Series in BioEngineering

The Series in Bioengineering serves as an information source for a professional audience in science and technology as well as for advanced students. It covers all applications of the physical sciences and technology to medicine and the life sciences. Its scope ranges from bioengineering, biomedical and clinical engineering to biophysics, biomechanics, biomaterials, and bioinformatics.

More information about this series at http://www.springer.com/series/10358

Costas Demetzos · Natassa Pippa
Editors

Thermodynamics and Biophysics of Biomedical Nanosystems

Applications and Practical Considerations

Springer

Editors
Costas Demetzos
School of Pharmacy
National and Kapodistrian University of Athens
Athens, Greece

Natassa Pippa
School of Pharmacy
National and Kapodistrian University of Athens
Athens, Greece

ISSN 2196-8861 ISSN 2196-887X (electronic)
Series in BioEngineering
ISBN 978-981-13-0988-5 ISBN 978-981-13-0989-2 (eBook)
https://doi.org/10.1007/978-981-13-0989-2

Library of Congress Control Number: 2018968396

This Springer imprint is published by the registered company Springer Nature Singapore Pte Ltd.
The registered company address is: 152 Beach Road, #21-01/04 Gateway East, Singapore 189721, Singapore

Preface

Thermodynamics and biophysics are considered as key elements for the development of innovative nanomedicines and for contributing to the evaluation process of biomedical nanoproducts. The book aims to highlight the recent advances of thermodynamics and biophysics in drug delivery nanosystems and in biomedical nanodevices. This multi-authored book is addressed to all those involved in recent advances of pharmaceutical nanotechnology and in nanomedicine. It aims to be an up-to-date book and to provide in-depth knowledge on bio-inspired nanotechnological systems for pharmaceutical applications. New aspects regarding the biophysics and thermodynamics as important elements for evaluating biomedical nanosystems as well as for correlating their physicochemical, biophysical and thermodynamical behaviour with those of a living organism form part of this book.

This approach is unique in bibliography, and we expect to shed light on the new and advanced physical properties of innovative nanomedicines which are considered to be important in terms of safety and effectiveness.

Additionally, thermal transitions that occur during nanosystem life and especially in nanosystems intended to be used in the field of health are extremely important. Their thermotropic behavior is related to nanoparticle dispersion and the thermodynamic behavior of the system changes according to the temperature. It is obvious that the methods used to study these changes contribute to a better understanding of nanoparticle stability as well as of the nanoparticles' rational design. According to ICTAC (International Confederation for Thermal Analysis and Calorimetry), thermal analysis can be defined as the gamut of analytical techniques where scientists can measure a property of the sample or its products in correlation to temperature, while the sample is submitted in a programmed thermal process, within a designated/controlled environment.

The book is divided into two main parts. Part I entitled: "Thermodynamic and Biophysics of Biomedical Systems" deals with the investigation of nano- and bio-systems with thermal analysis techniques.

Chapter "Introducing Thermodynamics and Biophysics in Health Sciences" discusses the importance of thermodynamics and biophysics in health sciences, especially in design and development medicines and innovative nanocarriers.

Methodological approaches for the characterization of the self-assembling behaviour in the pharmaceutical field are described in Chapter "Methodological Approaches for the Characterization of the Self-assembling Behaviour in the Pharmaceutical Field". This chapter reviews the common available techniques employed to characterize the self-assembling behaviour of drug delivery systems in a temperature or concentration-dependent manner. A particular focus will be placed on the calorimetric techniques (as differential scanning calorimetry, DSC, or isothermal titration calorimetry, ITC), but also on alternative methodologies such as high-resolution ultrasonic spectroscopy (HR-US). Special attention is given to bioresponsive nanomaterials and functionalized, coated, layered or multi-walled colloidal particles.

The phase transition in biological membranes is presented in Chapter "Phase Transitions in Biological Membranes", especially the experimental findings and the thermodynamics that describes the control of the membrane function. According to the author: "The thermodynamics of membranes brings about a coherent picture in which all phenomena are coupled in a relatively simple framework. The emergence of pulses and channel-like features in membranes and the coupling to the timescales are an immediate consequence. Such an approach is in strong contrast to the usual attempt to provide separate theories for each phenomenon". This chapter is useful to understand the membrane functionality on the basis of thermodynamics and biophysics.

Isothermal titration calorimetry (ITC) is a well-established technique that allows an accurate and precise determination of binding equilibrium constants; this technique is analyzed in depth in Chapter "Isothermal Titration Calorimetry: A Powerful Tool for the Characterization of Molecular Interactions". ITC plays an important role in biology, biochemistry and medicinal chemistry, providing researchers with important information on the structure, stability and functionality of biological and synthetic molecules. This chapter demonstrates the power and versatility of ITC in providing accurate, rapid and label-free measurement of the thermodynamics of molecular interactions and focuses on recent studies employing ITC to investigate compounds of great biotechnological interest.

The development of powerful and sensitive methods for rigorous monitoring of organismal and cellular iron in health and disease is described in Chapter "Dynamics of Iron Homeostasis in Health and Disease: Molecular Mechanisms and Methods for Iron Determination". While the biochemistry and genetic pathways involved in iron homeostasis have been investigated for several decades and our current understanding of the role of iron metabolism in biological systems has advanced enormously over the past years, plenty fundamental questions remain elusive. Generation of transgenic animal models has proven invaluable for the elucidation of key proteins and mechanisms that regulate iron homeostasis at the systemic and cellular level, as introduced in this chapter. Biophysical techniques for determining iron-containing complexes and the transition of iron oxidative states are also analyzed.

Part II entitled: "Applications and Practical Considerations" deals with the application of thermal analysis techniques in design and development of nanosystems, as well as in understanding the behavior of biological systems.

DNA nanotechnology and the thermodynamics of multivalent binding are presented in Chapter "Achieving Selective Targeting Using Engineered Nanomaterials". The most recent advances in nanomaterial design leading to vectors capable of selectively targeting biological surfaces are discussed in depth too.

Chapter "PAMAM and PPI Dendrimers in Biophysical and Thermodynamic Studies on the Delivery of Therapeutic Nucleotides, Nucleosides and Nucleobase Derivatives for Anticancer Applications" includes the current research, which is focused on the biochemical, biophysical and thermodynamic characterization of the interactions between dendrimers and their useful ligands like nucleotides, nucleosides and nucleobase derivatives. More data are also available for the biological assessment of their in vitro and in vivo activity. Available data concerning the formation and stability of complexes between dendrimer nanocarriers and (i) nucleoside/nucleotide, (ii) purine and pyrimidine antimetabolites for their potential application in innovative therapies are also analyzed with several examples from recent literature. In Chapter "Thermal Analysis of Glass-Ceramics and Composites in Biomedical and Dental Sciences", the longest established measurement techniques of glass-ceramics and composites in biomedical and dental sciences are briefly discussed. Examples from thermogravimetry (TG), differential scanning calorimetry (DSC), differential thermal analysis (DTA) and dynamic mechanical analysis (DMA) are given, and the experimental data are presented. The added value of each technique is also underlined.

Furthermore, differential scanning calorimetry (DSC) is applied to indicate their thermotropic behavior and thermal stability, providing useful information, in order to optimize the quality and therapeutic efficiency of the liposomal formulations. The aim of Chapter "Differential Scanning Calorimetry (DSC): An Invaluable Tool for the Thermal Evaluation of Advanced Chimeric Liposomal Drug Delivery Nanosystems" is to prove, through different literature examples of chimeric liposomal systems, the utility of the DSC technique upon the characterization of their thermotropic behavior which is strictly correlated with the interactions and cooperativity of the different biomaterials, as well as to demonstrate in what way it predicts the efficacy of the examined liposomal platforms. DSC has been quite frequently applied in pharmaceutical research for scanning the thermal behavior of the samples and for recording the difference between the heat flows, while it provides quick and accurate information about the physical and energetic parts of a material, especially for chimeric advanced drug delivery platforms.

Chapter "Drug-Membrane Interactions in the Renin Angiotensin System" reviews the contribution of the various biophysical techniques and molecular dynamics on drug–membrane interactions on the renin–angiotensin system (RAS), which plays a key role in the pathophysiology of the cardiovascular and renal system. Studies described in the manuscript mainly refer to drugs acting on the G protein-coupled receptor (GPCR) and AT1 by the peptide Angiotensin II (AII). AT1 receptor; however, it is envisaged that the discussed concepts of drug interaction can apply also to other GPCRs. The development of elegant experiments that promote the knowledge of the mode of drug–membrane interactions is desirable. Such experiments will aid the drug design and the development of safer drugs.

In Chapter "Biophysical Characterization of Polysialic Acid—Membrane Nanosystems, the biophysical characterization of polysialic acid–membrane nanosystems is presented. Polysialic acid (polySia) is a long, membrane-bound, polyanionic polymer (with the degree of polymerization, DP, up to 400) of negatively charged sialic acid monomers. Biological roles of polySia are based on its ability to modulate repulsive and attractive interactions, and its ability to modulate membrane surface charge density, pH at the membrane surface and membrane potentials. Several sophisticated techniques are presented in order to fully analyze the biophysical behavior of polySia encapsulated in different nanosystems (i.e. liposomes and micelles) and their biomedical applications.

Chapter "Self-assembling of Thermo-Responsive Block Copolymers: Structural, Thermal and Dielectric Investigations" provides a comprehensive overview on recent investigations on the micellar aggregation and the thermoresponsive behavior of amphiphilic model polymers. Thermoresponsive polymers have attracted much research attention because of their potential applications, which include rheological control additives, thermal affinity separation, controlled drug release, gene therapy and regenerative medicine. On the other hand, they represent model systems for many biological systems, for example for the investigation of the interaction between peptide-like groups and solvents and, thus, for the study of protein stability in aqueous solutions.

Deals with the sol-gel synthesis and characterization of hybrid materials for biomedical applications. Several techniques are analyzed in this chapter. The Chapter "Sol-Gel Synthesis and Characterization of Hybrid Materials for Biomedical Applications" describes results suggest that the polymer incorporated in the inorganic matrix favors the preparation of biomaterials with improved biocompatibility.

We would like to express our gratitude to all authors for their important contributions. Their expertise in the fields of thermodynamics and biophysics makes this book a comprehensive and valuable tool for all scientists working in nanosciences and biosciences.

Athens, Greece

Prof. Costas Demetzos
Dr. Natassa Pippa

Contents

About the Editors

Costas Demetzos Since 1991, Professor Demetzos' scientific and teaching work is related to Pharmaceutical Nanotechnology. The innovation of his research, apart from being in numerous publications and announcements in international scientific journals and conferences, has led to awards of international and national patents in Greece and in many European countries, USA, Australia, having as main objects the nanotechnological drug delivery systems and nanodevises. His research work has international recognition and there are more than 4000 cross references related to his research (h index 32). The total number of publications in international journals is greater than 240, while the total number of announcements is greater than 300. Being a scientific director he participates in national and European research programs. He has concluded many scientific collaborations internationally and within the Greek scientific community. Prof. Costas Demetzos is well known for his research activity in the field of Pharmaceutical Nanotechnology and for this reason he has been and still is invited for lectures and speeches in international scientific conferences, while he collaborates with a great number of scientists in European level, in the field of Nanotechnology. He acts as a member of the editorial board of scientific journals and has also been awarded in pharmaceutical congresses. He is the author of related University notes, textbooks and monographs. He was responsible for dozens of PhD students and their theses, Master's Diplomas and Undergraduate Dissertations. Also, he has been invited to teach Pharmaceutical Nanotechnology in other Greek Universities. It must be mentioned that he participates as a member of Organizing Committees but also as a president of International and European scientific Conferences, while being an evaluator in many committees. He is the chairman of the Hellenic Pharmaceutical Society (HPS), from 2008. He was also an elected member of the ExCo of EUFEPS (Network Coordinator) (2014-2016).

Natassa Pippa is post-doctoral researcher in Department of Pharmaceutical Technology, National and Kapodistrian University of Athens. She has completed her PhD in 2015 from Faculty of Pharmacy, National and Kapodistrian University of Athens, under the guidance of the expert of Pharmaceutical Nanotechnology, Prof. Costas Demetzos. She developed bio-inspired delivery systems, especially chimeric advanced Drug Delivery nano Systems (aDDnSs), which are complex self-assembled soft nanostructures, act as facilitators for the development of innovative therapeutic devices, with added value and respond to the requirements of modern trends in the therapeutics. During her PhD, she participated in the Research Programme "NANOMACRO: Functional Self-assembled Nanostructures from Block Copolymers and Proteins" (under the guidance of Dr. Stergios Pispas) in Theoretical and Physical Chemistry Institute, National Hellenic Research Foundation. She has published more than 60 peer –review papers and 6 chapters in Books. She has been selected as speaker in National and International Conferences and presented more than 70 posters. She has been awarded in Congresses for her oral and poster presentations. She also participated in several research programs with pharmaceutical industries in order to develop medicines, cosmetics and food supplements. She also received two scholarships for post-doctoral research in Greece and in France. She also received a Scholarship from French Government: "Séjours cientifique de haut niveau" (SSHN) in the Laboratoire de Chimie des Polymères Organiques (LCPO), University of Bordeaux (France) under the supervision of Prof. Dr. Sébastien Lecommandoux. Finally, she was awarded by International Association of Advanced Materials with the prestigious "International Association of Advanced Materials Scientist Medal (IAAM Scientist medal) for the year 2016" due to her contribution in the field of "Advanced Materials Science and Technology" (at European Advanced Materials Congress 2016 on 24th August 2016 Stockholm, Sweden).

Introducing Thermodynamics and Biophysics in Health Sciences

Costas Demetzos and Natassa Pippa

Abstract Biophysics and thermodynamics are effectively distributed in life sciences guided by the fields of biology and physics, respectively. The main concern raised regarding the usefulness of such scientific fields on the development and evaluation of therapeutic products as part of the research in the Pharmaceutical Industry and as part of the regulatory processes in the regulatory agencies. This short chapter deals with the benefits of using biophysics and thermodynamics in health sciences, in addition to the expanded physicochemical characterization as well to morphological issues. The following two questions are addressed in this article:

- Why should biophysics and thermodynamics be applied in the research and development processes of innovative therapeutic products?
- Why should these approaches be part of the dossier to the regulatory agencies by the stakeholders?

1 Thermodynamics and Biophysics in Biomaterials and Self-assembled Structures

Physical laws and their applications in health sciences demand an in-depth knowledge of physics and biology in order to develop new and innovative medicines. In the field of pharmaceutics, thermodynamics and biophysics are considered as ultimate scientific tools in order to effectively develop and evaluate innovative therapeutics, diagnostics and imaging agents. By researching nature, we manage to disclose its '*encrypted natural codes*', that are considered as dynamic tools that could provide new insights not only for expanding our scientific horizons, but also for applying new findings to health sciences. It is obvious that the properties and the functions of new synthetic and biocompatible materials emerge in the development process of

C. Demetzos (✉) · N. Pippa
Section of Pharmaceutical Technology, Department of Pharmacy,
School of Health Sciences, National and Kapodistrian University of Athens,
Panepistimioupolis, Zografou, 15771 Athens, Greece
e-mail: demetzos@pharm.uoa.gr

C. Demetzos and N. Pippa (eds.), *Thermodynamics and Biophysics of Biomedical Nanosystems*, Series in BioEngineering, https://doi.org/10.1007/978-981-13-0989-2_1

self-assembled structures in a variety of dimensions (i.e. micro, nano, etc.) and could be correlated with the functionality of living cells. Moreover, bio-colloids and the bio-interfaces which are discovered based on the laws and principles of biology and physics, are directly related with the health sciences, providing evidence and new technological outcomes that could be applied in biological and in therapeutic applications (i.e. implants, films etc.). It is of importance to point out that the self-assembly process of matter and of biomaterials is a field that gained attention by significant research. The spontaneous ($\Delta G < 0$) (1) processes of biomaterials to minimize their free energy by increasing the attractive molecular interactions, is a process driven by the thermodynamics and biophysics. The intramolecular interactions promote the lower free energy of the system and consequently the spontaneous self-assembly process. The thermodynamics process of self-assembly can be represented by Eq. (1) and is characterized as an equilibrium process. This process maintained the balance between attractive and repulsive forces which are crucial to retain the equilibrium between the biomolecules that contribute to the self-assembled structures

$$\Delta G = \Delta H - T\Delta S \quad (1)$$

It should be mentioned that in biology the self-assembly process is entropy driven, contrary to the enthalpy driven self-assembly process, in the field of nanomaterials [1, 2]. The reason that entropy is the driving force to produce self-assembled biostructures is because such biological structures and objects allow higher degree of freedom and are able to be arranged in such a polymorphic forms to be considered functional for the biosystem. The morphogenesis of biostructures, which is driven by quantum mechanics, is a quite complicated approach which is guided by thermodynamics in order to produce stable biological object and biological 'apartments'. Moreover, mathematical features and properties such as dimension, topology, asymmetry, the interfacial phenomena between bio-or artificial objects as well as the defects during the formation process in self-assembled biosystems, are important features and could be clarified by using biophysics and thermodynamics as complementary tools. More specifically, the cell membranes' integrity as well as their irregularities that promote several vital functions and clustering effects seem to play an important role in the cell life and should be kept into consideration during the development process of therapeutic agents.

Thermodynamics belongs to physics. The scientists have much appreciated the knowledge that emerged from the heat transfer process which reflects the thermodynamic laws and principles. It is well documented that changes in nature and in biological systems are governed by reduction of free energy, and thermal transitions are considered as crucial in the evolution process. The living cell phase transitions and the functionality of their membranes are related with their metastable phases and equilibrium process that are responsible for life organization and for life evolution. Bio-thermodynamics has emerged as a new and attractive approach for studying and evaluating the functionality and behaviour of artificial self-assembled and bio-inspired structures [3]. Biophysics is a scientific field that deals with the study of physical phenomena that have occurred in biological systems. It contributes to the study of living organisms and systems that are composed of biological objects like, proteins, lipids, carbohydrates etc. The biological functions and the behaviour of biological structures are influenced by the physical factors and laws. It acts as a complementary field to chemistry, biology, mathematics and engineering. Luigi Galvani (1737–1798), was the first one to assign biophysics as an experimental approach to study the static electricity in the muscles of a frog [4]. Biomolecules are important structures that contribute to the structure and organization of the high functionality of living organisms. It is of importance to point out that the functionality and the stability of natural elements is a result of the evolution process of organisms and of their thermodynamical profile.

2 Thermal Analysis Techniques in Design and Development of Medicines

Techniques of Thermal Analysis are considered as one of the most popular in material sciences. Even though most thermal analysis methods can deal with samples as solids, semi-solids or liquids, solid-state characterization, could apply to a majority of the applications in pharmaceutical research providing laboratories with applications that are of great importance for almost every Pharmaceutical field [5]. Thermal analysis can also provide detailed information on the stability, polymorphism, purity and interactions that concern the excipients and the active substance of the pharmaceutical products over time and during their development process. The stability and the thermodynamic profile of biological macromolecules including physical products are extensively studied by thermal analysis techniques [6, 7]. The biophysical profile of innovative biomaterials that are able to self-assemble in order to be used as excipients in therapeutics, suggests that the physical laws in biological substrates can affect their functionality, their properties and ultimately their survival. From pharmaceutical point of view, the compatibility of the excipients with Active Pharmaceutical Ingredients (APIs) are extensively quantified by thermal analysis techniques, especially by Differential Scanning Calorimetry [8, 9]. The importance of compatibility assessment for selection of excipients in specific unit operations such as milling, and grinding are also discussed extensively in the literature [8–13]. Estimation of drug-excipient interactions is a crucial step in pre-formulation and formulation studies of drug delivery design and development to achieve consistent stability, bioavailability and manufacturability of solid dosage forms [8–13]. These interactions can affect drug product stability in physical aspects such as organoleptic changes and dissolution slowdown, or chemically by causing drug degradation [8–13]. Differential Scanning Calorimetry (DSC) is used in the earliest stage of drug development to determine the physical behavior (polymorphism, purity, transition temperature etc.) and contributes to the development of pharmaceutical formulations and advanced drug delivery carriers based on nanotechnology [8–13].

3 Liquid Crystalline Phases: Trying to Understudy the Physics of Bio-Inspired Systems

The liquid crystalline state of matter which has been extensively studied by the scientific community, leads phase transitions and phase separations, metastable phases and thermodynamic equilibrium processes through the system free energy change (ΔG) and promotes biophysical changes that are in direct correlation with the functions of artificial cell organelles, such as cell membranes. However, artificial cell membranes like nanoparticulate systems such as liposomes, micelles, dendrimers, etc. behave as biological membranes and natural objects in terms of their metastability, structural polymorphism and thermotropic behaviour. Pharmaceutical products in therapeutics, in diagnosis and in imaging as in colloidal state, can affect the functionality of artificial systems by giving added value to human health. However, biophysics is considered as an important tool in order to study and understand the functions and behaviour of self-assembled organization levels of biological objects and of artificial platforms. As an example of the biophysical approach in the development of drug delivery nano-systems, we can refer to the phase transitions of the liquid crystalline phase that are present in the cells and correlate with their functionality, stability, and effectiveness.

It is obvious that such natural processes driven by thermodynamics and biophysics are very complicated and in scientific fields such as therapeutics and health sciences, we need open minded experts that can introduce new insights in the analysis and research and development process of innovative therapeutic products.

The organic substances that lead the pharmaceutical synthetic approaches to developing new bioactive molecules (i.e. drugs), should be checked for their purity and stability. Solid-solid transitions and solid-state reactions drive the thermal changes which occur in bioactive molecules. However, we need techniques that provide simple ways to identify such transitions and to prove the quality of the starting materials and of the final health product. In the past (1780), Lavoisier, (*A. L. Lavoisier, P. S. Laplace, Mem. R. Acad. Sci. Paris, 1784, 355*) was the first one to measure the quantities of heat energy and to correlate such quantities with the changes of state and of the chemical reactions. Therefore, he had the honor to become the father of thermochemistry. Heat transfer has been mathematically expressed by Fourier (*J. B. Fourier, Theorie Analytique de la Chaleur, Paris, 1822*) who worked on the equations that describe the transfer of heat and the transitions of the materials. These approaches became fundamental in studying biomaterials and synthetic molecules that are pharmacologically active. However, studies on the biomaterials' transitions and on their physical or chemical changes have been recognized as crucial in the research and development process of drugs and excipients. Moreover, the material science and the precise study of natural processes, have led the scientific community to produce technological platforms that are able to mimic the natural approaches, properties and functions. The self-assemble process is well recognized as the driving force for producing biomimetic and smart biodevices that are referred to as innovative excipients. These innovative self-assembled excipients, due to their unique physical properties, need more precise techniques that will be able to identify their thermotropic and biophysical changes which should accurately be measured. It is clear, that research and development of new medicines (drug + excipients) needs more appropriate techniques that should embrace thermodynamics and thermal analysis techniques as well as biophysical approaches. In the nineteenth century Joule's pioneer work revealed the strong relationship between heat and work; nowadays thermal analysis techniques lead the laws of electrical heating. In 1964, Watson et al., talked about the Differential Scanning Calorimetry (DSC) to quantitatively measure the transitions of materials and the heat transfer. Today, two types of DSC have been developed [14]. The first is the *power-compensated DSC* and the second one is the *Heat flux DSC*. The thermal changes that are well recognized in materials and biomaterials, are the following: melting, crystallinity, boiling, dehydration, solid-solid transition, sublimation, and glass transition. DSC could be used as a thermal analysis technique to identify the exothermic or endothermic phenomena and to quantitatively measure these thermal changes. The identification of the crystallinity and of the amorphous nature of bioactive substances as well as of the excipients, are recognized as crucial in the development process of medicines in Pharmaceutical Industry, because of the sensitivity of the pre- and final formulation processes of the bioactive substances in such thermal changes that affect the therapeutic outcome [4].

3.1 The Metastable Phases

Metastability and polymorphism, regarding pharmaceutical dosage forms, are considered as highly important features. They are related to the self-life of the final pharmaceutical product, contributing to its effectiveness. We must point out that not only the bioactive substances' polymorphic forms play a key role to the properties of the final product, but the excipients or innovative self-assembled structures, that are used as dispersed media, should be checked for metastable phases during the pre-formulation process, as well as during the formulation process. Metastable phases and polymorphism are considered as the '*holy grail*' that provides increasing effect of the drug stability and drug solubility. It is also well recognized that the crystallinity and the crystal form of a bioactive substance affects its physical properties (i.e. solubility) and consequently its therapeutic profile. The metastable phases of the self-assembled innovative excipients are highly appreciated as the leading medium that can affect the pharmacokinetics of the bioactive compound (i.e. drug). The polymorphism is well correlated with the crystal form of the bioactive substance and based on the physical properties of the polymorphic from the substance could be *enantiotropic* or *monotropic*. In the case of monotropic form a transition to a more stable form would be expected. This transition of the metastable form should be kept into account during the development and evaluation process of a medicine, especially in the case of self-assembled excipients that comprise of a plethora of metastable phases [4]. However, polymorphism and metastable phases in Pharmaceutical dosage forms should be researched in depth, in order to avoid physical and thermotropic '*threats*' that could affect the effectiveness of the final pharmaceutical product. In addition, the liquid crystalline phase of matter is well recognized as the state of matter that incorporates plethora of polymorphic forms. Transition upon heating and the ordering dynamic process due to the micro-environmental transitions is highly appreciated. The thermotropic behavior of liquid crystals and their lyotropism depend on the temperature and on their concentration in the media in which they exist. The lyotropic liquid crystals are considered as an amazing state of mater that could be used as innovative excipient to formulate biomimetic and smart, regarding their functionality, drug carriers. The complexity of the dynamic ordering of such drug carriers, due to their molecular arrangement, is a unique property of natural complex systems. However, by '*copying*' such biomolecular platforms based on their lyotropic effects and polymorphism complexity, we can produce artificial cell's membranes as technological platforms that are able to transport bioactive molecules to the target tissues. These complex systems should be studied by approaching the polymorphic forms using thermodynamics and thermal analysis techniques which are considered as excellent tools in the development and evaluation process of innovative drug delivery systems. We must point out that such studies on the structural polymorphism of self-assembled liquid crystalline structures, is considered as unique and can solve solubility and pharmacokinetic problems of innovative pharmaceutical product. The thermodynamic stability and consequently the self-life of the final pharmaceutical dosage form, is well recognized since the product (medicinal, diagnostic or imaging agent) may become less effective by lowering the beneficial or present adverse reactions. However, the stability of the bulk materials that are in storage in Pharmaceutical Industry as well as the steps in pre-formulation and in formulation processes should be checked by using thermal analysis techniques in order to identify polymorphism, chemical changes, decomposition etc. In complex dynamic systems in which more than one bioactive molecule co-exists, it is obvious, that thermodynamics become complicated and we need to plot the corresponding phase diagram that is an elegant approach to solve problems regarding the mixing and compatibility of the substances.

4 Nanosimilars: The Role of Thermodynamics for the Approval of the Next-Generation Nanomedicines

According to Ehmann et al. (2015) "*Over the last three decades many first-generation nanomedicines have successfully entered routine clinical use and it is now important for medicines regulatory agencies to consider the mechanisms needed to ensure safe introduction of 'follow-on' nanomedicine products, 'nanosimilars'. Moreover, drug regulators need to ensure that 'next'-generation nanomedicines enter clinical development and consequently the market in a safe and timely way for the benefit of public health.*" [15]. Additionally, a recently published work by Wibroe et al. [16] showed that an integrated biophysical toolbox for analysis and evaluation of physical and morphological properties of nanoparticulate vehicles of doxorubicin, may help to produce efficient therapeutic products. By looking carefully, the results from the aforementioned study, it becomes obvious that nanoparticulate drug delivery systems like liposomes, could benefit by using physical techniques in order to figure out differences of their physical and morphological properties that may be driven by the thermal changes during the development process into the Pharmaceutical Industry (Figs. 1 and 2; Tables 1 and 2). From regulatory point of view, the scientific discussion on the *nanosimilarity* started in 11th October 2011 when The EMA has published a reflection paper on general issues for consideration regarding the parental administration of nanomedicines with surface coating (EMA/325027/2013). This reflection paper highlights the effect of the coating on the stability, the pharmacokinetics and the biodistribution of the medicine, the potential interaction of the coating with other molecules in the biological environment and the potential consequences of such interactions. According to the EMA/CHMP/SWP/100094/2011: "variation in mean/median size and size distribution and/or the accuracy of methods employed for nano-sizing may result in the generic product displaying different physicochemical properties leading to a different biopharmaceutical profile in respect of pharmacokinetics and biodistribution. This has the potential to significantly impact on safety/efficacy in comparison to the reference product."

In other words, the challenge, when studying nanoparticulate drug delivery systems, is that the nanostructures do not follow the classic laws of physics because of their scale invariance; instead, they follow the principles of quantum mechanics. Mesoscale is an attractive field to study the thermodynamics and biophysics of new therapeutic products that are following quantum laws and statistics. However, nanothermodynamics could be the key to disclose the so-called '*silence functionality*' [3] which is related with their biophysical and thermodynamical profile. This approach could be attractive regarding the formation of an integrated regulatory framework that is able to facilitate the approval process of innovative therapeutic products, overcoming the existing regulatory difficulties.

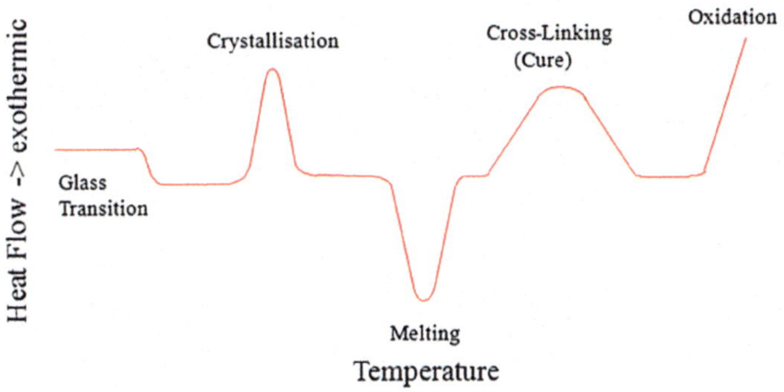

Fig. 1 A typical thermogram

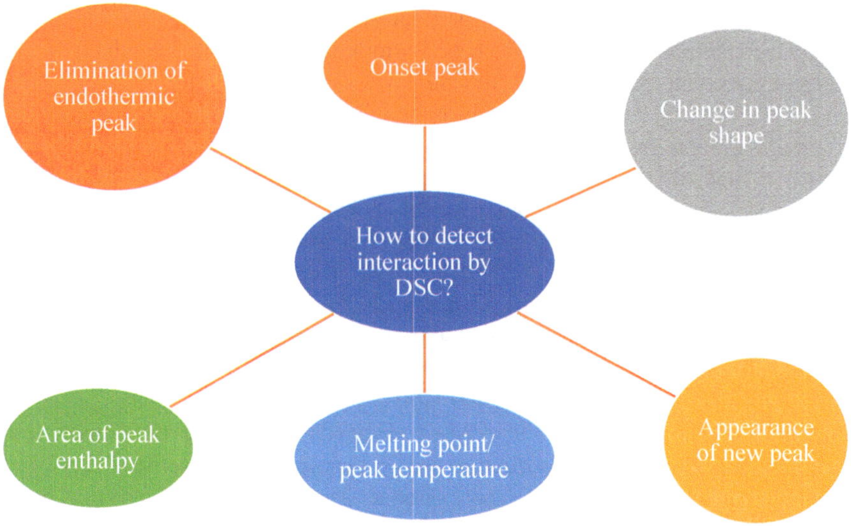

Fig. 2 The detection of drugs-excipients compatibility by using differential scanning calorimetry

Table 1 The advantages and the limitations of thermal analysis techniques

Advantages
Thermal analysis advantages in comparison to other analytical techniques are summarized in the following points:
The samples properties can study in a wide temperature range, using various heating and cooling programs
The sample can be in any form: liquid, solid (amorphous or crystalline) or gel, using various crucibles/sample holders made from different materials (e.g.: platinum, aluminum, etc.)
Smaller sample quantity is required i.e. 0.1–10 μg. Non-homogenous samples in combination with the small quantities required for analysis can cause problems related to the sample representativeness
Sample preparation is simple and refers to grinding and homogeneity of solids
The atmosphere around the sample is defined from the analyst
The required time of analysis is ranged between few minutes to few hours
The acquisition and function cost of thermal analysis tools is small
Limitations
They are non-selective techniques
Sensitivity and accuracy in quantity measurements of thermal analysis techniques do not exceed ±2%

Table 2 Industrial applications of thermal analysis

Industry	Transition	Purpose
Pharmaceutical	Tg	Collapse or storage temperature, amorphous content, self-life, impurities
	Cp	Processing conditions (temperature, pressure)
	Tm	Polymorphic forms, impurities, quality control
Polymers	Tg	Indicator of material properties, quality control, effect of additives
	Tm	Polymer processing, heat history
	Exotherm	Reactions rate, curing of materials, residual cure
	Cp	Energy needed to process
	Tc	Recrystallization times, kinetics
FoodTechnology	Tg	Storage temperature, properties, self-life, impurities
	Tm	Processing temperature

For example, nanosimilars which are the off-patent copies of nanomedicines should be studied in the light of biophysics and thermodynamics laws and principles. Nanosimilars are the follow-up 'copies' of prototype nanomedicines that present characteristics between generic medicines and biosimilars. Due to the complex nature of nanomedicines bioequivalence studies could not be applied and effective analytical techniques should be employed in order to achieve comparable data between nanomedicine prototype and its copied product. Nanosimilarity of liposomal products that are self-assembled nanostructures, is a demand and thermal analysis techniques should be reconsidered as effective tools in order to establish appropriate protocols regarding their polymorphism, metastability and morphological changes. It is of importance to point out that the compatibility between innovative excipients such as self-assembled nanostructures with the bioactive molecules is essential to prove the effectiveness of the nanosimilars. Such an approach incorporates thermodynamics and biophysics as the highly accurate and precise scientific paths for their evaluation. These two scientific approaches and their evolutionary added editions could be efficiently incorporated in a new regulatory framework for the approval of mesoscale dimensional therapeutic products and could become helper media for safer pharmaceutical outcomes.

5 Regulatory and Concluding Remarks

The regulatory environment is a dynamic system and is upgraded according to new scientific and social events which give rise to new regulatory guidance and revisions to that already exist. It is well documented in the case of nanotechnological products that the regulatory frame work is very fluid and it is in a dynamic continuing reconstruction. Small changes or differences, even in batch to batch production process in Pharmaceutical Industry of nanotechnological products, or in therapeutic agents, in which the nano-prefix governs their regulatory status, may affect their biological and therapeutic performance [14–18]. However, it is a challenge of the regulatory parties to disclose the ability of well-established techniques like thermal analysis or fractal analysis to accurate and precise characterization the thermal effects, the phase transitions as well as the metastability process and morphology, of innovative drug carriers, that may affect the stability of the therapeutic product and consequently its safety and effectiveness. In our point of view, biophysics and thermodynamics, and in more particular nano-thermodynamics, should be constitute an integral part of in depth studies of medicines and health products. The scientific community and the regulatory agencies should play an important role in promoting such scientific directions to stakeholders in order to produce effective and safe therapeutic products.

References

1. Fraden, S., Maret, G., Gaspar, D.L.D., Meyer, R.B.: The isotropic-nematic phase transition and angular correlations in isotropic suspensions of tobacco mosaic virus. Phys. Rev. Lett. **63**, 2068–2071 (1989)
2. Fraden, S., Maret, G., Caspar, D.L.: Angular correlation and the isotropic-nematic phase transition in suspensions of tobacco mosaic virus. Phys. Rev. E Stat. Phys. Plasmas Fluids Relat. Inter. discio. Topics **48**(4), 2837–2876 (1993)
3. Demetzos, C.: Biophysics and thermodynamics: the scientific building block of bio-inspired drug delivery nano systems. AAPS PharmSciTech **16**(3), 491–495 (2015)
4. Demetzos, C.: Pharmaceutical Nanotechnology: Fundamentals and Practical Applications. Springer, Berlin (2016)
5. Demetzos, C.: Differential Scanning Calorimetry (DSC): a tool to study the thermal behavior of lipid bilayers and liposomal stability. J. Liposome Res. **18**, 159–173 (2008)
6. Todinova, S., Raynova, Y., Idakieva, K.: Calorimetric study of *Helix aspersa* maxima hemocyanin isoforms. J. Anal. Methods Chem. (2018). https://doi.org/10.1155/2018/8450792
7. Idakieva, K., Parvanova, K., Todinova, S.: Differential scanning calorimetry of the irreversible denaturation of Rapanathomasiana (marine snail, Gastropod) hemocyanin. Biochim. Biophys. Acta **1748**(1), 50–56 (2005)
8. Chavan, R.B., Shastri, N.R.: Polymorphic transformation as a result of atovaquone incompatibility with selected excipients. J. Therm. Anal. Calorim. **131**(3), 2129–2139 (2018)
9. Ding, T., Chan, L., Zhai, L.H., Fu, Y., Sun, B.W.: Compatibility study of rivaroxaban and its pharmaceutical excipients. J. Therm. Anal. Calorim. **130**(3), 1569–1573 (2017)
10. Gressl, C., Brunsteiner, M., Davis, A., Landis, M., Pencheva, K., Scrivens, G., Sluggett, G.W., Wood, G.P.F., Gruber-Woelfler, H., Khinast, J.G., Paudel, A.: Drug-excipient interactions in the solid state: the role of different stress factors. Mol. Pharm. **14**(12), 4560–4571 (2017)
11. Narang, A.S., Desai, D., Badawy, S.: Impact of excipient interactions on solid dosage form stability. Pharm. Res. **29**(10), 2660–2683 (2012)
12. Panakanti, R., Narang, A.S.: Impact of excipient interactions on drug bioavailability from solid forms. Pharm. Res. **29**(10), 2639–2659 (2012)
13. Chadha, R., Bhandari, S.: Drug-excipient compatibility screening—role of thermo-analytical and spectroscopic techniques. J. Pharm. Biomed. Anal. **2014**(87), 82–97 (2014)
14. Watson, E.S., O'Neil, M.J., Justin, J., Brenner, N.: A differential scanning calorimeter for quantitative differential thermal analysis. Anal. Chem. **36**(7), 1233–1238 (1964)

15. Ehmann, F., Sakai-Kato, K., Duncan, R., Pérez, Hernán, de la Ossa, D., Pita, R., Vidal, J.M., Kohli, A., Tothfalusi, L., Sanh, A., Tinton, S., Robert, J.L., Silva, Lima B., Amati, M.P.: Next-generation nanomedicines and nanosimilars: EU regulators' initiatives relating to the development and evaluation of nanomedicines. Nanomedicine (Lond). **8**(5), 849–856 (2013)
16. Wibroe, P.P., Ahmadvand, D., Oghanian, M.A., Yaghmur, A., Mofhimi, S.M.: An integrated assessment of morphology, size, and complement activation of the PEGylated liposomal doxorubicin products Doxil®, Caelyx®, DOXOrubicin, and SinaDoxosome. J. Control Release **221**, 1–8 (2016)
17. Hussaarts, L., Mühlebach, S., Shah, V.P., McNeil, S., Borchard, G., Flühmann, B., Weinstein, V., Neervannan, S., Griffiths, E., Jiang, W., Wolff-Hilz, E., Crommelin, D.J.A., de Vlieger, J.S.B.: Equivalence of complex drug products: advances in and challenges for current regulatory frameworks. Ann. N. Y. Acad. Sci. **1407**(1), 39–49 (2017)
18. Zheng, N., Sun, D.D., Zou, P., Jaing, W.: Scientific and regulatory considerations for generic complex drug products containing nanomaterials. AAPS J. **19**(3), 619–631 (2017)

Methodological Approaches for the Characterization of the Self-assembling Behaviour in the Pharmaceutical Field

Diego Romano Perinelli and Giulia Bonacucina

Abstract Self-assembling is a thermodynamic process in which single components aggregate into large ordered structure as a function of temperature and concentration. Self-assembling has been longer exploited in the pharmaceutical field for the development of colloidal delivery systems for therapeutic, theranostic or preventive medicine purposes. In fact, the design of new polymeric or lipid self-assembling systems represents an advance in material science and nanomedicine, giving rise to a plenty of drug or functional substance carriers as polymeric or surfactants micelles, liposomes, vesicles and biomembranes. Recently, the growing interest in bioresponsive nanomaterials, able to respond to different stimuli in vivo (pH or temperature), has greatly influenced the architecture of these self-assembling systems, leading to the formulation of functionalized, coated, layered or multi-walled colloidal particles. This improvement has definitely made not effortless the chemical-physical characterization of the self-assembling process in the pharmaceutical field. The present chapter is aimed to review the common available techniques employed to characterize the self-assembling behaviour of drug delivery systems in a temperature or concentration dependent manner. A particular focus will be placed on the calorimetric techniques (as differential scanning calorimetry, DSC or isothermal titration calorimetry, ITC), but also on alternative methodologies as high-resolution ultrasonic spectroscopy (HR-US).

Keywords Ultrasound spectroscopy · Differential scanning calorimetry · Microcalorimetry · Isothermal titration calorimetry · Micellization · Liposomes

1 Self-assembly: A Ubiquitous Phenomenon

Self-assembly is a naturally occurred phenomenon, which can be observed everywhere in nature both at a macroscopic and a microscopic level. Despite scientist as well as common people have experienced self-assembly in their routinary life since

D. R. Perinelli · G. Bonacucina (✉)
School of Pharmacy, University of Camerino, Via Gentile III da Varano, 62032 Camerino, MC, Italy
e-mail: giulia.boncucina@unicam.it

C. Demetzos and N. Pippa (eds.), *Thermodynamics and Biophysics of Biomedical Nanosystems*, Series in BioEngineering, https://doi.org/10.1007/978-981-13-0989-2_2

time began, it is only in the last few decades that self-assembly has struck a new pivotal episode in its long-term history. This "new era" has been accompanied by the remarkable and even more numerous discoveries in the nanoscopic field occurred in the last century, so that, nowadays, the self-assembling concept is inseparably linked to all is related to nanotechnology. From this perspective, self-assembling configures itself as a boundary as well as a comprehensive concept among chemical, biological and material science disciplines in the world of nanosystems.

1.1 Definition and Thermodynamic Basis of Self-assembly

Apart from nanotechnology, in its broader sense, self-assembling refers to any self-induced structural organization of materials on all scales from molecules to galaxies. The peculiar characteristic of any self-assembling system, independently from its dimension, is to be a spontaneous and a reversible process. The key principle, upon this process, is the formation of thermodynamically favourable intermolecular secondary interactions among the "building blocks", which are able to drive the aggregation of pre-existing, single and disordered components into more ordered structures. In other words, self-assembly is a process able to make order among single or ensembles of molecules generating larger structures, which generally are in an equilibrium, or at least, in a metastable states. All secondary interactions including hydrogen bonds, ionic bonds, van der Waals bonds and hydrophobic interactions can be potentially involved, depending on the chemical structure of the "building blocks". The collective contribution of each of the mentioned interactions determines the formation of thermodynamically favourite larger and ordered structures. A different nomenclature has been used to categorize self-assembling processes. Generally, we refers to the "molecular self-assembly", when molecules or macromolecules are involved. Anyway, depending on the size of the "building blocks", also "colloidal" or meso-macroscopic self-assembly has to be considered. Yet, "molecular self-assembly" has been classified as "intermolecular" when the secondary interactions involve different molecules. This is the common meaning through which self-assembly is referred to and, for this reason, it is often called simply as 'self-assembly". Secondary interactions can be also formed intramolecularly, especially for macromolecules. In this case, the "building blocks" are the repeating units forming the macromolecules. This "intramolecular self-assembling" process naturally occurs in biological macromolecules, as proteins, and it is generally called as folding. Until 2000, only "static self-assembly" processes in which the ordered state occurs when the system is in an equilibrium state (at a constant temperature pressure and volume), without dissipating energy, have been investigated. By then, starting from the work of Grzybowski et al., dynamic self-assembly has been also considered and the formation of ordered supramolecular structures at non-equilibrium conditions have been experimentally demonstrated [1, 2].

2 Pharmaceutical and Biomedical Applications of Self-assembling

As for other technological sectors (e.g. electronics, optics), self-assembling has recently made a tremendous impact in the pharmaceutical and biomedical fields. In fact, in this research area, supramolecular self-assembling of functional nanomaterials have widely investigated and exploited for different purposes including tissue engineering, diagnostics and drug delivery [3–6]. The flourishing of new smart self-assembling nanomaterials in these fields has arisen from the acquired knowledge upon the chemical features of the "building blocks, determining the formation of secondary interaction, which can drive the self-aggregating process. The challenge of tuning the functional/structural properties in designing new "intelligent" self-assembling nanomaterials relies on the more or less successful attempt in mimicking processes occurring in nature in living systems.

For instance, synthetic or natural macromolecular materials can be employed for the preparation of in situ self-assembling hydrogels. They are administered by injection as a liquid and form a semi-solid hydrogel in situ, triggered by microenvironmental stimuli as temperature or pH. The formulation can be easily prepared and administered and can find a potential use both as cell scaffold in tissue engineering and as carriers in drug delivery.

A large variety of self-assembled nanocarriers as micelles, microemulsions, hydrogels, vesicles (liposomes, niosomes, cubosomes) and liquid crystals dispersion has drawn much attention for drug delivery. On a case-by-case basis, these nanocarriers resulted to be effective in improving the solubilisation, enhancing the targeting, reducing the toxicity and controlling the release of the encapsulated drug.

All these materials, commonly employed for biomedical and pharmaceutical self-assembly applications, can be categorized into: polymeric, lipidic, inorganic and biological nanosystems.

2.1 *Self-assembled Polymeric Nanosystems*

In the past decades, the use of both natural and synthetic polymers have marked an impressive incentive in the growing of the self-assembly field. The reasons for that can be attributed to the versatile chemical structures of the monomeric units composing the polymers giving a wide spectrum of possible inter-/intra molecular interactions, driving the self-assembling process. Among all polymers, as regards self-assembly, the interest is mainly focused on the amphiphilic polymers, made up of hydrophobic and hydrophilic units. As a function of the architecture of the monomeric units and the molar mass and chemical composition of the polymer, these systems can yield several self-assembled morphologies as spherical or worm-like micelles (also called filomicelles) or polymer vesicles (polymersomes). With the respect to aggregates resulting from the self-assembling of small molecules as surfactants (micelles) or phospholipids (vesicles), supramolecular self-assemblies derived from polymers have some advantages as a high colloidal stability and a large efficiency of solubilisation/encapsulation for both molecular compounds and/or macromolecules.

2.2 Self-assembled Lipidic Nanosystems

Lipid-based materials, able to spontaneously self-assembly, comprise surfactants phospholipids and other amphiphilic lipids as ceramides, gangliosides and cerebrosides. The structures of their nanoassemblies (micelles, bilayer and vesicles) is mainly determined by the chemical structure of the lipid described by the geometric factor referred to as the "critical packing parameter" *p*. It is a dimensionless parameter calculated according to the following equation:

$$p = v/a_0 l_c \tag{1}$$

where v is the volume of the hydrophobic tail, a_0 is the cross-sectional area occupied by the hydrophilic group at the aggregate interface and l_c is the critical length of the hydrophobic chain in the core of the aggregate. If the p value is less than $^1/_3$, the amphiphilic lipid is expected to form spherical micellar structures. If this value is above $^1/_3$ but below ½, it would be formed cylindrical or elongated micelles. For ½ $< p >$ 1, the lipid molecule self-assemblies into lamellar structure as vesicles sand liposomes. With $p > 1$, inverted assemblies as reverted micelles will be favourite.

2.3 Self-assembled Inorganic Nanosystems

Inorganic self-assembly generally refers to two distinctive group of materials as metal nanoparticles (e.g. gold or silver nanoparticles) and inorganic nanowires or nanotubes (e.g. carbon nanotubes). They mainly find application in diagnostic for the design of sensing probe of nanometric dimensions (optical, magnetic, electrochemical or amperometric sensors) [5, 7].

Self-assembly of inorganic and organic components into complex nanocomposites is also well-known [8].

2.4 Self-assembled Macromolecular (Biological) Nanosystems

Biological macromolecules as protein, peptides and nucleic acid naturally undergo a self-assembly process in vivo, and, since they are generally biocompatible, biodegradable, and, in some cases bioactive as well, they can offer itself as potential "building blocks" for the design of nanosystems in biomedical and pharmaceutical fields. Among all, peptides are the most utilized. In fact, biological peptides as well as synthetic analogues with different amino acid sequences self-aggregates into a plethora of ordered nanostructures (as cylindrical or elongated micelles, nanotubes or vesicles), which have been exploited in several fields, including drug or gene delivery. Peptides have been also employed to functionalize and to decorate onto the surface nanosystems formed by other materials (as synthetic polymers, lipids or metals) to enhance targeting properties as well as bioactivity.

3 Characterization of the Self-assembling Behaviour of Pharmaceutical and Biomedical Nanosystems

Several techniques are routinely employed for the self-assembling behaviour of nanosystems of diverse nature (polymeric, lipid, protein). These include spectroscopic, microscopy, scattering, calorimetric methodologies, each of them are able to highlight a different aspect of the process. The present chapter will focus on calorimetry and high-resolution spectroscopy techniques, since they allow a straightforward characterization of the self-assembling process from a temperature dependent and concentration dependent manner.

3.1 Ultrasound Spectroscopy Techniques

Ultrasound spectroscopy is a powerful analytical method for material characterization, which can find several applications in different technological fields as food chemistry, agriculture, pharmaceutics, cosmetics and, in general all related to material science. Despite several works have reported the use of this technique for the characterization of disperse systems such as micelles, emulsions and microemulsions, ultrasound spectroscopy is still underestimated in the pharmaceutical and biomedical fields. It is a spectroscopic technique that employs sound pressure waves with a frequency above the audible range for humans (~20 kHz) to study structural properties of a material in a fast, non-destructive and reliable manner. An ultrasound wave is generated from a motion of particles or masses, which causes an oscillatory and repetitive propagation. They are produced, for experimental purposes from piezoelectric elements able to convert electric energy into mechanical energy. Ultrasound waves at different energies can be generated by different experimental apparatus and be employed for different purposes. Low-intensity at high frequency (>1 MHz) ultrasound waves are used in the pharmaceutical field for particle size of a solid or liquid disperse phase, enabling reliable measurements also by analysing concentrated and optically opaque samples without needing of any dilution. Instead, high-intensity ultrasound at low frequency (20–100 kHz) are used to investigate structural properties of materials as a function of concentration or temperature [9].

3.1.1 Theory

According to a general definition, a wave can be described as a disturbance that travels through a medium, transporting energy from its source without transporting matter [10]. Waves can also be described in terms of pressure variation in the medium. In fact, when an excess of pressure is generated in a fluid (as air or a liquid), it tends to expand in the vicinity until all wave energy is dissipated.

As, acoustic waves are defined as longitudinal waves that propagate by means of adiabatic compression and decompression.

In this regards, ultrasounds can be also defined as waves generating an acoustic pressure above the highest pressure sensed by the human ear. Ultrasound waves propagate longitudinally through a compressional motion, which can be periodic or pulsed.

When the ultrasound wave is generated at regular intervals and with the same energy (that is its propagation occurs at a constant speed) we are in presence of a periodic wave:

$$y(x, t) = a \cos 2\pi f (x/c - t)$$

where: a is the peak amplitude, f is the frequency, c is the speed of the wave in the medium and x and t are the spatial and temporal coordinates.

Ultrasound spectroscopy, as the other spectroscopic technique, is based on the variation of the properties of ultrasound waves when they pass through the material. In fact, while travelling through a system, ultrasound wave loses part of its energy and it changes its velocity of propagation, depending on the structure of the system itself.

Each ultrasound wave can be defined in terms of the ultrasound parameters as sound speed and attenuation.

Sound speed is defined as the velocity of propagation of the ultrasound wave in the material. Being the ultrasound wave a succession of compression and depression in the medium of propagation, it depends from the elasticity and density of the medium as expressed by the Laplace equation:

$$U = \frac{1}{\sqrt{\beta\rho}} \quad (2)$$

where ρ is the density and β is the compressibility of the medium, representing the relative change of the medium volume per unit of pressure applied by the ultrasonic wave.

Ultrasonic attenuation, instead, is referred to the decrease in fluctuation amplitude consequent to the loss of energy occurring when the ultrasound wave travels through the material. In an ideal isotropic material, the amplitude of the sound wave, which passes through the material, remains constant and, its energy is conserved. On the other side, any discontinuity inside the materials can produce ultrasound attenuation.

The loss of ultrasound wave energy can occur through six different mechanism of interaction with the system: (a) viscous: if the discontinuities of the sample have different viscosity producing shear waves in the acoustic pressure field; (b) thermal: related to the temperature gradients across the material; (c) scattering: related to diffusion phenomena correlated to the interaction of the ultrasound wave with the discontinuities of material; (d) intrinsic: due to the intrinsic dissipation of wave energy during ultrasound propagation; (e) structural: caused by the oscillation of the disperse phase in the acoustic filed and (f) electrokinetic: related to ultrasound/electric double layer interactions.

Since the contribution of electrokinetic component is negligible in term of total attenuation, the total attenuation (α) can be calculated by the sum of the first five contributions:

$$\mathrm{A_{tot}} = up\alpha_{\mathrm{vis}} + \alpha_{\mathrm{th}} + \alpha_{\mathrm{sc}} + \alpha_{\mathrm{int}} + \alpha_{\mathrm{st}} \tag{3}$$

Acoustic attenuation coefficient α is dependent on the square of the ultrasound frequency and its decay is related to the properties of the medium by the relation:

$$\alpha = \frac{2\omega^2\eta}{3\rho\, c} \tag{4}$$

where ω is the angular frequency, η is the dynamic viscosity, ρ is the density and c is sound speed.

3.1.2 Applications of Ultrasound Spectroscopy for the Characterization of Self-assembling in Pharmaceutical and Biomedical Field

Ultrasound spectroscopy has been successfully employed to study the self-assembling behaviour, in a concentration dependent or temperature dependant manner, of a plenty of colloidal systems, investigated for their application in pharmaceutical and biomedical fields. They include surfactants, amphiphilic copolymers, microemulsions and liposomes [9].

A summary in the main applications of ultrasound techniques in the field is reported in Table 1.

Micellization is a spontaneous and thermodynamically-favoured phenomenon, which involves the self-self-aggregation of amphiphilic compounds as surfactants, amphiphilic block copolymers and amphiphilic peptide (also called as unimers) into larger supramolecular colloidal structure named micelles. They can have different dimensions, generally ranging between 2 and 15 nm for micelles formed by surfactants and between 10 and 50 nm for polymeric micelles. Despite the large variety of possible shapes for these aggregates (spherical, elongated, cylindrical, rod-like, worm-like), all micelles have the same general structure in aqueous dispersion: an internal core made of the hydrophobic portion of the unimer is surrounded by a shell, composed of the polar head of the amphiphiles, which remain in contact with

Table 1 Recent applications of ultrasound spectroscopy for the characterization of the self-assembling behaviour of colloidal dispersions for pharmaceutical and biomedical applications

Ultrasound spectroscopy		
Colloidal system	Application	References
Surfactants	CMC determination	[11–15]
Amphiphilic copolymers	CMT determination	[16–18]
Polymers	Gelation	[16, 19]
Liposomes	Sol-gel transition	[20, 21]

aqueous medium. Micellization is a concentration and temperature dependent process. Any micellization process is characterized by a minimum concentration of amphiphiles required for self-aggregation, termed as critical micelle concentration (CMC) and a minimum temperature above which micelle start to form, termed as critical micelle temperature (CMT). CMC can be determined using several experimental techniques including tensiometry, conductimetry, fluorescence spectroscopy, dynamic light scattering [22–24]. To this end, any technique able to detect a marked variation in a measured parameter related to the chemical-physical properties of the sample below and above CMC, as ultrasound velocity, can be potentially employed. As for many others chemical-physical properties of amphiphilic dispersions, ultrasound velocity change as a function of concentration at a different rate below and above CMC. As shown in Fig. 1, sound speed (U) or ΔU linearly increases over concentration below CMC. According to the equilibrium model for micelle formation, below the CMC, sound speed is only determined by surfactants in the unimer state and can be calculated from the following equation:

$$U_{C<cmc} = V_{unim} - \frac{K_{unim}}{2\beta_0} - \frac{1}{2\rho_0} \tag{5}$$

where V_{unim} and K_{unim} are respectively the specific volume and the compressibility of surfactants at the unimeric state and β_0 is the coefficient of adiabatic compressibility.

Above CMC, instead, the total sound speed is given by the sum of two contributions, surfactants in the unimer state, for which the concentration is constant and equal to CMC and surfactants in the micellar states:

$$U_{C>cmc} = V_{mic} - \frac{K_{mic}}{2\beta_0} - \frac{1}{2\rho_0} - \frac{CMC(2\beta_0(V_{mic} - V_{unim}) - (K_{mic} - K_{unim}))}{2\beta_0 C} \tag{6}$$

where V_{mic} and K_{mic} are respectively the specific volume and the compressibility of surfactants at the micellar state [11].

Below CMC, the increase in sound speed is linear since proportional to V_{unim}, which is dependent on surfactant concentration. Above CMC, the formation of micelles are consequent to a large loss of water bound to the amphiphile, leading to

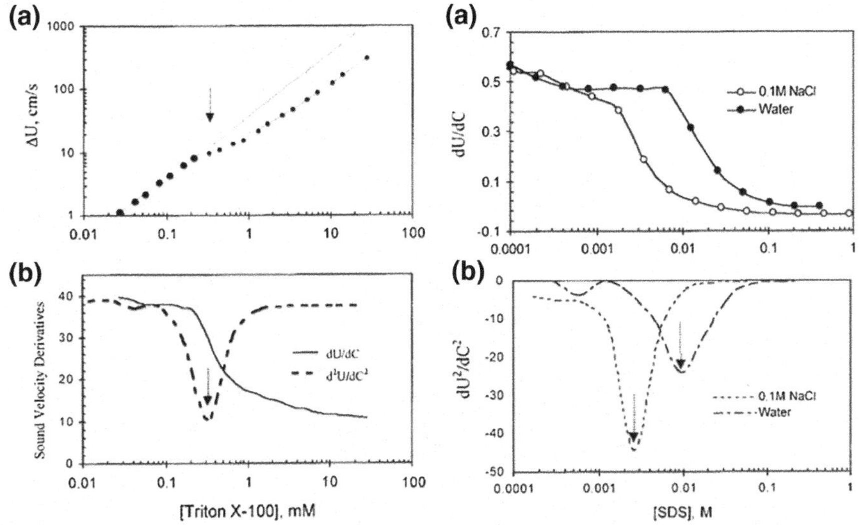

Fig. 1 Variation of ultrasound speed versus Triton X concentration and its derivative functions (on the left). First and second derivatives of sound speed versus SDS concentration in water (on the right). Adapted from [11]

an increase in compressibility and a slower increase of sound speed over surfactant concentrations.

CMC can be extrapolated from the peak value of the second derivate relative to the plot sound speed versus surfactants concentration. Sometimes, for an unambiguous determination of CMC, the second derivate function can be fitted using a gaussian model. In this case, the CMC value can be considered as the peak of the gaussian fit.

Being K_{mic} a term of Eq. 3, the variation in sound speed above CMC is also dependent on the intrinsic structure of the micellar aggregates.

Fluctuations in the ultrasound attenuation parameters close to CMC values have been theoretically postulated but they have been never exploited in experimental studies to determine CMC values for amphiphilic compounds [14].

Critical micelle temperature (CMT), is the minimum temperature at which micelles can form. Consequently, below CMT, there is no measurable CMC value. For common surfactants (e.g. sodium dodecyl sulphate), CMT is also referred as kraft temperature and determines the solubility of the amphiphile in water. In fact, below kraft temperature, surfactants are in a crystalline form and display a low solubility. At kraft temperature the solubility is equal to CMC, while above CMC, it sharply increases thanks to the contribution of the micellar component.

For amphiphilic block copolymers, instead, CMT is strongly dependent on concentration, in fact, CMT decreases by increasing copolymer concentration. The concentration- and temperature dependent self-aggregation behaviour has been well investigated on poloxamers, particularly Poloxamer 407. Several techniques have been employed including low energy high resolution ultrasound spectroscopy [16, 18, 25].

In this case, both sound speed and attenuation parameters resulted to be sensitive in highlighting the CMT of Poloxamer 407 as a function of concentration. The figure shows the variation of sound speed (normalized by the concentration) as a function of temperature for different Poloxamer 407 aqueous dispersions (10, 15, 20, 25 and 30% w/w). Overall, sound speed decreases over temperatures, but all plots have a characteristic profile in which three different trends of the signal can be identified. Indeed, a decrease in sound speed proportional to the increase of temperature can be recognised at the beginning and at the end of the heating ramp. These linear parts of the signal are linked together by a temperature range in which signal deviates from linearity. The linear trend is related to the overall effect of temperature on the system in terms of compressibility variations in the unimer or micellar state over temperates, while the middle non linear part is related to the micellization process. Particularly, the deviation of the signal depends on a modification in the amount of water bound to the copolymer (Fig. 2).

As such, micellization is a desolvation process, leading to an increment on the free water in the sample, determining an increase in the total compressibility and a reduction of sound speed.

Also in this case, the temperature of the transition can be determined by the peak value of the first derivate function, obtained from the raw signal.

Differently from CMC determination, the parameter attenuation was also successfully employed for the determination of CMT of Poloxamer 407. As described above, attenuation of sound speed is depending by several mechanisms, which take into account any variation in the size, thermal and viscous properties of the sample, which can result from any transition of the material, as micellization. Particularly, the measured attenuation was found to increase close to CMT when unimers start to

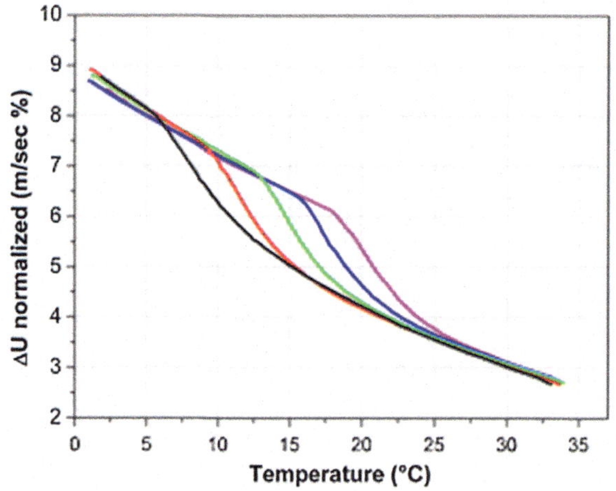

Fig. 2 Effect of temperature on the normalized relative ultrasound speed measure for aqueous dispersion of Poloxamer 407 in water at different concentrations (30, 25, 20, 15, and 10% w/w). Reprinted from [16]

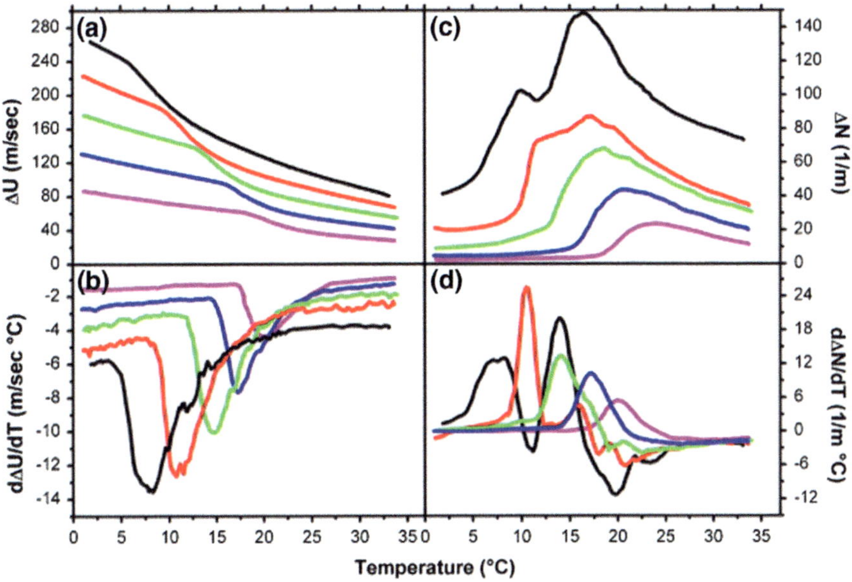

Fig. 3 Effect of temperature on the relative ultrasound speed (**a**); first derivate of relative ultrasound speed (**b**); relative attenuation (**c**); and first derivate of relative attenuation (**d**) as measured for aqueous dispersions of Poloxamer 407 at different concentrations (30, 25, 20, 15, and 10% w/w). Reprinted from [16]

aggregates into micelles (Fig. 3) as a consequence of a higher heterogenicity of the sample resulted from the coexistence of unimers and micelles in the samples. CMT can be considered as the maximum value in the plot attenuation versus temperature. Above CMT, attenuation decreases but it does not reach again the initial value of the unimer state. This denotes the different properties of the dispersion at the micellar state in comparison to the unimer state, correlated also to structural characteristic of the polymeric micelles.

Ultrasound spectroscopy, can represent also a valuable analytical tool to investigate micellar properties at different conditions. Information on micellar structure can be obtained by considering both the sound speed and attenuation profiles. Again, for Poloxamer 407, high resolution ultrasonic spectroscopy was employed to investigate the influence of salts, as potassium phosphate (KH_2PO_4/K_2HPO_4) commonly used for the preparation of buffer system, on the micellization behaviour of this amphiphilic block copolymer [18]. Interestingly by looking at the sound speed profiles (Figure), the curves (after normalization by concentration) are superimposed only below CMT when unimers are present, while the decrease in sound speed after CMT is proportional on the investigated salt concentrations (0, 100 and 200 mM). This suggests that the presence of phosphate salts influences only the hydration state of the micelles, but it does not affect the hydration state of the unimers. The lower sound speed values observed above CMT for sample prepared at high ionic strength

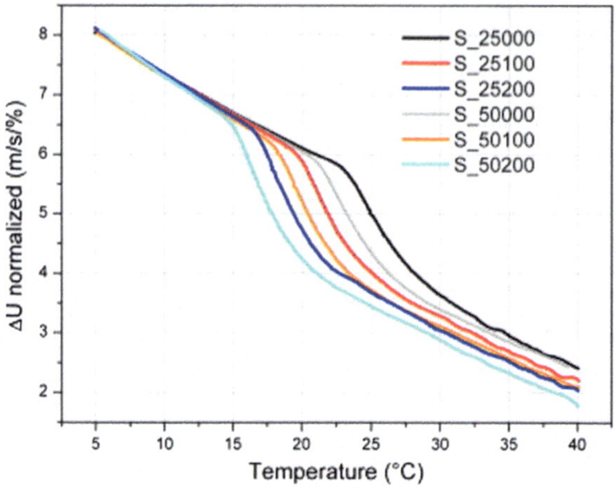

Fig. 4 Effect of temperature on the normalized ultrasound speed (on the left) and normalized ultrasound attenuation (on the right9 for Poloxamer 407 at different concentrations (5 and 2.5% w/w) in water or buffer phosphate (100 and 200 mM). Reprinted from [18]

(200 mM phosphate salts) is related to the higher compressibility of the systems related to the larger amount of free water, suggesting that water-water interactions in the medium are determinant as well as polymer-water interaction in driving the micellization process. In such a way, the presence of phosphate salts acts as a structuring agent for polymeric micelles, despite the less ordered state of the medium resulted from the increased number of free water molecules (Fig. 4).

Ultrasound spectroscopy has been also applied for the characterization of other nanosystems of great interest in the pharmaceutical or biomedical fields as liposomes. Liposomes are unilamellar vesicles of size generally ranging from 50 to 200 nm, resulting from the self-assembling in aqueous media of phospholipids into lipid bilayer. Despite the limited number of studies reported in the literature, ultrasound spectroscopy resulted to be very effective in the investigation of the thermal behaviour of liposomes. Particularly, reliable determinations of the main phase transition temperature, comparable to that obtained with classical techniques as calorimetry, was achieved by both analysing sound speed and attenuation parameters. The phase transition temperature (Tm) is defined as the temperature at which phospholipid change their physical state from an ordered gel phase to a liquid disordered crystalline phase. This transition can be detected by ultrasounds spectroscopy since the phospholipid bilayer modifies its internal organization from a state in which phospholipids are fully extended and closely packed (T < Tm) to a state in which phospholipids are relatively randomly oriented in a short-range disordered phase (T > Tm). In other words, the crystallinity degree of the lipid bilayer is modified, thereby affecting the total volume compressibility (influencing sound speed parameter) or the molecular relaxation component of total attenuation:

$$\alpha = \alpha_{mr} + \alpha_s \quad (7)$$

where α is the total attenuation, α_{mr} is the molecular relaxation component of attenuation and α_s is the scattering component of the attenuation depending on size and concentration of the dispersed colloidal particles. As for micellization, thermal phase transition can be highlighted in sound speed versus temperature plots as a change in the linearity of the signal, resulting in a more or less pronounced drop of sound speed in the vicinity of the transition. In the same way, phase transition of liposomal membrane appears in the attenuation versus concentration plots as a deviation from the baseline (Fig. 5). During the transition, attenuation increases up to reaching a maximum, and then, decreases, returning to the baseline values. Generally, the baseline tends to decrease slightly with temperature also above the phase transition at temperatures at which the thermal process can be considered completed. A larger sensitivity in the attenuation parameter was observed for loaded liposomes, as indicated by the higher measured peak values.

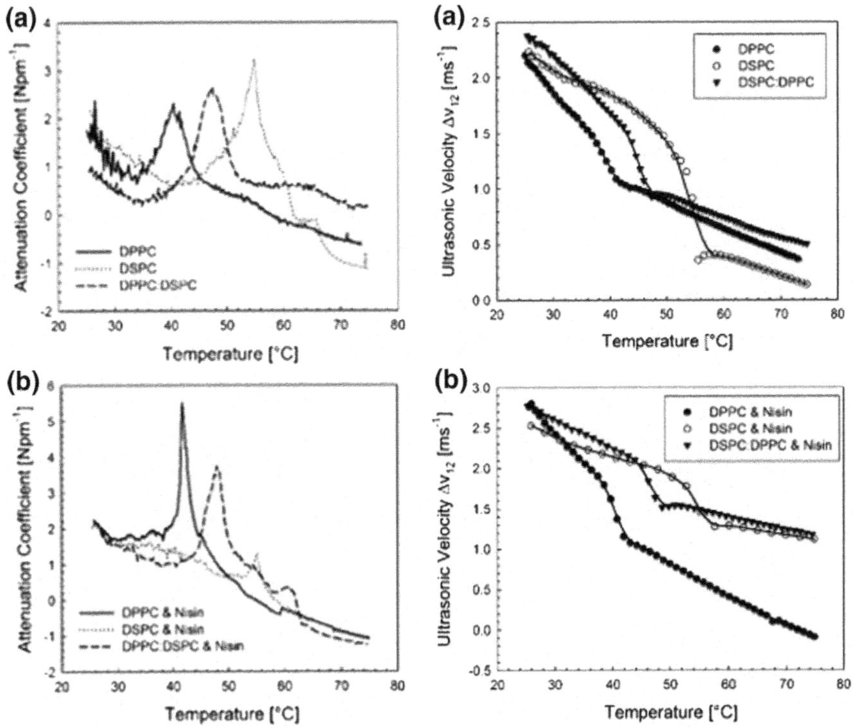

Fig. 5 Ultrasound attenuation coefficient (on the left) and ultrasound velocity (on the right) as a function of temperature for empty or nisin loaded (0.025 mg/mL) DPPC, DSPC and DPPC: DSPC 1:1 liposomes. Adapted from [20]

Figure 5 also shows the variation of ultrasound velocity over temperature for the analysed liposomal dispersions. As for micellization, before and after the phase transition, the decrease in sound speed with temperature is nearly linear; despite for pure phospholipids the slope prior to the transition was larger than that after the transition. The sol-gel transition from ultrasound speed measured can be identified as a stepwise decrease of the signal, and the exact measured values can be determined from peak value of the first derivate function.

High-resolution ultrasound spectroscopy is effective also for the characterization of thermal behaviour of complex liposome membranes, composed of a mixture of phospholipids, including PEGylated phospholipids. The thermal transition of the commercial formulation Doxil®, containing the anticancer drug doxorubicin, in comparison to the same formulation without the drug (placebo) are evident by taking into account both parameters as ultrasound velocity and attenuation (Fig. 6). In particular, attenuation was found to be highly sensitive to the presence of the drug. Loaded liposomes, in fact, showed a larger variation in the measured attenuation and velocity than placebo during the transition [21].

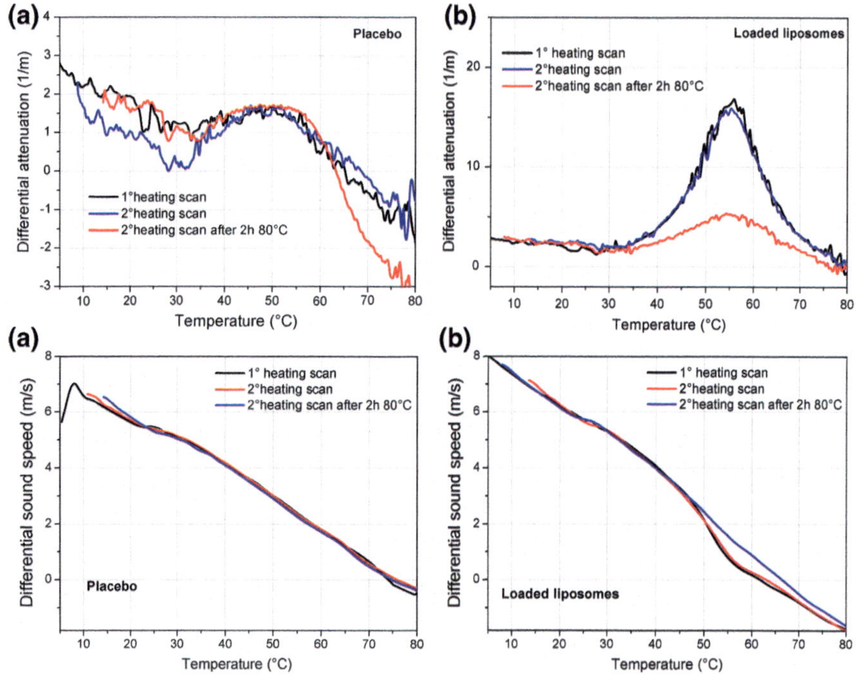

Fig. 6 Variation of the ultrasound parameter differential attenuation (on the top) and differential sound speed (on the bottom) as a function of temperature for unloaded PEGylated liposomes and doxorubicin loaded liposomes (Doxil®). Adapted from [21]

3.2 *Calorimetric Techniques*

Calorimetry has been a leading technique in the last decades for the characterization of the thermal behaviour of numerous single component or hybrid self-assembling materials with potential applications in the pharmaceutical or biomedical field. Calorimetry refers to a family of techniques, which are able to detect any transition involving heat exchange between the material and the environment. Since self-assembling is a thermodynamic phenomenon, actually any self-aggregation process can be followed by these techniques. On these premises, calorimetry configures as a versatile methodology, which can be exploited to investigate different aspects of self-assembling from a chemical-physical point of view.

3.2.1 Theory

Calorimetry techniques can be sub-grouped in two main category: differential scanning calorimetry (DSC) and isothermal titration calorimetry (ITC). DSC techniques measure the heat exchange between sample and reference when a specific thermal programme is applied. In other words, energy changes, together with the temperature at which they occur, are registered by the instrument, when sample is subjected to an heating ramp, a cooling ramp or an isotherm. In ITC technique, instead, samples are kept isothermally and heat variation after the addition of a further component is measured. Measurements are performed using instruments called calorimeters, made up of one or two adiabatic chambers (furnace) in which are placed two holders, one for sample and another for reference.

The main parameter measured by DSC is the heat flow, which is the flow of energy acquired or released by the sample as a function of temperature and time according to the equation:

$$\frac{dQ}{dt} = C_p \frac{dT}{dt} + f(T, t) \tag{8}$$

where dQ/dt is the total heat flow, C_p is sample heat capacity; dT/dt is the heating rate and f (T, t) is the kinetic dependence (temperature and time) of the measured heat flow.

For DSC analyses, the difference between the heat flow of the sample and the reference, maintained at the same temperature during the thermal programme, is measured. Since *Cp* values of the sample and reference are not the same, the energy required to maintain both at the same temperature is different, but this difference is constant whether no transition in the material occurs. In case of a thermal transition, energy can be adsorbed (endothermic) or released (exothermic) by the sample, determining a variation in the energy required to maintain sample and reference at the same temperature. The obtained signals are called thermograms from which transition temperature and the associated enthalpy can be calculated from the peak value and the area under the curve, respectively [26].

For ICT, the difference between the heat flow of the sample and the reference at a constant temperature is measured after the addition of a solution containing a second component. If this molecule interacts with the sample, there is a variation of the measured heat flow. The signal is the recording of consecutive peaks, after each addition of the component, whose the height progressively decreases until the saturation level has been reached. The area of each peak is then integrated and plotted versus the molar ratio of the two interacting components. The resulting curve can be fitted by a model from which several thermodynamic parameters, including the affinity constant (K_D), are derived [27].

3.2.2 Applications of Calorimetry for the Characterization of Self-assembling in Pharmaceutical and Biomedical Field

ITC and differential scanning calorimetry techniques have been widely exploited to study self-assembling behaviour. Particularly, ITC, by measuring the thermodynamic parameters at a constant temperature, is suitable for building up isotherms useful for studying the self-assembling process from a concentration dependent manner, while DSC techniques, by performing temperature scanning cycles on the samples, are effective in investigating the temperature dependent behaviour.

ITC is a recently emerged approach for the determination of CMC of surfactants of different nature as ionic and non-ionic synthetic amphiphiles and amphiphilic peptides [36, 41–43]. This technique is also very sensitive for the investigation of the aggregation of mixed system [37, 44]. ITC has the advantage over the other methodologies to provide directly the CMC together with thermodynamic parameters as the enthalpy of micellization. The CMC determination thought ITC is carried out by titrating with water or buffer a concentrated surfactant solution (at least 15–20 times above CMC) in the calorimetric chamber. The normalization by molarity of the integrated titration peaks is the thermodynamic parameter ΔH^0_{dil}. This parameter, when plotted against surfactants concentration, gives the profile, called demicellization isotherm, reported in Fig. 7.

This profile is dependent on different factors such as surfactant concentration, aggregation number, counterion, temperature, and micellar shape, but, in any case, three different regions can be identified. Two plateaux, in which ΔH^0_{dil} is almost constant, are characteristic of the micellar and unimer states and are linked by an intermediate region in which the recorded enthalpy sharply decreases or increases. From this plot, CMC value can be calculated from the maximum or minimum value of the first derivate.

The variation of Gibbs energy during the micellization process $\left(\Delta G^0_{mic}\right)$ can be calculated from CMC values for ionic or non ionic surfactants according to the following equations:

$$\Delta G^0_{mic} = RT Ln \chi_{CMC} \quad \text{for non ionic surfactants}$$
$$\Delta G^0_{mic} = (2 - \alpha) RT Ln \chi_{CMC} \quad \text{for ionic surfactants}$$

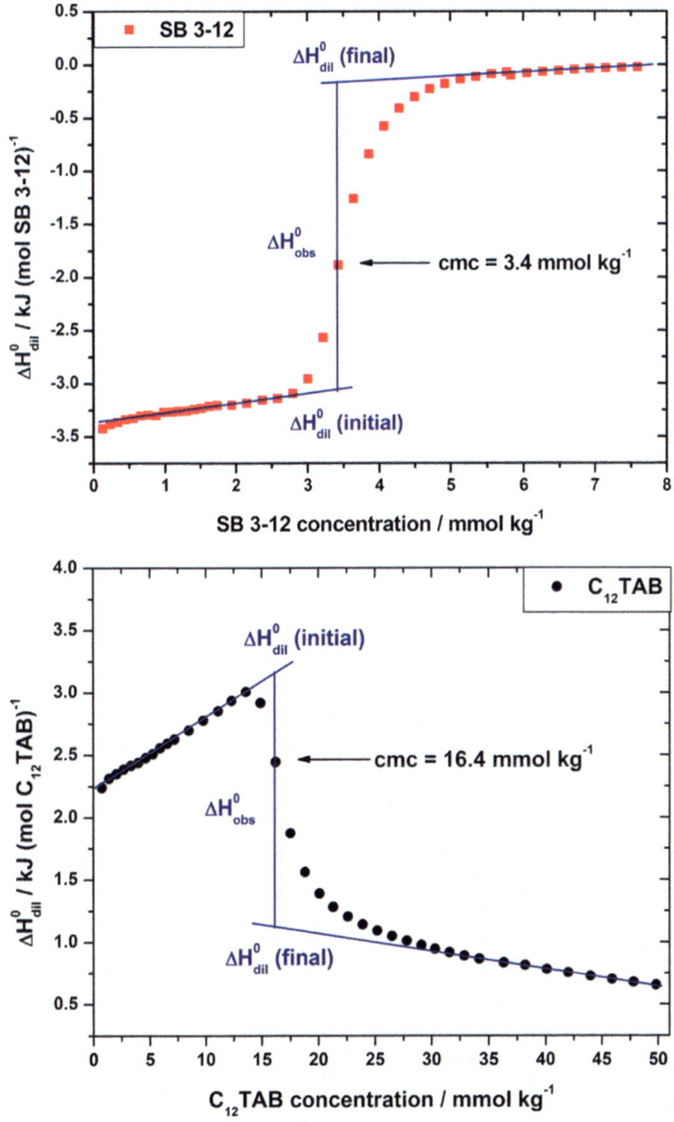

Fig. 7 Calorimetric titration curves from ITC measurements obtained for two different surfactants: SB 3-12, a zwitterionic surfactant and C12TAB, a cationic surfactant at 25 °C. Reprinted from [45]

where R is the gas constant, T is the Kelvin temperature, α is the ionization degree and χ_{CMC} is the mole fraction of monomers at CMC.

From the same plot, the micellization enthalpy can also be extrapolated ΔH^0_{mic}. This can be performed in two steps. Firstly, the enthalpy variation between the micellar and unimer state $\left(\Delta H^0_{Obs}\right)$ is calculated, then this value is normalized for the total amount of surfactant used for the titration (C_T) according to the equation [45]:

$$\Delta H^0_{mic} = \Delta H^0_{Obs} \frac{C_t}{C_t - cmc} \tag{9}$$

The versatility of ITC to investigate the interaction between molecular or macromolecular entities, providing an exhaustive overview about thermodynamics of the process, is well known. Over decades, this technique has been exploited to characterize colloidal mixed systems of different nature (e.g. lipid/polymeric, protein/peptides, inorganic/polymeric), intended for the most diverse applications.

Particularly, ITC is relevant to determine the binding stoichiometry and to identify the presence of one or more binding sites, allowing, at the same time the calculation of the isothermal association parameters. Actually, ITC is the most sensitive method currently available for the determination of the affinity or association constant (K) and the enthalpy variation associated with the interaction (ΔH), reflecting the heat released or absorbed.

For the experiments, one of the interacting species at fixed concentration is titrated, at constant temperature, with consecutive aliquots of the ligand up to saturation of the binding. The heat changes after each addition are recorded and the integrated area is plotted against the molar ratio of the reactants. The obtained profile can be fitted with various models to extrapolate thermodynamic parameters related to the association process (Fig. 8) [46, 47].

One possible application is the study of the association behaviour of the drug with colloidal carriers, in drug delivery field.

In this regards, many studies have been conducted on cyclodextrin-guest drug interactions [18]. These studies have highlighted as different drugs, sometimes with a similar chemical structure, can bind cyclodextrins with a diverse stoichiometry. For instance, β-cyclodextrin can form inclusion complexes with tropane alkaloids as homotropine hydrobromide, scopolamine and atropine sulfate. However, a different binding stoichiometry was found for the three alkaloids (1:1 for scopolamine and atropine sulfate and 1:2 for homatropine hydrobromide) [49].

Another example is offered by the investigation conducted to study drug partitioning phenomena in self assembling [50, 51]. To identify the nature and the strength of the drug with the carrier is a knowledge essential for a rational design of drug delivery systems.

Differential scanning calorimetry techniques include the traditional DSC, and its variants as micro- or nano-calorimetry. Micro- and nano-calorimetry differs mainly from traditional DSC for a larger sensitivity in the measured heat flow. In fact, the sensitivity of microcalorimetry and nanocalorimetry, depending on the instrument, ranging from 1 mW to below 1 μW. With the respect to traditional DSC, these tech-

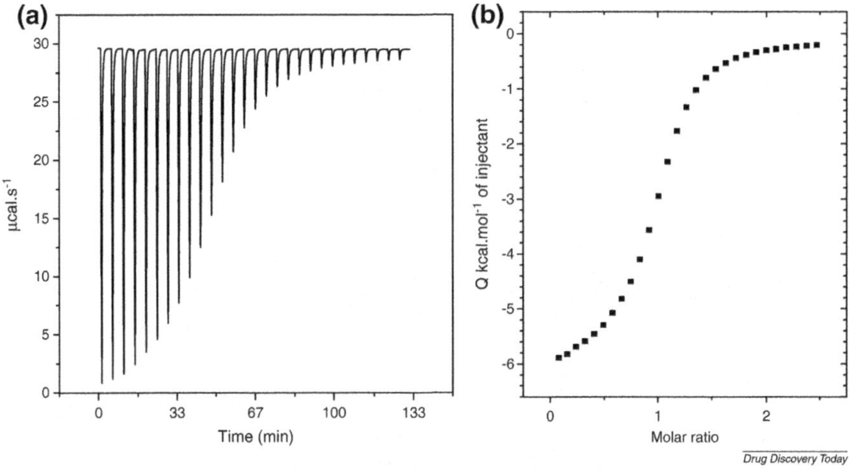

Fig. 8 Isothermal titration calorimetry data obtained from the binding interaction of 1-adamantylamine (1.3 mM) to b-CD (15 mM) in water at 25 °C. **a** Exothermic titration peak after each injection of 10 mL aliquots of b-CD in 1-adamantylamine solution. **b** Binding curve built up from the integrated heat data versus reagent molar ratio. Reprinted from [48]

niques are more accurate and sometimes, the only able to detect reliably transitions related to the self-assembling of colloidal systems as a function of temperature.

As such, differential scanning calorimetry is one of the most common technique to study micellization of amphiphlic compounds as surfactants or amphiphilic copolymers, since it is able to detect the endothermic transition associated to the desolvation of the unimer into micelle during the process. By thermal analysis, it is possible to calculate the onset of the transition, that is the temperature at which the aggregation into micelle starts, the peak temperature, that is the temperature at which the conversion unimer into micelles is faster and the area, that is the enthalpy associated with the transitions. Several examples are reported in the literature and, some recent studies are also summarized in Table 2. One case study is represented by poloxamer 407. Poloxamer 407 is a triblock copolymer of poly(ethylene oxide) (PEO) and poly(propylene oxide) (PPO), whose self-assembling properties (micellization and also gelation) are concentration and temperature dependent. Thermograms related to poloxamer 407 dispersion in water at different concentrations (30, 25, 20, 15, 10% w/w) are shown in Fig. 9. Each dispersion shows a broad endothermic peak, which is related to the micellization process. The peak of the trace can be considered as the CMT. The calculated CMT values decrease from around 30 °C to around 6 °C, by decreasing poloxamer 407 concentration from 30 to 10% w/w. Even, the enthalpy associated to the transition (ΔH_{cmc}) decreases with poloxamer 407 concentration from 6.5 to 2.5 J/g calculated respect to the sample weight. Moreover, for the dispersions with concentration higher than 15% w/w, another less energetic transition is

Table 2 Recent applications of ultrasound spectroscopy for the characterization of the self-assembling behaviour of colloidal dispersions for pharmaceutical and biomedical applications

	Calorimetry		
	Colloidal system	Application	References
Differential scanning calorimetry	Surfactants	Kraft temperature	[28, 29]
	Amphiphilic copolymers	CMT determination	[16, 18]
		Gelation	[16]
	Liposomes	Sol-gel transition	[21, 30]
ITC	Surfactants	CMC determination	[28, 31–33]
	Mixed systems	Drug-carrier interactions	[34, 35]
	Peptides	Self-assembling	[36–38]
	Nanoparticles	Self-assembling	[39, 40]

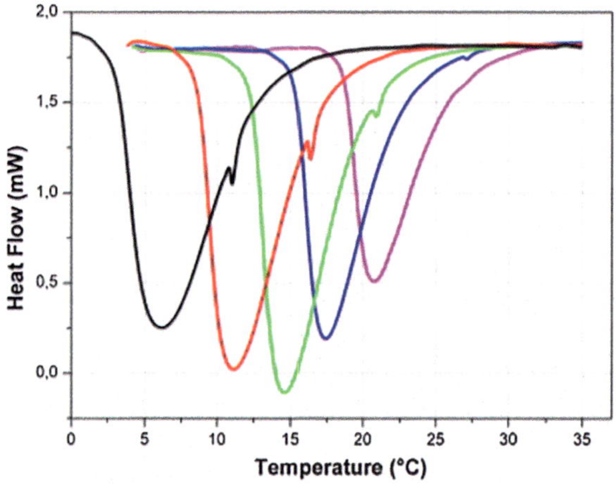

Fig. 9 Heat flow (exo up) versus temperature traces measured during heating ramp of aqueous dispersions of Poloxamer 407 at different concentrations (30, 25, 20, 15 and 10% w/w) obtained from mDSC. Reprinted from [16]

detectable. In fact, a smaller peak at temperature higher than CMT but superimposed to the end of the main transition (micellization process) can be observed. This secondary transition has been recognised as the rearrangement of polymeric micelles into a 3-D network (gelation) [16].

Recently, microcalorimetry has emerged as a useful technique able to highlight thermal behaviour of PEGylated liposomal dispersions. Wei et al. have reported for the first time thermograms obtained from the marketed liposomal formulation Doxil® in comparison to the same unloaded formulation. Microcalorimetry was able to detect two distinctive endothermic transitions, one centred at around 52 °C related to the phase transition of phospholipids composing the liposomal membrane (fully hydro-

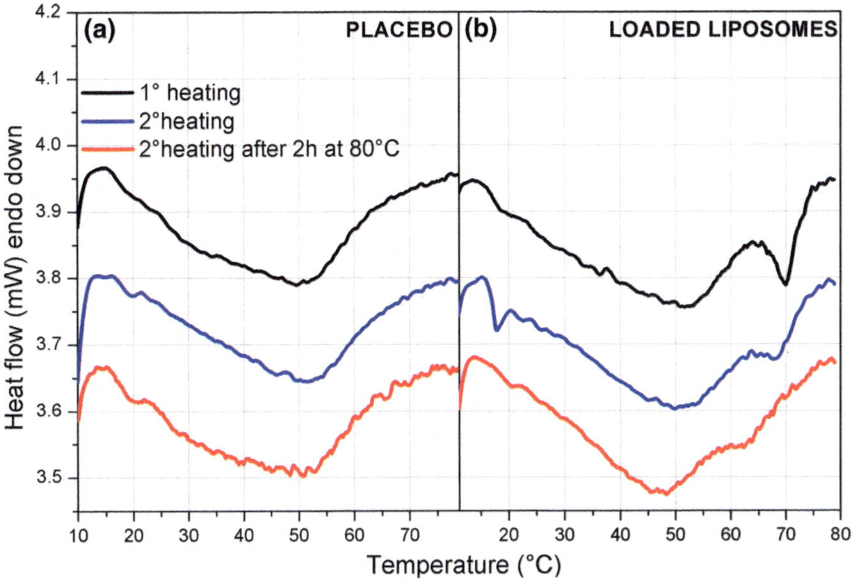

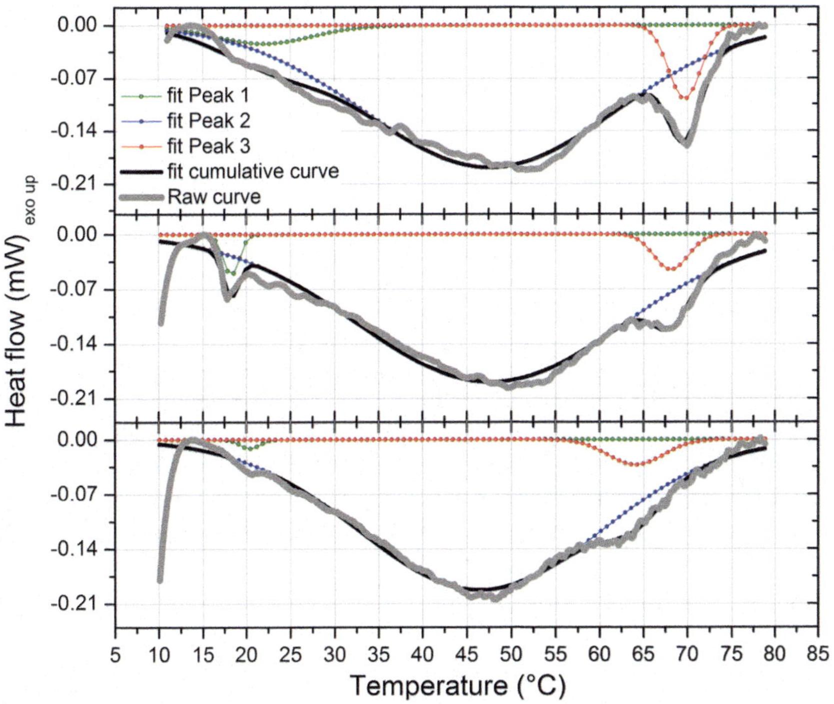

Fig. 10 mDSC traces of unloaded PEGylated liposomes and doxorubicin loaded liposomes (Doxil®) after different thermal treatments (1° heating, consecutive 2° heating, 2° heating performed after an isotherm for 2 h at 80 °C) (on the top). Deconvolution analysis of the thermograms obtained from the analysis by mDSC of doxorubicin loaded liposomes (Doxil®) for the different thermal treatments (on the bottom). Adapted from [21]

genated soy phospholipids HSPC, cholesterol and PEGylated phosphoethanolamide) and another one at around 70–72 °C attributed by the author to the melting of the intraliposomal doxorubicin. According to microcalorimetry results, the authors have concluded that the formulation is stable after one heating cycle 15 °C–90 °C–15 °C both as regards phospholipids and drug transitions [30]. A more detailed investigation of the thermal properties of Doxil® in comparison to the unloaded liposomes, was performed by Perinelli et al. [21], who have analysed different thermal programmes.

Figure 10 shows mDSC traces for Doxil® and unloaded liposomes. Differently from pure phospholipids, for which the sol-gel transition appears as sharp endotherms, in this case a broader trace in the temperature range 20°–60 °C was observed, as a result of the overlapping contribution of each lipid in the system. Interestingly, thermal transition at around 72 °C related to the loaded doxorubicin has emerged as sensitive to the different thermal programme applied. In fact, after heating the sample at a temperature higher than that of the doxorubicin transition (80 °C) for 2 h, the signal related to doxorubicin melting almost disappeared in the second heating scan. A deconvolution analysis on the thermograms was perform to discriminate between the contribution of each transition (Fig. 10). This analysis highlighted the effect of the analysed heating treatments. Particularly, after a heating for 2 h at 80 °C the transition related to doxorubicin shifted at a lower temperature from around 70 °C down to 63°–64 °C. Moreover, the area decreased from 6.6 to 3.6 (J/g of solution) and the peak became broader with a temperature interval at half height of endotherm increasing from 5 to 7.5 °C.

4 Conclusions

Different methodological approaches have been employed to characterize the self-assembling behaviour of materials of different nature (polymers, lipids, peptides or mixed systems). Every methodology such as particle size analysis, spectroscopic analysis or thermal behaviour characterization are able to investigate different aspects of the self-assembling process. Particularly, calorimetry and high-resolution ultrasound spectroscopy are very sensitive to highlight the behaviour of self-assembling systems from a temperature dependent or concentration dependent manner. These techniques alone, together or in combination with other methodologies represent a valuable tool in the field of the nanoscale materials for pharmaceutical or biomedical applications.

References

1. Grzybowski, B.A., Whitesides, G.M.: Dynamic aggregation of chiral spinners. Science **296**, 718–721 (2002). https://doi.org/10.1126/science.1068130

2. Grzybowski, B.A., Fitzner, K., Paczesny, J., Granick, S.: From dynamic self-assembly to networked chemical systems. Chem. Soc. Rev. **46**, 5647–5678 (2017). https://doi.org/10.1039/C7CS00089H
3. Peck, M., Dusserre, N., McAllister, T.N., L'Heureux, N.: Tissue engineering by self-assembly. Mater. Today **14**, 218–224 (2011). https://doi.org/10.1016/S1369-7021(11)70117-1
4. Verma, G., Hassan, P.A.: Self assembled materials: design strategies and drug delivery perspectives. Phys. Chem. Chem. Phys. **15**, 17016 (2013). https://doi.org/10.1039/c3cp51207j
5. Busseron, E., Ruff, Y., Moulin, E., Giuseppone, N.: Supramolecular self-assemblies as functional nanomaterials. Nanoscale **5**, 7098 (2013). https://doi.org/10.1039/c3nr02176a
6. Cheetham, A.G., Chakroun, R.W., Ma, W., Cui, H.: Self-assembling prodrugs. Chem. Soc. Rev. **46**, 6638–6663 (2017). https://doi.org/10.1039/C7CS00521K
7. Depero, L.E., Lucia Curri, M.: Inorganic self-assembly. Curr. Opin. Solid State Mater. Sci. **8**, 103–109 (2004). https://doi.org/10.1016/J.COSSMS.2004.01.006
8. Cong, H.-P., Yu, S.-H.: Self-assembly of functionalized inorganic–organic hybrids. Curr. Opin. Colloid Interface Sci. **14**, 71–80 (2009). https://doi.org/10.1016/j.cocis.2008.09.003
9. Bonacucina, G., Perinelli, D.R., Cespi, M., Casettari, L., Cossi, R., Blasi, P., Palmieri, G.F.: Acoustic spectroscopy: a powerful analytical method for the pharmaceutical field? Int. J. Pharm. **503**, 174–195 (2016). https://doi.org/10.1016/j.ijpharm.2016.03.009
10. Breazeale, M.A., McPherson, M.: Physical acoustics. In: Handbook of Acoustics. Springer, New York (2007)
11. Priev, A., Zalipsky, S., Cohen, R., Barenholz, Y.: Determination of critical micelle concentration of lipopolymers and other amphiphiles: comparison of sound velocity and fluorescent measurements. Langmuir **18**, 612–617 (2002). https://doi.org/10.1021/LA0110085
12. Savaroglu, G., Genc, L.: Determination of micelle formation of ketorolac tromethamine in aqueous media by acoustic measurements. Thermochim. Acta **552**, 5–9 (2013). https://doi.org/10.1016/J.TCA.2012.11.008
13. Savaroglu, G., Yurt, A.: Determination of the second critical micelle concentration of benzyldimethyltridecylazanium chloride in aqueous solution by acoustic and conductometric measurements. J. Chem. Thermodyn. **43**, 1552–1556 (2011). https://doi.org/10.1016/J.JCT.2011.05.011
14. Bhattacharjee, J.K., Kaatze, U.: Fluctuations near the critical micelle concentration. II. Ultrasonic attenuation spectra and scaling. J. Phys. Chem. B **117**, 3798–3805 (2013). https://doi.org/10.1021/jp401120x
15. Bhattacharjee, J.K., Kaatze, U.: Fluctuations near the critical micelle concentration I Premicellar aggregation, relaxation rate, and isentropic compressibility. J. Phys. Chem. B **117**, 3790–3797 (2013). https://doi.org/10.1021/jp4011185
16. Cespi, M., Bonacucina, G., Mencarelli, G., Pucciarelli, S., Giorgioni, G., Palmieri, G.F.: Monitoring the aggregation behaviour of self-assembling polymers through high-resolution ultrasonic spectroscopy. Int. J. Pharm. **388**, 274–279 (2010). https://doi.org/10.1016/j.ijpharm.2009.12.053
17. Farrugia, M., Morgan, S.P., Alexander, C., Mather, M.L.: Ultrasonic monitoring of drug loaded Pluronic F127 micellular hydrogel phase behaviour. Mater. Sci. Eng. C **34**, 280–286 (2014). https://doi.org/10.1016/J.MSEC.2013.09.018
18. Perinelli, D.R., Cespi, M., Pucciarelli, S., Casettari, L., Palmieri, G.F., Bonacucina, G.: Effect of phosphate buffer on the micellisation process of Poloxamer 407: microcalorimetry, acoustic spectroscopy and dynamic light scattering (DLS) studies. Colloids Surf. A Physicochem. Eng. Aspects **436**, 123–129 (2013). https://doi.org/10.1016/j.colsurfa.2013.06.002
19. Inoue, T., Norisuye, T., Sugita, K., Nakanishi, H., Tran-Cong-Miyata, Q.: Size distribution and elastic properties of thermo-responsive polymer gel microparticles in suspension probed by ultrasonic spectroscopy. Ultrasonics **82**, 31–38 (2018). https://doi.org/10.1016/J.ULTRAS.2017.07.007
20. Taylor, T.M., Davidson, P.M., Bruce, B.D., Weiss, J.: Ultrasonic spectroscopy and differential scanning calorimetry of liposomal-encapsulated nisin. J. Agric. Food Chem. **53**, 8722–8728 (2005). https://doi.org/10.1021/jf050726k

21. Perinelli, D.R., Cespi, M., Bonacucina, G., Rendina, F., Palmieri, G.F.: Heating treatments affect the thermal behaviour of doxorubicin loaded in PEGylated liposomes. Int. J. Pharm. **534**, 81–88 (2017). https://doi.org/10.1016/J.IJPHARM.2017.09.069
22. Dominguez, A., Fernandez, A., Gonzalez, N., Iglesias, E., Montenegro, L.: Determination of critical micelle concentration of some surfactants by three techniques. J. Chem. Educ. **74**, 1227 (1997). https://doi.org/10.1021/ed074p1227
23. Aguiar, J., Carpena, P., Molina-Bolívar, J.A., Carnero Ruiz, C.: On the determination of the critical micelle concentration by the pyrene 1:3 ratio method. J. Colloid Interface Sci. **258**, 116–122 (2003). https://doi.org/10.1016/s0021-9797(02)00082-6
24. Patist, A., Bhagwat, S.S., Penfield, K.W., Aikens, P., Shah, D.O.: On the measurement of critical micelle concentrations of pure and technical-grade nonionic surfactants. J. Surf. Deterg. **3**, 53–58 (2000). https://doi.org/10.1007/s11743-000-0113-4
25. Bonacucina, G., Misici-Falzi, M., Cespi, M., Palmieri, G.F.: Characterization of micellar systems by the use of acoustic spectroscopy. J. Pharm. Sci. **97**, 2217–2227 (2008). https://doi.org/10.1002/jps.21156
26. Schick, C., Lexa, D., Leibowitz, L., Schick, C., Lexa, D., Leibowitz, L.: Differential scanning calorimetry and differential thermal analysis. In: Characterization of Materials. Wiley, Hoboken, NJ, USA (2012). ISBN 9780471266969
27. del Río, J.M., Grolier, J.-P.E.: Chapter 4. Isothermal titration calorimetry. In: Enthalpy and Internal Energy, pp. 96–132. Royal Society of Chemistry, Cambridge (2017)
28. Bouchal, R., Hamel, A., Hesemann, P., In, M., Prelot, B., Zajac, J.: Micellization behavior of long-chain substituted alkyl guanidinium surfactants. Int. J. Mol. Sci. **17**, 223 (2016). https://doi.org/10.3390/ijms17020223
29. Lu, H., Pezron, I., Gaudin, T., Drelich, A.: Non-equilibrium micelles formed by sugar-based surfactants under their Krafft temperature. Colloids Surf. A Physicochem. Eng. Aspects **540**, 167–176 (2018). https://doi.org/10.1016/J.COLSURFA.2017.12.053
30. Wei, X., Cohen, R., Barenholz, Y.: Insights into composition/structure/function relationships of Doxil® gained from "high-sensitivity" differential scanning calorimetry. Eur. J. Pharm. Biopharm. **104**, 260–270 (2016). https://doi.org/10.1016/j.ejpb.2016.04.011
31. Šarac, B., Medoš, Ž., Cognigni, A., Bica, K., Chen, L.-J., Bešter-Rogač, M.: Thermodynamic study for micellization of imidazolium based surface active ionic liquids in water: effect of alkyl chain length and anions. Colloids Surf. A Physicochem. Eng. Aspects **532**, 609–617 (2017). https://doi.org/10.1016/J.COLSURFA.2017.01.062
32. Krouská, J., Pekař, M., Klučáková, M., Šarac, B., Bešter-Rogač, M.: Study of interactions between hyaluronan and cationic surfactants by means of calorimetry, turbidimetry, potentiometry and conductometry. Carbohydr. Polym. **157**, 1837–1843 (2017). https://doi.org/10.1016/J.CARBPOL.2016.11.069
33. Bai, G., Wang, Y., Ding, Y., Zhuo, K., Wang, J., Bastos, M.: Thermodynamics of self-assembling of mixture of a cationic gemini surfactant and sodium dodecylsulfate in aqueous solution: calorimetry, conductivity and surface pressure measurements. J. Chem. Thermodyn. **94**, 221–229 (2016). https://doi.org/10.1016/J.JCT.2015.11.017
34. Perspicace, S., Rufer, A.C., Thoma, R., Mueller, F., Hennig, M., Ceccarelli, S., Schulz-Gasch, T., Seelig, J.: Isothermal titration calorimetry with micelles: thermodynamics of inhibitor binding to carnitine palmitoyltransferase 2 membrane protein. FEBS Open Bio **3**, 204–211 (2013). https://doi.org/10.1016/j.fob.2013.04.003
35. Judy, E., Pagariya, D., Kishore, N.: Drug partitioning in micellar media and its implications in rational drug design: insights with streptomycin. Langmuir **34**, 3467–3848 (2018). https://doi.org/10.1021/acs.langmuir.7b04346
36. Kabiri, M., Unsworth, L.D.: Application of isothermal titration calorimetry for characterizing thermodynamic parameters of biomolecular interactions: peptide self-assembly and protein adsorption case studies. Biomacromolecules **15**, 3463–3473 (2014). https://doi.org/10.1021/bm5004515
37. Sikorska, E., Wyrzykowski, D., Szutkowski, K., Greber, K., Lubecka, E.A., Zhukov, I.: Thermodynamics, size, and dynamics of zwitterionic dodecylphosphocholine and anionic sodium

dodecyl sulfate mixed micelles. J. Therm. Anal. Calorim. **123**, 511–523 (2016). https://doi.org/10.1007/s10973-015-4918-0
38. Tiné, M.R., Alderighi, M., Duce, C., Ghezzi, L., Solaro, R.: Effect of temperature on self-assembly of an ionic tetrapeptide. J. Therm. Anal. Calorim. **103**, 75–80 (2011). https://doi.org/10.1007/s10973-010-1060-x
39. Vargas, C., Arenas, R.C., Frotscher, E., Keller, S.: Nanoparticle self-assembly in mixtures of phospholipids with styrene/maleic acid copolymers or fluorinated surfactants. Nanoscale **7**, 20685–20696 (2015). https://doi.org/10.1039/C5NR06353A
40. Loosli, F., Vitorazi, L., Berret, J.-F., Stoll, S.: Isothermal titration calorimetry as a powerful tool to quantify and better understand agglomeration mechanisms during interaction processes between TiO_2 nanoparticles and humic acids. Environ. Sci. Nano **2**, 541–550 (2015). https://doi.org/10.1039/C5EN00139K
41. Krouská, J., Pekař, M., Klučáková, M., Šarac, B., Bešter-Rogač, M.: Study of interactions between hyaluronan and cationic surfactants by means of calorimetry, turbidimetry, potentiometry and conductometry. Carbohydr. Polym. **157**, 1837–1843 (2017). https://doi.org/10.1016/J.CARBPOL.2016.11.069
42. Sastry, N.V., Singh, D.K., Trivedi, P.A.: Studies on micellar behavior of PEO-PBO or PEO-PBO-PEO copolymers, or surface active amphiphilic ionic liquids in aqueous media and exploration of the micellar solutions for solubilization of dexamethasone and its delayed release. J. Surf. Deterg. **21**, 65–79 (2018). https://doi.org/10.1002/jsde.12020
43. Šarac, B., Medoš, Ž., Cognigni, A., Bica, K., Chen, L.-J., Bešter-Rogač, M.: Thermodynamic study for micellization of imidazolium based surface active ionic liquids in water: Effect of alkyl chain length and anions. Colloids Surf. A Physicochem. Eng. Aspects **532**, 609–617 (2017). https://doi.org/10.1016/J.COLSURFA.2017.01.062
44. Patel, S.G., Bummer, P.M.: Thermodynamics of aggregate formation between a non-ionic polymer and ionic surfactants: an isothermal titration calorimetric study. Int. J. Pharm. **516**, 131–143 (2017). https://doi.org/10.1016/J.IJPHARM.2016.10.053
45. Loh, W., Brinatti, C., Tam, K.C.: Use of isothermal titration calorimetry to study surfactant aggregation in colloidal systems. Biochim. Biophys. Acta Gen. Subj. **1860**, 999–1016 (2016). https://doi.org/10.1016/j.bbagen.2015.10.003
46. Koper, G.J.M., Minkenberg, C.B., Upton, I.S., van Esch, J.H., Sudhölter, E.J.R.: Quantitatively interpreting thermal behavior of self-associating systems. J. Phys. Chem. B **113**, 15597–15601 (2009). https://doi.org/10.1021/jp909153n
47. Herrera, I., Winnik, M.A.: Differential binding models for isothermal titration calorimetry: moving beyond the Wiseman isotherm. J. Phys. Chem. B **117**, 8659–8672 (2013). https://doi.org/10.1021/jp311812a
48. Bouchemal, K., Mazzaferro, S.: How to conduct and interpret ITC experiments accurately for cyclodextrin–guest interactions. Drug Discov. Today **17**, 623–629 (2012). https://doi.org/10.1016/J.DRUDIS.2012.01.023
49. Wszelaka-Rylik, M., Gierycz, P.: Isothermal titration calorimetry (ITC) study of natural cyclodextrins inclusion complexes with tropane alkaloids. J. Therm. Anal. Calorim. **121**, 1359–1364 (2015). https://doi.org/10.1007/s10973-015-4658-1
50. Choudhary, S., Talele, P., Kishore, N.: Thermodynamic insights into drug–surfactant interactions: study of the interactions of naporxen, diclofenac sodium, neomycin, and lincomycin with hexadecytrimethylammonium bromide by using isothermal titration calorimetry. Colloids Surf. B Biointerfaces **132**, 313–321 (2015). https://doi.org/10.1016/j.colsurfb.2015.05.031
51. Mukhija, A., Kishore, N.: Partitioning of drugs in micelles and effect on micellization: physicochemical insights with tryptophan and diclofenac sodium. Colloids Surf. A Physicochem. Eng. Aspects **513**, 204–214 (2017). https://doi.org/10.1016/J.COLSURFA.2016.10.044

Phase Transitions in Biological Membranes

Thomas Heimburg

Abstract Native membranes of biological cells display melting transitions of their lipids at a temperature of 10–20 °C below body temperature. Such transitions can be observed in various bacterial cells, in nerves, in cancer cells, but also in lung surfactant. It seems as if the presence of transitions slightly below physiological temperature is a generic property of most cells. They are important because they influence many physical properties of the membranes. At the transition temperature, membranes display a larger permeability that is accompanied by ion-channel-like phenomena even in the complete absence of proteins. Membranes are softer, which implies that phenomena such as endocytosis and exocytosis are facilitated. Mechanical signal propagation phenomena related to nerve pulses are strongly enhanced. The position of transitions can be affected by changes in temperature, pressure, pH and salt concentration or by the presence of anesthetics. Thus, even at physiological temperature, these transitions are of relevance. Their position and thereby the physical properties of the membrane can be controlled by changes in the intensive thermodynamic variables. Here, we review some of the experimental findings and the thermodynamics that describes the control of the membrane function.

1 Introduction

It is a remarkable fact that most biomolecules display order transitions close to physiological conditions. DNA unfolds in a temperature regime above 65 °C depending on the fraction of GC pairs, salt concentration and pH [1, 2], and on the binding of proteins. Protein-unfolding temperature are also found above body temperature and range from slightly above physiological temperature to above 100 °C. At low temperatures, one finds the interesting phenomenon of cold unfolding [3], which has been postulated being the cause for the inactivation of many enzymes at low temperature. If physiological temperature is assumed to be around 310 K, the distance to the temperature of unfolding is small on the absolute temperature scale (e.g., typically

T. Heimburg (✉)
Niels Bohr Institute, University of Copenhagen, Blegdamsvej 17, 2100 Copenhagen, Denmark
e-mail: theimbu@nbi.ku.dk

C. Demetzos and N. Pippa (eds.), *Thermodynamics and Biophysics of Biomedical Nanosystems*, Series in BioEngineering, https://doi.org/10.1007/978-981-13-0989-2_3

only a few percent). Therefore, transitions can be induced by relatively small changes in transition enthalpy. As a consequence, proteins can be denatured by pressure [4], by changes in pH, by addition of small hydrogen-bonding molecules such as urea, or by voltage gradients. Thus, the temperature of denaturation of both DNA and proteins depends on the thermodynamic variables temperature, pressure, the chemical potentials of protons, ions and small molecules - and the electrostatic potential. It is clear that both DNA and protein function depend on variations of these variables and that the state of these molecules can be controlled by changing those variables from the outside.

Lipid membranes share many of the above features. They display melting of the lipid chains that depends on lipid composition, temperature, hydrostatic and lateral pressure, on ion concentration, pH and the binding or insertion of ligands like proteins or anesthetic drugs. It is surprisingly unknown that also biological membranes display melting transitions at temperatures of about 10–20° below body temperature [5, 6]. In contrast to proteins, membranes are quasi two-dimensional and macroscopic. They can support phenomena on scales that are larger than individual molecules. The transitions serve as a functional switch that can be controlled sensitively from the outside.

The membrane transitions and their important functional implications are the topic of this review.

2 Membrane Melting

Artificial membranes

In the lipid melting transition, membranes change their enthalpy H and entropy S, their volume V, their area A and their thickness D [7]. The most prominent change on the molecular scale is the loss of order in the lipid hydrocarbon chains. It is mostly due to rotations around the C–C bonds of the chains [8]. Furthermore, one finds a solid-liquid transition of the crystalline lipid arrangement within the membrane plane, i.e., the lipids can arrange on lattices at low temperature and have more random arrangements at high temperature, with important implications for the solubility of other molecules in the membrane. Consequently, the transitions are denoted as solid-ordered to liquid-disordered transitions (or more trivially as transitions from 'gel' and 'fluid'). Figure 1 shows the change in order, and the associated heat capacity (c_p-) profile of dipalmitoyl phosphatidylcholine (DPPC) - one of the most frequently studied synthetic lipids. The area under the heat capacity profile is the heat of melting, ΔH. The melting entropy can be deduced from these data because $\Delta S = \Delta H/T_m$ [8]. Functions such as the heat capacity are called susceptibilities. They indicate how sensitive an extensive property of the membrane reacts to changes in an intensive variable (such as temperature, pressure, pH or voltage). Beyond the heat capacity, other examples for susceptibilities are volume and area compressibility [7], and capacitance [9]. They generally display maxima in the transition, which implies that one can most sensitively adjust the state of the membrane by subtle changes in experimental conditions. It is unlikely that biology does not make use of such an

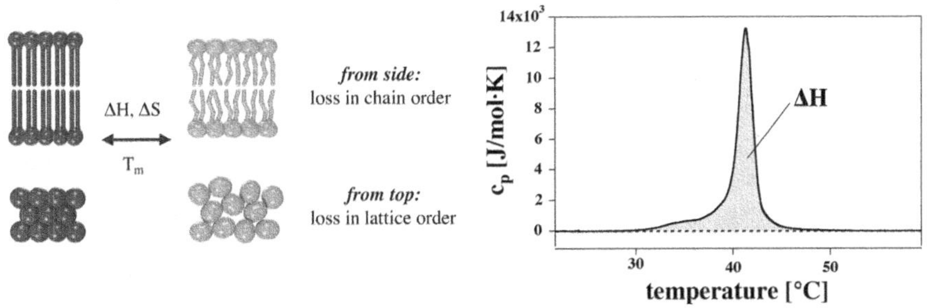

Fig. 1 Lipid membrane melting. Left: Schematic drawing of the changes in chain and lattice order in the melting transition of lipids. Right: Heat capacity profile of the artificial lipid dipalmitoyl phosphatidylcholine (DPPC). The melting transition can be found at 42 °C. The grey-shaded area indicates the melting enthalpy, i.e., the heat of melting. From Ref. [8]

important feature. The transition temperature of pure lipids depends on chain length and chain saturation. Unsaturated chains display lower melting points. Therefore, the fraction of unsaturated chains provides an estimate for the transition temperature. Another factor for the melting behavior is the headgroup. Of the uncharged lipid headgroups, phosphatidylethanolamines display higher melting temperatures as phosphatidylcholines. There exist also negatively charged lipid headgroups such as phosphatidylserine, phosphatidylglycerol or phosphatidylinositol. This renders transition temperatures sensitive to surface-active enzymes such as phospholipase A_1, A_2 [10, 11], phospholipases C and D, which alter the chain or headgroup structure of the lipids. The melting temperatures of charged lipids display a strong dependence on pH and divalent cations such as calcium, discussed below. The more charged a membrane is, the lower its melting temperature. Consequently, lowering of pH [12], increase in calcium concentration and the binding of basic proteins [13] increases the melting temperature of biomembranes.

In the transition, lipid mixtures can phase-separate into phases or domains of different composition and properties [14]. This is shown in Fig. 2. It is a plausible assumption that the frequently discussed 'rafts' [15] in biological membranes (often considered as functional platforms) are just a consequence of domain formation following the rules of phase diagrams [16, 17]. Further, as will be shown below, the elastic constants of lipid membranes are especially high in the transition. Within the transition range, membranes are soft, easily compressible, and very flexible (Sect. 6). They also display a maximum in permeability (Sect. 8). Therefore, the general assumption that lipid membranes are insulators for ions is incorrect close to transitions. The properties of the membranes are also obviously dependent on variables such as temperature, pressure, pH, calcium concentration, and voltage. Therefore, the function of the lipid membrane can be fine-tuned by adjusting these parameters.

Biological membranes

It is much less known that biological membranes also display transitions. There exists quite some literature from the 1970s when the interest in membrane thermodynamics

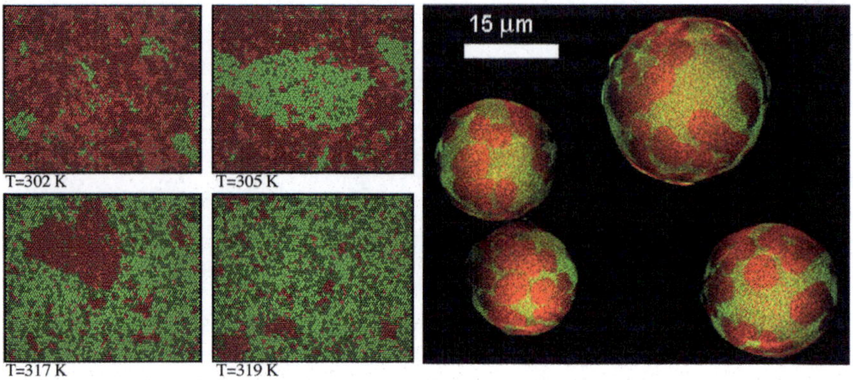

Fig. 2 Domain formation in lipid mixtures. Left: Monte-Carlo simulation at four different temperatures. Right: Domain formation in vesicles of a lipid mixture in the transition range as measured by confocal fluorescence microscopy. In both panels, red represents solid domains and green represents liquid domains. From Ref. [18]

was very high. After the finding of channel proteins and receptors, the interest shifted away from membrane function to more protein-based mechanisms. This led to a very unfortunate neglect of the very important role of the lipid membrane itself.

On average, biological membranes contain about 50% proteins and 50% lipids by mass. This includes the extra-membranous parts of the proteins. Therefore, in the membrane plane itself the lipids outweigh the proteins. There exist some exceptions with extremely high protein content: the purple membrane is effectively a 2D-crystal of bacteriorhodopsin with a lipid:protein ratio of about 1:4. But even in the purple membrane, lipid transitions have been described [19]. In contrast, myelin membranes and lung surfactant display low lipid:protein ratios of about 4:1. For lung surfactant, melting transitions are pronounced and well described [20]. The myelin is discussed below.

Most biomembranes (including all their membrane proteins) display transitions around 10–20° below body temperature. Figure 3 shows the melting profile of porcine spine membranes (adapted from Ref. [6]). One can recognize a pronounce maximum at 23.3 °C (grey-shaded peak) corresponding to lipid melting, and two more maxima corresponding to protein unfolding at 56.5 and 79 °C, respectively. The latter peaks disappear in a second heating scan because the heat unfolding of proteins is not reversible. One can distinguish the lipid melting from protein unfolding peaks because [21]

- in contrast to the lipid melting peaks, the protein unfolding peaks are not reversible and disappear in a second heating scan.
- lipid and protein unfolding peaks display very different pressure dependence.
- lipid extracts from the native membranes (in the absence of proteins) display transitions in the proximity of the lipid peak found in the recordings of native membranes.

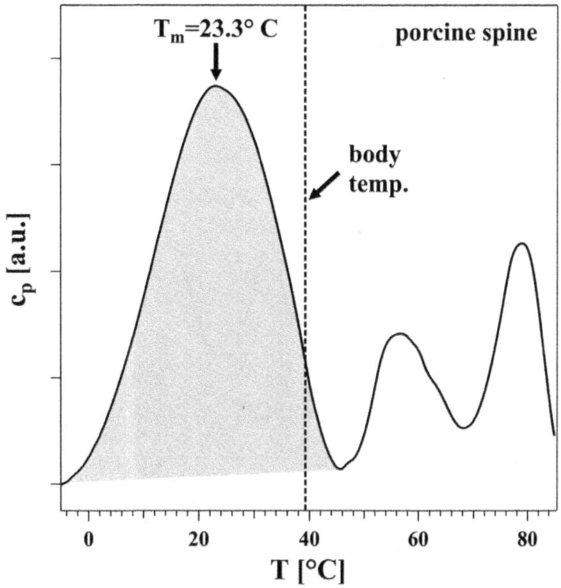

Fig. 3 Heat capacity profile of membranes from the spinal cord of pigs. The grey-shaded peak represents lipid melting. The dashed line represents the body temperature of pigs (∼39.3 °C). The peaks above body temperature represent protein unfolding. Adapted from Ref. [6]

Similar melting profiles have been reported for lung surfactant [5, 6, 20], *E.coli* membranes [5], *Bacillus subtilis* membranes [5] and various species of cancer cells [22]. Further, such transitions have been found in native membranes from chicken spine, rat spine and goat spine (unpublished data from our lab). In all of these preparations, the lipid melting peak was found about 10–20 °C below body temperature.

The assumption that transitions in biomembranes represent a meaningful feature of practically all cells that is actively maintained by the cell will guide the below considerations.

Adaptation

Many organisms do not live at constant temperature or pressure, e.g., bacteria or cold-blooded animals. The membranes of their cells have the possibility to adapt to growth conditions. Figure 4 (left) shows the melting peaks of *E. coli* membranes grown at two different temperatures, 15 and 37 °C, respectively [8]. One can recognize that the protein peaks recorded under different experimental conditions are found at the same temperatures indicating that protein structures are unaffected by growth temperature. However, the lipid peak (grey-shaded) adjusts to growth conditions. Since the lipids have no genome, this is most likely due to an adaption of the lipid composition to growth condition (see, e.g., [23–26]) and not due to selection of a particular strand.

Membrane composition also adapts to growth pressure. Some bacteria live at very high pressures up to 1100 bars in the deep sea. One example is the barophilic deep sea bacterium CNPT3 (Fig. 4, right). As a consequence of adaptation, the ratio of unsaturated to saturated lipids changes linearly with growth pressure [27]. At higher pressures, one finds a higher fraction of unsaturated lipids, which display lower melting temperatures as saturated lipids. Since pressure increases transition

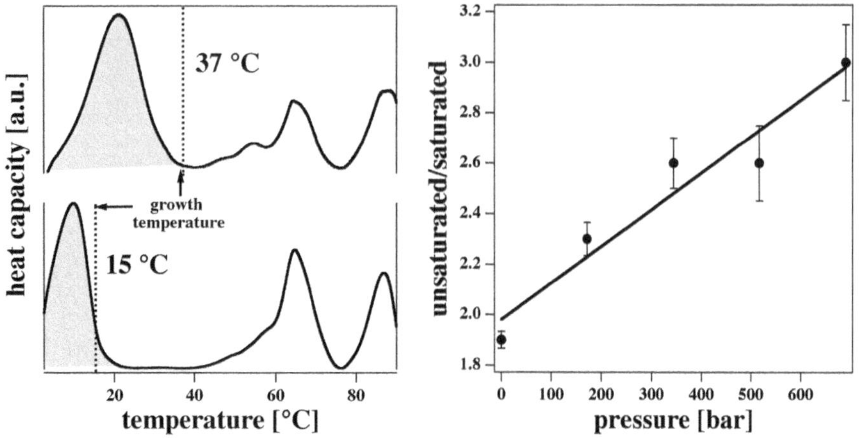

Fig. 4 Adaptation. Left: Heat capacity profiles of *E. coli* membranes grown at 37 °C (top) and 15 °C (bottom). The dashed lines represent growth temperature. The lipid melting peak (grey-shaded) adapt to growth temperature. Protein unfolding peaks remain unaltered. From [8]. Right: Ratio of lipids with low and high melting temperatures (unsaturated versus saturated) as a function of pressure of the barophilic deep sea bacterium CNPT3. Adapted from Ref. [27]

temperatures due to the excess volume of the lipids in transitions, the adaptation of lipid composition in barophilic bacteria compensates for the pressure changes such that the function of the membrane is maintained.

It has further been shown that the lipid composition of *E. coli* adapts to the presence of organic solvents (among those the anesthetic chloroform) in the growth medium [28]. This indicates that cells are 'eager' to maintain a particular state of the membrane. The physiological temperature is practically always found at the upper end of the melting transition (an exception being the purple membrane [19]), and changes in the physiological parameters are compensated by adaptation of the lipid membrane composition. The functional purpose of the transitions will be discussed below.

3 Influence of Anesthetic Drugs on Membrane Transitions

General anesthetics represent a class of small drugs with a surprisingly different chemical structures, ranging from the noble gas Xenon (Xe), laughing gas (N_2O), diethylether ($(C_2H_5)_2O$) and many alcohols to more complex molecules such as propofol ($C_{12}H_{18}O$) or thiopental ($C_{11}H_{18}N_2O_2S$). Surprisingly, they all have something in common. Despite their very different chemical nature, they display generic properties summarized by the famous Meyer-Overton correlation [29]. It states that the solubility of anesthetic drugs in membranes is inversely proportional to the critical dose that causes anesthesia. It is shown for numerous volatile anesthetics in Fig. 5 [29]. The correlation can be written as [30]

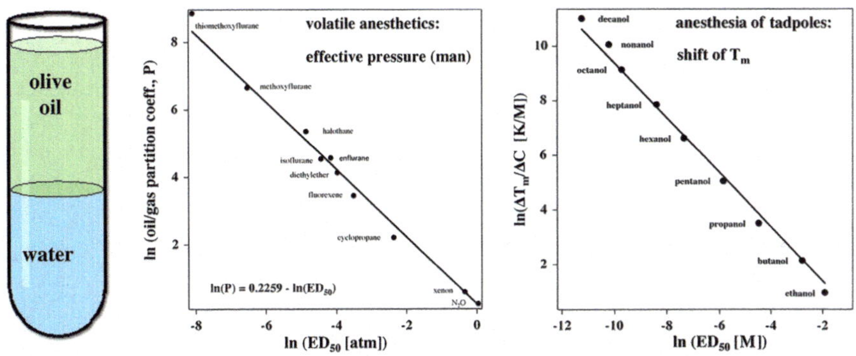

Fig. 5 The Meyer-Overton correlation. Left: The partition coefficient of an anesthetic drug represents the ratio of the concentration of an anesthetic in equal amounts of oil and water. Center: When plotting the partition coefficient of general anesthetics against the critical dose of an anesthetic, ED_{50}, one obtains a straight line with a slope of -1 (in a double-logarithmic plot). Data adapted from Ref. [29]. Right: The reduction of the membrane melting temperature as a function of the critical anesthetic dose for a series of alcohols. One also finds a slope of -1. Adapted from Ref. [31]

$$P \cdot [ED_{50}] = \text{const.} \tag{1}$$

Here, P is the oil/gas or the oil/water partition coefficient (i.e., the solubility of the drug in the lipid membrane), and $[ED_{50}]$ is the effective dose of anesthetics where 50% of all individuals are anesthetized. This correlation is remarkable but not well understood. Overton [32] himself noted that the complete unspecificity of this correlation indicates that its origin is not based on molecular chemistry but rather on generic physical laws. An interpretation of the Meyer-Overton finding is complicated by the fact that there exist membrane-soluble molecules that do not serve as general anesthetics, e.g., cholesterol [33, 34] or long-chained alcohols [35–39]. Thus, it is not exactly clear what distinguishes an anesthetic drug from other molecules. For this reason, the Meyer-Overton correlation has to be considered an empirical finding without much explanatory power. However, after a slight modification of the original assumption it obtains a profound physical meaning, and the requirement for an anesthetic drug become evident.

There exists a generic physical-chemistry law called 'freezing-point depression' law. Some substances such as NaCl reduce the melting point of ice proportional to their molar concentration. This effect exists also for other salts, sugars or alcohols and is independent of chemical structure of the antifreeze agent. The lowering of the melting point of ice, ΔT_m, is given by

$$\Delta T_m = -\frac{RT_m^2}{\Delta H} x_A, \tag{2}$$

where T_m is the melting temperature of ice, ΔH is the heat of melting of ice, R is the gas constant, and x_A is the molar fraction of the molecule A in water. When deriving this law, the main assumption is that the molecule A is ideally soluble in water but absolutely insoluble in ice.

It has been shown that this correlation also applies to anesthetics and the melting points of lipid membranes [30, 39]. In Eq. (2), ΔH is the heat of membrane melting, and x_A is the molar fraction of anesthetics dissolved in the membrane. General anesthetics lower the melting point of lipid membranes. One can calculate the lowering of the membrane melting temperature by postulating that the drugs are soluble in the liquid membrane but not in the crystalline solid membrane - in exact analogy to the freezing-point depression law. Interestingly, this finding is not only true for general anesthetics (e.g., octanol) but also for barbiturates (e.g., pentobarbital) and local anesthetics (such as lidocaine of bupivacaine) [39]. This is shown in Fig. 6.

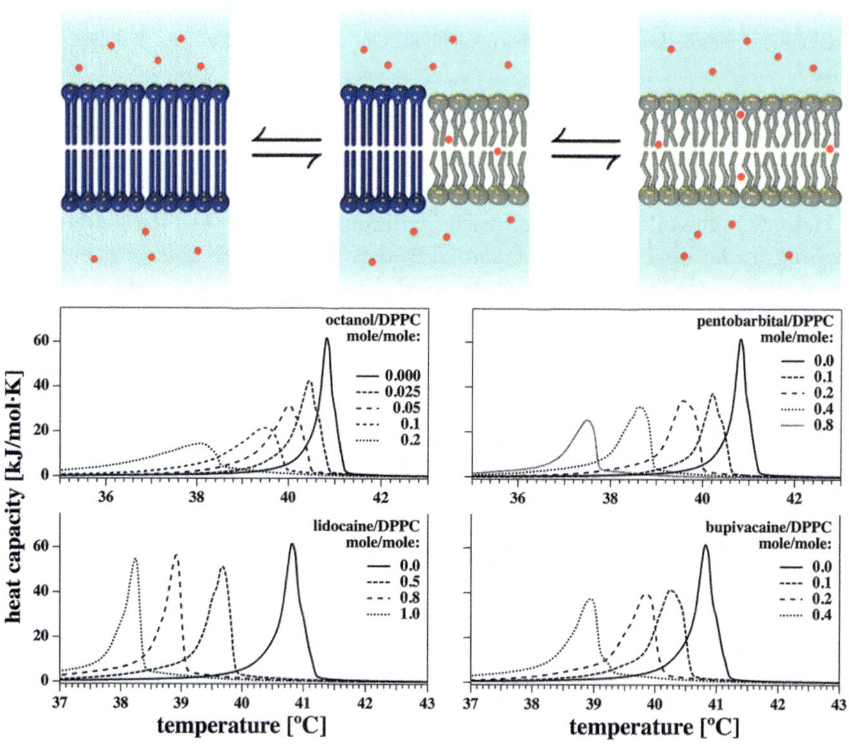

Fig. 6 Anesthesia. Top: The melting-point depression law requires that anesthetic drugs (red dots) are ideally soluble in the liquid membrane but insoluble in the solid membrane. This is shown for three different temperatures (different liquid fractions). Bottom: Melting point depression for four different anesthetics at different concentrations. The alcohol octanol and the barbiturate pentobarbital are general anesthetics, and lidocaine and bupivacaine are local anesthetics. From Ref. [39]

Therefore, the Meyer-Overton correlation can be reformulated: "The effectiveness of an anesthetic drug is proportional to its liquid phase solubility, P_f, while it is insoluble in the solid membrane". Since $P_f \cdot [ED_{50}]$ corresponds the concentration of the anesthetic in the fluid membrane, which is constant at critical dose (Eq. 1), one finds that the molar fraction of anesthetics in a membrane at critical dose is 2.6 mol% [30] for tadpoles. Under these (and only under these) circumstances, the change of the melting temperature upon increase of drug concentration, $d\Delta T_m/d[C]$, is inversely proportional to the critical dose of anesthetics, ED_{50}, or

$$\frac{d\Delta T_m}{d[C]} \cdot [ED_{50}] = \text{const.}, \tag{3}$$

From Eq. 3 combined with the Meyer-Overton correlation for tadpole anesthesia, one can estimate the change in melting temperature at critical dose and finds $\Delta T_m = -0.6°$ [30], completely independent of the chemical nature of the general anesthetic.

As mentioned above, the melting point depression is also valid for local anesthetics. This implies that local anesthetics have an identical influence on the thermodynamics of cell membranes as general anesthetics [39]. Further, the melting-point depression law defines clear criteria for what is an anesthetic and what is not. Required is ideal solubility in the fluid membrane, and no solubility at all in the gel membrane. Only under these conditions the maximum melting-point depression can be achieved. Therefore, whenever the parameters of a particular drug fall on a linear line defined by $dT_m/d[C]$ versus the partition coefficient ED_{50} (Fig. 5, right), the drug has a high potential of being a good anesthetic. These criteria are not fulfilled by molecules such as cholesterol and long-chain alcohols [37–39]. Therefore, they are no good anesthetics. However, the criteria are fulfilled for all drugs that follow the Meyer-Overton correlation [30], and for the barbiturates and local anesthetics investigated by us [39] (Fig. 6).

4 Pressure Reveral of Anesthesia

It has been shown by various authors that general anesthesia can be reversed by hydrostatic pressure [40–43]. This surprising effect can be understood when considering that the melting temperature of lipid membranes depends on pressure, and that the melting-point depression of general and local anesthetics can be reversed by hydrostatic pressure [42, 44–47]. At the melting temperature of a lipid membrane, the molar free energy difference between liquid and solid membrane is zero ($\Delta\mu_0 = \Delta H - T \cdot \Delta S = 0$). This leads to an equation for the melting temperature

$$T_m = \frac{\Delta H}{\Delta S}. \tag{4}$$

The enthalpy is given by $\Delta H = \Delta E + p\Delta V$ and is therefore pressure dependent. This can be written as $\Delta H^{p} = \Delta H^{1bar} \cdot (1 + \gamma_V \Delta p)$, where Δp is the pressure difference relative to 1 bar, and $\gamma_V = 7.8 \times 10^{-10}$ m^2/N is an empirical parameter measured for various artificial lipids and lung surfactant [20]. The change in melting temperature at pressure p is now given by ΔT_m^p:

$$\Delta T_m^p = \gamma_V \cdot \Delta p \cdot T_m^{1bar} . \tag{5}$$

This implies that it increases linearly with pressure. The influence of pressure on lipid melting transitions has been studied in detail for artificial lipids in [20, 48]. However, the effect also exists in biological membranes, such as lung surfactant [20] or *E. coli*-membranes [30], shown in Fig. 7 (center). The effect of pressure on biological preparations and on artificial membranes displays a very similar order of magnitude. Using Eqs. 2 and 5, one can calculate the pressure necessary to reverse general anesthesia. One obtains

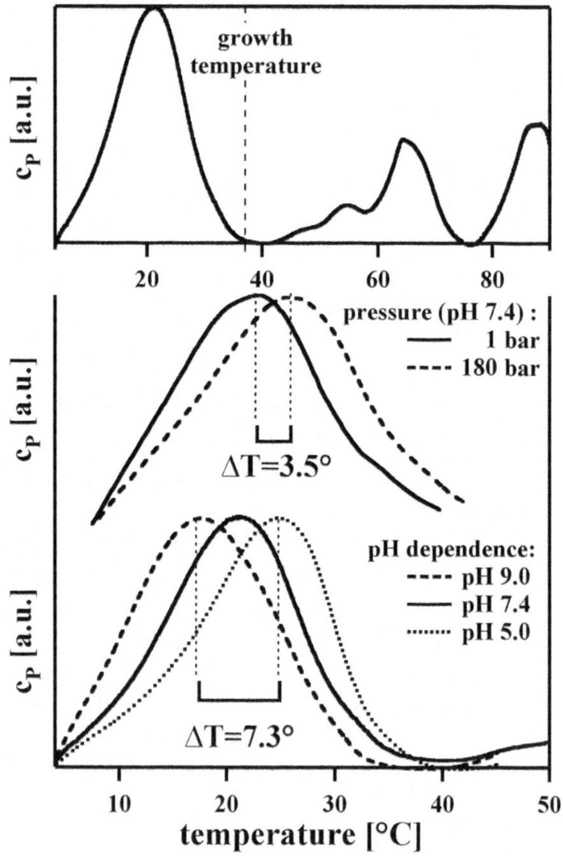

Fig. 7 Top: Melting profile of *E. coli* membranes. Center: Pressure dependence of the lipid meting peak of *E. coli* membranes. Bottom: pH dependence of the lipid melting peak of *E. coli* membranes. Adapted from Ref. [30]

$$\Delta p = \frac{RT_m^{1bar}}{\gamma_V \Delta H^{1bar}} x_A . \tag{6}$$

At this pressure, the melting point depression caused by anesthesia has exactly the same magnitude but opposite sign as the pressure induced shift. For the parameters of the artificial lipid dipalmitoyl phosphatidycholine ($T_m = 314$ K and $\Delta H \approx 35$ kJ/mol and a critical fraction of 2.6 mol%), this happens for approximately 25 bars. Values in the literature are of the same order but vary a lot. Johnson and Flagler [41] found that tadpoles anesthesia in 3–6 % of ethanol (3–6 times of critical doseED_{50}) could be reversed with 140–350 bars of pressure. Our calculation above would lead to a pressure of 75–150 bars of pressure. This is remarkably close taking into account that this calculation is based on the parameters of an artificial lipid while the pressure reversal was measured on tadpoles.

5 Influencing the Melting Transition by pH, Sodium, Potassium, Calcium and Other Charged Ligands

Electrostatics and the influence of ions

Biological membranes contain negatively charged lipids, which are usually asymmetrically distributed over the inner and outer monolayer of the membrane [49, 50]. This implies that the membranes possess both an electrical potential, and a net polarization across the membrane. The negative charges lead to repulsive forces in the membrane plane that cause a lowering of the melting temperature of the membrane [51]. The electrostatic potential can be partially shielded by protons (H^+), Na^+- or K^+-ions and Ca^{2+}-ions. Therefore, an increase in NaCl (or $CaCl_2$) concentration, or a lowering of the pH leads to an increase of the melting temperature [12] (see Fig. 8). An influence of the melting transition can also be induced by the binding of basic proteins such as cytochrome c [51].

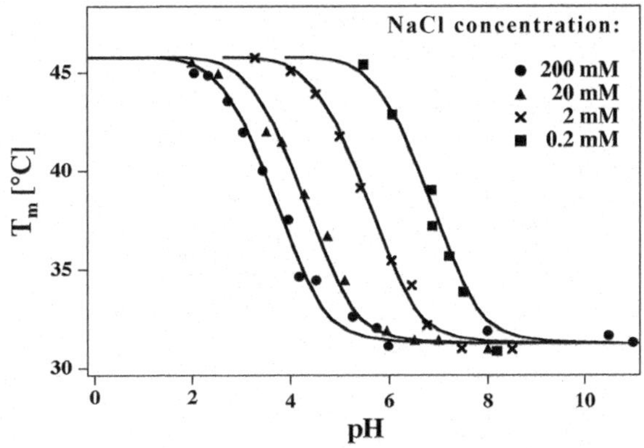

Fig. 8 The melting temperature of methylphosphatidic acid (MPA) as a function of pH and NaCl concentration. Adapted from Ref. [12, 52]

Influence of pH on anesthesia

This influence of ions and large molecules on the electrostatics of membranes, and the resulting changes in transition temperature of membranes is likely to have an influence on anesthesia as described in Sects. 3 and 4. Since lowering of pH increases and anesthetic drugs decrease the transition temperature in membranes, it is likely the decreasing pH reverses the anesthetic effect. In fact, it is known that inflammation leads to the failure of anesthesia. The pH of inflamed tissue is lowered by about 0.5 units [53] which is believed to be the cause for the failure. It has been shown in [30] that the increase of the transition caused by a pH-drop of this order is sufficient to reverse the melting-point depression caused by anesthetics (Fig. 7, bottom).

6 Elastic Constants and Time Scales

Elastic constants

In the transition of a membrane, its enthalpy, volume and area are changing. The temperature dependence of both volume and area is given by [7, 20, 54]

$$\Delta V(T) = \gamma_V \cdot \Delta H(T) \quad \text{and} \quad \Delta A(T) = \gamma_A \cdot \Delta H(T) \tag{7}$$

This is an empirical finding from experiment. Simultaneously, it is known from the fluctuation dissipation theorem that the heat capacity is proportional to the mean square fluctuations in enthalpy [7],

$$\Delta c_p = \frac{\langle \Delta H^2 \rangle - \langle \Delta H \rangle^2}{RT^2}, \tag{8}$$

while the isothermal volume and area compressibilities are proportional to the mean square fluctuations in volume and area [7], respectively

$$\Delta \kappa_T^V = \frac{\langle \Delta V^2 \rangle - \langle \Delta V \rangle^2}{\langle V \rangle R \cdot T} \quad \text{and} \quad \Delta \kappa_T^A = \frac{\langle \Delta A^2 \rangle - \langle \Delta A \rangle^2}{\langle A \rangle R \cdot T}. \tag{9}$$

From Eqs.(7)–(9) follows that

$$\Delta \kappa_T^V = \frac{\gamma_V^2 \cdot T}{\langle V \rangle} \Delta c_p \quad \text{and} \quad \Delta \kappa_T^A = \frac{\gamma_A^2 \cdot T}{\langle A \rangle} \Delta c_p, \tag{10}$$

i.e., volume and area compressibility are proportional to the heat capacity, and can be determined from calorimetric data. A similar relationship can be derived for the bending elasticity [7]:

$$\kappa_{bend} = \frac{16 \gamma_A^2 \cdot T}{\langle D \rangle^2 \langle A \rangle} \Delta c_p \tag{11}$$

This implies that membranes are much softer and more flexible close to a transition.

Relaxation times

Further, by a slightly more sophisticated argument it can be shown that the relaxation time τ is proportional to the heat capacity [55, 56]

$$\tau = \frac{T^2}{L} \Delta c_p \tag{12}$$

where L is a constant. This relation describes the time it takes for a membrane to equilibrate after a perturbation. Due to arguments originating from Lars Onsager [57], it is also the lifetime of the fluctuations. We show in Fig. 9 that the relaxation time is roughly proportional to the heat capacity. At the melting transition, fluctuations are especially slow, an effect known as critical slowing-down. Addition of drugs such as octanol (a general anesthetic), serotonine (a neurotransmitter) and gramicidin A (an antibiotic peptide) all influence the heat capacity profile and thereby influence relaxation times by a purely physical mechanism (Fig. 9). We will show the effect of this below in the section treating the lifetime of lipid pores (lipid channels) in membranes.

Summarizing, both volume and area compressibility, bending elasticity and relaxation time are proportional to the heat capacity and can therefore be estimated from the calorimetric experiment. All of these functions display maxima in the tran-

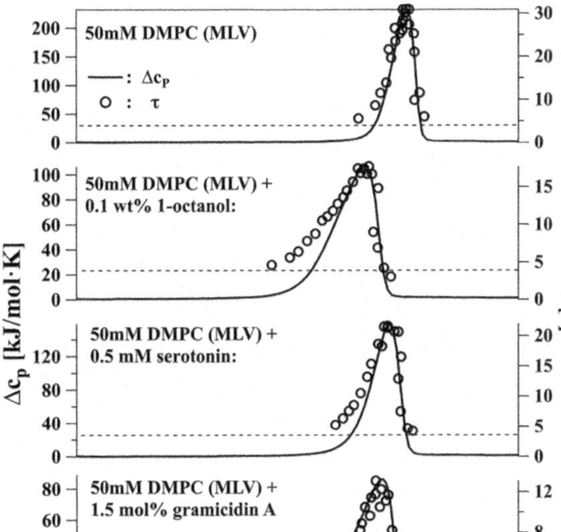

Fig. 9 Relaxation time scale of membranes in the absence and presence of the general anesthetic octanol, the neurotransmitter serotonine, and the antibiotic peptide gramicidin A, as compared to the heat capacity (solid lines). The relaxation times are at maximum in the transition. Adapted from Ref. [55]

sition. Thus, it is clear that the melting transitions have a profound influence on the mechanical properties of membranes, and on the equilibration timescales.

7 Voltage, Capacitance and Transitions

As mentioned in the previous section, lipid membranes change their dimension in the melting transition. For instance, the lipid DPPC decreases its thickness D from 4.79 nm to 3.92 n, and increases its area A from 0.474 nm^2 (at 25°) to 0.629 nm^2 (at 50°) [7]. At the transition itself, this corresponds to a reduction in relative thickness of -16.3% and and increase in relative area of 24.6%. This has a profound influence on the membrane capacitance [9]. Its value is given by

$$C_m = \varepsilon_0 \varepsilon \frac{A}{D} \tag{13}$$

where $\varepsilon_0 = 8.854 \times 10^{-12}$ F/m is the vacuum permittivity, and $\varepsilon \approx 3$ is the dielectric constant of the membrane core. With the above constants, the capacitance is about 0.6 μF/cm^2 at 25 °C, very close to reported values in the biological literature. However, the membrane capacitance of DPPC increases by about 50% in the transition. Figure 10 shows a measurement of the capacitance of a POPE:POPC = 8:2 lipid mixture as a function of temperature [58]. The capacitance changes by roughly 100% upon going through the transition. While this effect is normally neglected, it represents a significant change that has to be taken into account in modeling of the electrical properties of biomembranes, e.g., of nerves.

The fact that different lipid phases display different capacitance renders the membrane transition voltage dependent. It has been shown in [9] that the

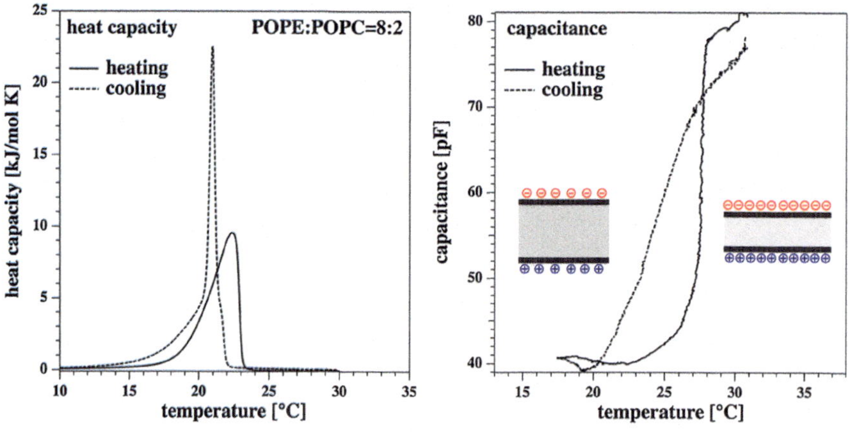

Fig. 10 Change in capacitance of an artificial DOPE:DOPC=8:2 membrane patch. From Ref. [58]

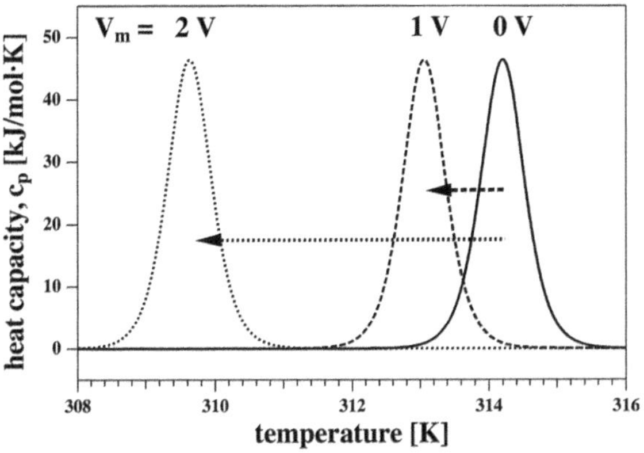

Fig. 11 Calculated change of the transition temperature of DPPC on voltage for $V_0 = 0\,\mathrm{V}$. From Ref. [9]

transition temperature displays a quadratic dependence on voltage on $V - V_0$, where V_0 is the equilibrium polarization of the membrane:

$$T_m(V) = T_{m,0}\left(1 + \alpha\,(V - V_0)^2\right), \tag{14}$$

where the coefficient $\alpha = -0.003634\ 1/\mathrm{V}^2$ was calculated for the lipid DPPC. For the case of $V_0 = 0$ V (no equilibrium polarization of the membrane), the dependence of the transition on the voltage is shown in Fig. 11. If $V_0 \neq 0$ V, the voltage dependence for positive and negative voltages is different.

8 Permeability and Channels

Close to transitions, artificial membranes are no electrical insulators but display a maximum in membrane permeability [59, 61, 62] (Fig. 12, left). This implies that

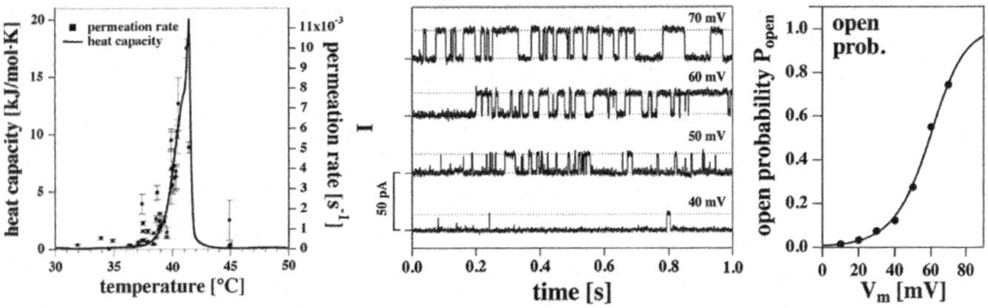

Fig. 12 Left: The permeability of an artificial lipid membrane for a fluorescence marker. From [59]. Center: The channels in a patch recording of synthetic lipid membrane are voltage gated. Right: Open-probability of lipid channels as a function of voltage. From Ref. [60]

close to transitions an important prerequisite of many membrane theories based on ion channel proteins (i.e., that ions do not pass through the lipid membrane) does not hold true. Further, it has been shown that close to transitions, artificial lipid membranes can display channel opening-events very similar to those reported for channel proteins (Fig. 12, center) [59, 63]. It is believed that lipid channels are related to transient opening of hydrophilic or hydrophobic pores in the membrane (for a review see [52, 64–66]). These lipid channels have been shown to display very similar single-channel conductances as proteins, and quite comparable open-lifetime distributions [60, 67]. They can be voltage-gated (Fig. 12, center) and current-voltage relations can display outward rectification [60, 66, 68]. This is particularly interesting because it is generally believed that both features are exclusively a property of proteins. It is clear from membrane experiments that this assumption is not correct.

Since the permeability is at a maximum close to a transition, it is not surprising that membrane-active drugs can influence the permeability of the biomembranes [59, 62]. Figure 13 displays a membrane measured at the transition temperature of a lipid mixture (dashed lines in the c_p profiles). Consecutive addition of a general anesthetics (octanol) lowers the melting temperatures due to melting-point depression and moves the heat capacity events towards lower temperatures - away from the maximum heat capacity situation. The right hand panels of Fig. 13 show the conduction events from electrophysiological patch-clamp recordings. Both, single-channel conductance, open-probabilities of lipid pores as well as the open lifetime of the lipid channels are reduced upon addition of the anesthetic octanol. It is worth-wile noting that the pure lipid membrane reacts in a very similar manner to the general anesthetic octanol as some protein channels, e.g., voltage-gated sodium channels [69] and the acetylcholine receptor [70]. The channels are apparently 'blocked'. However, in the

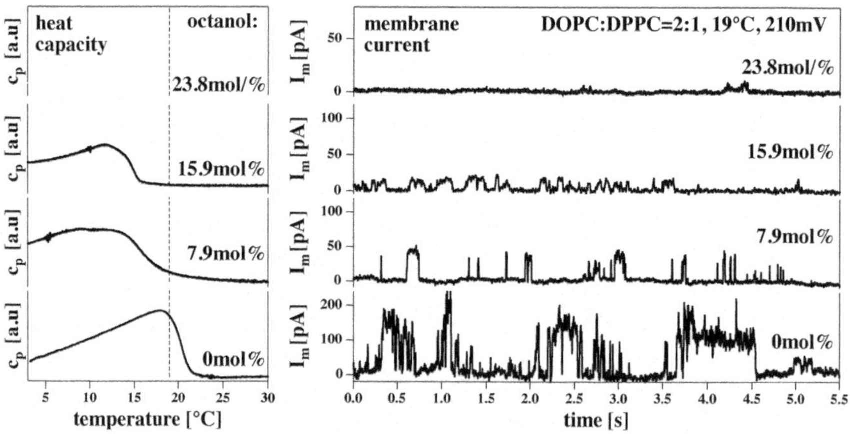

Fig. 13 The influence of anesthetics on the channel activity of a pure lipid membrane close to its transition.The vertical dashed lines indicate experimental temperature. Left: Heat capacity profiles. Right: Channel traces for different amounts of anesthetics. The channel conductance changes and the open probability decreases until the lipid channels are completely "blocked". From Ref. [59]

case of a pure lipid membrane this cannot hold as an explanation since there are no molecular channels that can be blocked. Obviously, there exists a purely physical mechanisms that can prevent conduction events, which is related to the shift of transitions. Summarizing, one has to be cautious when channel-events recorded in electrophysiological experiments are interpreted: They could originate from fluctuating pores in lipid membranes, and the "blocking" of such channels by drugs can be due to the thermodynamics of the molecules dissolving in the membranes [52, 59].

The concepts presented here also seem to apply to protein channels reconstituted in lipid membranes. Seeger et al. [71] demonstrated that the KcsA potassium channel displays a maximum conductance when the host lipid has a transition, and that the open lifetime of the channel is also at maximum. This indicates that the properties of the protein channel are closely tied to the thermodynamics of the surrounding lipid membrane (see also [66]).

9 Pulses Along Membranes

In the model for the nervous impulse established by Hodgkin and Huxley [72, 73], the axon is considered a cylindrical membrane. Cable theory is modified by employing a voltage-dependent membrane conductance for potassium and sodium. This effectively leads to the following differential equation for the time and position dependence of the voltage in an action potential:

$$\frac{a}{2R_i}\frac{\partial^2 V}{\partial x^2} = C_m \frac{\partial V}{\partial t} + g_K(V,t)\cdot(V - E_K) + g_{Na}(V,t)(V - E_{Na}) + \text{leak currents}. \tag{15}$$

Here, all complications from the voltage-gating of ion channel proteins are hidden in the complicated time and voltage-dependent conductances for potassium and sodium, g_K and g_{Na}, respectively. These were parametrized by [72] from experiments on squid axons [74] using many fit constants. The capacitance C_m is assumed being constant. E_K and E_{Na} are the Nernst potentials that reflect the potassium and sodium concentrations inside and outside of the neuron. R_i is the internal specific resistance of the cytoplasm, and a is the cable radius. Equation (15) displays pulse-like solutions, which are generally considered to be the explanation for the action potential.

However, the presence of a phase transition in neural membranes gives rise to a very interesting related phenomenon: The possibility of propagating solitary pulses in cylindrical membranes. These pulses are electromechanical in nature and resemble the propagation of sound. They consist of a variation in membrane area density, $\Delta\rho$. They are the consequence of the change of the elastic constants close to transitions, and the frequency dependence of the sound velocity (known as dispersion). The differential equation for the membrane density changes can be written as (technical details can be found in [5, 76])

$$\frac{\partial^2 \Delta\rho}{\partial t^2} = \frac{\partial}{\partial x}\left((c_0^2 + p\Delta\rho + q\Delta\rho^2)\frac{\partial \Delta\rho}{\partial x}\right) - h \cdot \frac{\partial^4 \Delta\rho}{\partial x^4} \tag{16}$$

Here, c_0 (speed of sound), p and q represent the elastic constants of the membrane close to a transition. They have been determined for artificial and biological membranes using the concepts leading to Eq. (10) [5]. The dispersion parameter h is responsible for describing the frequency dependence of the speed of sound. The theory behind Eq. (16) is often called the "soliton"-theory. While this differential equation has a somewhat different structure, it also possesses pulse-like solutions. During the pulse, the membrane is transiently shifted from a liquid to a solid state and back. This is schematically demonstrated in Fig. 14 (top) where it is shown that a solid region (red) travels in a liquid environment (green). This goes along with changes in membranes thickness and capacitance. We have measured thickness changes by atomic force microscopy. Figure 14 (bottom left) shows the thickness changes in a lobster axon and the corresponding voltage changes. Both mechanical and voltage changes are exactly in phase [75]. In Fig. 14 (bottom, right) a solution of Eq. (16) is given. The density pulses compare well with the shape of both the voltage signal and the thickness change of the membrane. Similar findings have been made by [77–80] on squid axons, garfish nerves and mammalian nerve terminals.

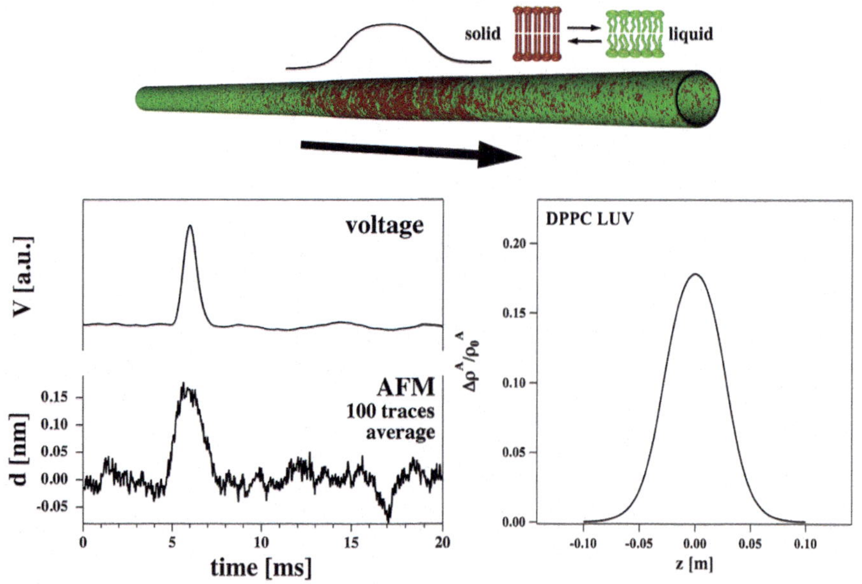

Fig. 14 The nerve pulse as a electromechanical pulse involving phase transitions.Top: Schematic drawing of a propagating solitary pulse that involves a transition (solid=red, liquid=green) From [75]. Bottom left: Mechanical pulse (changes in membrane thickness) measured by atomic force microscopy as compared to the voltage pulse. Adapted from Ref. [75]. Right: Calculated pulse using the elastic parameters of membranes of the lipid DPPC. Adapted from Ref. [5]

There exists no reasonable doubt that the action potential possesses a mechanical component. Therefore, the action potential couples to the thermodynamics of the membrane. Equation (16) describes many non-electrical features of nerves that are not contained in the Hodgkin-Huxley theory, including the mechanical changes, the length contraction, the reversible heat absorption observed in experiments (e.g., [81]) and the observation that two colliding nerve pulses may pass through each other [75, 82].

10 Influence of the Thermodynamic Variables on the Excitability of Nerves

The beauty of a thermodynamic model for the nerve pulse related to membrane transitions lies in the possibility to influence the nerve pulse by changes in the thermodynamic variables. E.g., anesthetics move the melting transition further away from physiological temperature, and thereby make it more difficult to push the membrane through is phase transition. This implies that it becomes more difficult to stimulate a solitary pulse as described in the previous section. Nevertheless, it is clear that while in the soliton model anesthetics will increase the stimulation threshold, they will not block a nerve pulse. This is conceptually shown in Fig. 15 (left). Here, the free energy difference between liquid and solid membrane is plotted versus the membrane density for different concentrations of anesthetics (remember that the amplitude of the soliton represents a change in membrane density). One finds sigmoidal profiles. This plot describes the free energy necessary to stimulate a solitary pulse and how it is influenced by anesthetics. We define the stimulation threshold as the half-maximum density. It corresponds to the position of the melting transition. One can recognize a significant shift in threshold for increasing concentration of the anesthetic drug. These theoretical predictions are compared to experiments on the human median nerve (Fig. 15 (right), adapted from Moldovan et al. [83]). In this experiment, the median nerve was locally anesthetized by high doses of lidocaine. The stimulus was administered be a current between two electrodes at the lower arm, and the response was measured in the thumb. The stimulus-response curve was recorded at different time after the experiment while the anesthetic concentration was gradually washed out. The stimulus-response profile continuously decreases its threshold, and returned to the original situation after about 24 h. After this time, the stimulus-response curve was the same as before the experiment. The threshold increases by a factor of 7–10 without any obvious blocking of the nerve. The maximum amplitude stays about the same, just as in the calculation in the left hand panel. Assuming that the action potential can be explained by the progagation of a soliton in the membrane, it becomes immediately clear why anesthesia can be reversed by pressure: Pressure reduces the threshold by moving the phase transition temperature closer to physiological temperature. Thereby, it counteracts the effect of anesthetics on the stimulation threshold.

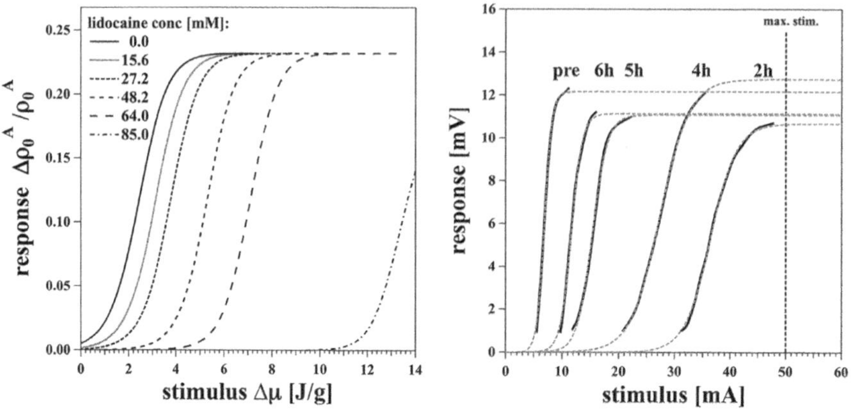

Fig. 15 Stimulus-response curves. Left: Calculated density changes as a function of stimulation free energy for various lidocaine concentrations. Increasing amount of the anesthetic increases the stimulation threshold. The maximum amplitude is conserved, and no blocking of the nerve occurs. Right: Stimulus-response profiles for the human median nerve at various times after administration of lidocain. Adapted from Ref. [6, 83]

11 Summary

In the past, the power of a thermodynamic treatment of biomembranes did not receive the appreciation it deserves. It is likely that many membrane-related properties are intrinsically related to the profound thermodynamic couplings described here. Treating phenomena such as propagating pulses, absorption of drugs, emergence of channels as independent processes will not elucidate the underlying mechanisms. In fact, separating these phenomena may even violate the second law of thermodynamics as shown in [84]. However, the beauty of the thermodynamic couplings recently began to attract more attention. In respect to biomembrane phenomena, it finds its application in the emergence of membrane pores as a consequence of thermal collisions, in propagating pulses as density variations close to regimes with large fluctuations, and the control of such processes by the variation of intensive variables including anesthetic drugs.

Here, we have reviewed the experimental evidences for transitions in biomembranes and the effect that transitions may have on heat capacity, elastic constants, relaxation lifetimes and on capacitance. A natural consequence of the transitions is the emergence of long-lived pores in the membrane. This effect makes the membrane most permeable in the transition. Various factors influence the transitions, e.g., pressure and voltage, but also the presence of drugs such as anesthetics. The latter lower transitions, while pressure increases them. Response functions (or susceptibilities) are derivatives of extensive thermodynamic variables with respect to intensive variables. These are often the properties accessible in an experiment. Examples are heat capacity (dH/dT), capacitance ($dq/d\Psi$), volume and area compressibility ($-V^{-1}dV/dp$ or $-A^{-1}dA/d\Pi$, respectively), the volume expansion coefficient (dV/dT) or the bending elasticity. All of them display maxima in transitions and are related to fluctuation relations. Changes in the intensive variables (temperature, voltage, hydrostatic and lateral pressure, and the chemical potential of membrane components) influence the position of transitions and thereby affect the susceptibilities.

It is very important to note that these couplings lead to correlations in many functions of the membranes. Anesthetics and pressure both influence elastic constants and fluctuation lifetimes. This also implies that pressure influences the effect of anesthetics. Similar couplings exist between capacitance, membrane thickness and electrical phenomena. They are all tied to the same thermodynamics, and by necessity a change in one intensive variable will effect the dependence of the membrane state on the change of other variables. Compressibilities are not independent of heat capacity, capacitance and fluctuation lifetime.

The thermodynamics of membranes brings about a coherent picture in which all phenomena are coupled in a relatively simple framework. The emergence of pulses and channel-like features in membranes, and the coupling to the time scales are an immediate consequence. Such an approach is in strong contrast to the usual attempt to provide separate theories for each phenomenon.

References

1. Marmur, J., Doty, P.: J. Mol. Biol. **5**, 109 (1962)
2. Manning, G.S.: Q. Rev. Biophys. **11**, 179 (1978)
3. Privalov, P.L.: Crit. Rev. Biochem. Mol. Biol. **25**, 281 (1990)
4. Ravindra, R., Winter, R.: Chem. Phys. Chem. **4**, 359 (2003)
5. Heimburg, T., Jackson, A.D.: Proc. Natl. Acad. Sci. USA **102**, 9790 (2005)
6. Wang, T., Mužić, T., Jackson, A.D., Heimburg, T.: Biochim. Biophys. Acta **1860**(2145–2153) (2018). https://doi.org/10.1016/j.bbamem.2018.04.003
7. Heimburg, T.: Biochim. Biophys. Acta **1415**, 147 (1998)
8. Heimburg, T.: Thermal Liophysics of Membranes. Wiley VCH, Berlin, Germany (2007)
9. Heimburg, T.: Biophys. J. **103**, 918 (2012)
10. Mouritsen, O.G., Andersen, T.L., Halperin, A., Hansen, P.L., Jakobsen, A.F., Jensen, U.B., Jensen, M.O., Jørgensen, K., Kaasgaard, T., Leidy, C., Simonsen, A.C., Peters, G.H., Weiss, M.: J. Phys.: Condens. Matter **18**, S1293 (2006)
11. Gudmand, M., Fidorra, M., Bjørnholm, T., Heimburg, T.: Biophys. J. **96**, 4598 (2009)
12. Träuble, H., Teubner, M., Woolley, P., Eibl, H.: Biophys. Chem. **4**, 319 (1976)
13. Heimburg, T., Biltonen, R.L.: Biophys. J. **70**, 84 (1996)
14. Lee, A.G.: Biochim. Biophys. Acta **472**, 285 (1977)
15. Simons, K., Ikonen, E.: Nature **387**, 569 (1997)
16. London, E.: Curr. Opin. Struct. Biol. **12**, 480 (2002)
17. Almeida, P.F.: Biophys. J. **106**, 1841 (2014)
18. Hac, A., Seeger, H., Fidorra, M., Heimburg, T.: Biophys. J. **88**, 317 (2005)
19. Jackson, M.B., Sturtevant, J.M.: Biochemistry **17**(Biochemistry, 17(5), 911–915.), 911 (1978)
20. Ebel, H., Grabitz, P., Heimburg, T.: J. Phys. Chem. B **105**, 7353 (2001)
21. Mužić, T., Madsen, S.B., Tounsi, F., Konrad, M., Heimburg, T.: In preparation (2018)
22. Højholt, K.L., Mužić, T., Jensen, S.D., Bilgin, M., Nylandsted, J., Heimburg, T., Frandsen, S.K., Gehl, J.: Submitted (2018)

23. Hazel, J.R.: Am. J. Physiol. Regul. Integr. Comp. Physiol. **287**, R633 (1979)
24. Hazel, J.R.: Ann. Rev. Physiol. **57**, 19 (1995)
25. Avery, S.V., Lloyd, D., Harwood, J.L.: Biochem. J. **312**, 811 (1995)
26. van de Vossenberg, J.L.C.M., Driessen, A.J.M., da Costa, M.S., Konings, W.N.: Biochim. Biophys. Acta **1419**, 97 (1999)
27. DeLong, E.F., Yayanos, A.A.: Science **228**, 1101 (1985)
28. Ingram, L.O.: Appl. Environ. Microbiol. **33**, 1233 (1977)
29. Overton, C.E.: Studies of Narcosis. Chapman and Hall, New York (1991) (English version of 'Studien der Narkose' from 1901)
30. Heimburg, T., Jackson, A.D.: Biophys. J. **92**, 3159 (2007)
31. Kharakoz, D.P.: Biosci. Rep. **210**, 801 (2001)
32. Overton, C.E.: Studien über die Narkose. Verlag Gustav Fischer., Jena, Germany (1901). (English translation: Studies of Narcosis, Chapman and Hall, 1991, R. Lipnick, Ed.)
33. Cantor, R.S.: J. Phys. Chem. B **101**, 1723 (1997)
34. Cantor, R.S.: Biochemistry **36**, 2339 (1997)
35. Pringle, M.J., Brown, K.B., Miller, K.W.: Mol. Pharmacol. **19**, 49 (1981)
36. Kaminoh, Y., Nishimura, S., Kamaya, H., Ueda, I.: Biochim. Biophys. Acta **1106**, 335 (1992)
37. Kamaya, H., Matubayasi, N., Ueda, I.: J. Phys. Chem. **88**, 797 (1984)
38. Kamaya, H., Ma, S.M., Lin, S.H.: J. Membr. Biol. **90**, 157 (1986)
39. Græsbøll, K., Sasse-Middelhoff, H., Heimburg, T.: Biophys. J. **106**, 2143 (2014)
40. Johnson, F.H., Brown, D., Marsland, D.: Science **95**, 200 (1942)
41. Johnson, F.H., Flagler, E.A.: Science **112**, 91 (1950)
42. Johnson, S., Miller, K.W.: Nature **228**, 75 (1970)
43. Halsey, M.J., Wardley-Smith, B.: Nature **257**, 811 (1975)
44. Trudell, J.R., Payan, D.G., Chin, J.H., Cohen, E.N.: Proc. Natl. Acad. Sci. USA **72**, 210 (1975)
45. Mountcastle, D.B., Biltonen, R.L., Halsey, M.J.: Proc. Natl. Acad. Sci. USA **75**, 4906 (1978)
46. Kamaya, H., Ueda, I., Moore, P.S., Eyring, H.: Biochim. Biophys. Acta **550**, 131 (1979)
47. Galla, H.J., Trudell, J.R.: Biochim. Biophys. Acta **599**, 336 (1980)
48. Winter, R., Pilgrim, W.C.: Ber. Bunsenges. Phys. Chem. **93**, 708 (1989)
49. Rothman, J.E., Lenard, J.: Science **195**, 743 (1977)
50. Rothman, J.E., Kennedy, E.P.: J. Mol. Biol. **110**, 603 (1977)
51. Heimburg, T., Marsh, D.: Biological membranes: a molecular perspective from computation and experiment. In: Merz, K.M., Roux, B. (eds.) pp. 405–462. Birkhäuser, Boston (1996)
52. Heimburg, T.: Biophys. Chem. **150**, 2 (2010)
53. Punnia-Moorthy, A.: J. Oral Pathol. **16**, 36 (1987)
54. Pedersen, U.R., Peters, G.H., Schrøder, T.B., Dyre, J.C.: J. Phys. Chem. B **114**(6), 2124 (2010)
55. Seeger, H.M., Gudmundsson, M.L., Heimburg, T.: J. Phys. Chem. B **111**, 13858 (2007)
56. Grabitz, P., Ivanova, V.P., Heimburg, T.: Biophys. J. **82**, 299 (2002)
57. Onsager, L.: Phys. Rev. **38**, 2265 (1931)
58. Zecchi, K.A., Mosgaard, L.D., Heimburg, T.: J. Phys. Conf. Ser. **780**, 012001 (2017)
59. Blicher, A., Wodzinska, K., Fidorra, M., Winterhalter, M., Heimburg, T.: Biophys. J. **96**, 4581 (2009)
60. Blicher, A., Heimburg, T.: PloS ONE **8**, e65707 (2013)
61. Papahadjopoulos, D., Jacobson, K., Nir, S., Isac, T.: Biochim. Biophys. Acta **311**, 330 (1973)
62. Sabra, M.C., Jørgensen, K., Mouritsen, O.G.: Biochim. Biophys. Acta **1282**, 85 (1996)
63. Laub, K.R., Witschas, K., Blicher, A., Madsen, S.B., Lückhoff, A., Heimburg, T.: Biochim. Biophys. Acta **1818**, 1 (2012)
64. Glaser, R.W., Leikin, S.L., Chernomordik, L.V., Pastushenko, V.F., Sokirko, A.I.: Biochim. Biophys. Acta **940**, 275 (1988)
65. Böckmann, R., de Groot, R., Kakorin, S., Neumann, E., Grubmüller, H.: Biophys. J. **95**, 1837 (2008)
66. Mosgaard, L.D., Heimburg, T.: Acc. Chem. Res. **46**, 2966 (2013)
67. Gallaher, J., Wodzinska, K., Heimburg, T., Bier, M.: Phys. Rev. E **81**, 061925 (2010)
68. Mosgaard, L.D., Zecchi, K.A., Heimburg, T.: Soft Matter **11**, 7899 (2015)

69. Horishita, T., Harris, R.A.: J. Pharm. Exp. Ther. **326**, 270 (2008)
70. Zuo, Y., Aistrup, G.L., Marszalec, W., Gillespie, A., Chavez-Noriega, L.E., Yeh, J.Z., Narahashi, T.: Mol. Pharmacol. **60**, 700 (2001)
71. Seeger, H.M., Aldrovandi, L., Alessandrini, A., Facci, P.P.: Biophys. J. **99**, 3675 (2010)
72. Hodgkin, A.L., Huxley, A.F.: J. Physiol. London **117**, 500 (1952)
73. Johnston, D., Wu, S.M.S.: Cellular Neurophysiology. MIT Press, Boston (1995)
74. Hodgkin, A.L., Huxley, A.F.: J. Physiol. London **116**, 449 (1952)
75. Gonzalez-Perez, A., Mosgaard, L.D., Budvytyte, R., Villagran Vargas, E., Jackson, A.D., Heimburg, T.: Biophys. Chem. **216**, 51 (2016)
76. Lautrup, B., Appali, R., Jackson, A.D., Heimburg, T.: Eur. Phys. J. E **34**, 57 (2011)
77. Iwasa, K., Tasaki, I.: Biochem. Biophys. Res. Comm. **95**, 1328 (1980)
78. Iwasa, K., Tasaki, I., Gibbons, R.C.: Science **210**, 338 (1980)
79. Tasaki, I., Kusano, K., Byrne, M.: Biophys. J. **55**, 1033 (1989)
80. Kim, G.H., Kosterin, P., Obaid, A., Salzberg, B.M.: Biophys. J. **92**, 3122 (2007)
81. Ritchie, J.M., Keynes, R.D.: Q. Rev. Biophys. **18**, 451 (1985)
82. Gonzalez-Perez, A., Budvytyte, R., Mosgaard, L.D., Nissen, S., Heimburg, T.: Phys. Rev. X **4**, 031047 (2014)
83. Moldovan, M., Lange, K.H.W., Aachmann-Andersen, N.J., Kjær, T.W., Olsen, N.V., Krarup, C.: J. Physiol. **592**, 2735 (2014)
84. Heimburg, T., Jackson, A.D.: Structure and dynamics of membranous interfaces. In: Nag, K. (ed.), pp. 317–339. Wiley (2008)

Isothermal Titration Calorimetry: A Powerful Tool for the Characterization of Molecular Interactions

Angelos Thanassoulas and George Nounesis

Abstract Isothermal titration calorimetry (ITC) is a well-established technique that allows the accurate and precise determination of binding equilibrium constants. It is able to provide detailed thermodynamic description of reacting systems without the need for van't Hoff analysis. ITC plays an important role in biology, biochemistry and medicinal chemistry, providing researchers with important information on the structure, stability and functionality of biological and synthetic molecules. This review demonstrates the power and versatility of ITC in providing accurate, rapid, and label-free measurement of the thermodynamics of molecular interactions. Moreover, this work focuses on recent studies employing ITC to investigate compounds of great biotechnological interest.

Keywords Isothermal titration calorimetry · Molecular recognition · Binding thermodynamics · Nanoparticles · Polyelectrolytes · Cyclodextrins · Liposomes · Rational drug design

1 Introduction

One of the most fundamental processes in biology, defined as molecular recognition, is the specific binding of molecules driven by noncovalent forces such as of hydrogen bonds, hydrophobic, van der Waals and charge-charge interactions. Networks of interacting molecules form intricate biological pathways that regulate all critical aspects of life, including homeostasis, signal transduction, energy production and gene expression.

To gain insight into cellular functions, many key-components of these networks, like proteins, peptides, lipids, carbohydrates and nucleic acids have been identified

A. Thanassoulas (✉) · G. Nounesis
Biomolecular Physics Laboratory, INRaSTES, National Centre for Scientific Research 'Demokritos', Athens, Greece
e-mail: athanas@rrp.demokritos.gr

G. Nounesis
e-mail: nounesis@rrp.demokritos.gr

C. Demetzos and N. Pippa (eds.), *Thermodynamics and Biophysics of Biomedical Nanosystems*, Series in BioEngineering, https://doi.org/10.1007/978-981-13-0989-2_4

and studied extensively. Over the past 60 years, rapid developments in genomics and proteomics allowed scientists to collect a large amount of structural information of ever-increasing detail on these biomolecules and their complexes. As of 2018, over 137,000 macromolecular structures are accessible via the Protein Data Bank repository (www.rcsb.org) [1] and new structures are being added every year at an accelerated rate. Knowledge of spatial orientation and surface complementarity of biological assemblies has been crucial for the in-depth study of many important cellular processes, allowing us to make useful connections about specific three-dimensional architectures and functionality. However, biological recognition is not governed by structural features alone; to explore fully the binding mechanisms and expand our understanding of physiological processes at the molecular level, we must be also able to describe how strongly and how fast the binding components interact. Any effort to modulate cellular activity by therapeutic molecules or bio-inspired nanomaterials requires a detailed information on the nature of the molecular forces involved. Thus, a combination of kinetic, structural and thermodynamic data is necessary for a rigorous description of any biological interaction.

Several biophysical techniques, including fluorescence spectroscopy, capillary electrophoresis, surface plasmon resonance and mass spectroscopy have been successfully employed over the years to study noncovalent molecular interactions. The strengths and weaknesses of the most common techniques in terms of sensitivity, sample concentration, throughput and ease of use are presented in Table 1. Isothermal titration calorimetry (ITC) stands out as a versatile and sensitive method that allows the rigorous characterization of molecular interactions without any need for modification or immobilization of the compounds involved.

ITC has near-universal applicability, allowing a non-invasive approach to quantify binding events in solution. Energy exchange with the environment in the form of heat is an inherit property of all binding reactions; as a result, the range of interactions that can be studied by ITC is vast. The most significant advantage of the technique is its unique ability to provide a model-independent measurement of the binding enthalpy change (ΔH_b), without the underlying assumptions of van't Hoff analysis used by other methods. In addition, with the use of an appropriate reaction model, important binding parameters including the binding constant (K_b), the binding entropy change (ΔS_b), the stoichiometry of the interaction (N) and the Gibbs free energy change (ΔG_b) can be determined from a limited set of experiments and calculations. This detailed thermodynamic description of the system facilitates the unraveling of the binding mechanism and the molecular forces governing the binding process.

Up until a few decades ago, only a few custom-made titration calorimeters were available worldwide, accessible only to pioneers that painstakingly developed methods and technologies that are still in use today. A prototype model of a titration calorimeter was presented by Christensen and Izatt in 1968. The sensitivity of the instrument was not sufficient to study biological interactions; however, the usefulness of the technique was demonstrated by measurements of binding enthalpies for simpler reactions [2]. In 1979, Langerman and Biltonen described instrumentation, concepts and methods that could be used in the calorimetric study of biological systems. The first commercially available titration calorimeters, designed specifically

Table 1 Biophysical techniques commonly used to analyze binding interactions

Technique	Advantages	Disadvantages	Daily throughput	Affinity range (M^{-1})	References
Surface plasmon resonance (SPR)	Small sample sizes, real time monitoring, high sensitivity	Need for immobilization, expensive, solvent interference in signal	High	10^3–10^{13}	[124, 125]
Nuclear magnetic resonance (NMR)	Reliable, high resolution, can provide structural information	Expensive, sample size restrictions, large amount of samples required	Low	10^3–10^6	[126, 127]
Analytical ultracentrifugation (AUC)	Small sample volume, limited to homomeric interactions	Cannot detect non-specific interactions, long experimental times, expensive	Low	10^3–10^8	[128, 129]
Fluorescence binding assays	Measurements in solution, easy to use, affordable	Requires presence of fluorescent groups, low resolution	Medium	10^4–10^{11}	[130, 131]
Capillary electrophoresis	Small sample requirements, no effect from impurities	Limited to low solution ionic strength	Low	10^2–10^9	[132, 133]
Mass spectroscopy	Low sample consumption, high resolution, high sensitivity, fast speed of analysis	Low affinity binders are difficult to detect, sample desalting is required, detergents are not tolerated	Very high	Up to 10^5	[134, 135]

(continued)

Table 1 (continued)

Technique	Advantages	Disadvantages	Daily throughput	Affinity range (M^{-1})	References
Isothermal titration calorimetry (ITC)	High sensitivity, direct enthalpy measurements, complete thermodynamic profile acquired, in solution measurements	High concentrations, requires samples of high purity	Low	10^3–10^9	[64, 136]

for biomolecular reactions, appeared a decade later [3]. Along with technological advancements, the sensitivity of titration calorimeters improved drastically. Modern isothermal titration calorimeters are able to detect very small amounts of heat and provide quantitative characterization of noncovalent interactions in conditions resembling those of biological systems.

As any other analytical technique, the use of ITC comes with certain drawbacks. ITC is a laborious and time-consuming method (low throughput). Compared to other techniques, it requires high concentrations of sample that needs to be of high purity (>95%). Despite these limitations, ITC is widely considered as the "gold standard" for the study of binding interaction. The popularity of the technique is evident by the number of publications generated each year (Fig. 1), covering disciplines including biochemistry, biophysics, structural biology, materials science, medicinal chemistry, supramolecular chemistry and food science.

This chapter aims to demonstrate the power of ITC for characterizing noncovalent molecular interactions and to highlight the suitability of the technique for pharmaceutical and biotechnological applications with selected examples from the literature.

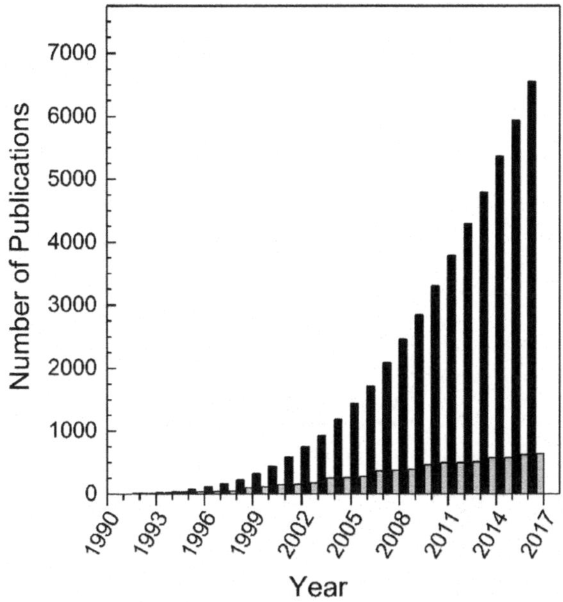

Fig. 1 Number of publications containing isothermal titration calorimetry data since 1990, sourced from the Web of Science citation database. Black bars correspond to the total number of available articles while grey bars represent the annual growth

2 Basic Concepts, Experimental Procedure and Data Analysis

2.1 Device Design and Experimental Procedure

Figure 2 shows the basic parts of a typical ITC instrument. Although significant steps have been made over the years towards scaled-down and ultrasensitive devices, the operating principles and general design remain unchanged.

Almost all ITC instruments operate in a differential mode. At the center of the device, symmetrically placed inside a jacket, there are two identical cells made of a heat-conductive corrosion-resistant metal. The sample cell is the vessel where the titration occurs, while the other cell serves throughout the experiment as a reference system with similar thermal behavior. Monitoring thermal fluxes as differential measurements ensures higher signal sensitivity and faster system responsiveness consequently the concept has been incorporated in most of the commercially available titration calorimeters. The temperature inside the jacket is constantly kept identical to that of the cells, functioning as an isothermal enclosure that prohibits any heat exchange with the environment. A series of thermoelectric sensors are attached to the cells and constantly provide temperature data to a temperature controller. The sensors detect temperature changes only in the active cell volume and not in the stem region that is used to collect the excess volume.

Separate heating and cooling systems in close proximity of the cells modulate their temperature. Some devices use a passive cooling system, while in other models

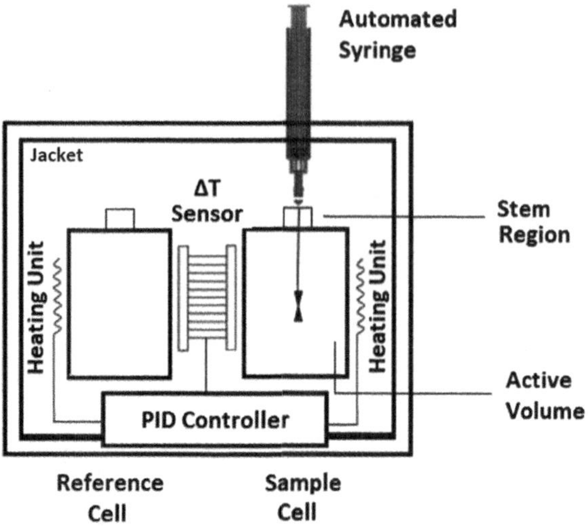

Fig. 2 Schematic representation of a typical isothermal titration calorimeter. Two identical cells are surrounded by a thermostated jacket that acts as an isothermal enclosure. A syringe is used for delivering small aliquots of a binding partner and also serves as a stirring device. Thermal sensors between the cells feed data to a temperature controller that adjusts the power supplied to the heating units to keep the temperature difference between the cells stable

it is actively adjusted by the temperature controller. Hence, in steady-state conditions, a certain amount of power needs to be supplied constantly to the cells in order to keep their temperature constant.

The direct observable in an ITC experiment is the differential power supply required to maintain the two cells at the same temperature as a function of time. When the sensors detect a small temperature gradient (ΔT) between the two cells, the power supply is properly adjusted to return the temperature back to the desired value (power compensation). The power supply difference between the two cells is often called differential power or cell feedback power (CFB). The CFB signal at the steady state before the injections occur consists the baseline of the titration. The baseline level can be adjusted by maintaining a small constant temperature difference ($\Delta T \sim 0.005$ °C) between the cells. This is important because near-zero CFB values destabilize the feedback circuit.

Two separate solutions of the reacting species are prepared. The solutions must be of high purity, known concentration and aggregate-free. The first reactant (titrand) is placed in the sample cell and the other (titrant) is loaded to the instrument's syringe. The syringe is then positioned on the top of the sample cell, with its lower tip submerged in the titrand solution. The syringe is computer-controlled and it can be programmed to execute automatically a series of small-volume injections at preset intervals with high accuracy. In addition, during the experiment, the syringe rotates at a constant speed, serving as a stirring device to ensure sample homogeneity and fast thermal equilibration.

For most applications, the heat capacity of the sample cell is effectively matched by filling the reference cell with deionized water. Filling the cell with the titration buffer is often avoided for cleaning purposes, without any measurable effect on data quality. However, in cases where the reacting solutions contain significant amounts of organic compounds, the reference cell should be filled with an identical solvent. The reference cell solution should be frequently replaced (i.e. every week), especially when working at higher-than-room temperatures.

The injection of solution into the sample cell is accompanied by a heat flow that tends to change the temperature of the sample cell. The heat sensors detect the imbalance and the feedback circuit minimizes the temperature difference between the cells by a proper adjustment of the power supply. Heat-generating reactions will reduce the power feed to the cells, while heat-absorbing reactions will result in an increased power feedback, until a new equilibrium is reached. The recorded heat change is proportional to the number of complex molecules formed after the injection. As the titration progresses, the number of binding sites that are available to the titrant is reduced. As a result, the observed heat effect becomes smaller with each subsequent injection until nearly all binding sites are occupied. After this saturation point is reached, only a small heat signal from mixing, dilution and friction phenomena is observed. The raw data of an ITC experiment is the plot of the CFB signal as a function of time, from the initial steady state to the end of the titration. This plot is the so-called thermogram of an ITC measurement (Fig. 3a).

ITC measurements are based on the first law of thermodynamics, which states that the change in the internal energy (ΔU) of a reacting system is equal to the difference between the total heat exchanged (Q) with its surroundings and the amount of work done by the system (W):

$$\Delta U = Q - W \tag{1}$$

In an ITC experiment, the reaction takes place at constant pressure (atmospheric pressure) and volume (active cell volume) and as a result, the total work is zero. Hence, for systems analyzed by ITC, it can be concluded that $\Delta U = Q$. For isobaric processes, ΔU is also equal to the total enthalpy change of the system (ΔH):

$$\Delta U = \Delta H \tag{2}$$

In this fashion, by measuring the total heat exchange for a given titration we can directly measure the enthalpy change of the occurring reaction without prior assumptions for the binding mechanism.

Measuring the binding enthalpy change of noncovalent interactions is not a trivial task. Noncovalent bonds are intrinsically weak, typically associated with total heat changes in the range of 4–20 kJ/mol upon formation. To make things even more challenging, during a titration experiment this small amount of heat is released or absorbed gradually. Nevertheless, modern ITC instruments are sensitive enough to detect changes of heat as small as 100 nJ.

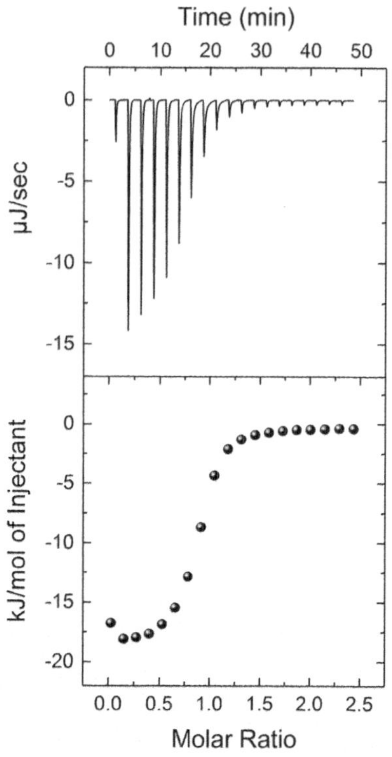

Fig. 3 (Upper panel) Change of power supply to the calorimetric cell during the titration of 10 mM of $CaCl_2$ (syringe) into 1 mM of EDTA (cell) at 25 °C in MES buffer (100 mM, pH 6.0). (Lower panel) Integration of the area under each injection, normalized per mol of injectant and plotted as a function of the $[Ca^{2+}]$:[EDTA] ratio at each point of the titration

Before the development of ITC, the only approach to calculate ΔH_b was to measure the binding constant of an interaction with a convenient method at several temperatures and then use the well-known van't Hoff equation:

$$\left(\frac{\partial \ln K_b}{\partial T}\right)_p = \frac{\Delta H_{vH}}{RT^2} \quad (3)$$

where K_b is the equilibrium constant, T is the absolute temperature and R is the gas constant. Although in theory the calorimetric and van't Hoff enthalpies are equivalent, there are reports of significant discrepancies for some systems [4–9].

2.2 *Design of an ITC Experiment*

To take full advantage of ITC as a powerful technique to investigate molecular interactions, solid experimental design and careful data analysis is required. There is no universal protocol for ITC measurements; for optimal results, each experimental setup should be based on the thermodynamic characteristics of the investigated system. In this section, we will present some general guidelines for a successful titration.

Selection of concentrations: The shape of the thermogram depends on both the K_b and the concentration of binding sites in the sample cell and can be described by the so-called Wiseman parameter (c) [3]:

$$c = N \cdot [A]_{tot} \cdot K_b \tag{4}$$

where N is the stoichiometry of the reaction (number of titrant molecules bound to one molecule of titrand when full saturation is reached) and $[A]_{tot}$ is the titrand concentration in the sample cell. The parameter c is dimensionless and can be used to determine the ideal concentrations for an ITC experiment. The c value of the reacting system is very important for an accurate determination of the binding constant. Very small c values lead to almost linear binding isotherms with poor ΔH_b or K_b resolution. On the other hand, large c values will result in step-like curves that cannot be used for K_b estimations. Moreover, for system with very high affinity, optimal concentrations will be so low that the heat produced will be lower than the sensitivity threshold of the instruments. For that reason, the largest binding constant that can be determined directly by an ITC experiment is around $10^9\ M^{-1}$.

Experience shows that selection of a c value between 5 and 50 leads to more accurate results. For interactions of known K_b and N the following inequality can be used to determine the optimal concentration range of the titrand:

$$\frac{5}{N \cdot K_b} \leq [A]_{tot} \leq \frac{50}{N \cdot K_b} \tag{5}$$

For interactions of unknown K_b and N, an initial titration under the assumptions $N = 1$ and $K_b = 10^5\ M^{-1}$ is designed and then the protocol is corrected based on the results of the experiment. Once the concentration of the titrand is selected, the following equation is typically used for calculating the titrant concentration in the syringe ($[T]_{syr}$):

$$[T]_{syr} = 15 \cdot N \cdot [A]_{tot} \tag{6}$$

Frequently, the intrinsic properties of the reacting species (i.e. low solubility, self-association etc.) or sample availability issues are prohibiting factors for working in the optimal range of concentrations. In this case, Eq. 5 is used for determining the concentrations based on the limitations imposed by the system.

The analysis of low c value systems is still possible if certain conditions are met. Turnbull and Daranas [10] demonstrated that it possible to extract meaningful data from systems of $c < 1$ by using a fixed N value when fitting the data to a binding model. This requires former knowledge of N, which can be provided by other methods or by the nature of the reaction in study. Sophisticated ITC protocols have also been developed for the indirect analysis of systems with $c > 1000$ or $c < 1$ [11–14].

In any case, for reliable measurements, the selected concentrations should produce heat signals of at least 3–5 μcal for the larger peaks of the titration. The accuracy of the ITC-derived parameters directly depends on how well the concentrations of the reacting species are determined. Hence, it is advisable to verify the concentrations used with an appropriate analytical method (i.e. UV/Vis absorbance).

Selection of injection sequence: In a conventional ITC experiment, the titrant solution is added to the reaction vessel in a stepwise manner as a sequence of small-volume injections; this approach ensures a strict control of the reactant concentrations in the cell for each point of the titration. Before an ITC measurement begin, the user must select the total number of injections, the time distance between two consecutive injections as well as the volume and duration of each injection.

The number of injections is equal to the number of points that can be fitted to an appropriate thermodynamic model. A common misconception is that one must have as many points as possible for better fitting results; by increasing the number of injections, one divides the finite amount of reaction heat into smaller and smaller portions, a practice that eventually leads to larger uncertainties and poor reproducibility. Tellinghuisen [15] provided a compelling case and mathematical foundation for using no more than 10 injections for standard ITC titrations. However, 20–25 injections is still the norm for most experiments.

Typical injection volumes range from 1 to 15 μL. Smaller injection volumes are inaccurate and significantly affected from the unavoidable diffusion of the titrant solution in the sample cell [16]. Larger injection volumes must be of very long duration to allow thermal equilibration of the sample cell solution before it is displaced from it as excess volume.

The duration of each injection depends on the injected volume and it should be such that proper mixing is achieved before significant volume displacements. On the other hand, injections with extremely long duration times suffer from titrant diffusion and the measured heat effect is underestimated. Consequently, a balancing act is required; for most reactions, 1–2 s duration for each μL of solution delivered will give reproducible results.

The time distance between two consecutive injections must be long enough to allow the occurring reaction to reach an equilibrium state and for the CFB signal to return to baseline levels. Typical spacing between injections is 300 s, while for reactions with very slow kinetics the waiting time can be longer than 600 s. Special attention is needed for systems with multiple binding sites that exhibit negative cooperativity. In these reactions, the binding process is slower at the middle of the experiment than at the beginning of it and the time distance needs to be adjusted accordingly.

The number of injections and the equilibration time between injections define the duration of the ITC experiment. A typical 20-injections titration with 300 s intervals between injections needs approximately two hours to finish (including the pre-titration equilibration period of the instrument).

A common problem in ITC experiments is that the first injection systematically produces a smaller than expected heat change (see Fig. 3, first injection). The prevailing explanation is that this phenomenon stems from the passive diffusion of the titrant during the long equilibration phase ($\geq$15 min) that precedes the injection sequence; however, it has been suggested that mechanical issues from the automated syringe motor could also give rise to this effect [17]. To circumvent this problem, it is common practice to include a small 1-μL injection at the beginning of the titration and then exclude the first point from data analysis.

An alternative approach that can be used in place of the conventional sequence of injections method is to deliver the syringe solution by a single continuous slow titration. The main advantage of this method is the increase rate of data acquisition, that significantly reduces the duration of the experiment and leads to more points in the binding isotherm for non-linear regression [18].

Selection of buffer and solution preparation: Buffer selection is very important, as it can have a significant effect on the ITC thermogram. Whenever the complex formation is accompanied by protonation or deprotonation of the reacting species, an equivalent number of will have to be taken up or released by the buffer to maintain the pH of the solution stable. This buffer activity is coupled to the binding reaction and will have a measurable heat effect (Q_{ion}), which depends only on the number of protons (n_p) exchanged per mole of titrant bound and the ionization enthalpy of the buffer (ΔH_{ion}). The net heat change of the binding reaction (Q_{bind}) can then be calculated from the experimentally determined heat (Q_{exp}) by the following equation:

$$Q_{bind} = Q_{exp} - (\Delta H_{ion} \cdot n_p) \tag{7}$$

Due to the complexity of the interactions, the n_p term is seldom known a priori. However, it can be calculated by repeating the binding experiment a few times in buffers with different ionization enthalpies; the collected Q_{exp} data are then plotted versus the ΔH_{ion} of the buffer and fitted linearly to Eq. 7 to estimate both Q_{bind} and n_p (Fig. 4). Knowledge of n_p can be very useful for molecular simulations of the binding reaction, especially in cases where no structural information is available. Moreover, buffer effects can be exploited to amplify the heat signal of a weak interaction by selecting a buffer with large ionization enthalpy. Nonetheless, for most titrations, a buffer with small ionization enthalpy ($\Delta H_{ion} \approx 0$) should be selected to avoid this kind of thermal interference. The ionization enthalpies and other important thermodynamic data for many buffer systems are readily available from previous studies [19, 20]. Table 2 contains ionization enthalpies and heat capacity changes for common biological buffers.

Mixing of solutions with very different chemical compositions can generate large heat signals of mixing or dilution that will mask any heat produced by a binding reaction. To minimize such heat signals, the buffer of the titrant and the titrand solutions should be identical in every aspect (pH, ionic strength, concentration, presence of co-solvents etc.). Hence, it is advisable to dialyze exhaustively the samples against large volumes of buffer or to directly dissolve them using a buffer solution of the same batch. To acquire data of high quality, the purity of samples must be higher than 95%. If a reducing agent must be added in the buffer, it is preferable to use

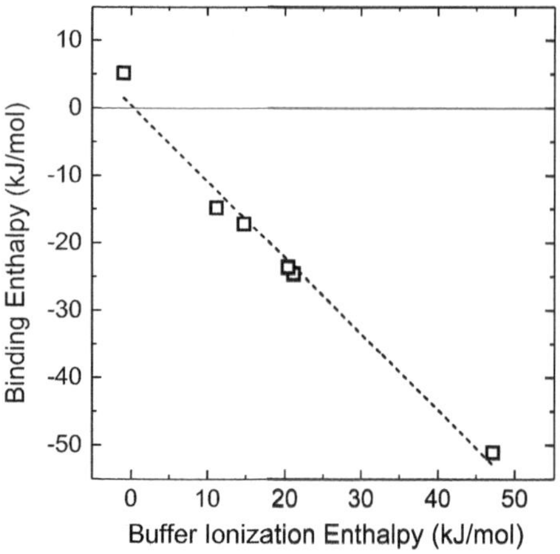

Fig. 4 Binding enthalpy (ΔH_b) versus buffer ionization enthalpy (ΔH_{ion}) plot (□) for the the titration of 10 mM of $CaCl_2$ (syringe) into 1 mM of EDTA (cell) at 25 °C in the following buffers: Tris ($\Delta H_{ion} = 47.5$ kJ/mol), MOPS ($\Delta H_{ion} = 21.1$ kJ/mol), PIPES ($\Delta H_{ion} = 11.2$ kJ/mol), MES ($\Delta H_{ion} = 14.8$ kJ/mol), acetate ($\Delta H_{ion} = -0.4$ kJ/mol) and HEPES ($\Delta H_{ion} = 20.4$ kJ/mol). The intercept of the linear fit (---) is 0.39 ± 1.7 kJ/mol and the slope is -1.1 ± 0.1 ($R^2 = 0.98$)

Table 2 Thermodynamic quantities for the ionization reactions of common buffers in water at T = 298.15 K and p = 0.1 MPa [19, 20]

Buffer	Equilibrium	pK_a	$\Delta_r H^\circ$ (kJ mol^{-1})	$\Delta_r C_p^\circ$ (J mol^{-1} K^{-1})
HEPES	$HL^{\pm} \rightleftarrows H^+ + L^-$, (HL: $C_8H_{18}N_2O_4S$)	7.6	20.4	47
Borate	$H_3BO_3 \rightleftarrows H^+ + H_2BO^{-3}$	9.2	13.8	≈−240
Acetate	$HL \rightleftarrows H^+ + L^-$, (HL: $C_2H_4O_2$)	4.8	−0.4	−142
MES	$HL^{\pm} \rightleftarrows H^+ + L^-$, (HL: $C_6H_{13}NO_4S$)	6.3	14.8	5
MOPS	$HL^{\pm} \rightleftarrows H^+ + L^-$, (HL: $C_7H_{15}NO_4S$)	7.2	21.1	25
PIPES	$HL^{\pm} \rightleftarrows H^+ + L^-$, (HL: $C_8H_{18}N_2O_6S_2$)	7.1	11.2	22
Tris	$HL^+ \rightleftarrows H^+ + L$, (HL: $C_4H_{11}NO_3$)	8.1	47.5	59
Phosphate	$H_2PO_4^- \rightleftarrows H^+ + HPO_2^-$	7.2	3.6	230

$\Delta_r H^\circ$ is the enthalpy of ionization and $\Delta_r C_p^\circ$ is the corresponding heat capacity change low concentrations of mercaptoethanol or tris(2-carboxyethyl)phosphine (TCEP), instead of dithiothreitol (DTT).

Before loading the samples for an ITC experiment, it important to thoroughly degass the solutions using a suitable vacuum chamber. Introduction of small air bubbles into the cell can generate a noisy power baseline and spurious heat signals that will affect the overall quality of the measurement. For the same reason, both solutions should be filtered or centrifuged to remove any precipitated material before filling the cell or the syringe. In case there is some precipitation, after the removal of the aggregates the concentration of the sample must be corrected.

Solution Placement: Another decision that needs to be made when designing an ITC experiment is to select which solution will be loaded in the syringe and which in the sample cell. Routinely, the compound with the lower molecular weight is filled the syringe and the larger compound is placed in the cell. Nevertheless, this is not always the optimal configuration for an ITC experiment; if a compound with significant population of oligomers is placed in the syringe, it will dissociate when diluted in the sample cell and will introduce an additional heat effect to the binding isotherm (dissociation heat). Considering that the cell solution is diluted to a lesser degree (around 20%), loading this compound in the cell is preferable in order to minimize these kind of heat contributions.

Another factor that can force the decision of solution placements is the concentrations needed for the titration. If there are solubility or availability issues with a specific sample, it is better to load it in the cell, where typically the concentration requirements are much lower than those needed for the syringe solution.

Selection of Temperature: The ITC's temperature operating range is 2–80 °C. The experimental temperature should be selected based on the properties of the system. If there are no system-specific requirements, the most common titration temperatures selected are 25 and ~37 °C. In general, ΔG_b, ΔS_b, ΔH_b and K_b exhibit temperature dependence and caution is required when comparing results from titrations at different experimental temperatures.

2.3 *Experimental Results and Data Analysis*

2.3.1 Raw Data Treatment

The experimentally determined heats of the binding isotherm ($DQ_{titration}$) unavoidably include heat contributions from the dilution of both the titrant and sample cell solutions, which need to be separated from the actual heat generated by the interaction ($DQ_{reaction}$). To achieve this, a series of control measurements is required: the dilution heat for the titrant ($DQ_{titrant}$), usually the most significant contribution, can be determined by replacing the sample cell solution with buffer and repeating the titration following the same protocol with the initial experiment. Likewise, the dilution heat of the titrand ($DQ_{titrand}$) is estimated by another experiment in which the syringe solution is the one replaced with buffer. For a thorough analysis, a final titration in which buffer is injected into buffer is used for calculating the buffer dilution heat (DQ_{buffer}). The DQ_{buffer} term arises from temperature differences between the syringe and the cell as well as friction heat from stirring and it can be considered as an instrument blank. The heat data from this set of titrations are used to correct the final binding isotherm of the reaction according to the following equation:

$$DQ_{reaction} = DQ_{titration} - DQ_{titrant} - DQ_{titrand} + DQ_{buffer} \tag{8}$$

The term DQ_{buffer} is added and not subtracted because buffer dilution heat is already included in both $DQ_{titrant}$ and $DQ_{titrand}$ terms and it is removed twice from $DQ_{titration.}$ For most reactions, the dilution heats of the titrand and the buffer are negligible and the corresponding control experiments are often omitted. Then, Eq. 8 is reduced to the equation:

$$DQ_{reaction} = DQ_{titration} - DQ_{titrant} \tag{9}$$

If the titrant is available in limited quantities or is too valuable to be used for a control experiment, it is possible to estimate the $DQ_{titrant}$ term from the final binding isotherm points of a fully saturated system. After the point of saturation is reached, any heat released upon injection is in essence the dilution heat of the titrant ($DQ_{reaction} \approx 0$). A linear fit of these last few points in the binding isotherm will provide a good approximation of $DQ_{titrant}$. However, it is strongly recommended to collect all necessary control experiments prior to data analysis for a more accurate calculation of the thermodynamic parameters.

2.3.2 Conversion of Raw Data to a Binding Isotherm

In an ideal situation, where the reacting system is always in thermal and thermodynamic equilibrium without any mechanical or electrical interference, the instrument's power baseline will always be constant. However, in reality the power baseline will always exhibits some drift and background noise. A very important step for calculating the binding isotherm from raw ITC data is to define the power baseline of the experiment. The baseline trace cannot be determined by a separate experiment and it is only intermittently observed in the thermogram. Hence, the parts of the baseline that are disrupted by the injection sequence must be extrapolated, a process that introduces a certain degree of ambiguity and affects the precision of the measurements [21]. Baseline approximations are generated automatically by the data analysis software provided by the device manufacturer and can be manually corrected if deemed necessary.

The heat change associated with each injection is calculated from the thermogram by integration of the area between the baseline and the corresponding CFB trace with respect to time. The lower time bound for the integration is a few seconds before the injection begins and the upper bound is a few seconds before the next injection is made. Each injection heat change is normalized as heat change per mole of titrant and plotted against the ratio of titrant to titrand concentrations in the active volume (molar ratio). The resulting curve is known as the binding isotherm of the titration (Fig. 3b).

2.3.3 Equilibrium Binding Models for ITC Data Analysis

The binding constant, binding enthalpy and stoichiometry of the interaction are determined by the best fit of the experimental data to an appropriate binding model. The equations describing these models derive from the law of mass action, based on the assumptions made for the binding mechanism and the equilibria involved. The dependent variable (i.e. heat change per injection) is expressed as a function of an independent variable (i.e. moles of titrant added) and the corresponding thermodynamic parameters (e.g. K_b, N and ΔH_b). The chosen model should be as simple as possible and compatible with the intrinsic properties of the system in study.

The fitting of a binding isotherm is an iterative process. At first, initial assumptions are made for the value of the fitted parameters. Based on these values, a curve is generated and compared to the experimental data. The deviation between experimental data and the curve is calculated and a suitable algorithm is used to adjust the value of the parameters in order to minimize the error. This procedure is repeated enough times until no further minimization is possible.

The simplest binding reaction is that of an analyte (A) with one available binding site, interacting with free titrant molecules (T) to form a stable complex (AT).

$$A + T \overset{K_b}{\Leftrightarrow} AT$$

The equilibrium association constant or binding constant (K_b) of the interaction is:

$$K_b = \frac{[AT]}{[A][T]} \tag{10}$$

where [A], [T] and [AT] are the concentrations of the corresponding species in the sample cell. Often, in biochemical or pharmacological studies, the reciprocally related dissociation constant (K_d) is used instead of K_b ($K_b = 1/K_d$). The dimension of K_b is M^{-1}.

If Q is the measured total heat change for the binding reaction (experimental observable), V is the active volume, $[A]_{tot}$ is the total titrand concentration, [A] is the unbound titrand concentration and [AT] is the concentration of the complex in the sample cell, then we have:

$$[A]_{tot} = [A] + [AT] \tag{11}$$

and

$$Q = \Delta H_b \cdot V \cdot [AT] \xrightarrow{Eq.10\ and\ Eq.11} \Delta H_b \cdot V \cdot [\mathrm{A}]_{\mathrm{tot}} \cdot \frac{K_b[T]}{1+K_b[T]} \tag{12}$$

If the analyte has N identical binding sites, then Eq. 11 can be rewritten in the form:

$$Q = \Delta H_b \cdot V \cdot [\mathrm{A}]_{\mathrm{tot}} \cdot \frac{N \cdot K_b[T]}{1 + K_b[T]} \tag{13}$$

In cases where the analyte contains N independent binding sites, each with different binding enthalpy (ΔH_b^i), binding constant (K_b^i) and stoichiometry (N^i), Eq. 12 is transformed to the following equation:

$$Q = V \cdot [\mathrm{A}]_{\mathrm{tot}} \cdot \sum_1^N \frac{N^i \cdot K_b^i[T] \cdot \Delta H_b^i}{1 + K_b^i[T]} \tag{14}$$

In order to be able to use these equations for nonlinear regression, we must determine the concentration of the unbound titrant in the active volume [T], at each point of the titration.

For systems with one binding site (N = 1):

$$[\mathrm{T}]_{\mathrm{tot}} = [\mathrm{T}] + [\mathrm{AT}] \tag{15}$$

where $[\mathrm{T}]_{\mathrm{tot}}$ is the total concentration of the titrant injected in the sample cell. The combination of Eqs. 10, 11 and 15 leads to:

$$[\mathrm{AT}]^2 + (-[\mathrm{A}]_{\mathrm{tot}} - [\mathrm{T}]_{\mathrm{tot}} - 1/\mathrm{K_b})[\mathrm{AT}] + [\mathrm{A}]_{\mathrm{tot}} \cdot [\mathrm{T}]_{\mathrm{tot}} = 0$$

or equivalently:

$$[\mathrm{AT}]^2 + \mathrm{b} \cdot [\mathrm{AT}] + \mathrm{c} = 0 \tag{16}$$

where $\mathrm{b} = -[\mathrm{A}]_{\mathrm{tot}} - [\mathrm{T}]_{\mathrm{tot}} - 1/\mathrm{K_b}$ and $\mathrm{c} = [\mathrm{A}]_{\mathrm{tot}} \cdot [\mathrm{T}]_{\mathrm{tot}}$. The solution of Eq. 15 is then:

$$[\mathrm{AT}] = \frac{-b - \sqrt{b^2 - 4c}}{2} \tag{17}$$

For systems with multiple identical binding sites (N > 1), $\mathrm{K_b}$ is equal to:

$$K_b = \frac{[AT]}{([\mathrm{A}]_{\mathrm{tot}} \cdot \mathrm{N} - [AT]) \cdot [T]} \tag{18}$$

Following a similar logic, we get to a solution identical to that of Eq. 17, with the following c and b values:

$$b = -N \cdot [A]_{tot} - [T]_{tot} - 1/K_b \text{ and } c = N \cdot [A]_{tot} \cdot [T]_{tot}.$$

Parameter c in both cases is equal to the Wiseman parameter of the system. Once [T] is determined, the experimental data can be fitted to Eq. 12 or 13. All concentration-dependent parameters are normalized per mole of injectant. If the thermodynamic data need to be presented in respect to the sample cell compound, then the analysis model must be modified accordingly [22].

The binding models described above are adequate for analyzing simple binding reactions. However, in some cases the nature of the reacting systems requires a different thermodynamic description. Binding models for complex interactions, including three independent set of sites [23], competitive binding [11, 12], multiple interactive binding sites [24], dimer dissociation [25] and combinations of them [26] are available in the literature. Model-free methodologies that utilize binding polynomials for the analysis of ITC data have also been described [27, 28].

Frequently, different binding models produce similar binding isotherms, making difficult to discriminate binding mechanisms. A useful test for the validity of the chosen binding model will be to repeat a second titration with reverse solution placement. If the model is correct, the calculated thermodynamic parameters from the two titrations should be the same [29].

2.3.4 Thermodynamics of Noncovalent Interactions

The measurement of binding affinity allows the calculation of the Gibbs free energy change of the interaction (ΔG_b) using the equation:

$$\Delta G_b = -RT \ln K_b \tag{19}$$

ΔG_b is the result of structural and energy differences between free and bound states for all reacting species. A reaction under thermodynamic control will spontaneously proceed to the direction in which $\Delta G_b < 0$, until a thermodynamic equilibrium is reached. When analyzed to enthalpic and entropic contributions ($\Delta G_b = \Delta H_b - T\Delta S_b$), ΔG_b can provide additional information for the binding interaction. The binding enthalpy change (ΔH_b) reflects variations on the interactions of all the atoms participating in the reacting system [30]. ΔH_b is the net result of the formation, disruption and destruction of many individual bonds; when the system gains energy through increased bonding, we have release of heat to the environment and a negative enthalpy change for the interaction ($\Delta H_b < 0$). The reaction is then termed exothermic and the observed ΔH_b is considered favorable, since it makes ΔG_b more negative. Analogously, when $\Delta H_b > 0$, we have decreased bonding when the complex is formed with a contemporary uptake of energy from the surrounding in the form of heat; the reaction is then described as endothermic and the observed ΔH_b is considered unfavorable.

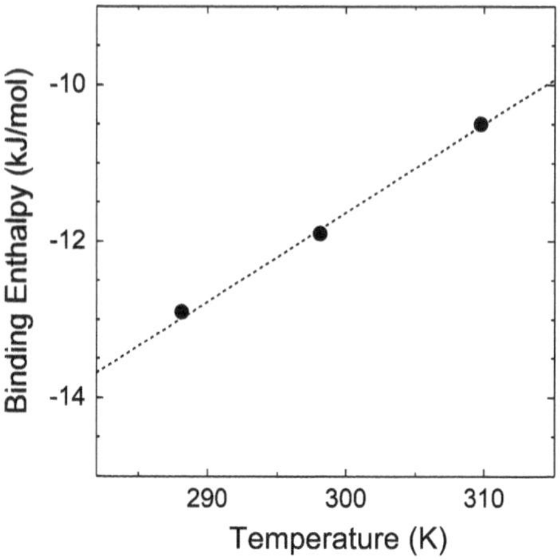

Fig. 5 Binding enthalpy versus temperature plot for the binding oxacillin to octakis(6-(2-aminoethylthio)-6-deoxy)-γ-cyclodextrin [123] at 288.15, 298.15 and 309.75 K. Dashed line represents the fit of the data to a simple linear model of the form: $\Delta H_b(T) = \Delta H° + \Delta C_p \cdot T$ ($R = 0.99$, $\Delta C_p = 0.11 \pm 0.02$ kJ/mol K, $\Delta H° = -45.76 \pm 7.56$ kJ/mol)

The binding entropy change (ΔS_b) is associated with changes in the order of the system. In general, a strongly interacting system tends to be more ordered than a weakly interacting system. Positive ΔS_b (increased disorder) leads to even more negative ΔG_b and it is favorable for the interaction. Parallel to ΔH_b, ΔS_b represents the sum of many positive or negative contributions: i.e., a positive solvation entropy change is associated with the burial on non-polar groups from the solvent, while conformational entropy penalties arise from the unavoidable loss of degrees of freedom upon complex formation. Additional contributions to the binding entropy change stem from the burial of polar surfaces and release of organized water molecules to the bulk [31].

Interactions with ΔH_b as the most significant contribution towards a negative ΔG_b are considered as enthalpy-driven, while interactions with a dominant entropic term are described as entropy-driven.

It is already described how ITC can directly measure the binding enthalpy of a given reaction at a specific temperature. If this process is repeated over a temperature range, the heat capacity change under constant pressure (ΔC_p) of the interaction can be determined, based on the following equation:

$$\Delta H_b(T) = \Delta H_b(T_0) + \Delta C_p \cdot (T - T_0) \tag{20}$$

where T_0 is an appropriate reference temperature. Consequently, if ΔH_b is plotted against temperature for a series of measurements and fitted using a linear model, the slope of the best fit will correspond to the ΔC_p of the system (Fig. 5). The use of Eq. 20 implies that ΔC_p is constant within the temperature range of the experiments. ΔC_p measures differences in the ability of the solution to absorb heat, reflecting changes in the molecular level on the order and structure of the system in study.

The fact that many reactions are accompanied by large changes in ΔC_p may have some practical implications for ITC experiments. For these systems, there is a temperature range where ΔH_b will be very small or zero. Consequently, no heat will be detected by the calorimeter. For this reason, in cases where no reaction is observed although it is expected, it is advisable to repeat the experiment at a new experimental temperature that differs at least 10–15 °C to that of the original experiment.

Heat capacity changes are of particular interest in the case of protein-protein and protein-peptide interactions. It has been established that there is a strong correlation between ΔC_p and the surface area screened from the solvent upon complex formation [32, 33].

Many noncovalent interactions are accompanied by characteristic changes in enthalpy, entropy and heat capacity that make them distinguishable from other heat-generating phenomena. These "thermodynamic signatures" are extremely useful for the interpretation of ITC data, by providing insight to the driving forces and mechanism of the interaction [34–40]. The thermodynamic profile of various processes is discussed below.

Hydrogen Bonds: Hydrogen bonds are very important for all processes involving aqueous solutions. Although weaker than covalent bonds, hydrogen bonds play a critical role in the stability, conformation and function of many molecular assemblies [41]. In solution, compounds with solvent-exposed hydrogen bonding groups interact with water molecules in their surroundings. When two molecules interact by forming hydrogen bonds at their binding interface, an equal number of hydrogen bonds with water molecules will have to be broken. Considering that hydrogen bonds are relatively weak and have more or less the same energy (around 20 kJ/mol), only a small change is observed in binding enthalpy. The properties of the hydrogen bonding groups and the solvent before and after the formation of the complex dictate the magnitude of the enthalpy change. Despite their small enthalpic contribution, hydrogen bonds significantly enhance the selectivity and affinity of interactions, with a single bond often leading to a 10- to 1000-fold increase in K_b [42].

Hydrophobic Interactions: By the term hydrophobic interaction, we describe the tendency of non-polar groups to aggregate in aqueous solutions to minimize their contact with the solvent. Non-polar surfaces exposed to the solvent cannot form hydrogen bonds with the surrounding water molecules. Consequently, water molecules near hydrophobic areas tend to make stronger hydrogen bonds among themselves, creating regions of "structured" water around them. Hiding non-polar surfaces from water leads to release of these organized molecules back to bulk solvent, a process characterized by a small enthalpy change ($\Delta H_b \sim 0$), a large favorable entropic term ($-T\Delta S_b < 0$) and a negative contribution to heat capacity change ($\Delta C_p < 0$) [43–46].

Electrostatic Interactions: Electrostatic forces originate from coulombic interactions among polar and charged chemical groups. The strength of these interactions is determined by the dielectric constant of the binding interface and the inverse distance between the charges. Electrostatic interactions in general are entropically driven processes, with a small binding enthalpy change [47]. The entropic gain originates from the release of water molecules solvating the charged groups. Electrostatic interactions have a varying effect on ΔC_p. Naturally, electrostatic interactions do not exhibit any binding specificity or selectivity.

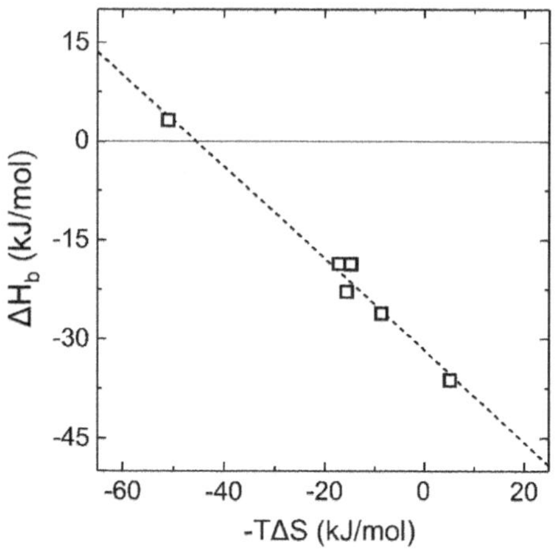

Fig. 6 Enthalpy-entropy compensation plot (□) for the the titration of 10 mM of $CaCl_2$ (syringe) into 1 mM of EDTA (cell) at 25 °C in Tris, MOPS, PIPES, MES, acetate and HEPES buffers. The intercept of the linear fit (---) is −31.7 ± 1.1 kJ/mol and the slope is −0.70 ± 0.04 ($R^2 = 0.98$)

Enthalpy-Entropy Compensation: Molecular interactions often yield compensating changes for $-T\Delta S_b$ and ΔH_b terms with no significant effect on ΔG_b (Fig. 6). This effect, better known as enthalpy-entropy compensation, makes hard to predict the effect of noncovalent interactions on binding affinity. This phenomenon is not limited in aqueous solutions and it appears be a universal property of systems stabilized by interconnecting networks of weak interactions [48–52]. The origin of the enthalpy-entropy compensation effect is a subject of debate for decades. The fact that both enthalpic and entropic contributions are controlled by ΔC_p may provide a mechanism for this correlation [53].

Thermodynamic parameters and structural changes: Empirical equations connecting important thermodynamic parameters (ΔH_b, ΔS_b and ΔC_p) with structural changes, based on protein folding/unfolding studies have been described in the literature [34, 54–56]. It was expected that these equations would have general use since molecular interactions are based on the same noncovalent forces. However, their applicability beyond protein-protein interactions is questionable, especially when the binding surface is relatively small [53]. Therefore, extreme caution is needed for the use of structural data to predict thermodynamic parameters and vice versa [57, 58]. The collection of thermodynamic data for interactions of pharmaceutical interest will eventually allow the development of similar equations for the binding of small ligands, a task of great importance for structure-based rational drug design [29, 59, 60].

3 Applications

ITC was originally developed to study the interactions of proteins and other biological macromolecules with small ligands. The number of protein-protein and protein-

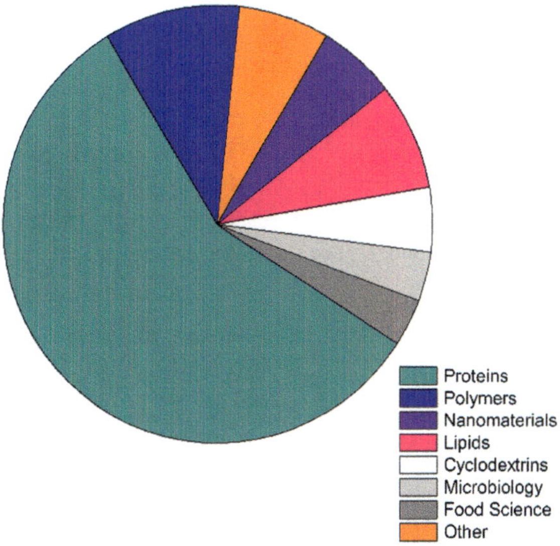

Fig. 7 Scientific articles containing isothermal titration calorimetry data in 2107, categorized in fields of research. Protein chemistry still dominates the number of publications (362), although the number of articles for synthetic molecules and lipid-based systems is constantly increasing

ligand interactions that have been investigated by ITC is truly impressive and there is a plethora of excellent reviews that summarize the most important methods and findings [61–68].

Even today, the majority of ITC-related publications still involve the study of protein interactions (Fig. 7). However, new application fields and protocols have emerged over the last decade to include the study of surfactant demicellization [69], membrane partitioning [70], food chemistry reactions [71, 72], antimicrobial agents [73, 74], degradation of pollutants [75] and enzyme kinetics [76–79]. In this section, we describe the advantages of including ITC in the drug optimization process and demonstrate the power of the technique for the characterization of novel drug delivery systems, nanomaterials and supramolecular assemblies.

3.1 The Value of ITC in Rational Drug Design

The development of a new drug is a laborious multi-step process which can take 12–15 years to complete with a total cost that often exceeds the $1,000,000,000 mark [80]. The first step in drug design is to identify ligands that exhibit affinity for a specific target of interest. This task is accomplished by screening compound libraries, often containing thousands of molecules, against a drug target of interest. Typically, initial hits involve compounds that bind with micromolar or even weaker dissociation constants (lead compounds). However, nanomolar and sub-nanomolar range affinities are necessary for the clinical use of a therapeutic compound. In addition, a successful drug should satisfy a wide range of physicochemical requirements,

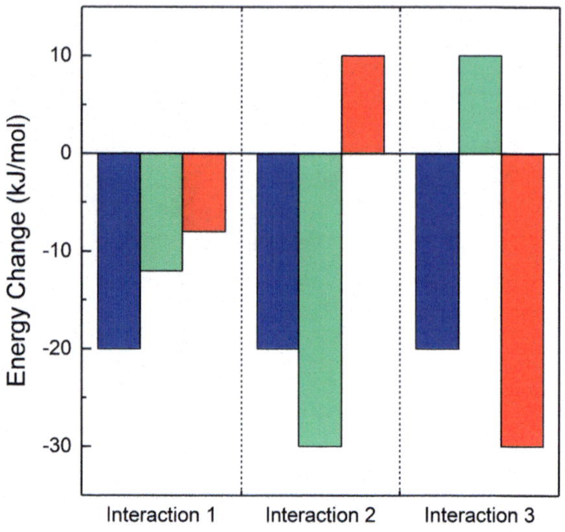

Fig. 8 Thermodynamic profiles for three different interactions with identical binding affinities. Interaction 1 has favorable enthalpic and entropic contributions; interaction 2 is an enthalpy-driven process and interaction 3 is an entropy-driven process. Blue bars represent the Gibbs free energy change (ΔG_b); green bars (ΔH_b) represent the binding enthalpy and red bars represent the contribution of the entropic term ($-T\Delta S_b$)

including binding selectivity, low toxicity, high solubility, membrane permeability etc. Therefore, the next step involves chemical optimization of the initial lead candidates for the development of molecules with a suitable pharmacokinetic profile (drug candidates).

Binding affinity is directly related to the free energy change of the interaction (Eq. 19). Therefore, different combinations of ΔH_b and ΔS_b contributions can result in identical binding affinities (Fig. 8). Enthalpic contributions in general provide a measure of the numerous noncovalent bonds (van der Waals, H-bonds, salt bridges etc.) created or destroyed at the binding interface upon complex formation. In a similar manner, the entropic term is the net effect of various smaller contributions: solvation entropy change is associated with the burial on non-polar groups from the solvent, while conformational entropy penalties arise from the unavoidable loss of degrees of freedom upon complex formation. Another phenomenon linked with the total entropy change is the burial of polar surfaces and release of organized water molecules from the hydration layer to the bulk [31]. Consequently, the thermodynamic signature of a binding event can provide valuable insight to the molecular mechanisms involved in the association. Enthalpy-driven interactions are based on the formation of stabilizing forces between chemical groups and complementary surfaces while entropy-driven interactions are dominated by the hydrophobic effect. Obviously, high binding affinity is achieved only when both enthalpic and entropic contributions are favorable [81]. As a result, the analysis of binding affinity in terms of enthalpy and entropy can be used to identify the best lead compounds among molecules with similar binding constants and to provide useful guidelines for the chemical alterations that would improve their effectiveness.

Even when detailed structural and thermodynamic data are available, lead optimization is a common bottleneck in drug development that many projects fail to overcome [81]. For example, the binding affinity of a compound for a hydrophobic binding pocket can be theoretically increased by introducing non-polar groups to its structure. However, this modification would decrease the solubility of compound and could have a negative effect on the surface complementarity and binding orientation of the molecule, leading to sub-optimal affinities. Therefore, a delicate balance of forces must be achieved for fabricating therapeutic agents with the desired set of properties.

ITC can be directly used to identify lead compounds by monitoring their affinity (K_b) for specific binding sites of interest. Furthermore, the thermodynamic signature (N, ΔH_b, ΔS_b, ΔC_p) of these interactions provides additional information on the complexation mechanism, which can be further exploited for the rational design of new compounds with optimized physicochemical characteristics. Describing an interaction not only in terms of affinity but in terms of ΔH_b as well is extremely helpful for identifying the most promising positive hits from a large collection of successfully screened compounds. Large favorable ΔH_b is a difficult optimization task that includes good shape complementarity, increased number of stabilizing bonds in the binding site interface and proper orientation of interacting groups. The ability to cull compounds that already exhibit large favorable ΔH_b can accelerate the optimization process considerably [60].

As mentioned previously, the effectiveness of a therapeutic molecule hinges on more than one properties. Naturally, a good drug should show high affinity for its target, but it should also be able to interact with serum proteins or biological membranes for an effective delivery to the site of interest. ITC can be employed to monitor these interactions as well. Another distinct advantage of the technique is the ability to measure accurately the stoichiometry of a binding interaction. Exact knowledge of N is an excellent method to monitor the effectiveness of a purification process or to assess sample consistency between batches. The ITC-derived stoichiometry is based on the "active" and not the total concentration of the reactants and therefore can be utilized as a sensitive quality indicator.

One of the main disadvantages for the use of ITC in drug development was the low throughput of the technique. Until recently, ITC was able to screen only 5–8 compounds on a daily basis, following protocols that required manual cleaning and loading of the device between experiments. However, technical advancements allowed the development of highly sophisticated instruments and the throughput of the technique was considerably increased. Nowadays, new instruments allow the conduct of 100 experiments per week by automated processes that require even smaller sample volumes [82]. Furthermore, advanced ITC protocols are now available for identifying unknown target proteins from a mixture of biomolecules for a given drug or a lead compound [83]. Although ITC is still a low throughput technique and not suitable as a primary screening method, it provides information-rich data that are invaluable for the identification and optimization of the most promising compounds.

3.2 Nanoparticle-Protein Interactions

The field of biotechnology has broadened considerably over the last decade, to include a wide range of novel nanomaterials to the traditional arsenal of lipid-based and polymer-based systems discuss above [84]. One of the most notable new approaches is the use of nanoparticles (NPs) for the development of bio-reactive assemblies with controlled surface characteristics.

Nanoparticles offer considerable scalability and versatility for the development of novel materials for problem-specific applications. Similar in size (<100 nm) to many biological components, their functionality can be exploited to control processes at the cellular level. However, their ability to penetrate cells and organelles can also be hazardous [85]. Hence, a deeper understanding of nanoparticle interactions with respect to their intrinsic properties is crucial in their design, applications and safe handling.

In principle, when introduced to biological fluids, NPs with the appropriate functionality will be able to readily interact with biomolecules due to their small dimensions and large surface-to-mass ratio. Since the absorbed biomolecules are mainly proteins, the term protein corona is often used to describe such complexes [86]. The corona composition is based on a dynamic equilibrium between the NP and the surroundings. When a NP is introduced to a new environment, the composition of the corona will change until a new equilibrium state is reached. Typical protein coronas in human plasma are consisted of proteins such as albumins, apolipoproteins and immunoglobins [87].

In parallel to many biological interfaces, the absorption of biomolecules to the nanoparticles surface is facilitated by several "weak" noncovalent forces such as van der Waals interactions, solvation effects, hydrogen bonds etc. As a result, the mechanism of the interaction depends on both the physicochemical properties of the reactants and the medium [88]. The release of the absorbed molecules can be achieved by disruption stabilizing forces between the nanoparticle and the compound, i.e. by changes in the pH or the ionic strength of the solution [89].

The absorption of biomolecules to the NP surface can promote the translocation of the assembly across membranes to specific cellular targets. On the other hand, absorption forces can also induce conformation changes to proteins, affecting the overall reactivity of the complex or even triggering immune system responses [90]. These adverse effects highlight the need to for a better understanding of the molecular mechanisms involved in NP interactions in order to engineer bio-reacting nanomaterials for future applications.

Absorbed serum proteins by functionalized NPs are frequently used as a model system to gain insight to the dynamics and energetics of such interactions. ITC is one of the few analytical methods that can provide detail information on NP-protein interactions and it has been used for the study a number of NP-protein systems.

Cedervall et al. [91] employed ITC to investigate the interaction of human serum albumin (HSA) with N-isopropylacrylamide (NIPAM): N-tert-butylacrylamide (BAM) copolymer nanoparticles. The absorption was exothermic with a binding isotherm that was adequately fitted by a single-set-of-sites thermodynamic model, indicating the absence of protein-protein interactions or binding cooperativity. The interactions were very strong ($K_b \sim 10^6\ M^{-1}$) and the binding affinity was not affected by the size or the composition of the nanoparticles, implying that the interactions are not specific. The stoichiometry of the interactions revealed that a large number of protein molecules could be absorbed to the NP surface. Small size NP (~70 nm) were able to interact with dozens of proteins, while large size NPs (~200 nm) needed more than 1000 protein molecules to reach saturation. Interestingly, the stoichiometry of the interaction increased with the NP hydrophobicity. This finding suggests that the hydrophobic BAM groups are the main protein binding sites and that hydrophobic effects are the driving force for protein absorption by NIPAM/ BAM copolymer nanoparticles.

Another study by Mandal et al. [88] investigated the interaction of BSA and HSA with carbon nanoparticles. The calorimetric profiles of the interactions were almost identical, showing a single binding event driven by equally favorable enthalpic and entropic contributions. The affinity of the NP-BSA and NP-HSA interactions is remarkably high ($\sim 2 \times 10^7\ M^{-1}$) compared to typical affinities for binding of small ligands to BSA and HSA (around $10^5\ M^{-1}$). The 40 nm carbon nanoparticles were able to interact with approximately 40 protein molecules, showing the same binding capacity for both BSA and HSA. Strong binding interactions on the highly curved surface of the NPs were expected to have an effect on the protein conformation. Circular dichroism data confirmed partial unfolding and significant alteration of protein secondary structure in the presence of carbon NPs. The authors suggested that the BSA and HSA unfolding on the NP surface would significantly change the transport properties of the complex, affecting in turn all the BSA- and HSA-mediated processes of a biological system.

A later publication by Fleischer and Payne [92] monitored the interactions of bovine serum albumin (BSA) with polystyrene NPs functionalized with either amine or carboxylate groups to provide a cationic or anionic surface charge respectively. ITC thermograms revealed that BSA is able to bind to both cationic and anionic NPs following a binding equilibrium that can be analyzed by a single set-of-sites model. The total enthalpy change of BSA binding was large, favorable and almost identical for both NPs, suggesting the presence of an extensive network of stabilizing forces between the NPs and the protein. However, the other parameters of the interaction were very different for the oppositely charged NPs. The binding constant was almost an order of magnitude greater for BSA adsorbed on anionic NP than that on their cationic counterpart. Moreover, the average number of BSA molecules absorbed on the carboxylate-modified NPs was much larger than that of the amine-modified NPs (871 and 27 respectively), even though they were approximately of the same size. The authors attributed the lower binding capacity of the cationic NPs to the loss of protein secondary structure, resulting in a less energetically favorable packing of BSA upon absorption on the positively charged surface. These differences in corona characteristics were very important for the binding of the BSA-NP complexes to cellular receptors. Although the same protein forms the corona, it was established that anionic NPs and cationic NPs bind to different cellular receptors, showing that the protein secondary structure is a key parameter that controls the NP interactions with the cell.

Eren et al. [93] investigated the interaction of lysozyme with silica NPs. The thermal profile of the interaction revealed the presence of two distinct modes of interaction that lead to lysozyme complexation. The binding isotherm was analyzed with the use of a sequential binding model that allowed the estimation of thermodynamic parameters for both processes. The first binding mode was of higher affinity, with favorable enthalpic and entropic terms and it was succeed by a weaker binding mode with a similar, although less pronounced, thermodynamic signature. The enthalpic change in both processes was attributed to the formation of multiple noncovalent interactions between the protein and the NP, whereas the entropic gain was associated with reorganization of the charged NP surface and release of solvent molecules and counterions from the binding interface to the bulk. Connecting ITC and zeta potential data, the authors proposed an interesting binding mechanism for the interaction of lysozyme with silica NPs. The first thermodynamic equilibrium involves binding of protein molecules to a positively charged NP. As the complexation progresses, more negatively charged protein molecules are bound and as a result the surface charge of NP becomes neutral. The stoichiometry ratio from ITC data confirm that this is the point where the second binding mode takes over, with the second thermodynamic equilibrium corresponding to protein binding events to a negatively charged surface. These energetically different thermodynamic equilibria appear as discrete binding events in the ITC thermogram.

3.3 Polyelectrolyte Interactions

Polyelectrolytes is a fascinating and rapidly growing research domain [94]. Charged polyelectrolytes can serve as model systems to investigate polyelectrolyte complexation events in biological systems, including the wrapping of DNA around histones to form nucleosomes. More importantly, it has been established that interactions between oppositely charged polyelectrolytes can lead to a wide variety of electrostatically assembled macromolecular architectures; for example, when combined with lipids, DNA can form characteristic shapes like rods or toroids [95–97]. Therefore, by introducing different chemical groups to the polyelectrolyte structures, we can create self-assembled complexes with many desired properties, conformations and functionalities. Appropriately, polyelectrolytes have many industrial and biotechnological applications including wastewater treatment, microencapsulation, protein separation, enzyme immobilization and drug delivery.

Despite their importance, our understanding of their behavior and functionality is far from complete. ITC is one of the few experimental techniques capable of monitoring their interactions without any significant limitation in respect to their charge, size and internal architecture. In addition, ITC can provide valuable data on the polyelectrolyte's aggregation state under various solvent conditions, by conducting proper dilution experiments [98–100]. However, studying these systems with

ITC can be very challenging, considering that the complexation between oppositely charged polyelectrolytes is often dominated by favorable entropic contributions, with a very small enthalpy change. The main source of heat during the complexation process is the release of counterions from the polyelectrolyte's hydration layer. Fortunately, modern ITC devices have become sensitive enough to follow this type of low-enthalpy reactions. As a result, during the last decade, ITC has been frequently employed to gain further insight into the mechanisms of various polyelectrolyte interactions. Some characteristic examples from the literature are discussed in brief below.

Feng et al. [101] studied the interactions between the strong polyelectrolytes poly(diallyldimethylammonium chloride) and potassium poly(vinyl sulfate) and between the weak polyelectrolytes polyvinylamine and carboxymethyl cellulose by ITC under various pH and ionic strength conditions. Strong polyelectrolytes were found to interact with high affinity through a relatively simple two-step process; the first step corresponds to the exothermic complexation of the two polyelectrolytes. The second step, which is also exothermic, is the result of aggregation and shrinking of the complex structure to a more compact state. Interestingly, the second process occurs near the point where the total charge of the sample cell solution is zero. The enthalpy of the interaction shows a strong dependence on the choice of buffer, a strong indication for the presence of a linked protonation/deprotonation equilibrium. However, ion specific effects from the buffer counterions could also play a significant role in these densely charged colloidal dispersions. The interaction between the two weak electrolytes was found to follow a different binding mechanism. Based on ITC data from experiments in different solution pH, the authors concluded that the formation of the polyelectrolyte complex at the 5.0–5.9 pH range induces the carboxymethyl cellulose to ionize further during the titration. This increased carboxyl ionization leads to more ionic crosslinks between the polyelectrolytes and the interaction appears to be quite strong. Outside this pH window, the ionization effect becomes less important and the interaction is described by a much smaller binding constant.

Another study by Maurstad et al. [102] investigated the interaction between xanthan, a bacterial polysaccharide, and a chitosan polymer. Thermodynamic analysis of ITC data provided a valuable insight to the mechanism of interaction. The small binding constant of the interaction is indicative of a non-specific, entropy-driven complexation process. Increasing the chain length of the chitosan led to a more complicated binding isotherm, suggesting that there was a secondary process after the formation of the complex. Atomic force microscopy topographs confirmed original ITC findings, showing that initial binding events were followed by aggregation when the concentration and chain length of chitosan were increased over a certain limit.

A recent study by Lounis et al. [103] used ITC to systematically investigate the interactions between linear poly(L-lysine) and oppositely charged acrylamide and 2-acrylamido-2-methyl-1-propanesulfonate copolymers at different ionic strength conditions. The ITC experimental results for these interactions were in very good agreement with previously published data obtained by other techniques. The thermodynamic analysis of the binding signature revealed that the complex formation is an entropy-driven process. Furthermore, the logarithm of the measured binding affinities demonstrated a linear dependence with the logarithm of the solution's ionic strength; taken together, these finding confirm that the binding mechanism is based on stabilizing electrostatic interactions between oppositely charged polyelectrolyte regions. The authors were able to estimate the number of counter ions release by combining ITC affinity data and the Manning theory [104].

One of the most interesting applications of polyelectrolyte systems is their use as drug stabilizing and delivery systems. Sideratou et al. [105] investigated the interaction between functionalized poly(L-lysine) dendrigrafts and insulin at physiological pH. Analysis of ITC data provided a clear picture for the binding mechanism of the dendrigrafts-insulin interaction. The final binding isotherm appears to be the result of two distinct thermodynamic equilibria: an endothermic process that is the major source of heat signals at low insulin concentrations (up to ~0.7 molar ratio) and an exothermic process that is predominant at higher molar ratio. The competition between these two opposite heat signals produces a characteristic V-shaped curve (Fig. 9). The endothermic trace is produced from the dissociation of insulin, which at high concentrations forms dimers or hexamers. When dendrigraft molecules are injected into the sample cell, the presence of the amino and/or guanidinium groups destabilizes the insulin oligomers, causing them to dissociate into monomers. The observed exothermic process results from the binding of insulin monomers to the functionalized dendrigraft by charge-pairing interactions. After a critical monomer concentration is reached in the solution, less and less dissociation events occur and the exothermic binding trace becomes dominant.

3.4 Interactions of Lipid-Based Systems

Some compounds do not have all the necessary properties to become efficient drugs. Problems like poor solubility, non-specific targeting or low permeability of biological barriers reduce their effective concentration significantly, leading to less than optimal results. One of the greatest challenges in the pharmaceutical industry is the development of new strategies and technologies to overcome these obstacles and enhance the bioavailability of the molecules in the area of interest.

The interaction of bioactive ligands with lipid bilayers is a critical step for their transportation through biological membranes to the intended target site. Most drugs will have to cross several membranes to reach the interior of the cell. Therefore, a detailed understanding of the interactions between drugs and biological membranes is very important for the formulation of effective therapeutic agents.

The thermodynamics of lipid interactions with other compounds depends on the structure and various physicochemical properties of the reacting species, including lipophilicity, charge, size and hydrogen bonding capacity. Lipids can assemble in a variety of morphologies, depending on their chemical structure, hydration state and solution conditions. Lipid are a very difficult system to work with; usually, the available sample quantities are limited and contain various impurities. Moreover, the

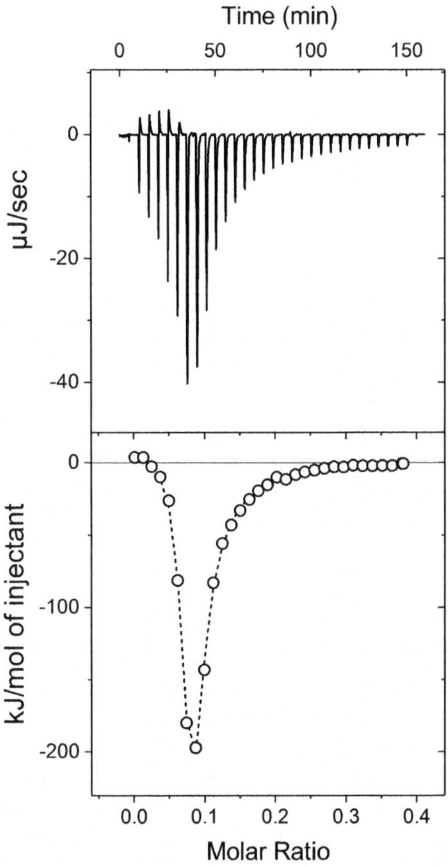

Fig. 9 Isothermal titration data for the interaction of 1 mM insulin with 0.5 mM poly(L-lysine) dendrigraft bearing 36 amino groups at the external surface at 25 °C in Tris buffer (10 mM, pH = 7.4). The upper panel shows the raw calorimetric data and the lower panel shows the corresponding integrated and normalized heats (d) given per mole of dendrigraft injected. Dashed line represent the non-linear least square fit of the ITC data to a two independent set-of-sites binding model

resulting suspensions are generally turbid, which complicates analysis by spectroscopic techniques. Opposite to specific binding, where the association requires the presence of well-defined binding sites, lipid interactions does not necessarily lead to the formation of complexes with certain stoichiometries.

In addition to their usefulness as model systems for biological membranes, lipid vesicles hold promise as efficient drug carriers, with numerous clinical applications [106, 107]. ITC is considered the method of choice for monitoring interactions between bioactive molecules (such as surfactants, peptides, proteins, drugs) and lipid membrane analogs like vesicles or micelles. Below we discuss in brief some interesting ITC studies on the interactions of lipid-based systems.

Ikonen et al. [108] investigated the thermodynamics of the binding and partitioning of three β-blockers (alprenolol, labetalol and propranolol) as well as that of the local anaesthetic tetracaine into liposomes composed of 1-Palmitoyl-2-oleoyl-sn-glycero-3-phosphocholine (POPC) and 1-palmitoyl-2-oleoyl-sn-glycero-3-[phospho-rac-(1-glycerol)] (POPG). A series of ITC experiments was performed in order to investigate thoroughly the effect of concentration, membrane curvature, temperature, and ionic strength to binding energetics. The interactions of all drugs were exothermic, with a dominant enthalpic term. The favorable enthalpy change can be explained by van der Waals interactions between the non-polar groups of the drugs and the long hydrocarbon chains of the lipids. Repeating the titrations with different concentrations of liposomes and drugs resulted in almost identical estimations for the thermodynamic parameters of the interaction, indicating that the partitioning of the compounds into liposomes is a thermodynamic equilibrium and not a kinetically controlled process. The effect of membrane curvature was examined by direct comparison of ITC titrations with liposomes of different diameter and it was found that there is no significant change to the binding parameters. Increase of the titration temperature resulted in more exothermic interactions, implying negative ΔC_p values for the binding processes. Negative ΔC_p values are often attributed to burial of hydrophobic areas, providing further evidence for an extended network of van der Waals interactions between the non-polar groups of the reactants. In conclusion, the effect of ionic strength on the affinity of the interactions was examined by carrying out ITC experiments at different salt concentrations. Measurable differences were detected for the binding constants of the interactions in low and high salt concentrations, providing evidence that electrostatic effects play an important role in the binding process. These electrostatic effects may include direct charge—charge interactions between the drugs and the liposomes as well as structural reorganization of the lipids in the negatively charged membrane.

Another publication by Al-Kaddah et al. [109] explored the interaction of the antibiotics gallidermin and vancomycin with bacterial membranes simulated by 1,2-dioleoyl-sn-glycero-3-phosphocholine liposomes (DOPC). To investigate the role of the cell wall precursor lipid II to the antibiotic interactions, pure DOPC liposomes and DOPC liposomes containing 0.1% (mol/mol) were prepared (DOPC + Lipid II). Both compounds were found to bind to DOPC + Lipid II liposomes with high affinity (Kb ~ 3×10^5 M^{-1}), displaying a typical sigmoidal ITC trace of exothermic peaks that can be analyzed by a single set-of -sites binding model. However, only vancomycin was able to interact with pure DOPC liposomes with similar affinity under identical experimental conditions. These results provide a clear indication for a direct interaction between gallidermin and Lipid II that facilitates the membrane insertion of the molecule. On the other hand, the vancomycin insertion is probably mediated by non-specific hydrophobic interactions with the membrane. These differences in binding behavior could also imply a different mechanism of antimicrobial activity for the two peptides.

Another interesting class of lipid-based molecules, which combine the aggregation characteristics of lipids and the specific functionalities of nucleosides, are the nucleolipids [110]. These hybrid moieties, composed of a lipid covalently linked to a nucleoside, show great promise as biocompatible nanocarriers for therapeutic compounds. A recent study [111] explored the interactions of thymidine- and adenine-3′-(1,2-dipalmitoyl-sn-glycero-3-phosphate) nucleolipids (abbreviated as diC16dT and diC16dA respectively) with the RNA analogs polyuridylic acid (polyU) and polyadenylic acid (polyA). Both nucleolipids were able to bind to their

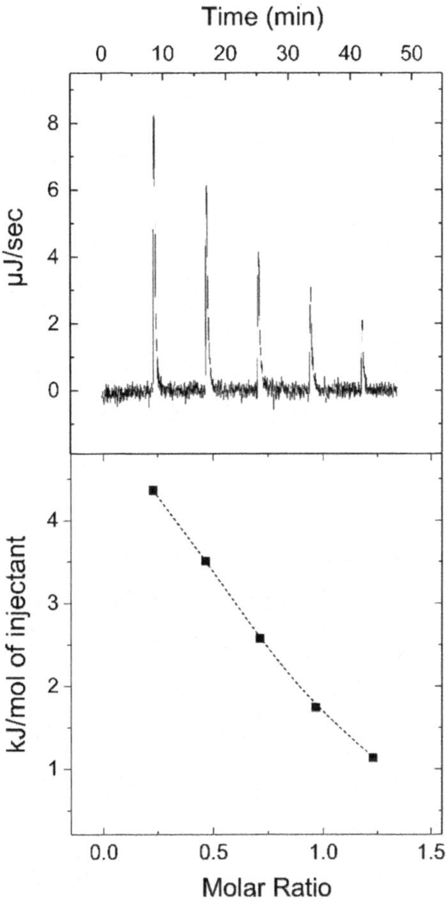

Fig. 10 Calorimetric data for the interaction of poly-A with diC16dT; Upper panel: Raw data for the titration of a 6 mM solution of poly-A [syringe] into 1 mM solution of diC16dT [cell]. Lower panel: Normalized injection heats, corrected for dilution effects and the fitting curve to the single set of sites model (dashed line). The binding constant is $K_b = 6660\ M^{-1}$ and the binding enthalpy is 5.6 kJ/mol

base-corresponding RNA analogs, however the interactions were relatively weak ($K_b < 10^3\ M^{-1}$) and non-specific, driven mainly by favorable entopic contributions (Fig. 10). The stoichiometry of the interactions ($N < 1$) also revealed that due to steric restrictions, only a fraction of the nucleolipid molecules are able to interact with the nucleic acid analogues. The conclusions of this study were used for an appropriate modification of the diC16dT and diC16dA nucleolipids, in order to increase the selectivity and affinity of the interactions [112]. The results were very encouraging, showing a tenfold increase in affinity for both compounds. Although the optimization of nucleolipids is far from complete, these studies demonstrate the usefulness of ITC in the development of new design strategies.

3.5 Cyclodextrin Interactions

For most drugs, a difficult balance needs to be achieved: they must be hydrophilic enough to dissolve in biological fluids and lipophilic enough to be able to penetrate all the biological membranes in their path. However, quite often, the most promising drug candidates exhibit poor water solubility [113] and they are classified as Class II (high permeability but low solubility) or Class IV (low permeability and low solubility) drugs in the FDA's biopharmaceutical classification system [114]. There are many ways to increase the apparent solubility of a compound, including salt formulation, particle size reduction, microemulsions etc. One very effective strategy to increase drug solubility is to create reversible complexes with a more hydrophilic carrier. Cyclodextrins are increasingly used for drug-delivery purposes, by virtue of their non-toxicity and effectiveness in enhancing the solubility, stability and bioavailability of otherwise poorly soluble drugs. Today, many therapeutic agents that contain natural or modified cyclodextrins are available in markets worldwide [115]. Cyclodextrins are naturally occurring cyclic oligosaccharides formed by six (α-CD), seven (β-CD) or eight (γ-CD) D-glucopyranosyl units organized in a hollow truncated cone architecture with amphiphilic properties; the exterior of the structure is hydrophilic, while the interior cavity surface is relatively hydrophobic. They are known for more than 100 years [116], but it is only recently that highly purified cyclodextrins have been incorporated in pharmaceutical formulations. In aqueous environments, CDs form inclusion complexes by incorporating hydrophobic molecules or hydrophobic parts of the molecule (depending on chemical structure and steric restrictions) into their central cavity, thereby increasing the solubility of the guest molecules and protecting them from their environment.

The rigidity of the structure and the specific dimensions of the hydrophobic cavity arise certain limitations on the potential guests that can be accommodated by native CDs. Chemical modifications of cyclodextrins have been used to enhance their functionality and in some cases to overcome molecular size restrictions by proper orientation of multiple CDs into new geometries that specifically match the structure of the guest. CD functionalization provide new opportunities for their use to complexation, delivery and controlled release of therapeutic molecules. Natural CDs are relatively large and contain a large number of hydrogen donors and acceptors; consequently, they do not readily permeate biological membranes [117]. However, modified cyclodextrins, such as methylated β-cyclodextrins, can enhance drug penetration through lipophilic barriers [118].

Most CD-drug inclusion complexes involve 1:1 stoichiometries (one molecule of drug per CD molecule) and binding constants in the range of 50–2000 M^{-1} [119]. The driving force for the inclusion can have many origins: inclusion of the drug in the central cavity displaces closely packed water molecules to the bulk with a favorable change in entropy. Electrostatic interactions, van der Waals interactions, hydrophobic effect, hydrogen-bond formation can also contribute to the energetics of the complexation process. Hence, CD-drug interactions are relatively weak interactions, which are often hard to study. ITC is an invaluable tool for studying the binding mechanism and the energetics of CD—drug inclusion, especially when combined with structural information from other techniques like NMR. In the following paragraphs, instructive examples of ITC studies on CD-drug interactions are discussed.

Segura-Sanchez et al. [120] used ITC to study the complexation mechanism of (+)-usnic acid (a molecule with antitumor activity), by native and modified CDs. The stoichiometry of all interactions was found to be 1:1 and the binding constants were in the range between 100 and 1100 M^{-1}, with γ-CD being the most suitable host of all the CDs of the study. The thermodynamic profile of the interactions suggests that the complex formation is enthalpically driven, stabilized by a combination of van der Waals forces and hydrogen bonds. To investigate further the complexation mechanism, the authors repeated the γ-CD experiments in different solution pH. The observed changes in binding energetics were correlated with various possible orientations and a planar configuration was finally proposed for the guest molecule. Molecular modeling simulations confirmed that the indicated binding topology was the most stable for this specific inclusion system.

Mazzaferro et al. [121] investigated the inclusion complexes of docetaxel (a potent anti-cancer drug) with various cyclodextrins. The solubility of the drug was increased over 5000 times when complexed with methyl-β-cyclodextrin (Me-β-CD), while the other cyclodextrins of the study offered less significant solubility enhancements. Two-dimensional 1H NMR, ITC, molecular docking simulations and circular dichroism spectroscopy were employed to investigate the binding of docetaxel to Me-β-CD. Analysis of ITC data revealed an unusual binding mechanism: the stoichiometry of the complex was found to be 2:1 (two Me-β-CDs bind one docetaxel molecule) and the experimental binding isotherm was indicative of an exothermic sequential binding process. The first step of the complex formation involves the exothermic binding of docetaxel to one Me-β-CD with an unfavorable entropic term. The exposed part of docetaxel then interacts with another Me-β-CD to form the final 2:1 complex, this time with a favorable entropy change. Molecular docking simulations explain the opposite entropic contributions of the sequential binding events by differences in the hydrophobic area of the two binding interfaces and the smaller flexibility of the already bound docetaxel molecule.

A study by Ignaczak et al. [122] evaluated the thermodynamic parameters of the complex formation between mianserin hydrochloride, a common antidepressant, with β-CD by ITC and quantum chemical methods. Titration data shown that the interaction is relatively strong (>1300 M^{-1}), with a small negative enthalpic term and a large favorable entropic change that is associated with release of high-ordered water molecules from the hydration layer to the bulk. Interestingly, the stoichiometry of the interaction suggests the formation of 2:1 complexes along with a smaller population of 1:1 β-CD—mianserin complexes. However, the binding isotherm was adequately described by a single step process, indicating that the 2:1 complex involves the synchronous binding of two β-CDs to a mianserin molecule. Quantum molecular dynamics calculations on both 1:1 and 2:1 structures confirmed that 2:1 complexes are energetically more favorable than their 1:1 counterparts.

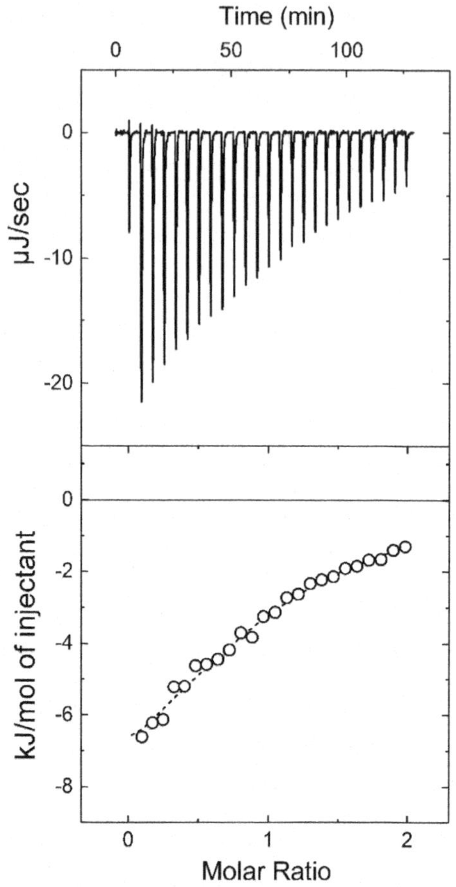

Fig. 11 (Upper panel) Change of power supply to the calorimetric cell during the titration of 10 mM of octakis(6-(2-aminoethylthio)-6-deoxy)-γ-cyclodextrin into 1 mM of various oxacillin at 25 °C in PBS buffer, after the subtraction of the appropriate reference experiments. (Lower panel) Integration of the area under each injection, normalized per mol of injectant and plotted as a function of the [Guest]:[Host] ratio at each point of the titration. Dashed line represent the non-linear least square fit of the ITC data to a simple 1:1 complexation thermodynamic model

A recent study by Agnes et al. used ITC to identify the optimal host for delivering penicillins to mammalian cells from a collection of anionic and cationic thioether-substituted-β- and γ-CD derivatives. The strongest interaction of the study was that of the positively charged octakis(6-(2-aminoethylthio)-6-deoxy)-γ-CD with oxacillin (Fig. 11). The binding constant was ~1700 M^{-1}, with almost equally favorable entropic and enthalpic terms. For this system, the binding stoichiometry suggested the presence of 1:1 complexes along with measurable population of 2:1 (host/guest) complexes, a result that was later confirmed by NMR data as well. The authors used this interaction to investigate the role of electrostatic effects in the β-lactam antibiotics inclusion mechanism by studying the effect of ionic strength on the binding of the oppositely charged molecules. No significant change in both enthalpy and binding affinity was detected for biologically relevant salt concentrations and it was concluded that the complex was stabilized though host-guest soft contacts rather than by electrostatic interactions.

4 Conclusions

Isothermal titration calorimetry is a well-established biophysical technique for studying the formation or dissociation of molecular complexes. Over the past years, ITC has developed from a convenient tool for studying macromolecular interactions with small ligands to a powerful and versatile method that is used in various research fields including biochemistry, biophysics, structural biology, materials science, medicinal chemistry, supramolecular chemistry and food science. The increased popularity of ITC in basic biochemical, biomedical biotechnology-oriented research lies in the ability to provide a complete thermodynamic characterization of noncovalent interactions and valuable insight to the underlying mechanisms that govern molecular recognition. Measuring the small amounts of heat released or absorbed upon binding, ITC is able to monitor molecular events that are invisible to other techniques such as binding to surfactants, cyclodextrin inclusions and nanoparticle corona formation. As titration calorimeters become more sensitive and widespread with time, the spectrum of ITC applications is expected to broaden even more in the following years.

References

1. Berman, H.M., et al.: The protein data bank. Nucleic Acids Res. **28**(1), 235–242 (2000)
2. Christensen, J.J., Johnston, H.D., Izatt, R.M.: An isothermal titration calorimeter. Rev. Sci. Instrum. **39**(9), 1356–1359 (1968)
3. Wiseman, T., et al.: Rapid measurement of binding constants and heats of binding using a new titration calorimeter. Anal. Biochem. **179**(1), 131–137 (1989)
4. Horn, J.R., Brandts, J.F., Murphy, K.P.: van't Hoff and calorimetric enthalpies II: effects of linked equilibria. Biochemistry **41**(23), 7501–7507 (2002)
5. Horn, J.R., et al.: van't Hoff and calorimetric enthalpies from isothermal titration calorimetry: are there significant discrepancies? Biochemistry **40**(6), 1774–1778 (2001)
6. Kantonen, S.A., Henriksen, N.M., Gilson, M.K.: Accounting for apparent deviations between calorimetric and van't Hoff enthalpies. Biochim. Biophys. Acta **1862**, 692–704 (2018)
7. Liu, Y., Sturtevant, J.M.: Significant discrepancies between van't Hoff and calorimetric enthalpies. II. Protein Sci. **4**(12), 2559–2561 (1995)
8. Liu, Y., Sturtevant, J.M.: Significant discrepancies between van't Hoff and calorimetric enthalpies. III. Biophys. Chem. **64**(1–3), 121–126 (1997)
9. Naghibi, H., Tamura, A., Sturtevant, J.M.: Significant discrepancies between van't Hoff and calorimetric enthalpies. Proc. Natl. Acad. Sci. U.S.A. **92**(12), 5597–5599 (1995)
10. Turnbull, W.B., Daranas, A.H.: On the value of c: can low affinity systems be studied by isothermal titration calorimetry? J. Am. Chem. Soc. **125**(48), 14859–14866 (2003)
11. Velazquez-Campoy, A., Freire, E.: Isothermal titration calorimetry to determine association constants for high-affinity ligands. Nat. Protoc. **1**(1), 186–191 (2006)
12. Khalifah, R.G., et al.: Thermodynamics of binding of the carbon dioxide-competitive inhibitor imidazole and related compounds to human carbonic anhydrase I: an isothermal titration calorimetry approach to studying weak binding by displacement with strong inhibitors. Biochemistry **32**(12), 3058–3066 (1993)
13. Zhang, Y.L., Zhang, Z.Y.: Low-affinity binding determined by titration calorimetry using a high-affinity coupling ligand: a thermodynamic study of ligand binding to protein tyrosine phosphatase 1B. Anal. Biochem. **261**(2), 139–148 (1998)

14. Sigurskjold, B.W.: Exact analysis of competition ligand binding by displacement isothermal titration calorimetry. Anal. Biochem. **277**(2), 260–266 (2000)
15. Tellinghuisen, J.: Designing isothermal titration calorimetry experiments for the study of 1:1 binding: problems with the "standard protocol". Anal. Biochem. **424**(2), 211–220 (2012)
16. Nilsson, S.O., Wadso, I.: A flow-microcalorimetric vessel for solution of small quantities of easily or slightly soluble liquids—solution of benzene in water at 298.15-K. J. Chem. Thermodyn. **16**(4), 317–330 (1984)
17. Mizoue, L.S., Tellinghuisen, J.: The role of backlash in the "first injection anomaly" in isothermal titration calorimetry. Anal. Biochem. **326**(1), 125–127 (2004)
18. Markova, N., Hallen, D.: The development of a continuous isothermal titration calorimetric method for equilibrium studies. Anal. Biochem. **331**(1), 77–88 (2004)
19. Christensen, J.J., Hansen, L.D., Izatt, R.M.: Handbook of Proton Ionization Heats and Related Thermodynamic Quantities. Wiley, New York (1976)
20. Goldberg, R.N., Kishore, N., Lennen, R.M.: Thermodynamic quantities for the ionization reactions of buffers. J. Phys. Chem. Ref. Data **31**(2), 231–370 (2002)
21. de Rivera, M.R., Socorro, F.: Baseline changes in an isothermal titration microcalorimeter. J. Therm. Anal. Calorim. **80**(3), 769–773 (2005)
22. Bhatnagar, R.S., Gordon, J.I.: Thermodynamic studies of myristoyl-coa—protein N-myristoyltransferase using isothermal titration calorimetry. Lipid Modif. Proteins **250**, 467–486 (1995)
23. Brautigam, C.A.: Fitting two- and three-site binding models to isothermal titration calorimetric data. Methods **76**, 124–136 (2015)
24. Brown, A.: Analysis of cooperativity by isothermal titration calorimetry. Int. J. Mol. Sci. **10**(8), 3457–3477 (2009)
25. McPhail, D., Cooper, A.: Thermodynamics and kinetics of dissociation of ligand-induced dimers of vancomycin antibiotics. J. Chem. Soc. Faraday Trans. **93**(13), 2283–2289 (1997)
26. Buurma, N.J., Haq, I.: Advances in the analysis of isothermal titration calorimetry data for ligand-DNA interactions. Methods **42**(2), 162–172 (2007)
27. Keeler, C., et al.: An explicit formulation approach for the analysis of calcium binding to EF-hand proteins using isothermal titration calorimetry. Biophys. J. **105**(12), 2843–2853 (2013)
28. Freire, E., Schon, A., Velazquez-Campoy, A.: Isothermal titration calorimetry: general formalism using binding polynomials. Methods Enzymol. **455**, 127–155 (2009)
29. Ladbury, J.E.: Calorimetry as a tool for understanding biomolecular interactions and an aid to drug design. Biochem. Soc. Trans. **38**(4), 888–893 (2010)
30. Fisher, H.F., Singh, N.: Calorimetric methods for interpreting protein-ligand interactions. Energ. Biol. Macromol. **259**, 194–221 (1995)
31. Ladbury, J.E., Chowdhry, B.Z.: Sensing the heat: the application of isothermal titration calorimetry to thermodynamic studies of biomolecular interactions. Chem. Biol. **3**(10), 791–801 (1996)
32. Livingstone, J.R., Spolar, R.S., Record Jr., M.T.: Contribution to the thermodynamics of protein folding from the reduction in water-accessible nonpolar surface area. Biochemistry **30**(17), 4237–4244 (1991)
33. Spolar, R.S., Livingstone, J.R., Record, M.T.: Use of liquid-hydrocarbon and amide transfer data to estimate contributions to thermodynamic functions of protein folding from the removal of nonpolar and polar surface from water. Biochemistry **31**(16), 3947–3955 (1992)
34. Sturtevant, J.M.: Heat capacity and entropy changes in processes involving proteins. Proc. Natl. Acad. Sci. U.S.A **74**(6), 2236–2240 (1977)
35. Gomez, J., Freire, E.: Thermodynamic mapping of the inhibitor site of the aspartic protease endothiapepsin. J. Mol. Biol. **252**(3), 337–350 (1995)
36. Haq, I., et al.: Specific binding of hoechst 33258 to the d(CGCAAATTTGCG)2 duplex: calorimetric and spectroscopic studies. J. Mol. Biol. **271**(2), 244–257 (1997)
37. Chaires, J.B.: Energetics of drug-DNA interactions. Biopolymers **44**(3), 201–215 (1997)
38. Cooper, A.: Thermodynamic analysis of biomolecular interactions. Curr. Opin. Chem. Biol. **3**(5), 557–563 (1999)

39. Holdgate, G.A., Ward, W.H.: Measurements of binding thermodynamics in drug discovery. Drug Discov. Today **10**(22), 1543–1550 (2005)
40. Spolar, R.S., Record Jr., M.T.: Coupling of local folding to site-specific binding of proteins to DNA. Science **263**(5148), 777–784 (1994)
41. Jeffrey, G.A., Saenger, W.: Hydrogen Bonding in Biological Structures. Springer, New York (1991)
42. Fersht, A.: Structure and Mechanism in Protein Science: A Guide to Enzyme Catalysis and Protein Folding, vol. xxi, 631 p. W.H. Freeman, New York (1999)
43. Scatena, L.F., Brown, M.G., Richmond, G.L.: Water at hydrophobic surfaces: weak hydrogen bonding and strong orientation effects. Science **292**(5518), 908–912 (2001)
44. Matulis, D.: Thermodynamics of the hydrophobic effect. III. Condensation and aggregation of alkanes, alcohols, and alkylamines. Biophys. Chem. **93**(1), 67–82 (2001)
45. Matulis, D., Bloomfield, V.A.: Thermodynamics of the hydrophobic effect. II. Calorimetric measurement of enthalpy, entropy, and heat capacity of aggregation of alkylamines and long aliphatic chains. Biophys. Chem. **93**(1), 53–65 (2001)
46. Matulis, D., Bloomfield, V.A.: Thermodynamics of the hydrophobic effect. I. Coupling of aggregation and pK(a) shifts in solutions of aliphatic amines. Biophys. Chem. **93**(1), 37–51 (2001)
47. Matulis, D., Rouzina, I., Bloomfield, V.A.: Thermodynamics of DNA binding and condensation: isothermal titration calorimetry and electrostatic mechanism. J. Mol. Biol. **296**(4), 1053–1063 (2000)
48. Ahmad, M., et al.: Enthalpy-entropy compensation upon molecular conformational changes. J. Chem. Theory Comput. **11**(4), 1410–1418 (2015)
49. Breiten, B., et al.: Water networks contribute to enthalpy/entropy compensation in protein-ligand binding. J. Am. Chem. Soc. **135**(41), 15579–15584 (2013)
50. Dragan, A.I., Read, C.M., Crane-Robinson, C.: Enthalpy-entropy compensation: the role of solvation. Eur. Biophys. J. **46**(4), 301–308 (2017)
51. Dunitz, J.D.: Win some, lose some: enthalpy-entropy compensation in weak intermolecular interactions. Chem. Biol. **2**(11), 709–712 (1995)
52. Lee, B.: Enthalpy-entropy compensation in the thermodynamics of hydrophobicity. Biophys. Chem. **51**(2–3), 271–277; discussion 277–278 (1994)
53. Holdgate, G.A.: Making cool drugs hot: isothermal titration calorimetry as a tool to study binding energetics. Biotechniques **31**(1), 164–166, 168, 170 passim (2001)
54. Baldwin, R.L.: Temperature dependence of the hydrophobic interaction in protein folding. Proc. Natl. Acad. Sci. U.S.A. **83**(21), 8069–8072 (1986)
55. Privalov, P.L., Gill, S.J.: Stability of protein structure and hydrophobic interaction. Adv. Protein Chem. **39**, 191–234 (1988)
56. Spolar, R.S., Ha, J.H., Record Jr., M.T.: Hydrophobic effect in protein folding and other noncovalent processes involving proteins. Proc. Natl. Acad. Sci. U.S.A. **86**(21), 8382–8385 (1989)
57. Kelley, R.F., O'Connell, M.P.: Thermodynamic analysis of an antibody functional epitope. Biochemistry **32**(27), 6828–6835 (1993)
58. Davies, T.G., Hubbard, R.E., Tame, J.R.: Relating structure to thermodynamics: the crystal structures and binding affinity of eight OppA-peptide complexes. Protein Sci. **8**(7), 1432–1444 (1999)
59. Chaires, J.B.: Calorimetry and thermodynamics in drug design. Ann. Rev. Biophys. **37**, 135–151 (2008)
60. Garbett, N.C., Chaires, J.B.: Thermodynamic studies for drug design and screening. Expert Opin. Drug Discov. **7**(4), 299–314 (2012)
61. Pierce, M.M., Raman, C.S., Nall, B.T.: Isothermal titration calorimetry of protein-protein interactions. Methods **19**(2), 213–221 (1999)
62. Freire, E., Mayorga, O.L., Straume, M.: Isothermal titration calorimetry. Anal. Chem. **62**(18), A950–A959 (1990)

63. Lewis, E.A., Murphy, K.P.: Isothermal titration calorimetry. Methods Mol. Biol. **305**, 1–16 (2005)
64. Leavitt, S., Freire, E.: Direct measurement of protein binding energetics by isothermal titration calorimetry. Curr. Opin. Struct. Biol. **11**(5), 560–566 (2001)
65. Ababou, A., Ladbury, J.E.: Survey of the year 2004: literature on applications of isothermal titration calorimetry. J. Mol. Recognit. **19**(1), 79–89 (2006)
66. Cliff, M.J., Ladbury, J.E.: A survey of the year 2002 literature on applications of isothermal titration calorimetry. J. Mol. Recogn. **16**(6), 383–391 (2003)
67. Ghai, R., Falconer, R.J., Collins, B.M.: Applications of isothermal titration calorimetry in pure and applied research–survey of the literature from 2010. J. Mol. Recogn. **25**(1), 32–52 (2012)
68. Falconer, R.J.: Applications of isothermal titration calorimetry—the research and technical developments from 2011 to 2015. J. Mol. Recogn. **29**(10), 504–515 (2016)
69. Majhi, P.R., Blume, A.: Thermodynamic characterization of temperature-induced micellization and demicellization of detergents studied by differential scanning calorimetry. Langmuir **17**(13), 3844–3851 (2001)
70. Vargas, C., Klingler, J., Keller, S.: Membrane partitioning and translocation studied by isothermal titration calorimetry. Methods Mol. Biol. **1033**, 253–271 (2013)
71. Poncet-Legrand, C., et al.: Interactions between flavan-3-ols and poly(L-proline) studied by isothermal titration calorimetry: effect of the tannin structure. J. Agric. Food Chem. **55**(22), 9235–9240 (2007)
72. Zheng, Y., et al.: Effect of pH on the complexation of kaempferol-4′-glucoside with three beta-cyclodextrin derivatives: isothermal titration calorimetry and spectroscopy study. J. Agric. Food Chem. **62**(1), 244–250 (2014)
73. Baldoni, D., et al.: Performance of microcalorimetry for early detection of methicillin resistance in clinical isolates of Staphylococcus aureus. J. Clin. Microbiol. **47**(3), 774–776 (2009)
74. Xi, L., et al.: Microcalorimetric study of Staphylococcus aureus growth affected by selenium compounds. Thermochim. Acta **387**(1), 57–61 (2002)
75. Mariana, F., et al.: Isothermal titration calorimetry—a new method for the quantification of microbial degradation of trace pollutants. J. Microbiol. Methods **82**(1), 42–48 (2010)
76. Hansen, L.D., et al.: Enzyme-catalyzed and binding reaction kinetics determined by titration calorimetry. Biochim. Biophys. Acta **1860**(5), 957–966 (2016)
77. Transtrum, M.K., Hansen, L.D., Quinn, C.: Enzyme kinetics determined by single-injection isothermal titration calorimetry. Methods **76**, 194–200 (2015)
78. Demarse, N.A., et al.: Determining enzyme kinetics via isothermal titration calorimetry. Methods Mol. Biol. **978**, 21–30 (2013)
79. Todd, M.J., Gomez, J.: Enzyme kinetics determined using calorimetry: a general assay for enzyme activity? Anal. Biochem. **296**(2), 179–187 (2001)
80. Hughes, J.P., et al.: Principles of early drug discovery. Br. J. Pharmacol. **162**(6), 1239–1249 (2011)
81. Ruben, A.J., Kiso, Y., Freire, E.: Overcoming roadblocks in lead optimization: a thermodynamic perspective. Chem. Biol. Drug Des. **67**(1), 2–4 (2006)
82. Torres, F.E., et al.: Higher throughput calorimetry: opportunities, approaches and challenges. Curr. Opin. Struct. Biol. **20**(5), 598–605 (2010)
83. Zhou, X., Kini, R.M., Sivaraman, J.: Application of isothermal titration calorimetry and column chromatography for identification of biomolecular targets. Nat. Protoc. **6**(2), 158–165 (2011)
84. Malmsten, M.: Soft drug delivery systems. Soft Matter **2**(9), 760–769 (2006)
85. Gwinn, M.R., Vallyathan, V.: Nanoparticles: health effects—pros and cons. Environ. Health Perspect. **114**(12), 1818–1825 (2006)
86. Cedervall, T., et al.: Detailed identification of plasma proteins adsorbed on copolymer nanoparticles. Angew. Chem. Int. Ed. Engl. **46**(30), 5754–5756 (2007)
87. Saptarshi, S.R., Duschl, A., Lopata, A.L.: Interaction of nanoparticles with proteins: relation to bio-reactivity of the nanoparticle. J. Nanobiotechnol. **11**, 26 (2013)

88. Mandal, S., et al.: Interaction of carbon nanoparticles to serum albumin: elucidation of the extent of perturbation of serum albumin conformations and thermodynamical parameters. J. Hazard. Mater. **248–249**, 238–245 (2013)
89. Malmsten, M.: Inorganic nanomaterials as delivery systems for proteins, peptides, DNA, and siRNA. Curr. Opin. Colloid Interface Sci. **18**(5), 468–480 (2013)
90. Cukalevski, R., et al.: Structural changes in apolipoproteins bound to nanoparticles. Langmuir **27**(23), 14360–14369 (2011)
91. Cedervall, T., et al.: Understanding the nanoparticle–protein corona using methods to quantify exchange rates and affinities of proteins for nanoparticles. Proc. Natl. Acad. Sci. **104**(7), 2050–2055 (2007)
92. Fleischer, C.C., Payne, C.K.: Secondary structure of corona proteins determines the cell surface receptors used by nanoparticles. J. Phys. Chem. B **118**(49), 14017–14026 (2014)
93. Eren, N.M., Narsimhan, G., Campanella, O.H.: Protein adsorption induced bridging flocculation: the dominant entropic pathway for nano-bio complexation. Nanoscale **8**(6), 3326–3336 (2016)
94. Decher, G., Schlenoff, J.B.: Multilayer thin Films: Sequential Assembly Of Nanocomposite Materials. 2nd compl. rev. and enl. edn. Wiley, Weinheim (2012)
95. Dan, N.: The structure of DNA complexes with cationic liposomes-cylindrical or flat bilayers? Biochim. Biophys. Acta **1369**(1), 34–38 (1998)
96. Golan, R., et al.: DNA toroids: stages in condensation. Biochemistry **38**(42), 14069–14076 (1999)
97. Vilfan, I.D., et al.: Time study of DNA condensate morphology: implications regarding the nucleation, growth, and equilibrium populations of toroids and rods. Biochemistry **45**(26), 8174–8183 (2006)
98. Perspicace, S., et al.: Isothermal titration calorimetry with micelles: Thermodynamics of inhibitor binding to carnitine palmitoyltransferase 2 membrane protein. FEBS Open Bio **3**, 204–211 (2013)
99. Loh, W., Brinatti, C., Tam, K.C.: Use of isothermal titration calorimetry to study surfactant aggregation in colloidal systems. Biochim. Biophys. Acta **1860**(5), 999–1016 (2016)
100. Marsh, D.: Thermodynamics of phospholipid self-assembly. Biophys. J. **102**(5), 1079–1087 (2012)
101. Feng, X., Leduc, M., Pelton, R.: Polyelectrolyte complex characterization with isothermal titration calorimetry and colloid titration. Colloids Surf. A **317**(1), 535–542 (2008)
102. Maurstad, G., Kitamura, S., Stokke, B.T.: Isothermal titration calorimetry study of the polyelectrolyte complexation of xanthan and chitosan samples of different degree of polymerization. Biopolymers **97**(1), 1–10 (2012)
103. Lounis, F.M., et al.: Interactions between oppositely charged polyelectrolytes by isothermal titration calorimetry: effect of ionic strength and charge density. J. Phys. Chem. B **121**(12), 2684–2694 (2017)
104. Manning, G.S.: Limiting laws and counterion condensation in polyelectrolyte solutions I. Colligative properties. J. Chem. Phys. **51**(3), 924–933 (1969)
105. Sideratou, Z., et al.: Arginine end-functionalized poly(L-lysine) dendrigrafts for the stabilization and controlled release of insulin. J. Colloid Interface Sci. **351**(2), 433–441 (2010)
106. Bulbake, U., et al.: Liposomal formulations in clinical use: an updated review. Pharmaceutics **9**(2), 12 (2017)
107. Sercombe, L., et al.: Advances and challenges of liposome assisted drug delivery. Front Pharmacol **6**, 286 (2015)
108. Ikonen, M., Murtomaki, L., Kontturi, K.: Microcalorimetric and zeta potential study on binding of drugs on liposomes. Colloids Surf B Biointerfaces **78**(2), 275–282 (2010)
109. Al-Kaddah, S., et al.: Analysis of membrane interactions of antibiotic peptides using ITC and biosensor measurements. Biophys. Chem. **152**(1), 145–152 (2010)
110. Allain, V., Bourgaux, C., Couvreur, P.: Self-assembled nucleolipids: from supramolecular structure to soft nucleic acid and drug delivery devices. Nucleic Acids Res. **40**(5), 1891–1903 (2012)

111. Thanassoulas, A., et al.: From nucleobases to nucleolipids: an ITC approach on the thermodynamics of their interactions in aqueous solutions. J. Phys. Chem. B **118**(24), 6570–6585 (2014)
112. Patwa, A., et al.: Tuning molecular interactions in lipid-oligonucleotides assemblies via locked nucleic acid (LNA)-based lipids. Org. Biomol. Chem. **11**(41), 7108–7112 (2013)
113. Lipinski, C.A.: Drug-like properties and the causes of poor solubility and poor permeability. J. Pharmacol. Toxicol. Methods **44**(1), 235–249 (2000)
114. Amidon, G.L., et al.: A theoretical basis for a biopharmaceutic drug classification: the correlation of in vitro drug product dissolution and in vivo bioavailability. Pharm. Res. **12**(3), 413–420 (1995)
115. Gidwani, B., Vyas, A.: A comprehensive review on cyclodextrin-based carriers for delivery of chemotherapeutic cytotoxic anticancer drugs. Biomed. Res. Int. **2015**, 198268 (2015)
116. Villiers, A.: Sur la fermentation de la fécule par l'action du ferment butyrique. Compt. Rend. Acad. Sci **112**, 536–538 (1891)
117. Lipinski, C.A., et al.: Experimental and computational approaches to estimate solubility and permeability in drug discovery and development settings. Adv. Drug Deliv. Rev. **46**(1–3), 3–26 (2001)
118. Marttin, E., Verhoef, J.C., Merkus, F.W.: Efficacy, safety and mechanism of cyclodextrins as absorption enhancers in nasal delivery of peptide and protein drugs. J. Drug Target. **6**(1), 17–36 (1998)
119. Connors, K.A.: The stability of cyclodextrin complexes in solution. Chem. Rev. **97**(5), 1325–1358 (1997)
120. Segura-Sanchez, F., et al.: Elucidation of the complexation mechanism between (+)-usnic acid and cyclodextrins studied by isothermal titration calorimetry and phase-solubility diagram experiments. J. Mol. Recogn. **22**(3), 232–241 (2009)
121. Mazzaferro, S., et al.: Bivalent sequential binding of docetaxel to methyl-beta-cyclodextrin. Int. J. Pharm. **416**(1), 171–180 (2011)
122. Ignaczak, A., Palecz, B., Belica-Pacha, S.: Quantum chemical study and isothermal titration calorimetry of beta-cyclodextrin complexes with mianserin in aqueous solution. Org. Biomol. Chem. **15**(5), 1209–1216 (2017)
123. Agnes, M., et al.: Designed positively charged cyclodextrin hosts with enhanced binding of penicillins as carriers for the delivery of antibiotics: the case of oxacillin. Int. J. Pharm. **531**(2), 480–491 (2017)
124. Nguyen, H.H., et al.: Surface plasmon resonance: a versatile technique for biosensor applications. Sens. (Basel) **15**(5), 10481–10510 (2015)
125. Piliarik, M., Vaisocherova, H., Homola, J.: Surface plasmon resonance biosensing. Methods Mol. Biol. **503**, 65–88 (2009)
126. Tyszka, J.M., Fraser, S.E., Jacobs, R.E.: Magnetic resonance microscopy: recent advances and applications. Curr. Opin. Biotechnol. **16**(1), 93–99 (2005)
127. Cala, O., Guilliere, F., Krimm, I.: NMR-based analysis of protein-ligand interactions. Anal. Bioanal. Chem. **406**(4), 943–956 (2014)
128. Lebowitz, J., Lewis, M.S., Schuck, P.: Modern analytical ultracentrifugation in protein science: a tutorial review. Protein Sci. **11**(9), 2067–2079 (2002)
129. Howlett, G.J., Minton, A.P., Rivas, G.: Analytical ultracentrifugation for the study of protein association and assembly. Curr. Opin. Chem. Biol. **10**(5), 430–436 (2006)
130. Rossi, A.M., Taylor, C.W.: Analysis of protein-ligand interactions by fluorescence polarization. Nat. Protoc. **6**(3), 365–387 (2011)
131. Hall, M.D., et al.: Fluorescence polarization assays in high-throughput screening and drug discovery: a review. Methods Appl. Fluoresci. **4**(2), 022001 (2016)
132. Heegaard, N.H., Nilsson, S., Guzman, N.A.: Affinity capillary electrophoresis: important application areas and some recent developments. J. Chromatogr. B Biomed. Sci. Appl. **715**(1), 29–54 (1998)
133. Albishri, H.M., et al.: Recent advances in affinity capillary electrophoresis for binding studies. Bioanalysis **6**(24), 3369–3392 (2014)

134. Hofstadler, S.A., Sannes-Lowery, K.A.: Applications of ESI-MS in drug discovery: interrogation of noncovalent complexes. Nat. Rev. Drug Discov. **5**(7), 585–595 (2006)
135. Vivat Hannah, V., et al.: Native MS: an 'ESI' way to support structure- and fragment-based drug discovery. Future Med. Chem. **2**(1), 35–50 (2010)
136. Freyer, M.W., Lewis, E.A.: Isothermal titration calorimetry: experimental design, data analysis, and probing macromolecule/ligand binding and kinetic interactions. Methods Cell Biol. **84**, 79–113 (2008)

Dynamics of Iron Homeostasis in Health and Disease: Molecular Mechanisms and Methods for Iron Determination

Christina Ploumi, Emmanouil Kyriakakis and Nektarios Tavernarakis

Abstract Iron is a versatile trace metal, indispensable for the survival of all living organisms. Despite its crucial role in vital biological processes, exceeded iron levels can be harmful for cellular and organismal homeostasis, due to iron's involvement in the generation of toxic hydroxyl radicals. As such, maintaining balanced iron levels is highly required in order for the organisms to avoid iron toxicity and at the same time preserve iron-dependent processes. This is achieved by the tight coordination of intricate systemic, cellular and subcellular mechanisms for iron absorption, excretion, utilization and storage. Those mechanisms decline during ageing, as well as in multiple human pathologies, leading to iron overload or deprivation, and eventually to death. To gain insight into how perturbations in iron homeostasis lead to disease, it is of great importance to use efficient methods for iron detection in distinct biological samples. Towards this direction, several biochemical and biophysical methods have been developed for the determination of iron and iron-containing compounds.

1 Introduction

Iron is one of the most important trace metals indispensable for life. Primarily found in the form of heme or iron-sulfur clusters (ISCs), it is highly required for numerous cellular and systemic biological processes. DNA replication and repair [1, 2], oxidative phosphorylation [3], and oxygen transport [4], are only some of the iron-dependent processes that highlight its crucial role in maintaining cellular and organismal homeostasis.

Iron's wide versatility mainly lies in its ability to easily cycle between two oxidation states: ferrus (Fe^{2+}) and ferric (Fe^{3+}) forms. Owing to this redox-state ability, excess iron can lead to the formation of hydroxyl radicals ($^{\bullet}OH$), through hydro-

C. Ploumi · N. Tavernarakis (✉)
Institute of Molecular Biology and Biotechnology (IMBB),
Foundation for Research and Technology-Hellas (FORTH), Heraklion, Crete, Greece
e-mail: tavernarakis@imbb.forth.gr

E. Kyriakakis
Biozentrum, University of Basel, Basel, Switzerland

C. Demetzos and N. Pippa (eds.), *Thermodynamics and Biophysics of Biomedical Nanosystems*, Series in BioEngineering, https://doi.org/10.1007/978-981-13-0989-2_5

$$Fe^{3+} + O_2^{\bullet -} \longleftrightarrow Fe^{2+} + O_2$$

$$Fe^{2+} + H_2O_2 \longrightarrow Fe^{3+} + {}^{-}OH + {}^{\bullet}OH \qquad \text{Fenton reaction}$$

$$O_2^{\bullet -} + H_2O_2 \longrightarrow O_2 + {}^{-}OH + {}^{\bullet}OH \qquad \text{Haber-Weiss reaction}$$

Fig. 1 Fenton and Haber-Weiss reactions that result in the generation of hydroxyl radicals in the presence of iron

gen peroxide (H_2O_2) reduction, a process widely known as Fenton reaction (Fig. 1) [5]. The resulted hydroxyl radicals are highly harmful, since they can cause severe oxidative damage and disruption of cellular constituents (including proteins, lipids and DNA), eventually leading to cell death. On the other hand, iron deprivation is also detrimental for life, since vital iron-dependent processes (i.e. oxygen distribution in tissues, cell proliferation and energy production) are perturbed. As such, maintaining balanced iron levels is of great importance, in order for the organisms to circumvent iron toxicity and at the same time ensure the function of significant iron-dependent processes. This is achieved by the existence of delicate mechanisms for the regulation of iron trafficking in tissues, cells and subcellular compartments. Particularly, there have been characterized mechanisms for iron sensing (Iron regulatory proteins—IRPs, hepcidin), acquisition (transferrins, divalent metal transporters, mitoferrins) excretion (ferroportin, multicopper oxidases) utilization (factors implicated in ISC and heme biogenesis), as well as storage (ferritins) [6, 7]. These molecular pathways have been found to be significantly dysregulated during ageing, and in various pathological conditions, including cancer, neurodegenerative and cardiovascular diseases [8]. In order to understand the exact molecular mechanisms through which iron dyshomeostasis leads to ageing and disease, it is critical to employ efficient techniques to determine iron quantification and distribution, as well as to assess its oxidative status in biological samples.

In this chapter we are going to analyze the mechanisms that govern iron homeostasis and discuss about how perturbations in these pathways lead to ageing and disease in humans. Finally, we are going to give an overview of the most common biochemical and biophysical methods, used to gain information concerning abundance, oxidation state and distribution of iron and iron-containing molecules.

2 Systemic Iron Homeostasis

The average amount of total iron in adults under physiological conditions is approximately 4 ± 1 g [9]. Iron absorption, storage, utilization and excretion need to be tightly regulated for the maintenance of optimal iron levels. So far, it is unclear whether specific mechanisms are required for iron excretion. Iron is mostly lost in uncontrolled ways and mainly through blood loss, menstruation, shedding from epithelial cells and sweat. Thus, in order to avoid perturbation of systemic iron homeostasis, processes controlling intestinal iron uptake and recycling of existing iron pools need to be well orchestrated.

As expected, differences in total levels as well as in the homeostatic mechanisms, exist between infants, children, adult male, pregnant, pre- and post-menopausal women, aged individuals and centenarians [10–13]. Even under physiological conditions, it is imperative for the body to respond fast according to iron demands. In embryos, optimal iron concentration is required for normal development. Iron reaches the fetus through the placenta; however, how iron reaches the fetal blood remain largely unknown [14]. After birth, dietary iron absorption by the enterocyte is required to maintain physiological iron levels [15]. Infants are born with approximately 270 mg of iron. Initially and during development, the demand of iron is high and is subsequently stabilized at around 1 mg/day for men and 2 mg/day for women, reflecting the higher iron amounts required due to menstrual bleeding and pregnancy [9, 15].

The only way to compensate iron loss and high iron demands is through diet. Dietary iron is classified as heme (or haem) and non-heme iron. Heme is more bioavailable and highly abundant in meat and seafood. Free heme is not present in circulation since it is cytotoxic and its acquisition must be tightly regulated. Yet, the specific pathways involved for heme iron uptake remain enigmatic. Heme is released from the hemoproteins in the acidic environment of the stomach with the assistance of proteolytic enzymes and iron is released from heme in enterocytes, in a process that requires heme oxygenase, an enzyme that catalyzes heme degradation [15, 16]. However, how cells internalize heme iron is less well understood. Heme carrier protein 1 (HCP1) and heme responsive gene-1 (HRG-1) were initially thought to play important role in heme uptake, however HCP1 is believed to be primarily a folate transporter and only involved in low-affinity heme-Fe uptake [17]. HRG (heme responsive-gene) proteins have been proposed to participate in heme homeostasis [18–20]. *C. elegans* are natural heme auxotrophs and thus, they acquire heme from their environment. Taking advantage of this unique characteristic, heme trafficking can be studied without the mechanisms of heme biogenesis to interfere. In an evolutionary conserved mechanism, in *C. elegans*, HRG-4 mediates heme uptake at the plasma membrane, whereas HRG-1 has been proposed to transport heme into the cytosol of macrophages via the endocytosis pathway [18, 19]. In a more recent study, an inter-tissue mechanism of heme regulation was investigated [20]. Intestinal HRG-7 functions as a secreted signaling factor during heme starvation in extra-intestinal tissues. More interestingly, with the involvement of Smad transcription factor 9 (SMA-9), HRG-7 levels are repressed via a neuron-derived DBL-1-dependent signal (member of the transforming growth factor beta (TGF-beta) superfamily), suggesting that an analogous cell non-autonomous evolutionary conserved mechanism may exist for systemic heme iron homeostasis. Despite the recent progress in the field, further investigation is required to fully understand heme regulation and trafficking.

Non-heme iron refers to the rest inorganic iron, deriving mainly from plants. It is mainly found in its oxidized less soluble (ferric) form and its bioavailability further declines in the presence of calcium and plant-derived compounds such as tannins, phenols and phytates. On the other side, ascorbic acid (vitamin C), vitamin A and citric acid may increase non-heme iron absorption [21]. Non-heme iron is also taken up on the apical brush border membrane of enterocytes in the small intestine, the duodenum and proximal jejunum via a distinct pathway. Dietary non-heme iron is initially reduced by reductases such as duodenal cytochrome B reductase to its ferrous form. Divalent metal transporter 1 (DMT1) at the apical membrane of intestinal epithelial cells is then responsible for ferrous iron acquisition, which is then stored within the iron-storage protein ferritin until needed, or exported through the basolateral epithelial membrane via ferroportin-1 (FPN1) into the blood stream [9, 15, 22, 23]. In cooperation with FPN1, hephaestin, which converts ferrous into ferric ions, enhances iron efflux. Then, plasma transferrin delivers iron to cells of all tissues for utilization [9]. Hepcidin a hepatocyte-secreted peptide hormone is considered the main regulator of systemic iron homeostasis. Hepcidin directly interacts with ferroportin, driving its internalization and degradation [24]. These events lead to decreased export of cellular iron and finally perturbed whole body iron homeostasis with implication in various iron-related diseases. Thus, current studies are focusing in the identification of novel hepcidin regulators as potential therapeutic targets of iron related diseases [25–28].

An underappreciated parameter that might play crucial role in iron uptake and whole body iron homeostasis is the existence of microbiota and most importantly of gut microflora, where iron acquisition happens. Iron is an essential mineral for pathogenic and non-pathogenic microbes. The role of iron in infection and immunity has been long appreciated and iron-withholding defense mechanisms exist to curtail iron as an innate immune mechanism against invading pathogens [29–32]. Microbes practically hijack iron from the host organism. Novel pharmacological approaches target iron acquisition from microbes to treat infectious diseases. On the other hand, the importance of symbiosis is increasingly appreciated. Interactions between host and microbiota may explain several phenomena that have so far eluded understanding. The importance of iron in bacterial growth and microflora composition under normal or pathological conditions is long acknowledged. However, how gut microflora may influence iron homeostasis is poorly studied. In a recent study, new aspects of the crosstalk between the microbiota and the intestinal epithelium in respect to iron sensing were revealed [33]. Gut microbiota altered the iron-related protein signature of mice, although they did not affect systemic iron load. Furthermore as discussed above, several proteins during various steps are required for efficient iron uptake. For instance, non-heme iron is first reduced to its ferrous form. Since, microbes contain evolutionary conserved mechanisms for iron homeostasis such as microbial ferric reductases, someone could assume that microbiota could provide important metabolically active material for more efficient iron uptake. Fungi, such as *Saccharomyces cerevisiae*, *Candida glabrata* and *Candida albicans* express cell surface ferric reductases and are capable of utilize iron from most ferric-containing molecules, including ferritin and transferrin [32]. Further investigation is required on how host-microflora interactions may influence iron homeostasis, with potential to provide answers to the so far unanswered questions.

3 Cellular Iron Homeostasis

In addition to the systemic control of iron abundance, cells have developed sophisticated mechanisms to maintain intracellular iron homeostasis (Fig. 2).

3.1 Cellular Iron Uptake and Utilization

The majority of mammalian cells incorporate blood-derived non-heme iron by using the Transferrin (Tf)-Transferrin Receptor (TfR) system. Each Tf molecule has the capacity to bind two ferric ions with high affinity. The binding of ferric ions to Tf is regulated by various factors, including temperature, pH and the presence of anionic species, like chloride [34–36]. Diferric Tf can be recognized by the specific receptor TfR1, which forms homodimers in the cell membrane of almost all mammalian cells. TfR2 is another TfR, which is predominantly expressed in hepatocytes and has a significantly lower affinity to diferric Tf [37, 38]. The entire Tf-TfR1 complex is subsequently internalized in cells, using the endocytotic machinery [39]. Importantly, the myotonic dystrophy kinase-related Cdc42-binding kinase alpha (MRCKα) participates in the recruitment of the actin cytoskeleton, which supports endocytosis [40]. The endosomal acidic environment, generated by a proton pump, promotes the

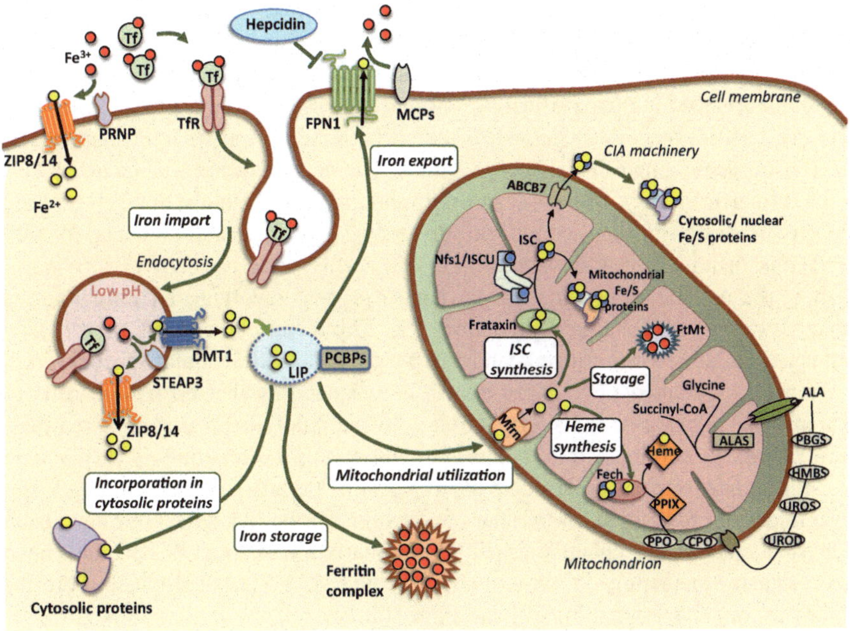

Fig. 2 Cellular mechanisms for iron homeostasis

dissociation of ferric ions from Tf, thus facilitating the recycling of apo-Tf back to the cell membrane. In addition to low pH, iron release from Tf is suggested to be enhanced by the reduction of ferric to ferrous ions. This process is mediated by six-transmembrane epithelial antigen of the prostate 3 (STEAP3), a ferrireductase resided at the endosomal membrane. Once reduced, ferrous ions are transported through DMT1 to the cytosol [41, 42].

Cells can also take up non-transferrin-bound iron (NTBI) from plasma. ZRT/IRT-like protein 8 (ZIP8) and ZIP14, members of the solute carrier 39 (SLC39) family can directly mediate the intracellular transport of plasma NTBI [43–45]. The prion protein (PRNP), localized in the cellular membrane is suggested to enhance the ZIP14-mediated NTBI transport due to its ferrireductase activity [46]. Apart from its role in NTBI transport, ZIP14 has also been implicated in transporting endosomal ferrous iron into the cytosol [47]. Although DMT1 and ZIP proteins seem to have similar roles in ferrous transport, their activity is highly pH-dependent. While ZIP8 and ZIP14 display maximal iron transport capacity at neutral pH [45, 48], DMT1 is more active in acidic environments (pH 5.5) [49]. This supports the hypothesis that ZIP8/14 may be the main NTBI transporters in the cell surface, as they directly communicate with the neutral pH-plasma, while DMT1 may be the main ferrous transporter in the acidic endosomes.

Cytosolic ferrous iron, released either by ZIP8/14 or DMT1 transporters, enters a low-molecular weight and chelatable labile iron pool (LIP). Due to the high redox activity of the ferrous state and the oxidative environment of the cytosol, LIP-derived iron is almost directly sequestered into cytosolic proteins, or delivered to mitochondria to support crucial iron-dependent processes. Cytosolic iron that is not utilized for iron-dependent processes can be stored into ferritin molecules. Their ferrireductase activity renders ferritins important factors of the cellular anti-oxidant response, as they transform the highly toxic ferrous ions into the less redox-active ferric form. Particularly, ferritins form shell-shaped 24-mers, comprised of ferritin light chain (FTL) and ferritin heavy chain (FTH) polypeptides. The composition of FTLs and FTHs in the 24mer may vary, and this variation modulates accordingly the iron-binding capacity of the entire complex. Intriguingly, each 24-mer can incorporate up to 4500 ferric ions [50]. Upon iron-depleted conditions, ferritin-incorporated iron is released to meet the cellular iron demands, through a process involving ferritinophagy, a selective type of autophagic degradation [51]. The remaining cytosolic ferrus iron that is neither stored nor utilized can be exported to the extracellular space through the hepcidin-regulated FPN1. Iron export is enhanced by the ferroxidase activity of multicopper-oxidases (MCPs). Although differentially expressed in human tissues, the MCPs ceruloplasmin, hephaestin and zyklopen are similarly involved in cytosolic iron export [52]. Members of the Poly-(rC)-binding protein (PCBP) seem to be important regulators for the fate of cytosolic ferrus iron, since they have been identified to act as iron chaperones [53]. Particularly, PCBP1 and PCBP2 have been involved in the metalation of iron-containing proteins [54, 55], but also in the delivery of ferrous ions to ferritins [56, 57] and FPN1 [58].

3.2 Regulation of Cellular Iron Homeostasis

The majority of the above-mentioned cellular mechanisms are under the tight control of the Iron Regulatory Protein (IRP) system. IRPs modulate the expression of iron-related genes to regulate iron import, export, storage and utilization. This is achieved by their ability to associate with specific stem-loop structures within the untranslated regions (UTRs) of the respective mRNAs, called iron-responsive elements (IREs). IRPs' binding to 5′-IREs blocks mRNA translation, whereas their binding to 3′-IREs induces it, through mRNA degradation blockage. As such, the expression of iron-related proteins is differentially regulated, depending on the IRE-type their mRNAs contain [59]. There have been identified two IRPs in mammals, IRP1 and IRP2, both of which have similar affinity to IRE-containing transcripts [60].

The RNA binding activity of IRPs is strongly dependent on the intracellular iron levels. Upon iron deprivation, the IRE-binding affinity of IRPs is highly induced leading to the translational suppression of the 5′IRE-containing mRNAs encoding for ferritins [61, 62] and FPN1 [22] and concurrently to the stabilization of the 3′IRE-containing mRNAs encoding for TfR1 [63], MRCKα [64] and DMT1 [65]. This homeostatic mechanism promotes iron uptake pathways and ensures that the remaining cytosolic iron will not be stored or exported, in order to be used for crucial iron-dependent processes. Under iron-replete conditions, IRPs' RNA-binding activity is strongly inhibited. When iron is intracellularly abundant, IRP1 is predominantly found assembled with a cubane type [4Fe–4S] of iron-sulfur cluster (ISC). The particular ISC assembly enables IRP1 to function as a cytosolic aconitase, and at the same time diminishes its affinity for IRE-containing transcripts [63]. A totally different inhibition mechanism has been characterized for IRP2, which lacks aconitase activity. Specifically, IRP2 is post-translationally regulated by an iron-dependent Ubiquitin proteasome mechanism. F-Box And Leucine Rich Repeat Protein 5 (FBXL5) is an enzyme 3 (E3) ubiquitin ligase, the activity of which depends on the formation of a [Fe–O–Fe] centre within its hemerythrin-binding domain. Upon iron-replete conditions, FBXL5 is activated, leading to the proteasomal degradation of its target IRP2 [66, 67]. Despite their similar RNA-binding affinity, several studies suggest that IRP2 responds better to varying iron concentrations, in comparison to IRP1 [68, 69]. This could be partly explained by the fact that the regulation of IRP2 is directly dependent on cytosolic iron and does not require ISC formation, like IRP1.

Another well characterized mechanism that orchestrates the expression of iron-related genes is the transcriptional regulation mediated by hypoxia inducible factors (HIFs) [70, 71]. The transcriptional activity of HIF1α and HIF2α is post-translationally regulated in an oxygen- and iron-dependent manner by specific hydroxylases, including prolyl-4-hydroxylase domain proteins (PHDs) [72] and the asparaginyl hydroxylase factor inhibiting HIF1 (FIH1) [73]. Upon iron- and oxygen-replete conditions, PHDs are fully functional and hydroxylate both HIF1α and HIF2α, leading to their proteasomal degradation. On the other hand, FIH1 directly inhibits HIFs' transcriptional activity by hydroxylating a conserved asparagine residue. Under hypoxic or iron-deprivation conditions, the function of both hydroxylases is blocked, leading to the enhancement of HIFs' transcriptional activity. Genes encoding for Tf, TfR1, DMT1, FPN and CP, are among the HIF targets implicated in the regulation of iron homeostasis [70, 74]. In a very recent study, they identified endonuclease Regnase-1, as novel regulator for HIF- mediated regulation of iron uptake. Regnase-1 promotes HIF2α activation, by destabilizing PHD3 mRNA [75]. Interestingly HIF2α expression is negatively regulated by the IRP/IRE system, since its mRNA contains a 5′-IRE. This is suggested to be a negative-feedback mechanism to prevent hemoglobin production, upon conditions that both heme's building blocks (oxygen, and iron) are deprived [76].

Hepcidin, a peptide hormone mainly synthesized in hepatocytes, is also considered to be an important regulator of both systemic and cellular iron homeostasis. Hepcidin directly interacts with FPN and promotes its internalization and subsequent degradation, thus blocking iron export from hepatocytes, macrophages and intestinal cells. Hepcidin's expression seems to be mainly regulated by systemic iron status. Increased Tf-TfR signaling, resulting from high extracellular iron levels (in the plasma or in the intestine) promotes the transcriptional activation of Hepcidin through BMP signaling. Hepcidin is also positively regulated by inflammatory cytokines, including interleukin-6 [77, 78].

4 Mitochondrial Iron Homeostasis

Mitochondria have a pivotal role in the maintenance of cellular and systemic iron homeostasis (Fig. 2). They are the lone sites for heme biogenesis, a vital prosthetic group, highly required for oxygen transport and mitochondrial respiration [79]. Additionally, mitochondria are essential for the biogenesis of ISCs. ISCs serve as cofactors of numerous enzymes, participating in indispensable cellular processes, including DNA replication, DNA repair and mitochondrial respiration [80]. Although there has been identified a cytosolic ISC assembly machinery, disruption of the mitochondrial ISC generation pathway, strongly inhibits the maturation of cytosolic and nuclear ISC-binding proteins [81]. This highlights the importance of the mitochondrial involvement for cellular and organismal homeostasis.

4.1 Mitochondrial Iron Import

A significant portion of cytosolic iron is transported into mitochondrial matrix for the generation of both heme and ISCs. There have been proposed several mechanisms that could mediate cytosolic iron transfer into the mitochondrial intermembrane space (IMS) [82]. One of the first theories suggests that LIP-derived cytosolic iron can be directly imported into mitochondria in a membrane potential-dependent and an ATP-independent process [83]. A growing amount of evidence supports the existence of a "kiss-and-run" mechanism, in which Tf-containing endosomes get in close proximity and fuse with the OMM, thus leading to the release of ferrus iron in the IMS [84, 85]. Alternatively, it has been suggested that cytosolic iron existed in a chelator-inaccessible state (likely bound in iron containing proteins) can be transferred across the OMM via protein-protein interactions with residual mitochondrial proteins. PCBPs, the cytosolic iron chaperones, could be involved in this process [53].

In all cases, the transport of IMS ferrous iron across the inner mitochondrial membrane (IMM) is mediated by mitoferrins (Mfrns), members of the mitochondrial solute carrier family. Mfrn2 is ubiquitously expressed in mammals, while the expression of its homolog Mfrn1 is restricted to hematopoietic tissues [86]. Mitochondrial iron import seems to be predominantly controlled by the post-translational regulation of Mfrns. Particularly, the protein levels of Mfrn1 (and not Mfrn2) are significantly increased in developing red blood cells, due to extension in the protein's half-life [87]. The increased stability of Mfrn1 is mediated by its physical interaction with ATP-binding cassette subfamily B member 10 (Abcb10), which is highly expressed during erythroid maturation. The specific interaction promotes mitochondrial iron uptake to support the increased demands for hemoglobin synthesis [88]. Concerning Mfrn2, there has not been reported any regulatory mechanism. Though, there must be a kind of regulation, since overexpression of Mfrn2 does not affect mitochondrial iron levels [87]. In addition to Mfrns, mitochondrial calcium uniporter is also involved in mitochondrial iron import [89, 90]. Once transported into the mitochondrial matrix, ferrous ions can be directly utilized for the biosynthesis of heme or ISCs, or they are stored for subsequent use.

4.2 Iron-Sulfur Cluster Biosynthesis

ISCs are essential cofactors, required for all living organisms, as they ensure the function of vital biological processes. These inorganic compounds consist of iron and sulfide ions, assembled in various combinations, the most common of which are the rhomboid [2Fe–2S] and the cubane [4Fe–4S] forms. Although it was initially believed that ISCs could be spontaneously formed in vivo, their biogenesis, as well as the maturation of ISC-containing proteins involve tightly regulated multistep processes, which are highly conserved from bacteria to yeast and humans [91].

In mammalian systems, as in almost all organisms, mitochondria are the major sites for ISC biogenesis. A critical initiation step for ISC formation is the abstraction of sulfur ions from cysteine. This is mediated by the homodimeric cysteine desulferase Nfs1 (Nitrogen Fixation 1 *S. Cerevisiae* Homolog), in complex with ISD11 which stabilizes and structurally supports Nfs1 activity [92]. The abstracted sulfur ions are subsequently transferred to the scaffold protein ISCU, which binds to each Nfs1 monomer. Frataxin, a mitochondrial matrix iron chaperone, likely donates ferrus ions into the ISCU complex, which serves as a platform for bringing ferrus and sulfur ions in close proximity [93]. Apart from its main role as a ferrus donor, frataxin enhances the reactivity of Nfs1 with cysteine's thiol side chains [94]. Ferrus and sulfur ions react in a process involving ferredoxins and ferredoxin reductases, thus leading to the formation of [2Fe–2S] ISCs. The newly synthesized rhomboid ISCs are sequestered to a mitochondrial chaperone complex, comprising the Heat Shock Protein Family A (Hsp70) Member 9 (HSPA9) and the co-chaperone DnaJ Homolog Subfamily C Member 20 (HSC20). This chaperone/co-chaperone system delivers [2Fe–2S] either to recipient mitochondrial ISC-binding apo-proteins, or to specific carrier proteins, such as glutaredoxin 5 (Grx5), that mediate the downstream ISC trafficking [95]. Particularly, Grx5 has been found to interact with and transfer ISCs to the ATP Binding Cassette Subfamily B Member 7 (ABCB7), an IMM transporter that mediates the export of ISCs to IMS [96]. Grx5 seems to be also involved in the formation of cubane [4Fe–4S] ISCs, by providing the rhomboid cluster to specific ISC proteins [97, 98].

In addition to the well-characterized mitochondrial pathway for ISC biogenesis, eukaryotic cells have developed a cytosolic ISC assembly (CIA) mechanism for the maturation of cytosolic and nuclear ISC-binding proteins [99]. Although there are cytosolic ISCU variants that could mediate de novo synthesis of ISCs in the cytosol, the CIA machinery requires intact mitochondrial ISC biogenesis [96, 100]. Remarkably, it has been suggested that a yet-unidentified ISC intermediate is exported to the cytosol in an ABCB7-dependent manner, in order to support CIA. Mutations in ABCB7 gene are the genetic cause of X-linked sideroblastic anemia with ataxia (XLSA/A), characterized by deficiency in the maturation of cytosolic ISC proteins, mitochondrial iron overload and defects in heme synthesis [81].

4.3 Heme Biosynthesis

Heme biogenesis is a multistep process, which involves eight sequential enzymatic reactions in both mitochondrial and cytosolic compartments [79]. The initial step is mediated by aminolevulinic acid synthase (ALAS), which catalyzes the condensation of glycine and succinyl-CoA to δ-aminolevulinic acid (ALA). The resulted ALA is exported to the cytoplasm through still not well-defined mechanisms. Although it was initially believed that ABCB10 is the transporter involved in ALA transport [101], recent studies rule out its involvement in this process and suggest that it is rather implicated in the regulation of hemoglobinization [102]. Cytosolic ALA is transformed to coproporphyrinogen III, through a 4-step enzymatic process, involving porphobilinogen synthase (PBGS), hydroxymethylbilane synthase (HMBS), uroporphyrinogen synthase (UROS) and uroporphyrinogen decarboxylase (UROD). The newly formed coproporphyrinogen III is transported to IMS, where it is converted to protoporphyrinogen IX by coproporphyrinogen oxidase (CPO). Protoporphyrinogen IX is then imported through ABCB6 to the mitochondrial matrix, and transformed by protoporphyrinogen oxidase (PPO) into protoporhyrin IX. Ferrochelatase (Fech) catalyzes the final step, by incorporating ferrous ions into protoporphyrin IX, leading to heme formation. Interestingly, Fech is an ISC-binding protein, highlighting the interconnection between heme and ISC biogenesis. The specific interplay explains the anemic phenotypes observed upon perturbation of ISC biogenesis [103].

4.4 Mitochondrial Iron Storage

Given that mitochondria are the principal sites for ROS production, increased levels of ferrous iron in mitochondrial matrix could lead to the formation of harmful hydroxyl radicals through Fenton reaction. To counteract iron toxicity, mammalian tissues with high metabolic demands express mitochondrial ferritins (FtMts) [104, 105]. FtMts are highly similar to cytosolic FTHs. They form shell-shaped oligomers in which they sequester ferrus ions and keep them in the less-redox active ferric form. Unlike cytosolic ferritins, FtMt expression is not regulated by the IRP/IRE system [106]. It was recently shown that its expression is controlled in the transcriptional level through a complex regulatory mechanism, involving multiple transcription factors and epigenetic changes [107, 108]. Interestingly, overexpression of FtMt in human lung cancer cell lines leads to a dose-dependent cytosolic iron depletion and concurrent iron overload in mitochondria [109].

Frataxin, the ferrous donor for the biosynthesis of ISCs is also suggested to act as an iron storage protein. Its yeast homolog Yfh1 is capable of forming 48-mers in which up to 2000 iron atoms can be stored [110]. The absence of frataxin multimers in mammalian systems could be explained by the redundant function of mitochondrial ferritins, which are not expressed in yeast [111]. In any case, independently of its capability to form multimers, frataxin can be classified as an iron storage protein, since it sequesters the redox-active ferrus ions in the matrix. Similar to Frataxin, Fech, the ferrous donor for heme biosynthesis can be also considered as an iron storage protein.

5 Iron Dyshomeostasis in Pathological Conditions and Ageing

Perturbations of systemic, cellular and subcellular iron levels can lead to life-threatening pathologies (summarized in Table 1) [80, 112–114]. Interestingly, iron dyshomeostasis is associated with ageing, the main risk factor for various diseases while iron accumulation appears to be a common feature during ageing [115–117]. However, how and whether loss of iron homeostasis influences normal ageing is not well known. Recent evidence shows that abundance of iron shortens the lifespan of animal models [117–120]. Knocking-down MCO-1, a *Drosophila melanogaster* ferroxidase, decreases iron content and protects from iron-induced toxicity [118]. Those flies live significantly longer compared to wild type animals in iron excess. Research on *C. elegans* has shown that iron influences longevity. Loss of ferritin leads to increased labile iron pool, ROS formation and rapid ageing [117]. Frataxin lesion initiates a pro-longevity iron starvation response in a mechanism that appears to involve induction of an evolutionary conserved pathway for mitophagy induction [119]. Thus, it is reasonable to hypothesize a direct and bidirectional association between iron metabolism and ageing. The mechanisms of iron accumulation during ageing and their involvement in the pathogenesis of various age-related disorders are only beginning to be elucidated. Deeper understanding of the underlying mechanisms that control iron homeostasis is expected to provide novel strategies against age-associated diseases and promote healthy ageing. Below, we mention only some of the diseases associated to iron dyshomeostasis.

5.1 *Friedrich's Ataxia*

Iron is required for normal brain function, but its accumulation has been observed in various diseases of the brain. Friedrich's ataxia (FA) is one of the best-studied neurological disorders of iron metabolism [121]. FA is an autosomal recessive inherited disease that causes progressive degeneration of sensory neurons in the dorsal root ganglia and symptoms begin to emerge early in life. Patients suffer from impaired muscle coordination (ataxia) and cardiomyopathy among others, but cognition is not affected. FA is characterized by iron accumulation in mitochondria due to dysfunction of iron-sulphur cluster biogenesis, caused by mutations in the frataxin gene [113, 122]. The majority of patients carry an unstable GAA repeat expansion that markedly reduces the expression of frataxin, whereas fewer patients have point mutations. Experimental models suggest that frataxin plays important roles in iron-dependent increase of ROS that affect Fe–S cluster proteins, in mitochondrial energy conversion and oxidative phosphorylation, highlighting the role of oxidative stress and normal mitochondrial function in FA [123]. Furthermore, a recent study has demonstrated that frataxin deficiency causes mitochondrial biogenesis defect in vivo, in vitro and in whole blood of patients [124]. Intriguingly, iron supplementation limits cardiac hypertrophy in mouse models [125]. Current understanding of the involved mechanisms allowed the assessment of genetic and cell-based treatment of FA. Ongoing approaches include increasing cellular frataxin levels and activity, improving mitochondrial function and decreasing oxidative stress, however no successful treatment is available owing to the complexity and inadequate understanding of ISC biogenesis and trafficking [126].

5.2 *Parkinson's Disease*

Parkinson's disease (PD) is a neurodegenerative disease of the central nervous system that affects predominantly dopaminergic neurons in the substantia nigra [127]. It is a progressive disorder that affects movement and early signs include tremors and trem-

Table 1 Pathologies associated with iron dyshomeostasis

Human disease	Involved gene	Encoded protein	Cellular localization	Relevant protein function	Relevance to iron homeostasis	References
Friedrich's ataxia	FXN/FRDA	Frataxin	Mitochondrial matrix (mMatrix)	Assembly of Fe–S cluster proteins	ISC biogenesis	[121–124]
Congenital sideroblastic anemias (CSAs):						
X-linked sideroblastic anemia (XLSA)	ALAS2	5′-Aminolevulinate synthase 2	Mitochondrial matrix	Generation of 5-aminolevulinic acid	Heme synthesis	[80]
XLSA with cerebellar ataxia	ABCB7	ATP binding cassette subfamily B member 7	Inner mitochondrial membrane (IMM)	Transporter	ISC biogenesis	[145–148]
Other types of CSAs	SLC25A38	Solute carrier family 25 member 38	IMM	Regulates mitochondrial glycine	Heme synthesis	[142]
	NDUFB11	NADH:Ubiquinone oxidoreductase subunit B11	mMatrix	Transfers e from NADH to Q10	Unclear	[143, 144]
	GLRX5	Glutaredoxin-related protein 5	mMatrix	Protein lipoylation	ISC and heme	[151, 152]
	ATP6	ATP synthase Fo subunit 6	mMatrix	ATP synthesis	Unclear	[153]
	TRNT1	tRNA-nucleotidyltransferase 1	mMatrix	tRNA Nucleotidyltransferase	Unclear	[155]

(continued)

Table 1 (continued)

Human disease	Involved gene	Encoded protein	Cellular localization	Relevant protein function	Relevance to iron homeostasis	References
	LARS2	Leucyl-tRNA synthetase	mMatrix	L-leucine to tRNA(Leu) catalysis	Unclear	[154]
	HSPA9	Mitochondrial 70 kDa heat shock protein	mMatrix	Chaperone	Unclear	[156]
Multiple respiratory chain complexes deficiency	LYRM4	ISD11	mMatrix	Forms a complex with the cysteine desulfurase NFS1	ISC biogenesis	[157]
Hereditary myopathy with lactic acidosis	ISCU	Iron-Sulfur cluster assembly enzyme	mMatrix	Scaffold protein for ISC assembly	ISC biogenesis	[158–160]
Multiple mitochondrial dysfunctions syndrome	NFU1	NFU1 iron–sulfur cluster scaffold	mMatrix	Scaffold proteins for ISC assembly	ISC biogenesis	[161, 162]
	BOLA3	BolA family member 3	mMatrix			

(continued)

Table 1 (continued)

Human disease	Involved gene	Encoded protein	Cellular localization	Relevant protein function	Relevance to iron homeostasis	References
Complex I deficiency	NUBPL	Nucleotide binding protein like	mMatrix	Assembly of the mitochondrial membrane respiratory chain NADH dehydrogenase (complex I)	ISC biogenesis	[163]
Wolfram syndrome type 2	CISD2	NAF-1/Miner 1	Outer mitochondrial membrane (OMM)	Fe–S cluster transport between mitochondria and cytosol (?)	ISC trafficking(?)	[164–166]
Parkinson's disease	SNCA	Alpha-synuclein	Cytoplasmic, found at tips of neuronal cells	Inhibit iron transporter recycling	Abnormal brain iron accumulation	[128–130]
Alzheimer's disease	APP	Amyloid precursor protein	Trans golgi netowork (TGN)	Physically interact with iron	Iron accumulation in senile plaques, neurofibrillary tangles	[134–139]
	MAPT	Tau proteins	Cytoplasmic	Physically interact with iron		
Diabetes	Adipokine genes	Adipokines		Systemic effects	Iron overload	[167–169]
Cardiovascular disease					Iron deficiency	[170–172]
Cancer	SLC40A1 HAMP TFRC STEAPs etc.	Ferroportin/SLC40A1 Hepcidin Transferrin receptor protein 1 STEAP metalloreductases etc.	Cell membrane Circulation Cell membrane Cytosolic	Iron export Regulates ferroportin Iron import Iron reduction	Low iron export and high import	[174, 176, 177, 179, 180]

bling. PD is also characterized by abnormal iron accumulation and mitochondrial defects, but its etiology remains enigmatic [112]. A hallmark of PD is the presence of Lewy bodies in the brain, formed by the abnormal accumulation of alpha-synuclein leading to brain cell death [127]. Interestingly, iron directly binds to alpha-synuclein, and accelerates a-synuclein aggregation, suggesting an important relevance to PD pathology [128, 129]. Inversely, alpha-synuclein dysregulates iron homeostasis by inhibiting SNX-3-retromer-mediated retrogate recycling of iron transporters [130]. Accumulating evidence suggests that macroautophagy plays an important role in dopaminergic neuron iron-dependent cell death [131, 132]. Overexpression of DMT1 in SH-SY5Y neuronal cells dramatically enhances ferrous uptake and promotes cell death. Chronic exposure to iron enhances oxidative stress and autophagy-induced cytotoxicity, but not apoptosis-induced cell death [132]. Whereas, in another setting, iron deposition increases alpha-synuclein levels and ROS generation and decreases cell viability by a mechanism that involves autophagy inhibition [131]. These seemingly opposite findings may suggest the importance of fine-tuning iron homeostasis, and that protective mechanisms may act as a double-edged sword according to intrinsic and extrinsic cues (high, prolonged stress versus low, acute stress).

5.3 *Alzheimer's Disease*

Alzheimer's disease (AD) is the most pervasive chronic neurodegenerative disease and its prevalence is expected to increase owing to population ageing [133]. It is characterized by progressive dementia and although pharmacological treatment may ameliorate some of the symptoms, there is currently no cure for AD. It has been identified as a protein folding disease, caused by accumulation of amyloid beta-containing extracellular plaques, and of neurofibrillary tangles made of abnormally folded tau protein in the brain. Interestingly, it has been demonstrated that senile plaques and neurofibrillary tangles contain elevated iron concentrations. However it remains unclear whether this accumulation is simply a repercussion of the disease or a significant causative factor, leading to neurodegeneration. Amyloid beta increases the levels of iron content and oxidative stress in neuronal cells overexpressing the Swedish mutant form of human β-amyloid precursor protein and in *C. elegans* [134]. The iron homeostasis disruption is probably associated with elevated expression of the iron transporter DMT1, but not TfR. Furthermore, iron has been found to associate with beta-amyloid and tau proteins and induce plaque and tangle formation [112, 135]. Iron overload stimulates amyloidogenic processing and alters neuronal signaling to increase amyloid beta plaque formation, leading to cognitive deterioration in transgenic mouse model of Alzheimer's disease [136]. In a *D. melanogaster* model of Alzheimer's disease, iron appears to induce ROS formation via Fenton chemistry and the resulted oxidative stress is essential for amyloid beta toxicity. Furthermore, treatment with clioquinol, an iron chelator, increases the lifespan of flies expressing amyloid beta [137]. Administration of iron in an Alzheimer's disease mouse model enhances APP cleavage and amyloid beta aggregate deposition, and impairs spatial learning and memory. Chelation of iron can prevent iron-induced amyloidogenic APP processing and reverse behavioral deficits [138]. Chelation of iron by deferoxamine can inhibit iron-induced hippocampal tau hyperphosphorylation in transgenic animals of Alzheimer disease, via CDK5 and GSK3β pathways [139]. Tau-knockout mice develop age-dependent brain atrophy, neuronal loss with concomitant cognitive deficits [140]. Iron is also elevated in brain regions, mediating age-dependent neurodegeneration. Oral supplementation of the iron chelator clioquinol is able to reverse these changes. Thus, we can surmise the importance of iron homeostasis in AD progression. Iron chelators may represent potential therapeutic agents for AD development. Further understanding of the role of iron in AD pathology might lead to novel pharmacological interventions.

5.4 Congenital Sideroblastic Anemias

The congenital sideroblastic anemias (CSAs) are a diverse group of inherited disorders characterized by pathological mitochondrial iron accumulation of erythroid precursors [141]. In previous years, the genetic basis of several distinctive forms of CSA has been elucidated. A common denominator is their involvement in subcellular iron metabolism. Most of them are associated with impaired heme synthesis or iron-sulfur cluster biogenesis.

X-linked sideroblastic anemia (XLSA) is the most common type of CSA. It is a genetic disorder associated with heme synthesis, due to a mutation in the erythroid-specific aminolevulinic acid synthase 2 (ALAS2) gene. Heme synthesis takes place in the mitochondrion with ALAS2 enzyme playing an important role in the initial step for the generation of ALA. Deficiency in ALAS2 results in disordered heme synthesis, iron-loaded mitochondria and potential death from hemochromatosis [80]. Mutation in mitochondrial carrier family transporter solute carrier 25A38 (SLC25A38) causes another form of CSA similar to XLSA [142]. SLC25A38 is involved in making glycine available for heme synthesis in the mitochondria suggesting its involvement in heme synthesis.

Two recent studies revealed a novel form of CSA caused by a mutation (p.F93del) in the respiratory complex 1 protein NDUFB11 [143, 144]. NDUFB11 is essential for mitochondrial oxidative phosphorylation, but its involvement in iron homeostasis is unclear. The p.F93del mutation causes respiratory insufficiency and impairs erythroid proliferation, suggesting the mechanism of anemia [143].

X-linked sideroblastic anemia with cerebellar ataxia (XLSA/A) is another type of CSA caused by mutations in the ABCB7 transporter [145–147]. ABCB7 is present in the inner mitochondrial membrane. Its direct role in heme synthesis in humans is not clear. Studies suggest that ABCB7 is associated with the maturation of Fe–S cluster containing proteins [80]. In mice, *Abcb7* is an essential gene, and participates in ISC biogenesis [148]. In *C. elegans*, depletion of ABCB7/ABTM-1 is lethal [149]. ABTM-1 lesion shows accumulation of ferric iron. Intriguingly, silencing of *abtm-1* induces developmental arrest and apoptosis and increases oxidative stress and lifespan. The paradoxical effects on lifespan may be attributed to the concept of hormesis, a phenomenon whereby favorable outcomes occur in response to low stress [150].

Mutation in the splice donor site of intron 1 of the mitochondrial glutaredoxin 5 (GLRX5) is the cause of another type of sideroplastic anemia [151, 152]. GLRX5 is required for ISC biogenesis and has an important role in erythropoiesis. More importantly, it was shown that deficiency of GLRX5 decreased proteins associated with heme, such as ferrochelatase, highlighting the tight association between heme and ISC synthesis, the two mitochondrial iron processing pathways that are merely thought to be independent [152].

Other mutations in various genes have been reported to be associated with sideroblastic anemia. These include mutations in the mitochondrial encoded ATP6 gene, the mitochondrial leucyl-tRNA synthetase LARS2 encoding gene, TRNT1 gene, which encodes a template-independent RNA polymerase for the maturation of cytosolic and mitochondrial tRNAs and the gene encoding for mitochondrial heat shock protein HSPA9 [153–156].

5.5 *Human Diseases Related to Gene Mutations Associated to ISC Biogenesis*

A homozygous mutation in LYRM4, encoding ISC biogenesis factor ISD11, was found to cause multiple respiratory chain complexes deficiency [157]. ISD11 forms a complex with the cysteine desulfurase NFS1. NFS1 is also important for the supply of inorganic sulfur for ISC biogenesis and its activity is lost in ISD1-deficient cells. Thus, the early step of ISC assembly is compromised in those patients. An intron mutation in ISCU gene, which results in a splicing defect, affecting both cytosolic and mitochondrial splice variants, causes hereditary myopathy with lactic acidosis [158–160]. Fatal mitochondrial diseases are associated with mutations in ISC scaffold genes NFU1 (alternative to ISCU) and BOLA3 [161, 162]. Mutations in nucleotide-binding protein-like (NUBPL), a Fe–S cluster protein that plays a critical role in the assembly of the mitochondrial membrane respiratory chain NADH dehydrogenase, cause complex I deficiency [163]. Recently, NEET proteins were described as a new class of iron–sulphur proteins with potential role in iron transport between mitochondria and the cytosol [164]. The class I NEET proteins are encoded by two genes, CISD1 and CISD2. Mutations in CISD2 gene have been linked to Wolfram syndrome type 2 [165, 166]. Treatment with the iron chelator Deferiprone is currently under evaluation. Development of appropriate models for these diseases and deeper investigation of the respective pathways and their exact role in ISC biogenesis may provide valuable information for the development of potential therapeutic strategies for multiple diseases associated to ISC biogenesis and trafficking.

5.6 *Other Human Pathologies Associated to Iron Dysregulation*

Human pathologies of the brain and different forms of anemia are the most well studied diseases related to iron. However, iron regulation has been proposed to be involved in far more maladies including cardiovascular disease (CVD), cancer, metabolic diseases and, as previously discussed, immunological and infectious disorders.

Diabetes is a group of metabolic diseases affecting an evergrowing proportion of world population. Iron overload has been associated with increased diabetes risk. However, the underlying mechanisms remain incompletely understood [167]. It is imperative to investigate the molecular mechanisms that are related to iron homeostasis and why it is a critical component of diabetes etiology. Recent studies show that iron levels may regulate adipokine expression and insulin sensitivity in adipocytes [168, 169]. Adipokines or adipose tissue-derived hormones are diverse proteins that may exert their effects in an autocrine, paracrine and most importantly in an endocrine fashion, affecting distal organs. The effect of iron on adipokine regulation may suggest a systemic effect of iron through release of adipokines from iron-sensing adipocytes. Potential molecular effectors for iron regulation in diabetes, apart from adipokines, include oxidative stress, ROS formation, hypoxia inducible factors, AMP-activated protein kinase (AMPK) and iron responsive elements [167].

Cardiovascular disease (CVD) is the leading cause of death in elderly and burdens the communities with high economic costs. Several studies have shown that both iron overload and iron deficiency can be detrimental in CVD patients [170–172]. For several years, the role of iron in CVD remains ambiguous and controversial, partially owing to the lack of reliable tools for monitoring systemic, cellular and subcellular iron levels and incomplete understanding of the relevant mechanisms. Future studies will aid in identifying the optimal treatment for patients, which should be determined according to particular cases.

Cancer is the second leading cause of death in developed countries and despite the recent advances in cancer research, a curative treatment with permanent results does not exist. Sustaining growth and proliferation are two of the hallmarks of cancer [173]. As already discussed, iron is vital for cell proliferation, metabolism and growth. Accumulating evidence implicates cellular iron changes as an imperative feature of cancer [113, 174, 175]. Cancer cells ingeniously utilize pathways of iron uptake, storage and efflux in order to increase intracellular metabolically-available iron for their survival and foster proliferation and growth. Most proteins implicated in iron metabolism, have now been associated to cancer development. FPN, the only iron efflux pump in vertebrates, is downregulated in cancer cells leading to elevated non-heme iron and accelerated proliferation, whereas FPN overexpression reduces breast tumor growth in mice [176]. Interestingly, the circulating levels of hepcidin are induced, potentially acting in a cell-non-autonomous fashion to induce FPN degradation and subsequent cellular iron levels [174]. Expression levels of TrF1 and some members of STEAP metalloreductases are also induced in various cancer types leading to increased iron uptake. A lot more proteins and pathways, associated to iron homeostasis, have been mechanistically linked to cancer, while novel strategies are now emerging towards cancer treatment. Iron chelators and interventions that block iron uptake and induce iron efflux (i.e. TfR overexpression) are now under investigation [174, 175]. Ferroptosis, an iron-dependent form of apoptosis, distinct from other well-known forms of cell death (including apoptosis and necrosis) is also involved in cancer development. Targeting the ferroptosis-related pathways is considered a promising strategy for therapeutic interventions [177–179]. Two distinct mechanisms associated to Glutathione Peroxidase 4 (GPX4) inhibition, have recently been implicated in ferroptosis, rendering GPX4 an essential regulator of ferroptotic cell death [177, 180]. Inactivation of the p53-mediated tumour suppression pathway, is the most common cause of tumorigenesis [181]. High levels of ROS can also initialize a p53-mediated ferroptosis, independenty of cell-cycle arrest, senescence or apoptosis, gradually leading to tumor growth suppression [179]. Additionally, the clinically approved anti-cancer drug, sorafenib, can initiate ferroptosis, supporting the hypothesis for the use of iron-dependent cell death mechanisms against cancer.

6 Biochemical and Biophysical Methods for Iron Detection

Multiple pathological conditions, including cancer and neurodegeneration, are characterized by impaired iron homeostasis. To better understand the iron-dependent molecular pathways that lead to disease, it is of great importance to use efficient methods to detect iron ions in distinct biological samples. The fact that iron can be found in numerous different forms (ferrus, ferric, ISCs, heme, etc.) renders its quantification quite difficult. In the following section we are going to discuss several biochemical and biophysical approaches, commonly used for iron determination (summarized in Table 2).

6.1 Colorimetric Assays

Colorimetric iron detection is based on the ability of several compounds to form colored complexes when interacting with iron. The majority of those compounds act as chelators, since they are bidentate ligands to ferrous ions. The most common chromogens used for colorimetric iron determination are Ferene S, Ferrozine and 2,2′-Bipyridyl [182–185]. The general mechanism involves the ligation of three chromogens to one ferrous ion, giving rise to a stable colored complex. The particular method is supposed to detect free chelatable ferrous iron. However, the measurement is highly dependent on the binding affinity of the respective ligands. Increased binding strength could lead to the chelation of ferrous ions that are already bound to endogenous proteins, rendering the accuracy of this method quiet questionable. To avoid those limitations, additional steps are required for the treatment of biological samples, prior to the colorimetric iron quantification [186]. Initially, a digestion treat-

Table 2 Common biophysical and biochemical methods for iron determination

Methods	Compounds used	Detectable Fe form	References
Colorimetric methods	Ferene-S	Fe^{2+}	[182]
	Ferrozine	Fe^{2+}	[183]
	2,2′-Bipyridyl	Fe^{2+}	[184]
	Ag nanoparticles (NALC–Ag NPs)	Fe^{3+}	[188]
	gold nanoparticles (AuNPs)	Fe^{3+}	[189]
Atomic spectroscopy methods	Atomic absorption Spectroscopy (AAS)	Total Fe	[190]
	Inductively Coupled plasma (ICP)-AAS	Total Fe	[191–193]
	ICP-Mass spectrometry (ICP-MS)	Both Fe isotopes (^{56}Fe, ^{57}Fe)	[191, 194–196]
In situ analyses	Perls' Prussian Blue (PPB)—potassium ferrocyanide	Fe^{3+} (but also Fe^{2+})	[200–204]
	Turnbull's blue	Fe^{2+}	[205–207]
	Electron energy loss spectroscopy (EELS)	Total Fe	[209–211]
	Confocal Raman microscopy	Heme iron	[212–214]
	Calcein AM	Labile iron pool (LIP)	[215]
	Fluorescein coumarin iron probe (FlCFe1)	Labile iron pool (LIP)	[216, 217]
	iron-caged luciferin-1 (ICL-1)	Labile iron pool (LIP)	[218]
	Fe-TRACER	Iron-containing proteins	[219]
	Synchrotron X-ray fluorescence (SXRF)	Total Fe	[220–224]
Biophysical approaches	Electron paramagnetic resonance (EPR)	Heme, [2Fe–2S], [4Fe–4S]	[229–234]
	UV-vis spectroscopy	Heme, [2Fe–2S], [4Fe–4S]	[241–244]

ment in acidic environment (by using hydrogen chloride or nitric acid) is required for the release of complexed iron ions. Subsequent addition of reducing agents, like ammonium acetate and sodium ascorbate, maintains the pH and reduces the iron ions to the ferrous state [187]. Following those treatments, the colorimetric compound is added in the sample, and after a short incubation time, the resulting absorbance is measured by using a spectrophotometer. Absorbance values are analogous to the total (ferrus and ferric) iron content of the respective sample. Interestingly, recent studies introduced two novel colorimetric methods for the determination of ferric ions, based on silver and gold nanoparticles respectively [188, 189].

In general, colorimetric iron determination is a simple and quite inexpensive technique, in comparison with the complex analytical spectrometric methods, that are described below. Though, there are still limitations that need to be taken under consideration. First, some chromogens, including Ferene S, can interfere with other metals, like copper, thus leading to underestimation of iron levels in the examined sample. Additionally, the total iron content is indirectly determined, after addition of the reducer, and may not represent the real iron levels. Another limitation is that colorimetric assays cannot simultaneously detect multiple metals in the same sample.

6.2 *Atomic Spectroscopy Methods*

Atomic absorption spectroscopy (AAS) is a highly sensitive technique, used for detection and quantitative analysis of chemical elements in biological samples. The particular method is initiated with the atomization of a given sample to its atomic constituents through heating. This is mediated by flame or electrothermal atomizers. Subsequent optical radiation produces light, part of which is absorbed by the atomic constituents. The emitted radiation then passes through a monochromator, which separates the element-specific radiation from the non-specific one. Then, a detector receives the separated wavelengths, which are further analyzed by a processor. AAS is considered a low-cost method, and it can analyze multiple elements of a given sample in a short period of time [190]. The major drawback of this technique is that it requires large amount of material, which is not always possible in biological samples.

Inductively coupled plasma (ICP) is a multi-elemental analytical technique, which is also used for determination of trace metals. Similar to AAS, the initial step is the atomization of the sample molecules to their atomic constituents. In ICP, the atomization is conducted by extremely high tempreratures (ranging from 6000 to 10,000 K), compared with those of AAS. This is achieved by a plasma torch, a device that receives Argon gas and leads to its ionization through the generation of an electromagnetic field. The generated electrons by colliding with the neutral Argon, are continuously moving within the circular magnetic field, generating intense heating. Subsequently, the sample is introduced into the excitation area of the plasma, after it has been converted to gas through a nebulizer. In the plasma, the sample molecules are atomized and become excitated. A detector is then used to "translate" the signal into information concerning element-identity.

According to the method used for detection, ICP can be distinguished into ICP-Optical emission spectroscopy (ICP-AES) and ICP-Mass spectrometry (ICP-MS). In ICP-AES, the detector receives the light emitted by the sample and a photomultiplier is used to convert it to electrical signals. The resulted electrical pattern is compared with known electrical patterns, finally leading to the simultaneous determination of multiple elements (including iron) in the given biological sample [191–193]. In ICP-MS, the ions generated in the plasma enter a mass-spectrometer, by passing through a series of cones. There, the ions are separated based on their mass-to-charge ratio, and subsequently an analyzer is used to determine the concentration of each element. Although its maintenance cost is extremely high, one of the major advantages of ICP-MS is that it can distinguish between different isotopes for each element. Additionally, ICP-MS is a highly sensitive method (compared to colorimetric assays and AAS) and does not require large sample amounts, as in the AAS [194–196]. In the case of iron determination, ICP-MS analysis sometimes leads to false-negative results, due to the signals obtained by Argon-containing molecules (Ar–O), which have the same molecular weight as the most common iron isotope (^{56}Fe). For this reason, there have been used specific chelator agents [197] as well as other gases, instead of Argon, for the generation of the plasma (like helium, hydrogen and ammonia) [198, 199].

6.3 *Methods for In Situ Iron Determination*

Although colorimetric assays and atomic spectroscopic methods can detect iron with relatively high sensitivity, they do not give any information concerning iron distribution in tissues, cells, or subcellular compartments. The starting biological material used for these methods has been subjected to homogenization, chemical treatments and sequential dilutions, which may interfere with the distribution and chemical properties of iron ions.

During the past decades, specific iron probes have been used for in situ iron determination in histological samples. Perls' Prussian Blue (PPB) is a common staining for ferric detection. The general principle of PPB's function is the reaction of the soluble potassium ferro-cyanide with ferric ions of the sample, giving rise to an insoluble bright blue pigment. This method is commonly used for diagnostic procedures due to its easy application and its relatively low cost. Importantly, it has been shown that ferrocyanide can also react with ferrous iron, producing a white pigment that is gradually transformed to blue, upon oxidation. As such, PPB can generally detect non-heme iron (both ferrus and ferric forms) [200–204]. Turnbull blue is a similar histochemical method, more rarely used, which uses the ferrus detector ferri-cyanide [205–207]. The sensitivity of both techniques can be enhanced by the addition of diaminobenzidine (DAB) [208].

Histochemical staining is commonly combined with electron microscopy, the high resolution of which can provide elemental detection even in subcellular compartments. Electron energy loss spectroscopy (EELS) is a high sensitive electron microscopy method, which allows quantitative determination of element distribution within subcellular compartments. EELS is based on the inelastic scattering of laser-beam electrons, when they interact with the atomic electrons of the specimen. The resulted energy-loss is used for the detection of specific elements in the sample. One of the major drawbacks of this method is the need for using extremely thin tissue sections [209–211].

Confocal Raman microscopy has been also used for in situ analysis of bioinorganic molecules. Its function is based on the inelastic scattering of monochromatic light (Raman scattering) when it passes through the specimen. Heme-containing proteins (including cytochromes) have several absorption bands in the visible region of the electromagnetic spectrum, and thus, they can be easily monitored via Confocal Raman microscopy [212–214].

Recent studies have suggested interesting fluorescent-based techniques for in situ iron determination. Fluorescein coumarin iron probe (FlCFe1) is an analog of the LIP-specific calcein AM [215], and was recently used for ratiometric determination of LIP within tumour spheroids [216, 217]. Another powerful method involves the bioluminescent probe iron-caged luciferin-1 (ICL-1). ICL-1 selectively reacts with ferrus ions and can be used for in vivo bioluminescence imaging of LIP [218]. Jiang et al. introduced a fluorescent-based method for specifically tracking iron-containing proteins. Fe-TRACER can successfully interact with known iron-containing proteins (as evidenced by X-ray diffraction), and thus, it can be easily used for imaging or proteomic analyses [219].

In addition to electron- and fluorescent microscopy, X-ray- based techniques have been successfully used for in situ iron determination. Specifically, Synchrotron X-ray fluorescence (SXRF) permits metal imaging with high lateral spatial resolution at cellular and subcellular level, requiring minimal sample pre-treatments. The method is based on the intrinsic properties of specific elements (including iron) to fluoresce, upon excitation of an X-ray beam and as such, it does not involve the use of external iron-specific probes. Importantly, due to the weak interaction of X-rays with matter, there is no need for chemical fixation and sectioning of the tissue samples [220–224].

6.4 *Biophysical Techniques for Determining Iron-Containing Complexes and the Transition of Iron Oxidative States*

As previously mentioned, iron has the ability to rapidly change its oxidative status by gaining or losing electrons. This transition ability is required for its function as a cofactor in the form of ISCs or heme. During the last decades, several biophysical methods have been developed for the investigation of iron's oxidative state and the determination of iron-containing complexes under normal or pathological conditions.

Electron Paramagnetic Resonance (EPR), alternatively known as Electron Spin Resonance (ESR), is a powerful spectroscopic method which allows the structural analysis of systems with unpaired electrons (paramagnetic systems). The EPR principle of function is based on the energy difference (ΔE) between the two possible spin states of an electron ($Ms = \pm 1/2$), in the presence of an external oscillating electromagnetic field of resonant energy equal to ΔE. The energy that is absorbed gives rise to a signal, which is monitored by a detector [225]. In the case of iron determination, the EPR signals may vary (different system spins) according to the number of iron atoms in the respective system (protein, ISC, heme, etc.) [226]. Remarkably, EPR is capable of distinguishing different oxidative states of iron [227, 228] and iron-containing complexes, including heme [229–231], rhomboid [2Fe–2S] [232] and cubane [4Fe–4S] ISCs [233, 234]. For the better determination of the charge-transfer transitions, EPR can be combined with Circular Dichroism (CD) spectroscopy [235–238]. Generally, EPR is a highly sensitive technique since it can detect iron species at really low concentrations. However, it is mostly used for iron determination in purified proteins, as cells or subcellular organelles have multiple metals that are EPR-detectable [239, 240].

UV-vis spectroscopy (UV/VIS) is a biophysical method which relies on the excitation of certain species when they absorb light in the ultraviolet to visible range. In several studies, UV/VIS has been used for determining the oxidation state of both heme [241, 242] and ISCs [243, 244] in iron-containing proteins.

7 Concluding Remarks and Future Directions

While the biochemistry and genetic pathways involved in iron homeostasis have been investigated for several decades and our current understanding for the role of iron metabolism in biological systems has advanced enormously over the past years, plenty fundamental questions remain elusive. Generation of transgenic animal models has been proven invaluable for the elucidation of key proteins and mechanisms that regulate iron homeostasis at the systemic and cellular level, as described in the previous sections. Novel iron-related proteins and related mechanisms are expected to be discovered soon and answer central questions. At the systemic level the discovery of hepcidin has been revolutionary, however how organ specific regulation is achieved is unknown. Furthermore, how iron crosses the placenta or the blood brain barrier remains a mystery. The reciprocal interaction of the microbiome and iron remains vague. A regulatory mechanism for systemic iron excretion is also unknown. At the cellular level, a key question is how intracellular trafficking and homeostasis of iron is achieved. Detailed mechanisms on mitochondrial iron efflux and intake need to be defined. These are only some of the key unresolved questions. The development of powerful and sensitive methods for rigorous monitoring of organismal and cellular iron in health and disease will facilitate bigger advancements in the field for the years to come. Current clinical interventions, such as iron chelation need to be evaluated further. Deeper understanding of the involved mechanisms will provide the possibility for new theurapeutic interventions for a plethora of iron-mediated diseases.

Acknowledgements CP is financially supported by General Secretariat for Research and Technology (GSRT), the Hellenic Foundation for Research and Innovation (HFRI) [Scholarship code: 1324], by grants from the European Research Council (ERC – GA695190 – MANNA, ERC – GA737599 – NeuronAgeScreen) and from the BIOIMAGING-GR-MIS5002755, which is co-financed by Greece and the European Union (European Regional Development Fund). E.K. is supported by the Stiftung für Herz- und Kreislaufkrankheiten.

References

1. Puig, S., Ramos-Alonso, L., Romero, A.M., Martinez-Pastor, M.T.: The elemental role of iron in DNA synthesis and repair. Metallomics Integr. Biometal Sci. **9**(11), 1483–1500 (2017). https://doi.org/10.1039/c7mt00116a
2. Veatch, J.R., McMurray, M.A., Nelson, Z.W., Gottschling, D.E.: Mitochondrial dysfunction leads to nuclear genome instability via an iron-sulfur cluster defect. Cell **137**(7), 1247–1258 (2009). https://doi.org/10.1016/j.cell.2009.04.014
3. Oexle, H., Gnaiger, E., Weiss, G.: Iron-dependent changes in cellular energy metabolism: influence on citric acid cycle and oxidative phosphorylation. Biochem. Biophys. Acta. **1413**(3), 99–107 (1999)
4. Coates, T.D.: Physiology and pathophysiology of iron in hemoglobin-associated diseases. Free Radical Biol. Med. **72**, 23–40 (2014). https://doi.org/10.1016/j.freeradbiomed.2014.03.039
5. Winterbourn, C.C.: Toxicity of iron and hydrogen peroxide: the Fenton reaction. Toxicol. Lett. **82–83**, 969–974 (1995)
6. Pantopoulos, K., Porwal, S.K., Tartakoff, A., Devireddy, L.: Mechanisms of mammalian iron homeostasis. Biochemistry **51**(29), 5705–5724 (2012). https://doi.org/10.1021/bi300752r
7. Lawen, A., Lane, D.J.: Mammalian iron homeostasis in health and disease: uptake, storage, transport, and molecular mechanisms of action. Antioxid. Redox Signal. **18**(18), 2473–2507 (2013). https://doi.org/10.1089/ars.2011.4271
8. Harigae, H.: Iron metabolism and related diseases: an overview. Int. J. Hematol. **107**(1), 5–6 (2018). https://doi.org/10.1007/s12185-017-2384-0
9. Anderson, G.J., Frazer, D.M.: Current understanding of iron homeostasis. Am. J. Clin. Nutr. **106**(Suppl 6), 1559S–1566S (2017). https://doi.org/10.3945/ajcn.117.155804
10. Cerami, C.: Iron Nutriture of the Fetus, Neonate, Infant, and Child. Annals of Nutrition and Metabolism **71**(suppl 3), 8–14 (2017)
11. Alwan, N.A., Hamamy, H.: Maternal iron status in pregnancy and long-term health outcomes in the offspring. J. Pediatr. Genet. **4**(2), 111–123 (2015). https://doi.org/10.1055/s-0035-1556742
12. Harrison-Findik, D.D.: Gender-related variations in iron metabolism and liver diseases. World J. Hepatol. **2**(8), 302–310 (2010). https://doi.org/10.4254/wjh.v2.i8.302
13. Joosten, E.: Iron deficiency anemia in older adults: a review. Geriatr. Gerontol. Int. **18**(3), 373–379 (2017). https://doi.org/10.1111/ggi.13194
14. Cao, C., Fleming, M.D.: The placenta: the forgotten essential organ of iron transport. Nutr. Rev. **74**(7), 421–431 (2016). https://doi.org/10.1093/nutrit/nuw009
15. Fuqua, B.K., Vulpe, C.D., Anderson, G.J.: Intestinal iron absorption. J. Trace Elem. Med Biol. **26**(2–3), 115–119 (2012). https://doi.org/10.1016/j.jtemb.2012.03.015
16. Kikuchi, G., Yoshida, T., Noguchi, M.: Heme oxygenase and heme degradation. Biochem. Biophys. Res. Commun. **338**(1), 558–567 (2005). https://doi.org/10.1016/j.bbrc.2005.08.020

17. Le Blanc, S., Garrick, M.D., Arredondo, M.: Heme carrier protein 1 transports heme and is involved in heme-Fe metabolism. Am. J. Physiol. Cell Physiol. **302**(12), C1780–C1785 (2012). https://doi.org/10.1152/ajpcell.00080.2012
18. White, C., Yuan, X., Schmidt, P.J., Bresciani, E., Samuel, T.K., Campagna, D., Hall, C., Bishop, K., Calicchio, M.L., Lapierre, A., Ward, D.M., Liu, P., Fleming, M.D., Hamza, I.: HRG1 is essential for heme transport from the phagolysosome of macrophages during erythrophagocytosis. Cell Metab. **17**(2), 261–270 (2013). https://doi.org/10.1016/j.cmet.2013.01.005
19. Rajagopal, A., Rao, A.U., Amigo, J., Tian, M., Upadhyay, S.K., Hall, C., Uhm, S., Mathew, M.K., Fleming, M.D., Paw, B.H., Krause, M., Hamza, I.: Haem homeostasis is regulated by the conserved and concerted functions of HRG-1 proteins. Nature **453**(7198), 1127–1131 (2008). https://doi.org/10.1038/nature06934
20. Sinclair, J., Pinter, K., Samuel, T., Beardsley, S., Yuan, X., Zhang, J., Meng, K., Yun, S., Krause, M., Hamza, I.: Inter-organ signalling by HRG-7 promotes systemic haem homeostasis. Nat. Cell Biol. **19**(7), 799–807 (2017). https://doi.org/10.1038/ncb3539
21. Lopez, M.A., Martos, F.C.: Iron availability: an updated review. Int. J. Food Sci. Nutr. **55**(8), 597–606 (2004). https://doi.org/10.1080/09637480500085820
22. McKie, A.T., Marciani, P., Rolfs, A., Brennan, K., Wehr, K., Barrow, D., Miret, S., Bomford, A., Peters, T.J., Farzaneh, F., Hediger, M.A., Hentze, M.W., Simpson, R.J.: A novel duodenal iron-regulated transporter, IREG1, implicated in the basolateral transfer of iron to the circulation. Mol. Cell **5**(2), 299–309 (2000)
23. Donovan, A., Brownlie, A., Zhou, Y., Shepard, J., Pratt, S.J., Moynihan, J., Paw, B.H., Drejer, A., Barut, B., Zapata, A., Law, T.C., Brugnara, C., Lux, S.E., Pinkus, G.S., Pinkus, J.L., Kingsley, P.D., Palis, J., Fleming, M.D., Andrews, N.C., Zon, L.I.: Positional cloning of zebrafish ferroportin1 identifies a conserved vertebrate iron exporter. Nature **403**(6771), 776–781 (2000). https://doi.org/10.1038/35001596
24. Nemeth, E., Tuttle, M.S., Powelson, J., Vaughn, M.B., Donovan, A., Ward, D.M., Ganz, T., Kaplan, J.: Hepcidin regulates cellular iron efflux by binding to ferroportin and inducing its internalization. Science **306**(5704), 2090–2093 (2004). https://doi.org/10.1126/science.1104742
25. Sangkhae, V., Nemeth, E.: Regulation of the iron homeostatic hormone hepcidin. Adv. Nutr. **8**(1), 126–136 (2017). https://doi.org/10.3945/an.116.013961
26. Casu, C., Nemeth, E., Rivella, S.: Hepcidin agonists as therapeutic tools. Blood **131**(16), 1790–1794 (2018). https://doi.org/10.1182/blood-2017-11-737411
27. Colucci, S., Pagani, A., Pettinato, M., Artuso, I., Nai, A., Camaschella, C., Silvestri, L.: The immunophilin FKBP12 inhibits hepcidin expression by binding the BMP type I receptor ALK2 in hepatocytes. Blood **130**(19), 2111–2120 (2017). https://doi.org/10.1182/blood-2017-04-780692
28. Pasricha, S.R., Lim, P.J., Duarte, T.L., Casu, C., Oosterhuis, D., Mleczko-Sanecka, K., Suciu, M., Da Silva, A.R., Al-Hourani, K., Arezes, J., McHugh, K., Gooding, S., Frost, J.N., Wray, K., Santos, A., Porto, G., Repapi, E., Gray, N., Draper, S.J., Ashley, N., Soilleux, E., Olinga, P., Muckenthaler, M.U., Hughes, J.R., Rivella, S., Milne, T.A., Armitage, A.E., Drakesmith, H.: Hepcidin is regulated by promoter-associated histone acetylation and HDAC3. Nat. Commun. **8**(1), 403 (2017). https://doi.org/10.1038/s41467-017-00500-z
29. Cassat, J.E., Skaar, E.P.: Iron in infection and immunity. Cell Host Microbe **13**(5), 509–519 (2013). https://doi.org/10.1016/j.chom.2013.04.010
30. Ganz, T., Nemeth, E.: Iron homeostasis in host defence and inflammation. Nat. Rev. Immunol. **15**(8), 500–510 (2015). https://doi.org/10.1038/nri3863
31. Michels, K., Nemeth, E., Ganz, T., Mehrad, B.: Hepcidin and host defense against infectious diseases. PLoS Pathog. **11**(8), e1004998 (2015). https://doi.org/10.1371/journal.ppat.1004998
32. Nevitt, T.: War-Fe-re: iron at the core of fungal virulence and host immunity. Biometals **24**(3), 547–558 (2011). https://doi.org/10.1007/s10534-011-9431-8

33. Deschemin, J.C., Noordine, M.L., Remot, A., Willemetz, A., Afif, C., Canonne-Hergaux, F., Langella, P., Karim, Z., Vaulont, S., Thomas, M., Nicolas, G.: The microbiota shifts the iron sensing of intestinal cells. FASEB J. **30**(1), 252–261 (2016). https://doi.org/10.1096/fj.15-276840
34. Hamilton, D.H., Turcot, I., Stintzi, A., Raymond, K.N.: Large cooperativity in the removal of iron from transferrin at physiological temperature and chloride ion concentration. J. Biol. Inorg. Chem. (JBIC) Publ. Soc. Biol. Inorg. Chem. **9**(8), 936–944 (2004). https://doi.org/10.1007/s00775-004-0592-6
35. He, Q.Y., Mason, A.B., Nguyen, V., MacGillivray, R.T., Woodworth, R.C.: The chloride effect is related to anion binding in determining the rate of iron release from the human transferrin N-lobe. Biochem. J. **350**(Pt 3), 909–915 (2000)
36. Gomme, P.T., McCann, K.B., Bertolini, J.: Transferrin: structure, function and potential therapeutic actions. Drug Discov. Today **10**(4), 267–273 (2005). https://doi.org/10.1016/S1359-6446(04)03333-1
37. Kawabata, H., Germain, R.S., Vuong, P.T., Nakamaki, T., Said, J.W., Koeffler, H.P.: Transferrin receptor 2-alpha supports cell growth both in iron-chelated cultured cells and in vivo. J. Biol. Chem. **275**(22), 16618–16625 (2000). https://doi.org/10.1074/jbc.M908846199
38. Kleven, M.D., Jue, S., Enns, C.A.: Transferrin receptors TfR1 and TfR2 bind transferrin through differing mechanisms. Biochemistry **57**(9), 1552–1559 (2018). https://doi.org/10.1021/acs.biochem.8b00006
39. Mayle, K.M., Le, A.M., Kamei, D.T.: The intracellular trafficking pathway of transferrin. Biochem. Biophys. Acta **1820**(3), 264–281 (1820). https://doi.org/10.1016/j.bbagen.2011.09.009
40. Cmejla, R., Ptackova, P., Petrak, J., Savvulidi, F., Cerny, J., Sebesta, O., Vyoral, D.: Human MRCKalpha is regulated by cellular iron levels and interferes with transferrin iron uptake. Biochem. Biophys. Res. Commun. **395**(2), 163–167 (2010). https://doi.org/10.1016/j.bbrc.2010.02.148
41. Fleming, M.D., Romano, M.A., Su, M.A., Garrick, L.M., Garrick, M.D., Andrews, N.C.: Nramp2 is mutated in the anemic Belgrade (b) rat: evidence of a role for Nramp2 in endosomal iron transport. Proc. Natl. Acad. Sci. U.S.A. **95**(3), 1148–1153 (1998)
42. Andrews, N.C.: The iron transporter DMT1. Int. J. Biochem. Cell Biol. **31**(10), 991–994 (1999)
43. Jenkitkasemwong, S., Wang, C.Y., Mackenzie, B., Knutson, M.D.: Physiologic implications of metal-ion transport by ZIP14 and ZIP8. Biometals Int. J. Role Metal Ions Biol. Biochem. Med. **25**(4), 643–655 (2012). https://doi.org/10.1007/s10534-012-9526-x
44. Liuzzi, J.P., Aydemir, F., Nam, H., Knutson, M.D., Cousins, R.J.: Zip14 (Slc39a14) mediates non-transferrin-bound iron uptake into cells. Proc. Natl. Acad. Sci. U.S.A. **103**(37), 13612–13617 (2006). https://doi.org/10.1073/pnas.0606424103
45. Wang, C.Y., Jenkitkasemwong, S., Duarte, S., Sparkman, B.K., Shawki, A., Mackenzie, B., Knutson, M.D.: ZIP8 is an iron and zinc transporter whose cell-surface expression is up-regulated by cellular iron loading. J. Biol. Chem. **287**(41), 34032–34043 (2012). https://doi.org/10.1074/jbc.M112.367284
46. Tripathi, A.K., Haldar, S., Qian, J., Beserra, A., Suda, S., Singh, A., Hopfer, U., Chen, S.G., Garrick, M.D., Turner, J.R., Knutson, M.D., Singh, N.: Prion protein functions as a ferrireductase partner for ZIP14 and DMT1. Free Radical Biol. Med. **84**, 322–330 (2015). https://doi.org/10.1016/j.freeradbiomed.2015.03.037
47. Zhao, N., Gao, J., Enns, C.A., Knutson, M.D.: ZRT/IRT-like protein 14 (ZIP14) promotes the cellular assimilation of iron from transferrin. J. Biol. Chem. **285**(42), 32141–32150 (2010). https://doi.org/10.1074/jbc.M110.143248
48. Pinilla-Tenas, J.J., Sparkman, B.K., Shawki, A., Illing, A.C., Mitchell, C.J., Zhao, N., Liuzzi, J.P., Cousins, R.J., Knutson, M.D., Mackenzie, B.: Zip14 is a complex broad-scope metal-ion transporter whose functional properties support roles in the cellular uptake of zinc and nontransferrin-bound iron. Am. J. Physiol. Cell Physiol. **301**(4), C862–C871 (2011). https://doi.org/10.1152/ajpcell.00479.2010

49. Gunshin, H., Mackenzie, B., Berger, U.V., Gunshin, Y., Romero, M.F., Boron, W.F., Nussberger, S., Gollan, J.L., Hediger, M.A.: Cloning and characterization of a mammalian proton-coupled metal-ion transporter. Nature **388**(6641), 482–488 (1997). https://doi.org/10.1038/41343
50. Harrison, P.M., Arosio, P.: The ferritins: molecular properties, iron storage function and cellular regulation. Biochem. Biophys. Acta **1275**(3), 161–203 (1996)
51. Mancias, J.D., Wang, X., Gygi, S.P., Harper, J.W., Kimmelman, A.C.: Quantitative proteomics identifies NCOA4 as the cargo receptor mediating ferritinophagy. Nature **509**(7498), 105–109 (2014). https://doi.org/10.1038/nature13148
52. Vashchenko, G., MacGillivray, R.T.: Multi-copper oxidases and human iron metabolism. Nutrients **5**(7), 2289–2313 (2013). https://doi.org/10.3390/nu5072289
53. Philpott, C.C., Ryu, M.S., Frey, A., Patel, S.: Cytosolic iron chaperones: proteins delivering iron cofactors in the cytosol of mammalian cells. J. Biol. Chem. **292**(31), 12764–12771 (2017). https://doi.org/10.1074/jbc.R117.791962
54. Nandal, A., Ruiz, J.C., Subramanian, P., Ghimire-Rijal, S., Sinnamon, R.A., Stemmler, T.L., Bruick, R.K., Philpott, C.C.: Activation of the HIF prolyl hydroxylase by the iron chaperones PCBP1 and PCBP2. Cell Metab. **14**(5), 647–657 (2011). https://doi.org/10.1016/j.cmet.2011.08.015
55. Frey, A.G., Nandal, A., Park, J.H., Smith, P.M., Yabe, T., Ryu, M.S., Ghosh, M.C., Lee, J., Rouault, T.A., Park, M.H., Philpott, C.C.: Iron chaperones PCBP1 and PCBP2 mediate the metallation of the dinuclear iron enzyme deoxyhypusine hydroxylase. Proc. Natl. Acad. Sci. U.S.A. **111**(22), 8031–8036 (2014). https://doi.org/10.1073/pnas.1402732111
56. Shi, H., Bencze, K.Z., Stemmler, T.L., Philpott, C.C.: A cytosolic iron chaperone that delivers iron to ferritin. Science **320**(5880), 1207–1210 (2008). https://doi.org/10.1126/science.1157643
57. Leidgens, S., Bullough, K.Z., Shi, H., Li, F., Shakoury-Elizeh, M., Yabe, T., Subramanian, P., Hsu, E., Natarajan, N., Nandal, A., Stemmler, T.L., Philpott, C.C.: Each member of the poly-r(C)-binding protein 1 (PCBP) family exhibits iron chaperone activity toward ferritin. J. Biol. Chem. **288**(24), 17791–17802 (2013). https://doi.org/10.1074/jbc.M113.460253
58. Yanatori, I., Yasui, Y., Tabuchi, M., Kishi, F.: Chaperone protein involved in transmembrane transport of iron. Biochem. J. **462**(1), 25–37 (2014). https://doi.org/10.1042/BJ20140225
59. Muckenthaler, M.U., Galy, B., Hentze, M.W.: Systemic iron homeostasis and the iron-responsive element/iron-regulatory protein (IRE/IRP) regulatory network. Annu. Rev. Nutr. **28**, 197–213 (2008). https://doi.org/10.1146/annurev.nutr.28.061807.155521
60. Wilkinson, N., Pantopoulos, K.: The IRP/IRE system in vivo: insights from mouse models. Front. Pharmacol. **5**, 176 (2014). https://doi.org/10.3389/fphar.2014.00176
61. Hentze, M.W., Caughman, S.W., Rouault, T.A., Barriocanal, J.G., Dancis, A., Harford, J.B., Klausner, R.D.: Identification of the iron-responsive element for the translational regulation of human ferritin mRNA. Science **238**(4833), 1570–1573 (1987)
62. Hentze, M.W., Rouault, T.A., Caughman, S.W., Dancis, A., Harford, J.B., Klausner, R.D.: A cis-acting element is necessary and sufficient for translational regulation of human ferritin expression in response to iron. Proc. Natl. Acad. Sci. U.S.A. **84**(19), 6730–6734 (1987)
63. Hentze, M.W., Kuhn, L.C.: Molecular control of vertebrate iron metabolism: mRNA-based regulatory circuits operated by iron, nitric oxide, and oxidative stress. Proc. Natl. Acad. Sci. U.S.A. **93**(16), 8175–8182 (1996)
64. Cmejla, R., Petrak, J., Cmejlova, J.: A novel iron responsive element in the 3′UTR of human MRCKalpha. Biochem. Biophys. Res. Commun. **341**(1), 158–166 (2006). https://doi.org/10.1016/j.bbrc.2005.12.155
65. Lee, P.L., Gelbart, T., West, C., Halloran, C., Beutler, E.: The human Nramp2 gene: characterization of the gene structure, alternative splicing, promoter region and polymorphisms. Blood Cells Mol. Dis. **24**(2), 199–215 (1998). https://doi.org/10.1006/bcmd.1998.0186
66. Salahudeen, A.A., Thompson, J.W., Ruiz, J.C., Ma, H.W., Kinch, L.N., Li, Q., Grishin, N.V., Bruick, R.K.: An E3 ligase possessing an iron-responsive hemerythrin domain is a regulator of iron homeostasis. Science **326**(5953), 722–726 (2009). https://doi.org/10.1126/science.1176326

67. Vashisht, A.A., Zumbrennen, K.B., Huang, X., Powers, D.N., Durazo, A., Sun, D., Bhaskaran, N., Persson, A., Uhlen, M., Sangfelt, O., Spruck, C., Leibold, E.A., Wohlschlegel, J.A.: Control of iron homeostasis by an iron-regulated ubiquitin ligase. Science **326**(5953), 718–721 (2009). https://doi.org/10.1126/science.1176333
68. Wang, W., Di, X., D'Agostino Jr., R.B., Torti, S.V., Torti, F.M.: Excess capacity of the iron regulatory protein system. J. Biol. Chem. **282**(34), 24650–24659 (2007). https://doi.org/10.1074/jbc.M703167200
69. Meyron-Holtz, E.G., Ghosh, M.C., Iwai, K., LaVaute, T., Brazzolotto, X., Berger, U.V., Land, W., Ollivierre-Wilson, H., Grinberg, A., Love, P., Rouault, T.A.: Genetic ablations of iron regulatory proteins 1 and 2 reveal why iron regulatory protein 2 dominates iron homeostasis. The EMBO journal **23**(2), 386–395 (2004). https://doi.org/10.1038/sj.emboj.7600041
70. Peyssonnaux, C., Nizet, V., Johnson, R.S.: Role of the hypoxia inducible factors HIF in iron metabolism. Cell Cycle **7**(1), 28–32 (2008). https://doi.org/10.4161/cc.7.1.5145
71. Shah, Y.M., Xie, L.: Hypoxia-inducible factors link iron homeostasis and erythropoiesis. Gastroenterology **146**(3), 630–642 (2014). https://doi.org/10.1053/j.gastro.2013.12.031
72. Mole, D.R.: Iron homeostasis and its interaction with prolyl hydroxylases. Antioxid. Redox Signal. **12**(4), 445–458 (2010). https://doi.org/10.1089/ars.2009.2790
73. Lando, D., Peet, D.J., Gorman, J.J., Whelan, D.A., Whitelaw, M.L., Bruick, R.K.: FIH-1 is an asparaginyl hydroxylase enzyme that regulates the transcriptional activity of hypoxia-inducible factor. Genes Dev. **16**(12), 1466–1471 (2002). https://doi.org/10.1101/gad.991402
74. Qian, Z.M., Wu, X.M., Fan, M., Yang, L., Du, F., Yung, W.H., Ke, Y.: Divalent metal transporter 1 is a hypoxia-inducible gene. J. Cell. Physiol. **226**(6), 1596–1603 (2011). https://doi.org/10.1002/jcp.22485
75. Yoshinaga, M., Nakatsuka, Y., Vandenbon, A., Ori, D., Uehata, T., Tsujimura, T., Suzuki, Y., Mino, T., Takeuchi, O.: Regnase-1 maintains iron homeostasis via the degradation of transferrin receptor 1 and prolyl-hydroxylase-domain-containing protein 3 mRNAs. Cell Rep. **19**(8), 1614–1630 (2017). https://doi.org/10.1016/j.celrep.2017.05.009
76. Sanchez, M., Galy, B., Muckenthaler, M.U., Hentze, M.W.: Iron-regulatory proteins limit hypoxia-inducible factor-2alpha expression in iron deficiency. Nat. Struct. Mol. Biol. **14**(5), 420–426 (2007). https://doi.org/10.1038/nsmb1222
77. Ganz, T., Nemeth, E.: Hepcidin and iron homeostasis. Biochem. Biophys. Acta **1823**(9), 1434–1443 (2012). https://doi.org/10.1016/j.bbamcr.2012.01.014
78. Reichert, C.O., da Cunha, J., Levy, D., Maselli, L.M.F., Bydlowski, S.P., Spada, C.: Hepcidin: homeostasis and diseases related to iron metabolism. Acta Haematol. **137**(4), 220–236 (2017). https://doi.org/10.1159/000471838
79. Hamza, I., Dailey, H.A.: One ring to rule them all: trafficking of heme and heme synthesis intermediates in the metazoans. Biochem. Biophys. Acta. **1823**(9), 1617–1632 (2012). https://doi.org/10.1016/j.bbamcr.2012.04.009
80. Huang, M.L., Lane, D.J., Richardson, D.R.: Mitochondrial mayhem: the mitochondrion as a modulator of iron metabolism and its role in disease. Antioxid. Redox Signal. **15**(12), 3003–3019 (2011). https://doi.org/10.1089/ars.2011.3921
81. Bekri, S., Kispal, G., Lange, H., Fitzsimons, E., Tolmie, J., Lill, R., Bishop, D.F.: Human ABC7 transporter: gene structure and mutation causing X-linked sideroblastic anemia with ataxia with disruption of cytosolic iron-sulfur protein maturation. Blood **96**(9), 3256–3264 (2000)
82. Lane, D.J., Merlot, A.M., Huang, M.L., Bae, D.H., Jansson, P.J., Sahni, S., Kalinowski, D.S., Richardson, D.R.: Cellular iron uptake, trafficking and metabolism: key molecules and mechanisms and their roles in disease. Biochem. Biophys. Acta. **1853**(5), 1130–1144 (2015). https://doi.org/10.1016/j.bbamcr.2015.01.021
83. Lange, H., Kispal, G., Lill, R.: Mechanism of iron transport to the site of heme synthesis inside yeast mitochondria. J. Biol. Chem. **274**(27), 18989–18996 (1999)
84. Zhang, A.S., Sheftel, A.D., Ponka, P.: Intracellular kinetics of iron in reticulocytes: evidence for endosome involvement in iron targeting to mitochondria. Blood **105**(1), 368–375 (2005). https://doi.org/10.1182/blood-2004-06-2226

85. Sheftel, A.D., Zhang, A.S., Brown, C., Shirihai, O.S., Ponka, P.: Direct interorganellar transfer of iron from endosome to mitochondrion. Blood **110**(1), 125–132 (2007). https://doi.org/10.1182/blood-2007-01-068148
86. Shaw, G.C., Cope, J.J., Li, L., Corson, K., Hersey, C., Ackermann, G.E., Gwynn, B., Lambert, A.J., Wingert, R.A., Traver, D., Trede, N.S., Barut, B.A., Zhou, Y., Minet, E., Donovan, A., Brownlie, A., Balzan, R., Weiss, M.J., Peters, L.L., Kaplan, J., Zon, L.I., Paw, B.H.: Mitoferrin is essential for erythroid iron assimilation. Nature **440**(7080), 96–100 (2006). https://doi.org/10.1038/nature04512
87. Paradkar, P.N., Zumbrennen, K.B., Paw, B.H., Ward, D.M., Kaplan, J.: Regulation of mitochondrial iron import through differential turnover of mitoferrin 1 and mitoferrin 2. Mol. Cell. Biol. **29**(4), 1007–1016 (2009). https://doi.org/10.1128/MCB.01685-08
88. Chen, W., Paradkar, P.N., Li, L., Pierce, E.L., Langer, N.B., Takahashi-Makise, N., Hyde, B.B., Shirihai, O.S., Ward, D.M., Kaplan, J., Paw, B.H.: Abcb10 physically interacts with mitoferrin-1 (Slc25a37) to enhance its stability and function in the erythroid mitochondria. Proc. Natl. Acad. Sci. U.S.A. **106**(38), 16263–16268 (2009). https://doi.org/10.1073/pnas.0904519106
89. Sripetchwandee, J., Sanit, J., Chattipakorn, N., Chattipakorn, S.C.: Mitochondrial calcium uniporter blocker effectively prevents brain mitochondrial dysfunction caused by iron overload. Life Sci. **92**(4–5), 298–304 (2013). https://doi.org/10.1016/j.lfs.2013.01.004
90. Zhang, X., Lemasters, J.J.: Translocation of iron from lysosomes to mitochondria during ischemia predisposes to injury after reperfusion in rat hepatocytes. Free Radical Biol. Med. **63**, 243–253 (2013). https://doi.org/10.1016/j.freeradbiomed.2013.05.004
91. Braymer, J.J., Lill, R.: Iron-sulfur cluster biogenesis and trafficking in mitochondria. J. Biol. Chem. **292**(31), 12754–12763 (2017). https://doi.org/10.1074/jbc.R117.787101
92. Adam, A.C., Bornhovd, C., Prokisch, H., Neupert, W., Hell, K.: The Nfs1 interacting protein Isd11 has an essential role in Fe/S cluster biogenesis in mitochondria. EMBO J. **25**(1), 174–183 (2006). https://doi.org/10.1038/sj.emboj.7600905
93. Yoon. T., Cowan. J.A.: Iron-sulfur cluster biosynthesis Characterization of frataxin as an iron donor for assembly of [2Fe–2S] clusters in ISU-type proteins. J. Am. Chem. Soc. **125**(20), 6078–6084 (2003). https://doi.org/10.1021/ja027967i
94. Parent, A., Elduque, X., Cornu, D., Belot, L., Le Caer, J.P., Grandas, A., Toledano, M.B., D'Autreaux, B.: Mammalian frataxin directly enhances sulfur transfer of NFS1 persulfide to both ISCU and free thiols. Nat. Commun. **6**, 5686 (2015). https://doi.org/10.1038/ncomms6686
95. Olive, J.A., Cowan, J.A.: Role of the HSPA9/HSC20 chaperone pair in promoting directional human iron-sulfur cluster exchange involving monothiol glutaredoxin 5. J. Inorg. Biochem. **184**, 100–107 (2018). https://doi.org/10.1016/j.jinorgbio.2018.04.007
96. Lill, R., Dutkiewicz, R., Freibert, S.A., Heidenreich, T., Mascarenhas, J., Netz, D.J., Paul, V.D., Pierik, A.J., Richter, N., Stumpfig, M., Srinivasan, V., Stehling, O., Muhlenhoff, U.: The role of mitochondria and the CIA machinery in the maturation of cytosolic and nuclear iron-sulfur proteins. Eur. J. Cell Biol. **94**(7–9), 280–291 (2015). https://doi.org/10.1016/j.ejcb.2015.05.002
97. Kim, K.D., Chung, W.H., Kim, H.J., Lee, K.C., Roe, J.H.: Monothiol glutaredoxin Grx5 interacts with Fe–S scaffold proteins Isa1 and Isa2 and supports Fe–S assembly and DNA integrity in mitochondria of fission yeast. Biochem. Biophys. Res. Commun. **392**(3), 467–472 (2010). https://doi.org/10.1016/j.bbrc.2010.01.051
98. Brancaccio, D., Gallo, A., Mikolajczyk, M., Zovo, K., Palumaa, P., Novellino, E., Piccioli, M., Ciofi-Baffoni, S., Banci, L.: Formation of [4Fe–4S] clusters in the mitochondrial iron-sulfur cluster assembly machinery. J. Am. Chem. Soc. **136**(46), 16240–16250 (2014). https://doi.org/10.1021/ja507822j
99. Netz, D.J., Mascarenhas, J., Stehling, O., Pierik, A.J., Lill, R.: Maturation of cytosolic and nuclear iron-sulfur proteins. Trends Cell Biol. **24**(5), 303–312 (2014). https://doi.org/10.1016/j.tcb.2013.11.005

100. Tong, W.H., Rouault, T.A.: Functions of mitochondrial ISCU and cytosolic ISCU in mammalian iron-sulfur cluster biogenesis and iron homeostasis. Cell Metab. **3**(3), 199–210 (2006). https://doi.org/10.1016/j.cmet.2006.02.003
101. Bayeva, M., Khechaduri, A., Wu, R., Burke, M.A., Wasserstrom, J.A., Singh, N., Liesa, M., Shirihai, O.S., Langer, N.B., Paw, B.H., Ardehali, H.: ATP-binding cassette B10 regulates early steps of heme synthesis. Circ. Res. **113**(3), 279–287 (2013). https://doi.org/10.1161/CIRCRESAHA.113.301552
102. Seguin, A., Takahashi-Makise, N., Yien, Y.Y., Huston, N.C., Whitman, J.C., Musso, G., Wallace, J.A., Bradley, T., Bergonia, H.A., Kafina, M.D., Matsumoto, M., Igarashi, K., Phillips, J.D., Paw, B.H., Kaplan, J., Ward, D.M.: Reductions in the mitochondrial ABC transporter Abcb10 affect the transcriptional profile of heme biosynthesis genes. J. Biol. Chem. **292**(39), 16284–16299 (2017). https://doi.org/10.1074/jbc.M117.797415
103. Lange, H., Muhlenhoff, U., Denzel, M., Kispal, G., Lill, R.: The heme synthesis defect of mutants impaired in mitochondrial iron-sulfur protein biogenesis is caused by reversible inhibition of ferrochelatase. J. Biol. Chem. **279**(28), 29101–29108 (2004). https://doi.org/10.1074/jbc.M403721200
104. Santambrogio, P., Biasiotto, G., Sanvito, F., Olivieri, S., Arosio, P., Levi, S.: Mitochondrial ferritin expression in adult mouse tissues. J. Histochem. Cytochem.: Official J. Histochem. Soc. **55**(11), 1129–1137 (2007). https://doi.org/10.1369/jhc.7A7273.2007
105. Gao, G., Chang, Y.Z.: Mitochondrial ferritin in the regulation of brain iron homeostasis and neurodegenerative diseases. Front. Pharmacol. **5**, 19 (2014). https://doi.org/10.3389/fphar.2014.00019
106. Drysdale, J., Arosio, P., Invernizzi, R., Cazzola, M., Volz, A., Corsi, B., Biasiotto, G., Levi, S.: Mitochondrial ferritin: a new player in iron metabolism. Blood Cells Mol. Dis. **29**(3), 376–383 (2002)
107. Guaraldo, M., Santambrogio, P., Rovelli, E., Di Savino, A., Saglio, G., Cittaro, D., Roetto, A., Levi, S.: Characterization of human mitochondrial ferritin promoter: identification of transcription factors and evidences of epigenetic control. Sci. Rep. **6**, 33432 (2016). https://doi.org/10.1038/srep33432
108. Wu, Q., Wu, W.S., Su, L., Zheng, X., Wu, W.Y., Santambrogio, P., Gou, Y.J., Hao, Q., Wang, P.N., Li, Y.R., Zhao, B.L., Nie, G., Levi, S., Chang, Y.Z.: Mitochondrial ferritin is a hypoxia-inducible factor 1alpha-inducible gene that protects from hypoxia-induced cell death in brain. Antioxid. Redox Sign. **30**(2), 198–212 (2019).https://doi.org/10.1089/ars.2017.7063
109. Nie, G., Sheftel, A.D., Kim, S.F., Ponka, P.: Overexpression of mitochondrial ferritin causes cytosolic iron depletion and changes cellular iron homeostasis. Blood **105**(5), 2161–2167 (2005). https://doi.org/10.1182/blood-2004-07-2722
110. Park, S., Gakh, O., O'Neill, H.A., Mangravita, A., Nichol, H., Ferreira, G.C., Isaya, G.: Yeast frataxin sequentially chaperones and stores iron by coupling protein assembly with iron oxidation. J. Biol. Chem. **278**(33), 31340–31351 (2003). https://doi.org/10.1074/jbc.M303158200
111. Sutak, R., Seguin, A., Garcia-Serres, R., Oddou, J.L., Dancis, A., Tachezy, J., Latour, J.M., Camadro, J.M., Lesuisse, E.: Human mitochondrial ferritin improves respiratory function in yeast mutants deficient in iron-sulfur cluster biogenesis, but is not a functional homologue of yeast frataxin. MicrobiologyOpen **1**(2), 95–104 (2012). https://doi.org/10.1002/mbo3.18
112. Hare, D., Ayton, S., Bush, A., Lei, P.: A delicate balance: iron metabolism and diseases of the brain. Front. Aging Neurosci. **5**, 34 (2013). https://doi.org/10.3389/fnagi.2013.00034
113. Gozzelino, R., Arosio, P.: Iron homeostasis in health and disease. Int. J. Mol. Sci. **17**(1), 130 (2016). https://doi.org/10.3390/ijms17010130
114. Apostolakis, S., Kypraiou, A.M.: Iron in neurodegenerative disorders: being in the wrong place at the wrong time? Rev. Neurosci. **28**(8), 893–911 (2017). https://doi.org/10.1515/revneuro-2017-0020
115. Bilgic, B., Pfefferbaum, A., Rohlfing, T., Sullivan, E.V., Adalsteinsson, E.: MRI estimates of brain iron concentration in normal aging using quantitative susceptibility mapping. Neuroimage **59**(3), 2625–2635 (2012). https://doi.org/10.1016/j.neuroimage.2011.08.077

116. Cook, C.I., Yu, B.P.: Iron accumulation in aging: modulation by dietary restriction. Mech. Ageing Dev. **102**(1), 1–13 (1998)
117. James, S.A., Roberts, B.R., Hare, D.J., de Jonge, M.D., Birchall, I.E., Jenkins, N.L., Cherny, R.A., Bush, A.I., McColl, G.: Direct in vivo imaging of ferrous iron dyshomeostasis in ageing Caenorhabditis elegans. Chem. Sci. **6**(5), 2952–2962 (2015). https://doi.org/10.1039/c5sc00233h
118. Lang, M., Braun, C.L., Kanost, M.R., Gorman, M.J.: Multicopper oxidase-1 is a ferroxidase essential for iron homeostasis in Drosophila melanogaster. Proc. Natl. Acad. Sci. U.S.A. **109**(33), 13337–13342 (2012). https://doi.org/10.1073/pnas.1208703109
119. Schiavi, A., Maglioni, S., Palikaras, K., Shaik, A., Strappazzon, F., Brinkmann, V., Torgovnick, A., Castelein, N., De Henau, S., Braeckman, B.P., Cecconi, F., Tavernarakis, N., Ventura, N.: Iron-starvation-induced mitophagy mediates lifespan extension upon mitochondrial stress in C. elegans. Curr Biol **25**(14), 1810–1822 (2015). https://doi.org/10.1016/j.cub.2015.05.059
120. Klang, I.M., Schilling, B., Sorensen, D.J., Sahu, A.K., Kapahi, P., Andersen, J.K., Swoboda, P., Killilea, D.W., Gibson, B.W., Lithgow, G.J.: Iron promotes protein insolubility and aging in C elegans. Aging (Albany NY) **6**(11), 975–991 (2014). https://doi.org/10.18632/aging.100689
121. Pandolfo, M.: Friedreich ataxia. Arch. Neurol. **65**(10), 1296–1303 (2008). https://doi.org/10.1001/archneur.65.10.1296
122. Campuzano, V., Montermini, L., Molto, M.D., Pianese, L., Cossee, M., Cavalcanti, F., Monros, E., Rodius, F., Duclos, F., Monticelli, A., Zara, F., Canizares, J., Koutnikova, H., Bidichandani, S.I., Gellera, C., Brice, A., Trouillas, P., De Michele, G., Filla, A., De Frutos, R., Palau, F., Patel, P.I., Di Donato, S., Mandel, J.L., Cocozza, S., Koenig, M., Pandolfo, M.: Friedreich's ataxia: autosomal recessive disease caused by an intronic GAA triplet repeat expansion. Science **271**(5254), 1423–1427 (1996)
123. Lupoli, F., Vannocci, T., Longo, G., Niccolai, N., Pastore, A.: The role of oxidative stress in Friedreich's ataxia. FEBS Lett. **592**(5), 718–727 (2018). https://doi.org/10.1002/1873-3468.12928
124. Jasoliya, M.J., McMackin, M.Z., Henderson, C.K., Perlman, S.L., Cortopassi, G.A.: Frataxin deficiency impairs mitochondrial biogenesis in cells, mice and humans. Hum. Mol. Genet. **26**(14), 2627–2633 (2017). https://doi.org/10.1093/hmg/ddx141
125. Whitnall, M., Suryo Rahmanto, Y., Huang, M.L., Saletta, F., Lok, H.C., Gutierrez, L., Lazaro, F.J., Fleming, A.J., St Pierre, T.G., Mikhael, M.R., Ponka, P., Richardson, D.R.: Identification of nonferritin mitochondrial iron deposits in a mouse model of Friedreich ataxia. Proc. Natl. Acad. Sci. U.S.A. **109**(50), 20590–20595 (2012). https://doi.org/10.1073/pnas.1215349109
126. Tai, G., Corben, L.A., Yiu, E.M., Milne, S.C., Delatycki, M.B.: Progress in the treatment of Friedreich ataxia. Neurol. Neurochir. Pol. **52**(2), 129–139 (2018). https://doi.org/10.1016/j.pjnns.2018.02.003
127. Michel, P.P., Hirsch, E.C., Hunot, S.: Understanding dopaminergic cell death pathways in Parkinson disease. Neuron **90**(4), 675–691 (2016). https://doi.org/10.1016/j.neuron.2016.03.038
128. Peng, Y., Wang, C., Xu, H.H., Liu, Y.N., Zhou, F.: Binding of alpha-synuclein with Fe(III) and with Fe(II) and biological implications of the resultant complexes. J. Inorg. Biochem. **104**(4), 365–370 (2010). https://doi.org/10.1016/j.jinorgbio.2009.11.005
129. Golts, N., Snyder, H., Frasier, M., Theisler, C., Choi, P., Wolozin, B.: Magnesium inhibits spontaneous and iron-induced aggregation of alpha-synuclein. J. Biol. Chem. **277**(18), 16116–16123 (2002). https://doi.org/10.1074/jbc.M107866200
130. Patel, D., Xu, C., Nagarajan, S., Liu, Z., Hemphill, W.O., Shi, R., Uversky, V.N., Caldwell, G.A., Caldwell, K.A., Witt, S.N.: Alpha-synuclein inhibits Snx3-retromer-mediated retrograde recycling of iron transporters in S. cerevisiae and C. elegans models of Parkinson's disease. Hum Mol Genet **27**(9), 1514–1532 (2018). https://doi.org/10.1093/hmg/ddy059
131. Wan, W., Jin, L., Wang, Z., Wang, L., Fei, G., Ye, F., Pan, X., Wang, C., Zhong, C.: Iron deposition leads to neuronal alpha-synuclein pathology by inducing autophagy dysfunction. Front. Neurol. **8**, 1 (2017). https://doi.org/10.3389/fneur.2017.00001

132. Chew, K.C., Ang, E.T., Tai, Y.K., Tsang, F., Lo, S.Q., Ong, E., Ong, W.Y., Shen, H.M., Lim, K.L., Dawson, V.L., Dawson, T.M., Soong, T.W.: Enhanced autophagy from chronic toxicity of iron and mutant A53T alpha-synuclein: implications for neuronal cell death in Parkinson disease. J. Biol. Chem. **286**(38), 33380–33389 (2011). https://doi.org/10.1074/jbc.M111.268409
133. Masters, C.L., Bateman, R., Blennow, K., Rowe, C.C., Sperling, R.A., Cummings, J.L.: Alzheimer's disease. Nat. Rev. Dis. Primers **1**, 15056 (2015). https://doi.org/10.1038/nrdp.2015.56
134. Wan, L., Nie, G., Zhang, J., Luo, Y., Zhang, P., Zhang, Z., Zhao, B.: beta-Amyloid peptide increases levels of iron content and oxidative stress in human cell and Caenorhabditis elegans models of Alzheimer disease. Free Radic. Biol. Med. **50**(1), 122–129 (2011). https://doi.org/10.1016/j.freeradbiomed.2010.10.707
135. Everett, J., Cespedes, E., Shelford, L.R., Exley, C., Collingwood, J.F., Dobson, J., van der Laan, G., Jenkins, C.A., Arenholz, E., Telling, N.D.: Ferrous iron formation following the co-aggregation of ferric iron and the Alzheimer's disease peptide beta-amyloid(1-42). J. R. Soc. Interface **11**(95), 20140165 (2014). https://doi.org/10.1098/rsif.2014.0165
136. Becerril-Ortega, J., Bordji, K., Freret, T., Rush, T., Buisson, A.: Iron overload accelerates neuronal amyloid-beta production and cognitive impairment in transgenic mice model of Alzheimer's disease. Neurobiol. Aging **35**(10), 2288–2301 (2014). https://doi.org/10.1016/j.neurobiolaging.2014.04.019
137. Rival, T., Page, R.M., Chandraratna, D.S., Sendall, T.J., Ryder, E., Liu, B., Lewis, H., Rosahl, T., Hider, R., Camargo, L.M., Shearman, M.S., Crowther, D.C., Lomas, D.A.: Fenton chemistry and oxidative stress mediate the toxicity of the beta-amyloid peptide in a Drosophila model of Alzheimer's disease. Eur. J. Neurosci. **29**(7), 1335–1347 (2009). https://doi.org/10.1111/j.1460-9568.2009.06701.x
138. Guo, C., Wang, T., Zheng, W., Shan, Z.Y., Teng, W.P., Wang, Z.Y.: Intranasal deferoxamine reverses iron-induced memory deficits and inhibits amyloidogenic APP processing in a transgenic mouse model of Alzheimer's disease. Neurobiol. Aging **34**(2), 562–575 (2013). https://doi.org/10.1016/j.neurobiolaging.2012.05.009
139. Guo, C., Wang, P., Zhong, M.L., Wang, T., Huang, X.S., Li, J.Y., Wang, Z.Y.: Deferoxamine inhibits iron induced hippocampal tau phosphorylation in the Alzheimer transgenic mouse brain. Neurochem. Int. **62**(2), 165–172 (2013). https://doi.org/10.1016/j.neuint.2012.12.005
140. Lei, P., Ayton, S., Finkelstein, D.I., Spoerri, L., Ciccotosto, G.D., Wright, D.K., Wong, B.X., Adlard, P.A., Cherny, R.A., Lam, L.Q., Roberts, B.R., Volitakis, I., Egan, G.F., McLean, C.A., Cappai, R., Duce, J.A., Bush, A.I.: Tau deficiency induces parkinsonism with dementia by impairing APP-mediated iron export. Nat. Med. **18**(2), 291–295 (2012). https://doi.org/10.1038/nm.2613
141. Fleming, M.D.: Congenital sideroblastic anemias: iron and heme lost in mitochondrial translation. Hematol. Am. Soc. Hematol. Educ. Program. **2011**, 525–531 (2011). https://doi.org/10.1182/asheducation-2011.1.525
142. Guernsey, D.L., Jiang, H., Campagna, D.R., Evans, S.C., Ferguson, M., Kellogg, M.D., Lachance, M., Matsuoka, M., Nightingale, M., Rideout, A., Saint-Amant, L., Schmidt, P.J., Orr, A., Bottomley, S.S., Fleming, M.D., Ludman, M., Dyack, S., Fernandez, C.V., Samuels, M.E.: Mutations in mitochondrial carrier family gene SLC25A38 cause nonsyndromic autosomal recessive congenital sideroblastic anemia. Nat. Genet. **41**(6), 651–653 (2009). https://doi.org/10.1038/ng.359
143. Lichtenstein, D.A., Crispin, A.W., Sendamarai, A.K., Campagna, D.R., Schmitz-Abe, K., Sousa, C.M., Kafina, M.D., Schmidt, P.J., Niemeyer, C.M., Porter, J., May, A., Patnaik, M.M., Heeney, M.M., Kimmelman, A., Bottomley, S.S., Paw, B.H., Markianos, K., Fleming, M.D.: A recurring mutation in the respiratory complex 1 protein NDUFB11 is responsible for a novel form of X-linked sideroblastic anemia. Blood **128**(15), 1913–1917 (2016). https://doi.org/10.1182/blood-2016-05-719062
144. Torraco, A., Bianchi, M., Verrigni, D., Gelmetti, V., Riley, L., Niceta, M., Martinelli, D., Montanari, A., Guo, Y., Rizza, T., Diodato, D., Di Nottia, M., Lucarelli, B., Sorrentino, F.,

Piemonte, F., Francisci, S., Tartaglia, M., Valente, E.M., Dionisi-Vici, C., Christodoulou, J., Bertini, E., Carrozzo, R.: A novel mutation in NDUFB11 unveils a new clinical phenotype associated with lactic acidosis and sideroblastic anemia. Clin. Genet. **91**(3), 441–447 (2017). https://doi.org/10.1111/cge.12790

145. D'Hooghe, M., Selleslag, D., Mortier, G., Van Coster, R., Vermeersch, P., Billiet, J., Bekri, S.: X-linked sideroblastic anemia and ataxia: a new family with identification of a fourth ABCB7 gene mutation. Eur. J. Paediatr. Neurol. **16**(6), 730–735 (2012). https://doi.org/10.1016/j.ejpn.2012.02.003
146. Shimada, Y., Okuno, S., Kawai, A., Shinomiya, H., Saito, A., Suzuki, M., Omori, Y., Nishino, N., Kanemoto, N., Fujiwara, T., Horie, M., Takahashi, E.: Cloning and chromosomal mapping of a novel ABC transporter gene (hABC7), a candidate for X-linked sideroblastic anemia with spinocerebellar ataxia. J. Hum. Genet. **43**(2), 115–122 (1998). https://doi.org/10.1007/s100380050051
147. Pondarre, C., Campagna, D.R., Antiochos, B., Sikorski, L., Mulhern, H., Fleming, M.D.: Abcb7, the gene responsible for X-linked sideroblastic anemia with ataxia, is essential for hematopoiesis. Blood **109**(8), 3567–3569 (2007). https://doi.org/10.1182/blood-2006-04-015768
148. Pondarre, C., Antiochos, B.B., Campagna, D.R., Clarke, S.L., Greer, E.L., Deck, K.M., McDonald, A., Han, A.P., Medlock, A., Kutok, J.L., Anderson, S.A., Eisenstein, R.S., Fleming, M.D.: The mitochondrial ATP-binding cassette transporter Abcb7 is essential in mice and participates in cytosolic iron-sulfur cluster biogenesis. Hum. Mol. Genet. **15**(6), 953–964 (2006). https://doi.org/10.1093/hmg/ddl012
149. Gonzalez-Cabo, P., Bolinches-Amoros, A., Cabello, J., Ros, S., Moreno, S., Baylis, H.A., Palau, F., Vazquez-Manrique, R.P.: Disruption of the ATP-binding cassette B7 (ABTM-1/ABCB7) induces oxidative stress and premature cell death in Caenorhabditis elegans. J. Biol. Chem. **286**(24), 21304–21314 (2011). https://doi.org/10.1074/jbc.M110.211201
150. Nikoletopoulou, V., Kyriakakis, E., Tavernarakis, N.: Cellular and molecular longevity pathways: the old and the new. Trends Endocrinol. Metab. **25**(4), 212–223 (2014). https://doi.org/10.1016/j.tem.2013.12.003
151. Wingert, R.A., Galloway, J.L., Barut, B., Foott, H., Fraenkel, P., Axe, J.L., Weber, G.J., Dooley, K., Davidson, A.J., Schmid, B., Paw, B.H., Shaw, G.C., Kingsley, P., Palis, J., Schubert, H., Chen, O., Kaplan, J., Zon, L.I., Tubingen Screen, C.: Deficiency of glutaredoxin 5 reveals Fe–S clusters are required for vertebrate haem synthesis. Nature **436**(7053), 1035–1039 (2005). https://doi.org/10.1038/nature03887
152. Ye, H., Jeong, S.Y., Ghosh, M.C., Kovtunovych, G., Silvestri, L., Ortillo, D., Uchida, N., Tisdale, J., Camaschella, C., Rouault, T.A.: Glutaredoxin 5 deficiency causes sideroblastic anemia by specifically impairing heme biosynthesis and depleting cytosolic iron in human erythroblasts. J. Clin. Invest. **120**(5), 1749–1761 (2010). https://doi.org/10.1172/JCI40372
153. Burrage, L.C., Tang, S., Wang, J., Donti, T.R., Walkiewicz, M., Luchak, J.M., Chen, L.C., Schmitt, E.S., Niu, Z., Erana, R., Hunter, J.V., Graham, B.H., Wong, L.J., Scaglia, F.: Mitochondrial myopathy, lactic acidosis, and sideroblastic anemia (MLASA) plus associated with a novel de novo mutation (m.8969G > A) in the mitochondrial encoded ATP6 gene. Mol Genet Metab **113**(3), 207–212 (2014). https://doi.org/10.1016/j.ymgme.2014.06.004
154. Riley, L.G., Rudinger-Thirion, J., Schmitz-Abe, K., Thorburn, D.R., Davis, R.L., Teo, J., Arbuckle, S., Cooper, S.T., Campagna, D.R., Frugier, M., Markianos, K., Sue, C.M., Fleming, M.D., Christodoulou, J.: LARS2 variants associated with hydrops, lactic acidosis, sideroblastic anemia, and multisystem failure. JIMD Rep. **28**, 49–57 (2016). https://doi.org/10.1007/8904_2015_515
155. Chakraborty, P.K., Schmitz-Abe, K., Kennedy, E.K., Mamady, H., Naas, T., Durie, D., Campagna, D.R., Lau, A., Sendamarai, A.K., Wiseman, D.H., May, A., Jolles, S., Connor, P., Powell, C., Heeney, M.M., Giardina, P.J., Klaassen, R.J., Kannengiesser, C., Thuret, I., Thompson, A.A., Marques, L., Hughes, S., Bonney, D.K., Bottomley, S.S., Wynn, R.F., Laxer, R.M., Minniti, C.P., Moppett, J., Bordon, V., Geraghty, M., Joyce, P.B., Markianos, K., Rudner, A.D., Holcik, M., Fleming, M.D.: Mutations in TRNT1 cause congenital sideroblastic anemia

with immunodeficiency, fevers, and developmental delay (SIFD). Blood **124**(18), 2867–2871 (2014). https://doi.org/10.1182/blood-2014-08-591370
156. Schmitz-Abe, K., Ciesielski, S.J., Schmidt, P.J., Campagna, D.R., Rahimov, F., Schilke, B.A., Cuijpers, M., Rieneck, K., Lausen, B., Linenberger, M.L., Sendamarai, A.K., Guo, C., Hofmann, I., Newburger, P.E., Matthews, D., Shimamura, A., Snijders, P.J., Towne, M.C., Niemeyer, C.M., Watson, H.G., Dziegiel, M.H., Heeney, M.M., May, A., Bottomley, S.S., Swinkels, D.W., Markianos, K., Craig, E.A., Fleming, M.D.: Congenital sideroblastic anemia due to mutations in the mitochondrial HSP70 homologue HSPA9. Blood **126**(25), 2734–2738 (2015). https://doi.org/10.1182/blood-2015-09-659854
157. Lim, S.C., Friemel, M., Marum, J.E., Tucker, E.J., Bruno, D.L., Riley, L.G., Christodoulou, J., Kirk, E.P., Boneh, A., DeGennaro, C.M., Springer, M., Mootha, V.K., Rouault, T.A., Leimkuhler, S., Thorburn, D.R., Compton, A.G.: Mutations in LYRM4, encoding iron-sulfur cluster biogenesis factor ISD11, cause deficiency of multiple respiratory chain complexes. Hum. Mol. Genet. **22**(22), 4460–4473 (2013). https://doi.org/10.1093/hmg/ddt295
158. Olsson, A., Lind, L., Thornell, L.E., Holmberg, M.: Myopathy with lactic acidosis is linked to chromosome 12q23.3-24.11 and caused by an intron mutation in the ISCU gene resulting in a splicing defect. Hum. Mol. Genet. **17**(11), 1666–1672 (2008). https://doi.org/10.1093/hmg/ddn057
159. Mochel, F., Knight, M.A., Tong, W.H., Hernandez, D., Ayyad, K., Taivassalo, T., Andersen, P.M., Singleton, A., Rouault, T.A., Fischbeck, K.H., Haller, R.G.: Splice mutation in the iron-sulfur cluster scaffold protein ISCU causes myopathy with exercise intolerance. Am. J. Hum. Genet. **82**(3), 652–660 (2008). https://doi.org/10.1016/j.ajhg.2007.12.012
160. Kollberg, G., Tulinius, M., Melberg, A., Darin, N., Andersen, O., Holmgren, D., Oldfors, A., Holme, E.: Clinical manifestation and a new ISCU mutation in iron-sulphur cluster deficiency myopathy. Brain **132**(Pt 8), 2170–2179 (2009). https://doi.org/10.1093/brain/awp152
161. Cameron, J.M., Janer, A., Levandovskiy, V., Mackay, N., Rouault, T.A., Tong, W.H., Ogilvie, I., Shoubridge, E.A., Robinson, B.H.: Mutations in iron-sulfur cluster scaffold genes NFU1 and BOLA3 cause a fatal deficiency of multiple respiratory chain and 2-oxoacid dehydrogenase enzymes. Am. J. Hum. Genet. **89**(4), 486–495 (2011). https://doi.org/10.1016/j.ajhg.2011.08.011
162. Navarro-Sastre, A., Tort, F., Stehling, O., Uzarska, M.A., Arranz, J.A., Del Toro, M., Labayru, M.T., Landa, J., Font, A., Garcia-Villoria, J., Merinero, B., Ugarte, M., Gutierrez-Solana, L.G., Campistol, J., Garcia-Cazorla, A., Vaquerizo, J., Riudor, E., Briones, P., Elpeleg, O., Ribes, A., Lill, R.: A fatal mitochondrial disease is associated with defective NFU1 function in the maturation of a subset of mitochondrial Fe–S proteins. Am. J. Hum. Genet. **89**(5), 656–667 (2011). https://doi.org/10.1016/j.ajhg.2011.10.005
163. Calvo, S.E., Tucker, E.J., Compton, A.G., Kirby, D.M., Crawford, G., Burtt, N.P., Rivas, M., Guiducci, C., Bruno, D.L., Goldberger, O.A., Redman, M.C., Wiltshire, E., Wilson, C.J., Altshuler, D., Gabriel, S.B., Daly, M.J., Thorburn, D.R., Mootha, V.K.: High-throughput, pooled sequencing identifies mutations in NUBPL and FOXRED1 in human complex I deficiency. Nat. Genet. **42**(10), 851–858 (2010). https://doi.org/10.1038/ng.659
164. Mittler, R., Darash-Yahana, M., Sohn, Y.S., Bai, F., Song, L., Cabantchik, I.Z., Jennings, P.A., Onuchic, J.N., Nechushtai, R.: NEET proteins: a new link between iron metabolism, reactive oxygen species, and cancer. Antioxid Redox Signal. (2018). https://doi.org/10.1089/ars.2018.7502
165. Amr, S., Heisey, C., Zhang, M., Xia, X.J., Shows, K.H., Ajlouni, K., Pandya, A., Satin, L.S., El-Shanti, H., Shiang, R.: A homozygous mutation in a novel zinc-finger protein, ERIS, is responsible for Wolfram syndrome 2. Am. J. Hum. Genet. **81**(4), 673–683 (2007). https://doi.org/10.1086/520961
166. Rouzier, C., Moore, D., Delorme, C., Lacas-Gervais, S., Ait-El-Mkadem, S., Fragaki, K., Burte, F., Serre, V., Bannwarth, S., Chaussenot, A., Catala, M., Yu-Wai-Man, P., Paquis-Flucklinger, V.: A novel CISD2 mutation associated with a classical Wolfram syndrome phenotype alters Ca^{2+} homeostasis and ER-mitochondria interactions. Hum. Mol. Genet. **26**(9), 1599–1611 (2017). https://doi.org/10.1093/hmg/ddx060

167. Simcox, J.A., McClain, D.A.: Iron and diabetes risk. Cell Metab. **17**(3), 329–341 (2013). https://doi.org/10.1016/j.cmet.2013.02.007
168. Gabrielsen, J.S., Gao, Y., Simcox, J.A., Huang, J., Thorup, D., Jones, D., Cooksey, R.C., Gabrielsen, D., Adams, T.D., Hunt, S.C., Hopkins, P.N., Cefalu, W.T., McClain, D.A.: Adipocyte iron regulates adiponectin and insulin sensitivity. J. Clin. Invest. **122**(10), 3529–3540 (2012). https://doi.org/10.1172/JCI44421
169. Gao, Y., Li, Z., Gabrielsen, J.S., Simcox, J.A., Lee, S.H., Jones, D., Cooksey, B., Stoddard, G., Cefalu, W.T., McClain, D.A.: Adipocyte iron regulates leptin and food intake. J. Clin. Invest. **125**(9), 3681–3691 (2015). https://doi.org/10.1172/JCI81860
170. von Haehling, S., Jankowska, E.A., van Veldhuisen, D.J., Ponikowski, P., Anker, S.D.: Iron deficiency and cardiovascular disease. Nat. Rev. Cardiol. **12**(11), 659–669 (2015). https://doi.org/10.1038/nrcardio.2015.109
171. Kraml, P.: The role of iron in the pathogenesis of atherosclerosis. Physiol. Res. **66**(Supplementum 1), S55–S67 (2017)
172. Sullivan, J.L.: Iron and the genetics of cardiovascular disease. Circulation **100**(12), 1260–1263 (1999)
173. Hanahan, D., Weinberg, R.A.: Hallmarks of cancer: the next generation. Cell **144**(5), 646–674 (2011). https://doi.org/10.1016/j.cell.2011.02.013
174. Torti, S.V., Torti, F.M.: Iron and cancer: more ore to be mined. Nat. Rev. Cancer **13**(5), 342–355 (2013). https://doi.org/10.1038/nrc3495
175. Bogdan, A.R., Miyazawa, M., Hashimoto, K., Tsuji, Y.: Regulators of iron homeostasis: new players in metabolism, cell death, and disease. Trends Biochem. Sci. **41**(3), 274–286 (2016). https://doi.org/10.1016/j.tibs.2015.11.012
176. Pinnix, Z.K., Miller, L.D., Wang, W., D'Agostino Jr., R., Kute, T., Willingham, M.C., Hatcher, H., Tesfay, L., Sui, G., Di, X., Torti, S.V., Torti, F.M.: Ferroportin and iron regulation in breast cancer progression and prognosis. Sci Transl Med **2**(43), 43–56 (2010). https://doi.org/10.1126/scisignal.3001127
177. Lu, B., Chen, X.B., Ying, M.D., He, Q.J., Cao, J., Yang, B.: The role of ferroptosis in cancer development and treatment response. Front. Pharmacol. **8**, 992 (2017). https://doi.org/10.3389/fphar.2017.00992
178. Dixon, S.J., Lemberg, K.M., Lamprecht, M.R., Skouta, R., Zaitsev, E.M., Gleason, C.E., Patel, D.N., Bauer, A.J., Cantley, A.M., Yang, W.S., Morrison 3rd, B., Stockwell, B.R.: Ferroptosis: an iron-dependent form of nonapoptotic cell death. Cell **149**(5), 1060–1072 (2012). https://doi.org/10.1016/j.cell.2012.03.042
179. Jiang, L., Kon, N., Li, T., Wang, S.J., Su, T., Hibshoosh, H., Baer, R., Gu, W.: Ferroptosis as a p53-mediated activity during tumour suppression. Nature **520**(7545), 57–62 (2015). https://doi.org/10.1038/nature14344
180. Yang, W.S., SriRamaratnam, R., Welsch, M.E., Shimada, K., Skouta, R., Viswanathan, V.S., Cheah, J.H., Clemons, P.A., Shamji, A.F., Clish, C.B., Brown, L.M., Girotti, A.W., Cornish, V.W., Schreiber, S.L., Stockwell, B.R.: Regulation of ferroptotic cancer cell death by GPX4. Cell **156**(1–2), 317–331 (2014). https://doi.org/10.1016/j.cell.2013.12.010
181. Muller, P.A., Vousden, K.H.: Mutant p53 in cancer: new functions and therapeutic opportunities. Cancer Cell **25**(3), 304–317 (2014). https://doi.org/10.1016/j.ccr.2014.01.021
182. Eskelinen, S., Haikonen, M., Raisanen, S.: Ferene-S as the chromogen for serum iron determinations. Scand. J. Clin. Lab. Invest. **43**(5), 453–455 (1983)
183. Stookey, L.L.: Ferrozine—a new spectrophotometric reagent for iron. Anal. Chem. **42**(7), 779–781 (1970). https://doi.org/10.1021/ac60289a016
184. Moss, M.L., Mellon, M.G.: Colorimetric determination of iron with 2,2′-bipyridyl and with 2,2′,2′-terpyridyl. Ind. Eng. Chem. Anal. Ed. **14**(11), 862–865 (1942). https://doi.org/10.1021/i560111a014
185. Goodwin, J.F., Murphy, B.: The colorimetric determination of iron in biological material with reference to its measurement during chelation therapy. Clinical chemistry **12**(2), 58–69 (1966)
186. Holmes-Hampton, G.P., Tong, W.H., Rouault, T.A.: Biochemical and biophysical methods for studying mitochondrial iron metabolism. Methods Enzymol. **547**, 275–307 (2014). https://doi.org/10.1016/B978-0-12-801415-8.00015-1

187. Fish, W.W.: Rapid colorimetric micromethod for the quantitation of complexed iron in biological samples. Methods Enzymol. **158**, 357–364 (1988)
188. Gao, X., Lu, Y., He, S., Li, X., Chen, W.: Colorimetric detection of iron ions (III) based on the highly sensitive plasmonic response of the N-acetyl-l-cysteine-stabilized silver nanoparticles. Anal. Chim. Acta **879**, 118–125 (2015). https://doi.org/10.1016/j.aca.2015.04.002
189. Kim, K., Nam, Y.S., Lee, Y., Lee, K.B.: Highly sensitive colorimetric assay for determining Fe(3+) based on gold nanoparticles conjugated with glycol chitosan. J. Anal. Methods Chem. **2017**, 3648564 (2017). https://doi.org/10.1155/2017/3648564
190. Jackson, K.W.: Electrothermal atomic absorption spectrometry and related techniques. Anal. Chem. **72**(12), 159R–167R (2000)
191. Payá-pérez, A., Sala, J., Mousty, F.: Comparison of ICP-AES and ICP-MS for the analysis of trace elements in soil extracts. Int. J. Environ. Anal. Chem. **51**(1–4), 223–230 (1993). https://doi.org/10.1080/03067319308027628
192. Çlolak, N., Gümgüm, B.: Determination of Iron by ICP-AES in stereoregular poly(propylene oxide) prepared by Pruitt-Baggett catalyst. Polym.-Plast. Technol. Eng. **33**(1), 105–109 (1994). https://doi.org/10.1080/03602559408010734
193. Bok-Badura, J., Jakobik-Kolon, A., Turek, M., Boncel, S., Karon, K.: A versatile method for direct determination of iron content in multi-wall carbon nanotubes by inductively coupled plasma atomic emission spectrometry with slurry sample introduction. RSC Adv. **5**(123), 101634–101640 (2015). https://doi.org/10.1039/c5ra22269a
194. Profrock, D., Prange, A.: Inductively coupled plasma-mass spectrometry (ICP-MS) for quantitative analysis in environmental and life sciences: a review of challenges, solutions, and trends. Appl. Spectrosc. **66**(8), 843–868 (2012). https://doi.org/10.1366/12-06681
195. Fiorito, V., Geninatti Crich, S., Silengo, L., Altruda, F., Aime, S., Tolosano, E.: Assessment of iron absorption in mice by ICP-MS measurements of (57)Fe levels. Eur. J. Nutr. **51**(7), 783–789 (2012). https://doi.org/10.1007/s00394-011-0256-6
196. Wheal, M.S., DeCourcy-Ireland, E., Bogard, J.R., Thilsted, S.H., Stangoulis, J.C.: Measurement of haem and total iron in fish, shrimp and prawn using ICP-MS: Implications for dietary iron intake calculations. Food Chem. **201**, 222–229 (2016). https://doi.org/10.1016/j.foodchem.2016.01.080
197. Segura, M., Madrid, Y., Camara, C.: Elimination of calcium and argon interferences in iron determination by ICP-MS using desferrioxamine chelating agent immobilized in sol-gel and cold plasma conditions. J. Anal. At. Spectrom. **18**(9), 1103–1108 (2003). https://doi.org/10.1039/b301719m
198. Hu, Q.-H.: Simultaneous separation and quantification of iron and transition species using LC-ICP-MS. Am. J. Anal. Chem. **02**(6), 675–682 (2011). https://doi.org/10.4236/ajac.2011.26077
199. Tanner, S.D., Baranov. V.: A dynamic reaction cell for inductively coupled plasma mass spectrometry (ICP-DRC-MS). II. Reduction Interferences Produced Cell **10** (1999). https://doi.org/10.1016/s1044-0305(99)00081-1
200. Nestor, S.L., Bancroft, J.D., Gamble, M.: Techn. Neuropathol. (2008)
201. Brunt, E.M., Olynyk, J.K., Britton, R.S., Janney, C.G., Di Bisceglie, A.M., Bacon, B.R.: Histological evaluation of iron in liver biopsies: relationship to HFE mutations. Am. J. Gastroenterol. **95**(7), 1788–1793 (2000). https://doi.org/10.1111/j.1572-0241.2000.02175.x
202. Deugnier Yves, M., Turlin, B., Moirand, R., Loréal, O., Brissot, P., Powell Lawrie, W., Summers Kim, M., Fletcher, L., Halliday June, W.: Differentiation between heterozygotes and homozygotes in genetic hemochromatosis by means of a histological hepatic iron index: a study of 192 cases. Hepatology **17**(1), 30–34 (1993). https://doi.org/10.1002/hep.1840170107
203. Ortega, L., Ladero, J.M., Carreras, M.P., Alvarez, T., Taxonera, C., Oliván, M.P., Sanz-Esponera, J., Díaz-Rubio, M.: A computer-assisted morphometric quantitative analysis of iron overload in liver biopsies. A comparison with histological and biochemical methods. Pathol. Res. Pract. **201**(10), 673–677 (2005). https://doi.org/10.1016/j.prp.2005.07.002
204. Hall, A.P., Davies, W., Stamp, K., Clamp, I., Bigley, A.: Comparison of computerized image analysis with traditional semiquantitative scoring of Perls' Prussian blue stained

hepatic iron deposition. Toxicol. Pathol. **41**(7), 992–1000 (2013). https://doi.org/10.1177/0192623313476576
205. Hansen, L.D., Litchman, W.M., Daub, G.H.: Turnbull's blue and Prussian blue: KFe(III)[Fe(II)(CN)6]. J. Chem. Educ. **46**(1), 46 (1969). https://doi.org/10.1021/ed046p46
206. Pavlishchuk Vitaly, V., Koval Iryna, A., Goreshnik, E., Addison Anthony, W., van Albada Gerard, A., Reedijk, J.: The first example of a true "Turnbull's Blue" family compound with trapped iron oxidation states. Eur. J. Inorg. Chem. **1**, 297–301 (2001). https://doi.org/10.1002/1099-0682(20011)2001:1%3c297:aid-ejic297%3e3.0.co;2-n
207. Reynolds, E.J.: LXIII—the composition of Prussian blue and Turnbull's blue. J. Chem. Soc. Trans. **51**(0), 644–646 (1887). https://doi.org/10.1039/ct8875100644
208. Meguro, R., Asano, Y., Odagiri, S., Li, C., Iwatsuki, H., Shoumura, K.: Nonheme-iron histochemistry for light and electron microscopy: a historical, theoretical and technical review. Arch. Histol. Cytol. **70**(1), 1–19 (2007)
209. Aronova, M.A., Leapman, R.D.: Elemental mapping by electron energy loss spectroscopy in biology. Methods Mol. Biol. **950**, 209–226 (2013). https://doi.org/10.1007/978-1-62703-137-0_13
210. Kapp, N., Studer, D., Gehr, P., Geiser, M.: Electron energy-loss spectroscopy as a tool for elemental analysis in biological specimens. In: Kuo, J. (ed.) Electron Microscopy: Methods and Protocols, pp. 431–447. Humana Press, Totowa, NJ (2007). https://doi.org/10.1007/978-1-59745-294-6_21
211. Leapman, R.D.: Detecting single atoms of calcium and iron in biological structures by electron energy-loss spectrum-imaging. J. Microsc. **210**(Pt 1), 5–15 (2003)
212. van Manen, H.-J., Kraan, Y.M., Roos, D., Otto, C.: Intracellular chemical imaging of heme-containing enzymes involved in innate immunity using resonance raman microscopy. J. Phys. Chem. B **108**(48), 18762–18771 (2004). https://doi.org/10.1021/jp046955b
213. Atkins, C.G., Buckley, K., Blades, M.W., Turner, R.F.B.: Raman spectroscopy of blood and blood components. Appl. Spectrosc. **71**(5), 767–793 (2017). https://doi.org/10.1177/0003702816686593
214. Wang, H., Lee, A.M.D., Lui, H., McLean, D.I., Zeng, H.: A method for accurate in vivo micro-Raman spectroscopic measurements under guidance of advanced microscopy imaging. Sci. Rep. **3**, 1890 (2013). https://doi.org/10.1038/srep01890
215. Al-Qenaei, A., Yiakouvaki, A., Reelfs, O., Santambrogio, P., Levi, S., Hall, N.D., Tyrrell, R.M., Pourzand, C.: Role of intracellular labile iron, ferritin, and antioxidant defence in resistance of chronically adapted Jurkat T cells to hydrogen peroxide. Free Radical Biol. Med. **68**, 87–100 (2014). https://doi.org/10.1016/j.freeradbiomed.2013.12.006
216. Carney, I.J., Kolanowski, J.L., Lim, Z., Chekroun, B., Torrisi, A.G., Hambley, T.W., New, E.J.: A ratiometric iron probe enables investigation of iron distribution within tumour spheroids. Metallomics: Integr. Biometal Sci. **10**(4), 553–556 (2018). https://doi.org/10.1039/c7mt00297a
217. Long, L., Wang, N., Han, Y., Huang, M., Yuan, X., Cao, S., Gong, A., Wang, K.: A coumarin-based fluorescent probe for monitoring labile ferrous iron in living systems. Analyst (2018). https://doi.org/10.1039/c8an00556g
218. Aron, A.T., Heffern, M.C., Lonergan, Z.R., Vander Wal, M.N., Blank, B.R., Spangler, B., Zhang, Y., Park, H.M., Stahl, A., Renslo, A.R., Skaar, E.P., Chang, C.J.: In vivo bioluminescence imaging of labile iron accumulation in a murine model of Acinetobacter baumannii infection. Proc. Natl. Acad. Sci. U.S.A. **114**(48), 12669–12674 (2017). https://doi.org/10.1073/pnas.1708747114
219. Jiang, N., Cheng, T., Wang, M., Chan, G.C., Jin, L., Li, H., Sun, H.: Tracking iron-associated proteomes in pathogens by a fluorescence approach. Metallomics: Integr. Biometal Sci. **10**(1), 77–82 (2018). https://doi.org/10.1039/c7mt00275k
220. Ralle, M., Lutsenko, S.: Quantitative imaging of metals in tissues. Biometals: Int. J. Role Metal Ions Biol. Biochem. Med. **22**(1), 197–205 (2009). https://doi.org/10.1007/s10534-008-9200-5

221. Collingwood, J.F., Davidson, M.R.: The role of iron in neurodegenerative disorders: insights and opportunities with synchrotron light. Front. Pharmacol. **5**, 191 (2014). https://doi.org/10.3389/fphar.2014.00191
222. Punshon, T., Ricachenevsky, F.K., Hindt, M., Socha, A.L., Zuber, H.: Methodological approaches for using synchrotron X-ray fluorescence (SXRF) imaging as a tool in ionomics: examples from Arabidopsis thaliana. Metallomics: Integr. Biometal Sci. **5**(9), 1133–1145 (2013). https://doi.org/10.1039/c3mt00120b
223. Kashiv, Y., Austin 2nd, J.R., Lai, B., Rose, V., Vogt, S., El-Muayed, M.: Imaging trace element distributions in single organelles and subcellular features. Sci. Rep. **6**, 21437 (2016). https://doi.org/10.1038/srep21437
224. Ducic, T., Barski, E., Salome, M., Koch, J.C., Bahr, M., Lingor, P.: X-ray fluorescence analysis of iron and manganese distribution in primary dopaminergic neurons. J. Neurochem. **124**(2), 250–261 (2013). https://doi.org/10.1111/jnc.12073
225. Henry, Y.A.: Basic EPR Methodology. In: Nitric Oxide Research from Chemistry to Biology, pp. 47–60. Springer, Boston, MA, US (1997). https://doi.org/10.1007/978-1-4613-1185-0_4
226. Taiwo, F.A.: Electron paramagnetic resonance spectroscopic studies of iron and copper proteins. Spectroscopy **17**(1) (2003). https://doi.org/10.1155/2003/673567
227. Telser, J., van Slageren, J., Vongtragool, S., Dressel, M., Reiff, W.M., Zvyagin, S.A., Ozarowski, A., Krzystek, J.: High-frequency/high-field EPR spectroscopy of the high-spin ferrous ion in hexaaqua complexes. Magn. Reson. Chem.: MRC **43**(Spec no), S130–S139 (2005). https://doi.org/10.1002/mrc.1689
228. Zhang, Y., Gan, Q.-F., Pavel, E.G., Sigal, E., Solomon, E.I.: EPR definition of the non-heme ferric active sites of mammalian 15-lipoxygenases: major spectral differences relative to human 5-lipoxygenases and plant lipoxygenases and their ligand field origin. J. Am. Chem. Soc. **117**(28), 7422–7427 (1995). https://doi.org/10.1021/ja00133a015
229. Brautigan, D.L., Feinberg, B.A., Hoffman, B.M., Margoliash, E., Preisach, J., Blumberg, W.E.: Multiple low spin forms of the cytochrome c ferrihemochrome. EPR spectra of various eukaryotic and prokaryotic cytochromes c. J. Biol. Chem. **252**(2), 574–582 (1977)
230. Bertrand, P., Janot, J.M., Benosman, H., Gayda, J.P., Labeyrie, F.: An EPR study of the interactions between heme and flavin in yeast flavocytochrome b2. Eur. Biophys. J. **14**(5), 273–278 (1987). https://doi.org/10.1007/bf00254891
231. Walker, F.A.: NMR and EPR spectroscopy of paramagnetic metalloporphyrins and heme proteins. In: Handbook of Porphyrin Science, pp. 1–337. World Scientific Publishing Company (2012). https://doi.org/10.1142/9789814307246_0001
232. Iwasaki, T., Samoilova, R.I., Kounosu, A., Ohmori, D., Dikanov, S.A.: Continuous-wave and pulsed EPR characterization of the [2Fe–2S](Cys)3(His)1 cluster in rat MitoNEET. J. Am. Chem. Soc. **131**(38), 13659–13667 (2009). https://doi.org/10.1021/ja903228w
233. Priem, A.H., Klaassen, A.A.K., Reijerse, E.J., Meyer, T.E., Luchinat, C., Capozzi, F., Dunham, W.R., Hagen, W.R.: EPR analysis of multiple forms of [4Fe–4S]3+ clusters in HiPIPs. J. Biol. Inorg. Chem. **10**(4), 417–424 (2005). https://doi.org/10.1007/s00775-005-0656-2
234. Bhave, D.P., Hong, J.A., Lee, M., Jiang, W., Krebs, C., Carroll, K.S.: Spectroscopic studies on the [4Fe–4S] cluster in adenosine 5′-phosphosulfate reductase from Mycobacterium tuberculosis. J. Biol. Chem. **286**(2), 1216–1226 (2011). https://doi.org/10.1074/jbc.M110.193722
235. Bonomi, F., Iametti, S., Morleo, A., Ta, D., Vickery, L.E.: Facilitated transfer of IscU–[2Fe2S] clusters by chaperone-mediated ligand exchange. Biochemistry **50**(44), 9641–9650 (2011). https://doi.org/10.1021/bi201123z
236. Chandramouli, K., Johnson, M.K.: HscA and HscB stimulate [2Fe–2S] cluster transfer from IscU to apoferredoxin in an ATP-dependent reaction. Biochemistry **45**(37), 11087–11095 (2006). https://doi.org/10.1021/bi061237w
237. Shakamuri, P., Zhang, B., Johnson, M.K.: Monothiol glutaredoxins function in storing and transporting [Fe2S2] clusters assembled on IscU scaffold proteins. J. Am. Chem. Soc. **134**(37), 15213–15216 (2012). https://doi.org/10.1021/ja306061x

238. Baker, T.M., Nakashige, T.G., Nolan, E.M., Neidig, M.L.: Magnetic circular dichroism studies of iron(ii) binding to human calprotectin. Chem. Sci. **8**(2), 1369–1377 (2017). https://doi.org/10.1039/c6sc03487j
239. Holmes-Hampton, G.P., Jhurry, N.D., McCormick, S.P., Lindahl, P.A.: Iron content of Saccharomyces cerevisiae cells grown under iron-deficient and iron-overload conditions. Biochemistry **52**(1), 105–114 (2013). https://doi.org/10.1021/bi3015339
240. Garber Morales, J., Holmes-Hampton, G.P., Miao, R., Guo, Y., Munck, E., Lindahl, P.A.: Biophysical characterization of iron in mitochondria isolated from respiring and fermenting yeast. Biochemistry **49**(26), 5436–5444 (2010). https://doi.org/10.1021/bi100558z
241. Nienhaus, K., Nienhaus, G.U.: Probing heme protein-ligand interactions by UV/visible absorption spectroscopy. Methods Mol. Biol. **305**, 215–242 (2005). https://doi.org/10.1385/1-59259-912-5:215
242. Berry, E.A., Trumpower, B.L.: Simultaneous determination of hemes a, b, and c from pyridine hemochrome spectra. Anal. Biochem. **161**(1), 1–15 (1987)
243. Mapolelo, D.T., Zhang, B., Naik, S.G., Huynh, B.H., Johnson, M.K.: Spectroscopic and functional characterization of iron-sulfur cluster-bound forms of Azotobacter vinelandii (Nif)IscA. Biochemistry **51**(41), 8071–8084 (2012). https://doi.org/10.1021/bi3006658
244. Albrecht, A.G., Netz, D.J., Miethke, M., Pierik, A.J., Burghaus, O., Peuckert, F., Lill, R., Marahiel, M.A.: SufU is an essential iron-sulfur cluster scaffold protein in Bacillus subtilis. J. Bacteriol. **192**(6), 1643–1651 (2010). https://doi.org/10.1128/JB.01536-09

Achieving Selective Targeting Using Engineered Nanomaterials

Roberta Lanfranco, Bortolo M. Mognetti and Gilles Bruylants

Abstract The development of Drug Delivery Systems (DDS) able to selectively deliver a controlled amount of a drug only to diseased cells would represent a dramatic development in nanomedicine. One of the multiple challenges still paving the way towards this goal is the elaboration of strategies that would allow targeting with extreme accuracy specific cells, as cancerous cells, among a large variety of closely related ones. In this work, we review the most recent nanotechnology applications aiming at controlling the selectivity of the interaction of delivery nanosystems with cells, with a focus on multivalent targeting. We briefly review thermodynamic models of multivalent interactions and highlight the challenges that still need to be addressed to transfer theoretical design principles into practical applications. In particular, suitable experimental systems based on multivalent models often require the control of the nanocarrier characteristics at the molecular level. Traditional delivery methods, however, fail to provide such degree of control. DNA nanotechnology is a growing field of nanoscience that has witnessed impressive developments in the past decades and has led to major advances in the fabrication of nanostructures and self-assembled systems. Relying on the possibility of controlling their molecular interactions by sequence design, nucleic acids can serve the drug delivery program by providing desired nanostructures with nearly atomic precision. In combination with the recent achievements in the research on DNA aptamers, short nucleic acid sequences isolated to interact selectively with a specific target, DNA nanotechnology is undoubtedly one of the most promising tools for the development of selective DDS.

R. Lanfranco
Biological and Soft Systems, Department of Physics,
University of Cambridge, Cambridge, UK

R. Lanfranco · G. Bruylants (✉)
Engineering of Molecular NanoSystems, Chemistry and Material Science Department,
Université libre de Bruxelles, Brussels, Belgium
e-mail: gbruylan@ulb.ac.be

B. M. Mognetti
Physics of Complex Systems and Statistical Mechanics, Department of Physics,
Université libre de Bruxelles, Brussels, Belgium

C. Demetzos and N. Pippa (eds.), *Thermodynamics and Biophysics of Biomedical Nanosystems*, Series in BioEngineering, https://doi.org/10.1007/978-981-13-0989-2_6

1 Introduction

Understanding and controlling biological systems, from the molecular interactions up to their functional behaviour, is one of the most relevant goals on which medicine, bio-engineering, physics, and chemistry have focused their strains in the last decades. Two distinct types of applications are fuelling this research area. On the one hand, the motivation is the production of smart sensors capable of rapidly detecting the presence of either hazardous compounds threatening the health of living beings or biological markers indicative of pathologies. This branch of research focuses its attention on developing proper techniques performing in a real and complex environments [207]. Several point-of-care devices have been proposed to monitor toxic substances [104, 105, 109, 113], replacing long and complex laboratory tests [1, 177, 190]. In the last years, this topic has been particularly successful in producing devices with low detection limits [176], till the single molecule level [12]. On the other hand, understanding the rules that govern molecular interactions in biological systems and lead to the extreme level of selectivity of biological entities such as antibodies, proteins or nucleic acids is also crucial to guide the development of new medical treatments and therapies as well as new nanomaterials capable of responding to different stimuli in complex biological environments and interacting only with selected targets.

In this chapter, we review the most recent advances in nanomaterial design leading to vectors capable of selectively targeting biological surfaces. We focus both on experimental results and theoretical modelling. Even if the scenario in which a carrier loaded with a drug would be programmed to target exogenous or diseased and release the carried compound only once these are reached is a dreaming and fascinating possibility, the issues linked to this process are still far from being solved [22, 51, 103, 198]. Overview all the problems still to overcome on the way leading to smart and efficient DDS is out of the scope of this chapter. Instead, here we want to focus on the importance to be selective in targeting while delivering a drug. The challenge in this field is to develop strategies allowing to target a specific entity to which deliver the drug in a sea of similar ones, for instance, a tumour cell among healthy ones.

After a brief introduction on engineered nanosystems, we introduce in Sect. 2 the strategies that can be implemented to reach selective and superselective targeting. Section 3 is devoted to the thermodynamics of multivalent binding, starting from the laws controlling ligand-receptor interactions and underlying selective targeting. As we are going to explain, even if several theoretical and computational works demonstrate that selectivity can be achieved, there is still a lack of experimental proofs-of-concepts backing theoretical predictions.

A way to fill this gap would be to bridge the field of DDS developments with that of DNA nanotechnology. Indeed, by taking advantages of the remarkable characteristics of nucleic acids to program and control the properties of nano- and micro-structures, we point out how it would be possible to exploit multivalent systems to succeed in developing selective targeting. Achievements in DNA nanotechnologies are reviewed in Sect. 4. Finally, in Sect. 5, we present novel results in the research about aptamers, short ssDNA able to bind a large variety of entities, and we point out how exploiting DNA nanotechnology could be one of the most promising routes towards the development of functional selective Drug Delivery nanovectors.

2 Engineered Nanosystems

Systems showing at least one characteristic dimension below 100 nm and conceived to achieve complex tasks are usually identified as engineered nanosystems. These systems are constituted of designed components that are programmed to self-assemble into specific structures. The recent advances in the design of new nanomaterials are positively impacting the fields of disease diagnosis, treatment, and prevention, which are essential steps towards the development of personalized nanomedicine. This wide spectrum of innovations can turn molecular discoveries into nanotechnological designs to target specific needs [103].

The engineering of nanosystems, by controlling and programming interactions can help the development of novel DDS, where high concentrations of a pharmaceutical compound would be transported to the desired site of action and delivered only at that location [203]. This scenario would avoid damaging healthy cells and would make a more parsimonious use of the, usually expensive, drug. Furthermore, the carrier can be often loaded with several compounds, allowing the simultaneous delivery of multiple therapeutic molecules in controlled ratios or the combination of therapeutic and diagnostic moieties, as a contrast agent. Drug delivery, in principle, should control both the rate at which a drug is released and its location. Successful delivery is conditional on solving many challenges including finding strategies for loading/releasing the drug, avoiding adsorption of plasma proteins or the clearance of the nanovector by the immune system [92]. These problematics are specific to the carrier that is used. For example, the internalization of a drug using a lipid vesicle can be more efficient compared to a polymeric particle [4], but the release of the therapeutic compound is more difficult to activate. Discussing and analysing all the aspects related to the control of DDS is not the main aim of this chapter. Instead, in this work we focus on understanding the possible ways to selectively target a specific entity, by controlling the functionalization of the drug carrier and we refer the reader to recent reviews in the literature for a more complete overview of the problems that have just been mentioned above [170, 197].

A simple approach to selective targeting consists in functionalizing the carriers with a ligand that binds strongly to a receptor, usually a membrane protein, expressed by the target cells. If this protein is expressed exclusively by these cells, this "active targeting" strategy allows the nanovectors to be selective. This strategy has proven to be effective in several cases, using, for example, monoclonal antibodies or Pattern Recognition Receptors agonists specific to the desired cell population [133, 172]. However, there are around 200 different cell types in the human body and more than 10,000 different microbial species (mostly bacteria), and most of these can exist in a variety of physiological states.

The fact that each cell can express many different surface receptors renders the identification of a receptor that is only present on the surfaces of the target cells extremely difficult, if not impossible [192]. The level of expression of surface receptors also varies dramatically between different organs. For example, it has been shown that the expression levels of the intracellular adhesion molecule-1 (ICAM-1) varies respectively from approximately 50 to 2000 proteins/μm^2 on cell surface in the lungs and the heart of C57BL female mice [151]. A scheme capable of sorting molecules based on receptor concentrations would then be desirable. Multivalent interactions feature interesting collective behaviours resulting from ensemble averages and provide properties to these systems that are not displayed by their monovalent constituents [119, 166]. Based on this principle ruling biological recognition processes, a second strategy that is currently envisaged consists in functionalizing the carriers with many ligands that weakly bind to their target receptors. This strategy should allow targeting only those cells expressing a concentration of receptors above a certain threshold. Several modelling efforts have been undertaken, aiming at assessing the influence of parameters, such as ligand binding affinity, ligand density, tether length and flexibility, tether valency (i.e., bi-, tri-, tetravalent tethers), delivery vector shape and substrate (cell surface) rigidity on the efficiency and selectivity of the recognition process [50, 61, 164, 171]. A complete mastering of the design of multi- or poly-valent interactions would allow reaching an extreme selectivity in the targeting of biological membranes [119].

3 Thermodynamics of Multivalent Binding

The physics of cell-cell adhesion has been extensively investigated starting from the pioneering works of Bell and collaborators [18, 42]. In these contributions, the strength of the adhesion between pairs of cells has been characterized using thermodynamic functions depending on the number of free/bound complexes featured by the system calculated starting from the binding affinity between the interacting moieties. Statistical approaches have also been used to study, for instance, endocytosis [50, 52, 69], targeting of multivalent probes [44, 71, 100, 110, 123], or synapse formation [38, 149, 153]. More recent developments have focused on the study of how membrane deformability alters the equilibrium constant of membrane-bound moieties [82, 168, 202].

Figure 1 explains the statistical mechanics approach to calculate effective interactions between functionalized particles or vesicles [7, 9, 13, 126, 123, 147, 154, 189]. This method derives the interaction free energy, F, using partition functions, Z, accounting for all microstates of the system specified by a given set of ligand-receptor complexes. The binding free energy, $\Delta G_{i,j}$, controls the probability of ligand i to bind to receptor j (see Fig. 1). $\Delta G_{i,j}$ comprises a term depending on the chemical details of the interacting groups, ΔG_0, and a configurational term, $\Delta G_{i,j,cnf}$, with $\Delta G_{i,j} = \Delta G_0 + \Delta G_{i,j,cnf}$. ΔG_0 can be obtained by measuring the association equilibrium constant of the interacting molecules (the “sticky” ends) free in solution or

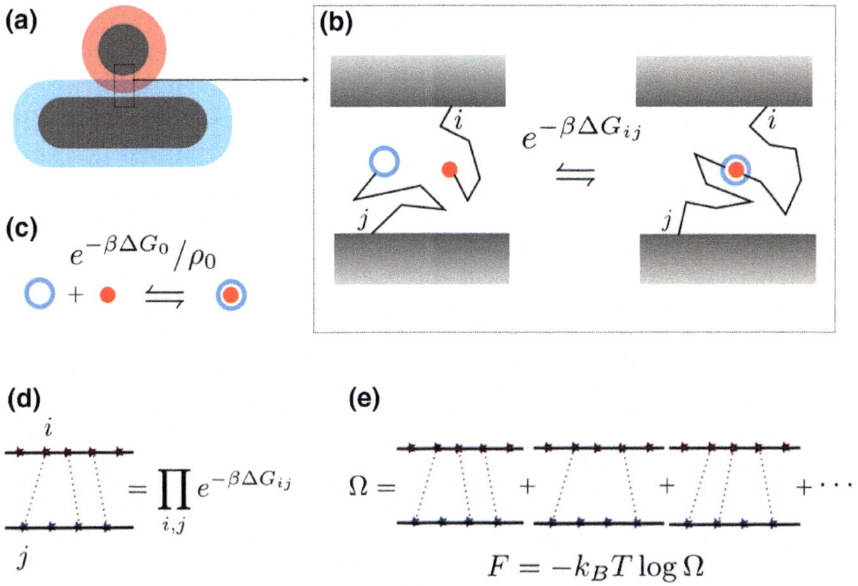

Fig. 1 Statistical mechanics modelling of multivalent interactions. A cell (carrying blue receptors) interacts with nanoparticles (carrying red ligands) through ligand-receptor complexes forming inter-particle bridges (**a** and **b**). Bridge formation is controlled by binding free energies, $\Delta G_{i,j}$, comprising configurational and chemical contributions (**b** and **c**, respectively). We assign a statistical weight to each microstate specified by a set of bridges (**d**), and calculate the multivalent free energy through an ensemble calculation (**e**)

using theoretical or empirical results like the nearest-neighbour rules in the case of nucleic acids [54, 162, 163, 205]. $\Delta G_{i,j,cnf}$ accounts for configurational constraints resulting from the fact that the interacting complexes are anchored to the facing surfaces, limiting the number of possible configurations of the tethered dimers as compared to the case of molecules free in solution (see Fig. 1). In particular, $\Delta G_{i,j,cnf}$ is a function of the distance between tethering points as well as of the physical properties of the spacers anchoring the "sticky" groups to the surfaces. Importantly, in most applications, $\Delta G_{i,j,cnf}$ cannot be neglected. For instance, considering ligands as thin rods of length L pivoting around the tethering points (as for the case of dsDNA spacers), a sound estimate of $\Delta G_{i,j,cnf}$ is given by $\Delta G_{i,j,cnf,a} = k_B T log(\rho_0 L^2 d)$, where ρ_0 is the standard concentration ($\rho_0 = 0.602\,\text{nm}^{-3}$) and d is the distance between tethering points [130]. Instead, in the case of fully flexible polymeric spacers made of N Kuhn segments of length a tipped by reactive units, we can write $\Delta G_{i,j,cnf,b} = k_B T\left[log(\rho_0 a^3) + 1.5d^2/(2Na^2)\right]$ [131, 189], where we have used Gaussian distributions for the end-to-end distance of ideal chains and have assumed a distance between tethering points comparable with the average end-to-end distance. Using $L = d = 20$ nm, $a = 5$ nm and $N = 16$ [154, 189] we obtain $\Delta G_{i,j,cnf,a} = 8.5\,k_B T$. This highlights that configurational effects in standard conditions are often comparable with ΔG_0 and should enter the modelling. When considering deformable units, like lipid vesicles, configurational terms should also account for the roughness of the substrate [82, 202].

Using $\Delta G_{i,j}$ we can calculate the probability of forming a ligand-receptor complex between i and j, pc_{ij}

$$pc_{ij} = pl_i * pr_j * \exp\left[-\beta \Delta G_{ij}\right] \tag{1}$$

In the previous expression, we assume that the probability of ligand i and receptor j to be free, prl, factorizes, $prl_{ij} = pl_i * pr_j$ where pl_i and pr_j are the probabilities of molecules i and j to be unbound. This hypothesis holds only in the limit of high coating densities [189]. Note how Eq. (1) represents a set of coupled equations for the ensemble of all pl_i and pr_j. This system can be solved using fixed-point iteration procedures as performed by Parolini et al. [143] and Bachmann et al. [14]. When the exact position of the tethering points is not known, as in most cases, Eq. (1) can be replaced by mean-field expressions for the densities of free/bound molecules. We refer to the work of Varilly et al. [189] for more details. Instead, here we review the case in which the tethering points are mobile, assumption that is appropriate in many biological systems and synthetic designs where lipid vesicles are functionalized with ligands carrying a hydrophobic head ensuring their attachment to the nanovector, as described in Sect. 4.3. In this case, the probability of making a complex, pc, is solely a function of the type of binding molecules (similarly for pl and pr), with the number of ligand-receptor complexes formed, nc, that reads as $nc = N_L * N_R * pc$, where N_L and N_R are the total numbers of ligands and receptors. After defining by n_L and n_R the number of free ligands and receptors ($n_L = N_L - nc, n_R = N_R - nc$), Eq. (1) becomes:

$$\begin{aligned} nc &= n_L * n_R * \exp[-\beta \Delta G] \\ \exp[-\beta \Delta G] &= \frac{1}{A_i * A_j} \int dr_i dr_j \exp\left[-\beta \Delta G_{ij}\right] \end{aligned} \tag{2}$$

In the previous equation, we calculate the interaction free energy, ΔG, by averaging ΔG_{ij} over the possible positions of the tethering points of molecules i and j running over the surfaces A_i and A_j. We also assume ideal ligands/receptors and neglect non-selective interactions. After dividing the first member of Eq. (2) by the total configurational spaces available to free/bound molecules, we recover chemical equilibrium equations. As compared to the fixed tethering case, for mobile tethers there is an extra entropic term associated to the fact that the tethering points of two binding molecules should be found in the particles' contact region.

The statistical mechanics approach is not limited to the calculation of the number of ligand/receptor complexes, but can also provide the free energy of the system. Interestingly, Angioletti-Uberti et al.[9] proposed a general expression for F that sidesteps explicit calculations of the partition function of the system, Z, usually done

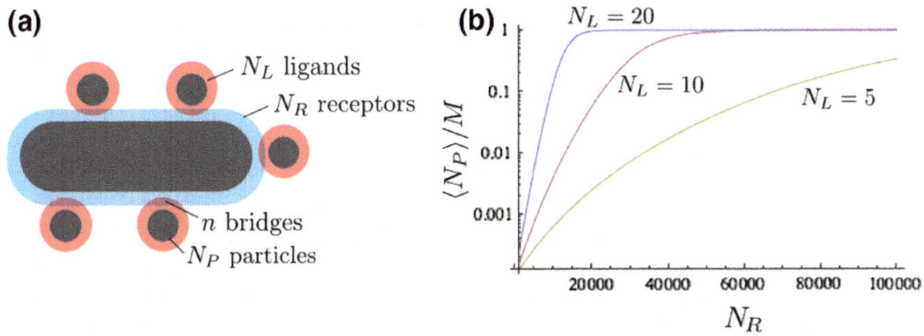

Fig. 2 **a** Targeting of a cell featuring N_R receptors by NPs functionalized by N_L ligands. **b** The average number of adsorbed particles as a function of the number of receptors for three different NPs. NPs with large amounts of ligands respond more sharply to changes in coating densities. We used $v_0 = 10^{-4}$, $M = 100$ and $\beta\Delta G = 10$

through thermodynamic integrations [154]. Instead, after calculating the probability of having free and interacting complexes, F reads as follows:

$$\beta F = \sum_i \log(pl_i) + \sum_j \log(pr_j) + \sum_{i,j} pc_{i,j} \tag{3}$$

The previous expression holds under the same conditions leading to Eq. (1). In particular, it assumes that interactions between different molecules not involved in the same complex are negligible. In case of mobile tethering points, we obtain:

$$\beta F = N_L log \frac{n_L}{N_L} + N_R log \frac{n_R}{N_R} + nc \tag{4}$$

We can easily generalize the previous expressions to systems featuring multiple types of ligands/receptors or more objects (like multiple NPs targeting a single cell). In particular, all different types of molecules on each particle contribute with a logarithmic term similar to the first two expressions of Eqs. (3) and (4), however a sum over all interacting complexes replaces the last terms of these equations. Di Michele et al. [126] recently extended Eq. (3) to systems featuring complexes made of more than two monomers. Another notable example involves fixed ligands against mobile receptors as relevant for studying nanoparticles targeting cells. A recent work of Di Michele et al. [52] has explicitly considered this case. Importantly, the authors also studied, for the first time, the effect of steric interactions between ligands/receptors on the adhesion free-energy.

In Fig. 2, we have considered the interaction between nanoparticles functionalized with N_L ligands and a cell featuring N_R receptors. When studying targeting of cells by small NPs or in experiments involving supported lipid bilayers it is more relevant to consider the target as a large surface featuring a constant density of receptors, ρ_0. This case is investigated in recent publications [52,171, 214].

We now use the theory outlined above to study targeting of a cell featuring N_R mobile receptors by nanoparticles functionalized by N_L ligands forming ligand-receptor complexes with binding free energy equal to $\Delta G_{i,j} = \Delta G$ (see Fig. 2a and e.g. [209]). We first consider configurations with N_P particles attached to the cell's surface. The chemical equilibrium equation, see Eq. (2), adapted to the present case becomes

$$nc = (N_L - nc)(N_R - nc * N_P)e^{-\beta \Delta G} \tag{5}$$

where nc is the number of bridges featured by each NP. By analytically solving the equation above we can calculate nc as a function of ΔG, N_L, N_R, and N_P. The multivalent free energy of the system, Eq. (4) then reads as

$$\beta F(N_P) = N_R log\left(1 - \frac{nc * N_P}{N_R}\right) + N_P N_L log\left(1 - \frac{nc}{N_L}\right) + N_P * nc \tag{6}$$

where nc is the solution of Eq. (5). Note that $F(N_P)$ is non-extensive in N_P given that all nanoparticles can bind the same set of receptors. We now allow attachment/detachment of NPs from the cell immersed in a bath of diluted NPs with density ρ. We use a cell model in which we divide the cell's surface into M spots containing at most one NP. If we neglect NP-NP interactions in bulk, the probability of finding N_P particles attached to the surface is

$$Prob(N_P) = \frac{1}{Z}\binom{M}{N_P} e^{-\beta F(N_P)} (v_0 \rho)^{N_P} \tag{7}$$

where v_0 is the configurational volume available to each bound NP and Z is the normalization factor. Using $Prob(N_P)$ we can calculate the average number of attached NPs as $\langle N_P \rangle = \sum_{N_P=0}^{M} Prob(N_P) N_P$. In Fig. 2b we study $\langle N_P \rangle$ and highlight superselective targeting against receptor densities for systems of NPs with many weak ligands [71, 123].

Although we recently witnessed advances in the modelling of multivalent interactions, many challenges are still ahead of us. A key point will be to probe the reliability of the equilibrium approach: under what conditions can we neglect ligand/receptor diffusivity when studying targeting? For instance, most of the multivalent theories with mobile tethering points described above have been validated using DNA linkers with a diffusion constant of ~1 μm^2/s. Membrane proteins in biology can be much slower [87]. After assessing the limits of validity of equilibrium approaches, it will be essential to develop theoretical approaches and efficient numerical methods for design purposes accounting for non-selective interactions. Probing the dynamics of the interactions will be unavoidable especially when aiming at designing complex functionalities and phase behaviours. Recent examples consist of competitions between inter-particle bridges and intra-particle loops [143]. Deformable vectors including polymeric materials are being used to deliver efficient drugs. We will need new simulation methods capable of screening the effect of invader deformability on targeting, selectivity and internalization. It is likely that such studies will not rely on brute force molecular dynamics simulations but rather on coarse-grained modelling coupled with efficient simulation schemes capable of tackling the length and the timescale of these approaches [140].

Experimental studies on multivalent binding remain challenging due to the complexity to achieve a precise control over parameters, such as surface ligand and receptor densities, ligand/receptor binding free energy, or to prevent non-specific interactions. Chemists have used the principles of multivalent interactions with various degrees of success regarding the increased stability of the multivalent complexes compared to the monomeric equivalents [101, 116, 152], highlighting the complexity of multi- or polyvalent ligand design. Whitesides and collaborators [120], in particular, developed a series of synthetic polymers containing benzylamine and sialic acid attached to a polyacrylamide backbone and obtained unprecedented inhibition of the adhesion of the attachment of influenza virus X-31 to mammalian cells. Kitov et al. [100] have proposed a model system based on the interaction between clustered multivalent oligosaccharides and the Shiga-like toxin Type 1, a homopentameric protein naturally biding carbohydrate, a trisaccharide known as P^k and issued from the Gb3 glycolipid. Each subunit presents three binding sites for P^k, with very different binding affinities. The authors conclude that both the avidity entropy and the cooperativity of binding are important for the enhancement of the multivalent complex stability.

In an interesting work Gestwicki et al. [71] compared the ability of multivalent ligands from five different structural classes, from small multivalent molecules to dendrimers, proteins, linear and polydispersed polymers, all containing mannose moieties, to inhibit and induce clustering of the concavilin A lectin. The authors studied the role of the ligand architecture on the avidity and kinetics of the interaction, but also on receptors clustering and the mean distance between the receptors in the clusters.

More recently, Dubacheva et al. [59] have realized quantitative studies of the interaction between hyaluronic acid functionalized with β-cyclodextrins and self-assembled monolayers (SAMs) of lipids functionalized with ferrocene receptors. This highly specific and well-defined model system allowed the authors to investigate the effect of receptor density on the multivalent avidity, showing regimes of superselectivity, demonstrating the possibility to target surfaces expressing a specific density of surface receptors.

It is interesting to point out that a key design rule for superselective systems is the use of weakly binding ligands. Curk et al. [44] have predicted that the ligand-receptor interaction free energy should be of the order of magnitude of the thermal energy, corresponding to a molar binding free energy of ~3 kJ/mol. Such value is much weaker than the binding free energies characterising the formation of complexes involving biological ligands [81], frequently used in DDS. In particular, experimentally, the binding strength cannot be reduced indefinitely, to avoid nonspecific interactions, but also because low ligand/receptor affinities need to be compensated by the formation of a large number of complexes [123]. In order to investigate the effect of binding affinity on the selectivity of the interaction, Dubacheva et al. [58] have compared targeting of surfaces functionalized with ferrocene or adamantane, for which β-cyclodextrin display respective dissociation constants of 10 and 200 μM. Both systems showed superselective responses but for receptor densities varying by more than one order of magnitude. These results indicate that ligands with tunable binding affinities are requested to develop superselective systems with a given receptor fingerprint.

These examples illustrate the difficulty of engineering systems in which all parameters controlling multivalent interactions are tunable. In the next sections, we review how DNA nanotechnology can provide better control over interactions.

4 Using DNA to Control Nanosystems

In the last decades, DNA has been one of the most investigated compounds used to control the properties of engineered nanosystems [3, 39, 74, 115, 117, 129, 199, 200, 206, 212] allowing the precise programming of their structures and functions. DNA is employed as building-block and glue to assemble nanosystems owing, respectively, to its peculiar structure and pairing selectivity properties. DNA is a polymer made from four basic repeating units called nucleotides: Adenine (A), Thymine (T), Guanine (G) and Cytosine (C), Fig. 3. The most common secondary structure of DNA is the well-known double helix formed by two DNA strands stabilised by hydrogen bonds between complementary nucleotides according to the canonical Watson-Crick base pairing rules (A with T and C with G) [196].

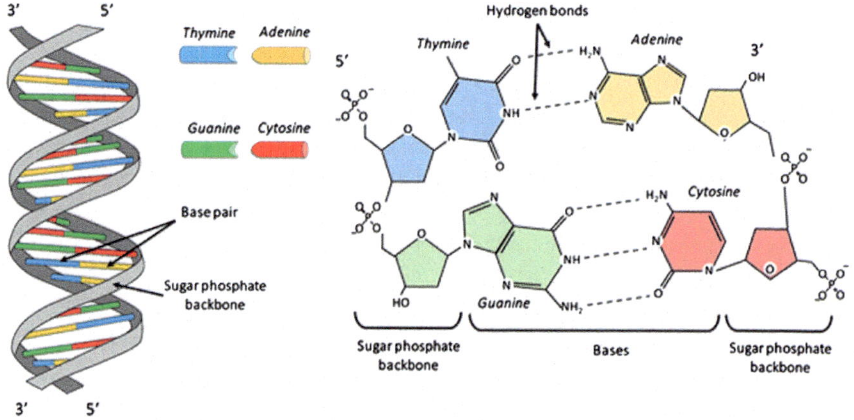

Fig. 3 Double-stranded DNA structure. Complementary single-stranded DNA oligomers pair to form a double helix. DNA oligomers are constituted by four monomeric units: Adenine (yellow), Thymine (blue), Guanine (green) and Cytosine (red). The double helix is stabilised by hydrogen bonds between complementary bases and stacking interactions (not shown)

Taking advantage of the selectivity of the Watson-Crick pairing, it is possible to design DNA sequences capable to form complex structures and materials. As illustrated in the next sections, these include gels, presented in Sect. 4.1, and 2- and 3-dimensional regular structures composed solely of nucleic acids, the so-called DNA origami, discussed in Sect. 4.2. In Sect. 4.3 we review how DNA has been used to direct the self-assembly of nano- and micro-particles, oil droplets and lipid vesicles into materials of defined architecture. Finally, we present in Sect. 4.4 some recent works about the use of DNA-based materials in DDS.

4.1 Materials Made of DNA

Due to its molecular programmability, DNA has been used to self-assemble liquids/gels/solids with physical properties unmatched by other materials. For instance, recent literature studied gels with limited valence made of DNA-nanostars [20], Fig. 4a. DNA-nano stars are constructs comprising 4 arms of double-stranded DNA self-assembled from 4 ssDNA oligomers of around 50 bases, Fig. 4b(i). The DNA sequences are designed so that a short ssDNA oligomer of 6 bases (sticky end) emanates from each arm. Gels are formed when complementary sticky-ends on different nanostars hybridize. In this system, self-assembly is a two steps process, Fig. 4a(ii), where the formation of the DNA nanostars happens at a temperature that is higher than the nanostars cross-linking. While the phase diagram resembles a standard fluid-liquid transition, the dynamics is very peculiar, as shown by DLS experiments, Fig. 4a(iii). At high temperature there is a single characteristic time related to the diffusivity of the nanostars. When temperature is lowered towards the critical temperature, the field autocorrelation function is characterized by two characteristic decay times, observation that is in disagreement with the theory of the dynamics of critical phenomena [79], which predicts a single decay time. The authors attribute this behaviour to the long lifetime of the bonds between DNA sticky ends that is larger than the time requested for structures made of a few DNA nanostars to diffuse over distances comparable to the correlation length even close to the critical temperature.

The work of Biffi et al. demonstrates how DNA nanostars are useful constructs to investigate the phase behaviour of gel-forming complexes in view of the possibility of adjusting the binding selectivity (via the DNA sequence), the strength of interactions (via the DNA length), and the valence. For this reason, DNA-nanostar systems have been extensively studied both experimentally [21, 26, 27, 66] and computationally [156, 158, 159].

Recently, Brady et al. [28] considered a system of DNA nanostars tipped by cholesterol groups, Fig. 4b(i). At high temperature, due to their hydrophobicity, cholesterol heads form micelles, Fig. 4b(ii). When the temperature is lowered, the micelles are cross-linked by DNA hybridizations leading to the formation of nanostars, a process that can be monitored by DLS, Fig. 4b(iii). This system is very efficient in yielding micron size crystals, Fig. 4b(iv), which internal structure depends on the character-

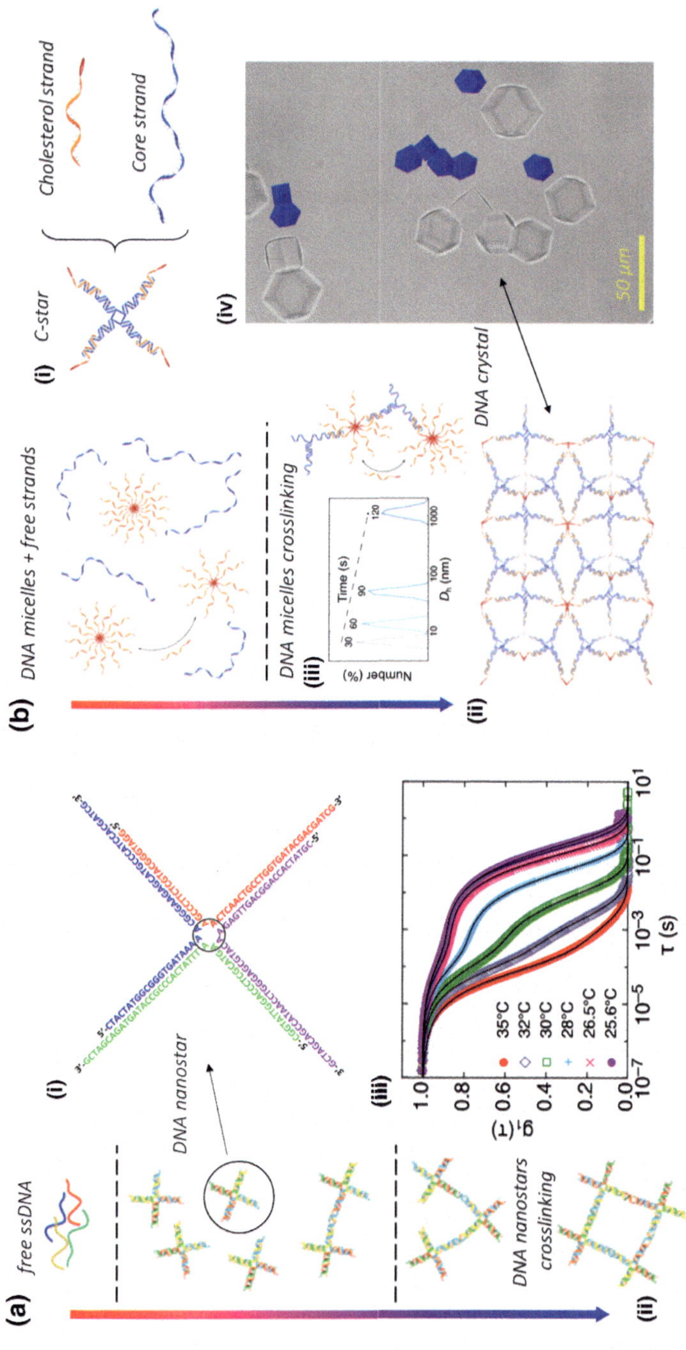
(a)
free ssDNA
DNA nanostar
DNA nanostars crosslinking
(i)
(ii)
(iii)
$g_1(\tau)$
τ (s)
35°C
32°C
30°C
28°C
26.5°C
25.6°C
(b)
DNA micelles + free strands
DNA micelles crosslinking
Time (s)
Number (%)
D_h (nm)
C-star
Cholesterol strand
Core strand
DNA crystal
(iv)
50 μm

◄**Fig. 4** DNA nanostars forming gels and crystals. **a** Biffi et al. [20] used four types of ssDNA oligomers to fabricate gels with a controlled valence using hierarchical self-assembly. When decreasing the temperature, DNA nanostars are formed following hybridization of the four arms (i). By further decreasing the temperature, nanostars are cross-linked by complementary sequences tipping the ends of the arms (ii). The field autocorrelation functions (iii) measured via dynamic light scattering show a two-step dynamic, the fastest being related to the diffusion of free nanostars, the slowest to the gel phase. **b** (i) Brady et al. [28] self-assembled micelles made of cholesterol modified ssDNA (in red) that were cross-linked by core ssDNA (in blue) while forming DNA stars (C-stars). At high temperature, cholesterol-tagged strands form micelles that coexist with core ssDNA. (ii) When the temperature is lowered, the core strands crosslink the micelles by hybridization of the ssDNA tails emanating from the cholesterol heads. (iii) Dynamic light scattering experiments show the size of the aggregates over time. (iv) Image in bright field microscopy that suggests self-assembly of extended crystalline networks as clarified by crystal reconstructions (in blue). *Credits* Figure **a**: adapted from [20]. Figure **b**: adapted from [28], further permissions should be directed to the ACS

istics of the cholesterol nanostars, as the number of arms and their length. Other studies about the use of crystals having different building blocks made of DNA fragments [94, 112, 111, 144, 199] have shown that the production of macromolecular, nano-structured crystalline materials is possible thanks to the nanometric control in building well-ordered self-assembled microstructures offered by nucleic acids.

DNA polymer can also be used to fabricate metamaterials. Usually, metamaterials are structurally engineered compounds that have unusual physical properties, such as negative refractive indices [25, 173]. To date, almost all metamaterials have been made from inorganic building blocks such as silicon and copper [114, 169]. Lee et al. [107] proposed a novel metamaterial formed of DNA strands elongated by the action of a polymerase enzyme and weaved non-covalently into a hydrogel. This material presents a liquid behaviour when dried and a solid shape when wetted. To fabricate this DNA meta-hydrogel, the authors took advantage of a special phage polymerase, φ29, which uses a single-stranded DNA as a template to elongate the primer. After the end of this procedure, a chain reaction was initiated, resulting in a DNA hydrogel. Unlike conventional hydrogels, which have amorphous internal structures [31, 142, 187], this meta-hydrogel has a hierarchical internal structure, as revealed by field-emission scanning electron microscopy. At the microscale, the densely packed DNA microstructures form a structure that resembles a bird's nest. These nest structures are monodisperse and woven together by DNA. The authors addressed to this particular structure the unusual solid/liquid properties of the material.

Short (less than 20-base pairs) dsDNA strands can also show liquid-crystalline phases [132], characterized by orientational order and liquid behaviour [49]. The isotropic-nematic transition depends on the DNA duplex length, concentration, and on the presence (or absence) of sticky parts between different duplexes [132], Fig. 5. Usually, the formation of liquid crystals is due to the rod shape of the molecules composing the sample [49]. In the case of DNA, the nematic transition is not only driven by the geometry of the dsDNA, but also by the end-to-end stacking between aligned helices [67] as shown in Fig. 5. This peculiar behaviour could have potential implications for the prebiotic chemical generation of complementarily H-bonded molecular assemblies. In fact, Fraccia et al. [68] have shown that, in the presence of appro-

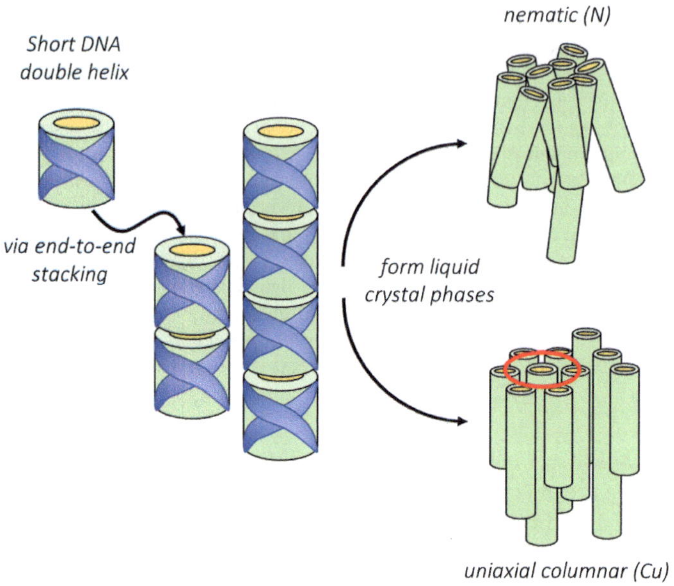

Fig. 5 DNA liquid crystals. Short DNA duplexes resemble cylinders and the hydrophobicity of the bases drives the formation of longer cylinders through head-to-head association. Nematic (N) to Columnar (C_U) phase transitions are observed when increasing DNA concentration

priate inorganic catalysts, head-to-head stacking enhances molecular concentration and promotes ligation in the liquid crystalline phase. Additionally, every ligation in the liquid-crystalline phase produces an extended complementary oligomer. This self-ordering behaviour of DNA could have implications correlated to the "origin of life", a field where researchers try to understand the impact of the molecular characteristics of simple biomolecules in building primordial forms of life as we know it [138, 137, 183].

In this section, we have presented some representative examples of how the peculiar selective nature of DNA can be exploited to build a wide variety of novel materials. The developments of new DNA materials is also boosting our understanding of the molecular properties of DNA.

4.2 DNA Origami

The main example of DNA engineering to self-assemble nanomaterials with pre-defined complex shapes is the DNA origami technology [161]. Owing to the strict configurational constraints enforced by the double helix along with the possibility to design hundreds of orthogonal sequences adopting pre-defined shapes, it is possible to build a large variety of DNA constructs with sizes ranging from tens of nanometers to microns [122]. In 2006, Rothemund [157] demonstrated the possibility of folding a 7-kilobase single-stranded scaffold, extracted from the virus M13mp18, by using over 200 short nucleotides (staples) to hold the long DNA precursors in place, Fig. 6b. Such technique allowed assembling flat aggregates resembling, for instance, stars or smiles, measuring hundreds of nanometers. Remarkably, once synthesized and mixed, the staple and scaffold strands self-assemble in a single step without requiring any external intervention.

Since the first appearance of DNA origami, different systems relying on similar principles have been proposed [19, 77, 161]. DNA origami have also motivated new methodological and theoretical developments aiming at understanding the physical principles controlling their self-assembly [85, 86, 139, 174, 182, 204, 208]. On the experimental side, Tikhomirov et al. [182] have recently shown how size limitation can be overcome (usually a single DNA origami has the size of 0.05 μm^2) by using a pre-assembled sets of tiles. The main advantage of using multistage protocols relies on the possibility to recycle the same type of DNA pattern to glue the edges of two neighbouring tiles (see edges having the same colour in Fig. 6b(i)) at different stages. The case of a three-stage process is reported in Fig. 6b(i). Following this method, the authors could control the relative position of the tiles and assemble complicate puzzles, Fig. 6b(ii) and c(iii).

New capabilities in DNA origami technology were achieved by using three-dimensional DNA bricks [73, 93]. These types of origami are made of DNA units that self-assemble according to selective inter-brick interactions, without the use of a scaffold. Gothelf et al. [73] used DNA bricks that resemble a two-stud LEGO brick composed by 32 nucleotides made of four regions of 8 nucleotides each. Two different bricks are connected by the hybridization of two 8-mers causing a 90° shift between the two units. These building blocks provide the possibility to rationally design 3D structures. More recently, Ong et al. [136] showed that, by using bricks made of four 13-mers and constraining the inter-brick bindings to remain perpendicular to each other, Fig. 6c(i), it is possible to self-assemble larger structures (0.1–1-gigadalton), as cuboids and symmetric tetramers, Fig. 6c(ii). Three-dimensional origami could serve as building blocks to assemble functional and responsive materials; in particular, they provide a DNA scaffold on which it is possible to position functional components at desired locations.

High costs related to DNA synthesis may hamper the use of origami in DD applications. Recently, Praetorius et al. [146] used bacteriophages to generate different single-stranded precursors in parallel. The authors produced all the ssDNA necessary to assemble multiple DNA origami using shaker-flask cultures and demonstrated end-to-end production of macroscopic amounts of a DNA origami nanorods using a liter-scale stirred-tank bioreactor.

A fundamental milestone that needs to be achieved to use DNA origami in DD applications is the understanding and modelling of how DNA origami self-assemble. This assembly is currently being investigated by several groups [85, 174] using simulations of nucleotide-level DNA models and more coarsened representations of DNA tiles [86], as well as theory [139]. Compared to classical schemes, the multicomponent nature of these systems challenges standard nucleation scenarios.

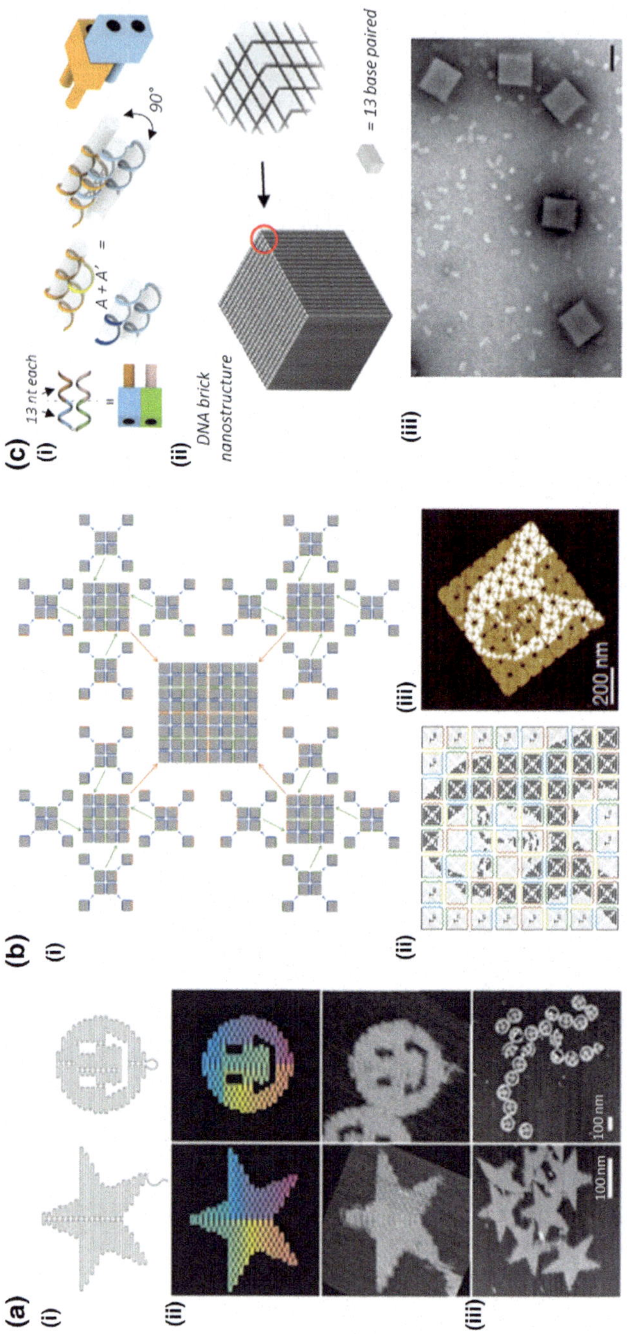
(a)
(i)
(ii)
(iii)
100 nm
100 nm
(b)
(i)
(ii)
(iii)
200 nm
(c)
(i)
13 nt each
A + A'
90°
(ii)
DNA brick nanostructure
= 13 base paired
(iii)

◀**Fig. 6** 2D and 3D DNA origami. Fig. **a**: folding of a 7-kilobases ssDNA using short ssDNA staples. (ii) Model of the folded backbone and (ii) colormap highlighting the route followed by the backbone red and purple label the first and last bases, respectively(iii) AFM images of the folded structures at two different magnifications. **b** DNA origami obtained via hierarchical self-assembly of DNA tiles fabricated using the procedure outlined in (**a**). (i) Three-stage self-assembly process that generates an 8×8 array of tiles. Blue, green and red edges (shown by arrows) are used to self-assemble arrays made of 2×2, 4×4, and 8×8 tiles, respectively. Panel (ii) and (iii) report a model and an AFM image of an 8×8 array in which the tiles form a prescribed puzzle. **c** Three-dimensional nanostructures self-assembled from DNA bricks. (i) Each brick is constituted by 52-mers composed of four 13-nt binding domains labelled as tails (blue and green), and heads (orange and pink). Each of these DNA bricks interacts with an adjacent brick through hybridization of one of its head domain with the complementary tail domain of another brick (A+A′ in blue and yellow). (ii) In this way, the formation of the 13-bp duplexes leads roughly to 1.25 turns of helix and to a 90° dihedral angle between two strands. All DNA bricks adopt an identical shape when incorporated into the desired structure. (iii) TEM Image of a 536 MDa brick cuboid (scale bar is 100 nm). Overall the figure shows how DNA origami technology allows assembling 2D (**a**-**b**) and 3D (**c**) nanomaterials with structural control at the single nucleotide level. *Credits* Figure **a**: adapted from [157]. Figure **b** adapted from [182] and Figure **c** adapted from [136]

4.3 Self-assembly of Hybrid Materials Driven by DNA Interactions

DNA has been largely used to drive the self-assembly of nanoparticles [89], soft vesicles [5, 145], and emulsions [90, 211]. DNA is easily conjugated with a wide variety of chemical functional groups that can be used to anchor the biopolymer to different types of surfaces [46, 117, 185].

Mirkin et al. [129] and Alivisatos et al. [3] developed a method to functionalize gold nanoparticles (NPs) with DNA oligonucleotides using thiol-gold bonds. Condensation of DNA coated NPs was driven by an oligonucleotide cross-linking pairs of particles through DNA hybridization of complementary sticky ends [129]. This aggregation process is reversible and mainly controlled by temperature. Since these works, many implementations of DNA functionalized gold NPs have been used to assemble composite materials [60, 134, 167, 213], featuring peculiar optical [55, 62, 178] and electrical properties [11, 23, 32, 41, 117, 175, 179, 199].

Extensive works to understand the design principles controlling the self-assembly of particles functionalized by sticky ends were performed. For example, Park et al. [141] have demonstrated that it is possible to build distinct and predictable crystal structures by tuning the size of the particles, the oligonucleotide sequence and length, as well as the degree of functionalization. Design principles were also provided by Macfarlane et al. [118] who were capable of self-assembling nine distinct types of crystal lattices, demonstrating the high degree of designability that can be achieved when using DNA [89].

Maye et al. [124] demonstrated the possibility of using DNA to direct the self-assembly of microparticles. The authors used both gold NPs, functionalized exploiting the thiol chemistry, and carboxylated polystyrene colloidal microspheres, functionalized with an amine modified ssDNA. They proposed a general route for regulating both micro- and nanoparticle self-assembly based on the fine balancing of the attractive forces generated by DNA hybridization with the steric repulsion provided by non-complementary DNA [135, 201, 210].

The biotin-avidin complex has also been largely used to functionalize microparticles with DNA. The most common strategy uses biotinylated DNA strands attached to avidinated surfaces [188]. For instance, functionalized microparticles [56, 188] were used to develop systematic experiments to understand key parameters controlling the self-assembly of DNA coated colloids. In particular, Dreyfus et al. [56] used polystyrene beads coated with three different types of ssDNA: two families of complementary sequences and an inert strand used to counterbalance the amount of grafted sticky strands, Fig. 7a(i). The authors confirmed that self-assembly is mainly controlled by temperature, Fig. 7a(ii), and clarified how both the configurational volume available to bound particles and the configurational entropy cost experienced by bound strands, see Sect. 3, are fundamental to predict the dissociation temperature.

More recently, Wang et al. developed a method employing DNA to engineer colloids featuring directional interactions [191]. The authors assembled particles presenting chemically distinct surface patches that mimic hybridized atomic orbitals, Fig. 7b. They functionalized the patches with DNA carrying sticky ends and forming highly directional, inter-particles bonds. These features allow the particles to self-assemble into structures with, for instance, triangular or tetrahedral symmetry, and should give access to a rich variety of new micro-structured materials [96]. Wang et al. [194] developed a new grafting strategy that led to higher functionalization densities using an alkyne-azide cycloaddition reaction which covalently grafts the DNA strands to the particle's surface. These DNA-coated colloids, having a higher degree of functionalization, can rearrange while annealing defects, enabling the growth of large crystals from of micrometer-sized DNA-coated colloids [195]. Other works that have investigated the fabrication of colloidal materials, such as colloidal crystals with diamond symmetry [193], mixtures of DNA-functionalized colloids bi-dispersed in size [128], DNA-mediated colloidal crystals [96] demonstrate that this research field is very proficient.

Recently, Angioletti-Uberti et al. [8] and Rogers and Manoharan [155] have demonstrated that it is possible to create colloidal systems that have the peculiarity of melting upon cooling. This inverse thermic behaviour can be achieved by designing sticky end sequences leading to competition between intra-particle loops and inter-particles bridges [8] or by DNA strand displacement reactions [70].

DNA functionalization can also be performed on non-solid particles, like oil droplets [90, 211] and lipid membranes [5, 17, 143, 171]. The importance of engineering fluidic particles is of fundamental importance in drug delivery applications. Indeed, for liquid particles, the whole interior volume can be loaded with the drug to be delivered, while, for solid particles, only the surface can be functionalised with the desired molecule. Moreover, lipid vesicles, such as liposomes, have been extensively used in DDS [4], mainly because their surface can in principle fuse with the cell membrane of the target cell. For these reasons, a systematic study of soft matter systems miming real cells is propaedeutic in understanding the behaviours of drug carriers in biological environments. An example is proposed in the work of Dave et al. [48]. The authors characterized DNA-functionalized liposomes as a function of lipid bilayer composition, DNA and suspension characteristics, such as size, charge, DNA length, DNA sequence and salt concentration, Fig. 7c. They highlighted that, while there are many similarities with the way solid particles interact, some important differences exist. For instance, in soft-particle systems, the inter-particle separation is not affecting the aggregation temperature, on the contrary to what is observed for solid particles, Fig. 7c(ii). The authors claimed that the control of "soft" systems is likely easier to achieve because these systems are less influenced by the specificities of both the lipid membrane and the linkers.

Others works involving interactions between lipid particles have enlightened how the membrane deformability plays a major role in multivalent recognition schemes [5, 13, 171]. Importantly, "soft" systems have been used to implement strand-displacement schemes, as we have already seen for solid colloids [70], that allow engineering the self-assembly kinetics [143].

4.4 Using DNA to Drive Selective DDS

The knowledge acquired in the control and programming of the bottom-up fabrication of customised structures using DNA has been employed to assemble scaffolds for drug delivery purposes [34, 83, 213]. The advantages of using DNA in biomedical applications rely on its biocompatibility as well as on the possibility to finely tune, for instance, the size of the delivery vector, which has been shown to be an important parameter in DD applications. Three years after the appearance of the first origami, Andersen et al. [6] proposed the first proof-of-principle of a DNA drug carrier. The authors engineered a square box, with tunable size and a foldable face (lid) allowing the internalisation of the drug, Fig. 8a(i) and (ii). The authors proved that it is possible to open the lid in a controlled way through the use of DNA "keys", Fig. 8a(iii). These carriers could be programmed to release their cargo remotely when triggered by specific conditions.

Another proof of the possibility of using DNA origami in drug delivery was provided by Jiang et al. [88], Fig. 8b. They used well-characterised 2D and 3D DNA origami resembling triangular structures [157] to deliver a well-known anti-cancer drug, Doxorubicin. Doxorubicin is an intercalating agent that can be loaded in the double helix structure with an efficiency of 50–60%. They also proved that their DNA origami structure is not cytotoxic, increases cell uptake, and circumvent cellular drug resistance, Fig. 8b(iii) and (iv).

Recently Banga et al. used DNA to develop carriers featuring selective properties [16]. They developed spherical nucleic acid nanostructures (SNAs), without the use of inorganic building-blocks, as nanoparticles, in contrast with most SNA systems [45]. The SNAs used in Banga et al. [16] consist of 30 nm liposomes functionalized by DNA strands modified with a tocopherol tail that intercalates into the phospholipid layer via hydrophobic effect, Fig. 8c. The authors demonstrated that the spherical nucleic acid architecture is a more reliable delivery construct than small unilamellar

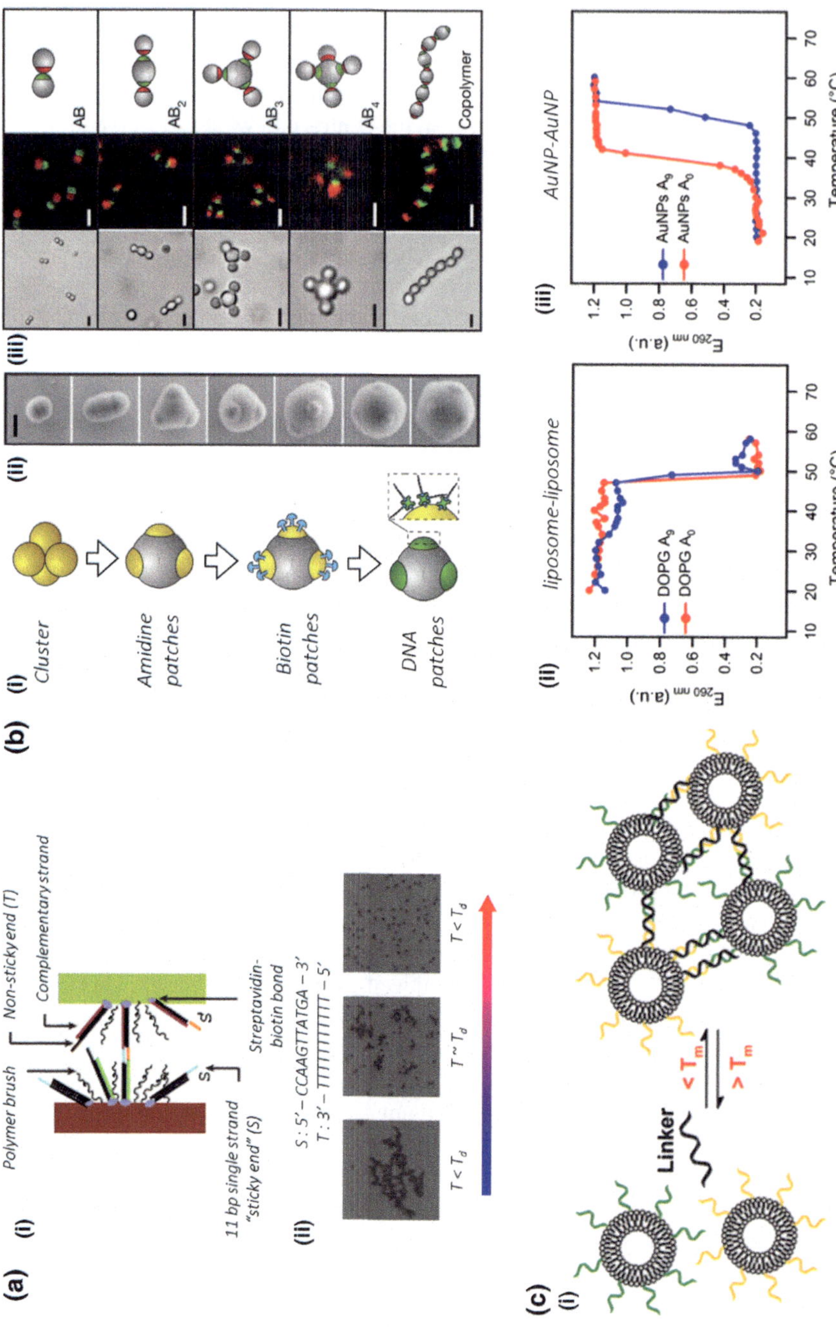
(a)
(i)
Polymer brush
Non-sticky end (T)
Complementary strand
11 bp single strand "sticky end" (S)
Streptavidin-biotin bond
(ii)
S : 5′ – CCAAGTTATGA – 3′
T : 3′ – TTTTTTTTTTT – 5′
T < T_d
T ~ T_d
T < T_d
(b)
(i)
Cluster
Amidine patches
Biotin patches
DNA patches
(ii)
(iii)
AB
AB2
AB3
AB4
Copolymer
(c)
(i)
Linker
< Tm
> Tm
(ii)
liposome-liposome
DOPG A9
DOPG A0
E260 nm (a.u.)
Temperature (°C)
(iii)
AuNP-AuNP
AuNPs A9
AuNPs A0
E260 nm (a.u.)
Temperature (°C)

◂**Fig. 7** DNA hybridization drives the self-assembly of functionalized colloids. **a** Dreyfus et al. [56] used two micrometre-sized particles coated with a mixture of sticky (S) and non-sticky (T) DNA strands grafted to their surface via streptavidin-biotin complex formation (i). At temperatures below the dissociation temperature (T_d) the colloids aggregate (ii). **b** Patchy micro-particles fabrication procedure. (i) A cluster of four amidinated polystyrene microspheres is glued by polymerized styrene that encapsulates the central part of the cluster leaving free the four extremities (patches). Directional interactions can be engineered by functionalizing the patches by streptavidin and biotinylated DNA as described in (**a**). Variations in the protocol allow engineering different types of patchy particles (ii). Directionality of particle-particle interactions can be probed using optical microscope (iii). Bright-field confocal fluorescence images and schematic representations show colloidal molecules self-assembled from patchy particles. According to the number of patches and their complementarity, monovalent, divalent, trivalent, tetravalent, and higher order assemblies can form. **c** DNA-functionalized DOPG liposomes assembled via DNA linkers. Liposomes self-assemble below their melting temperature (T_m) (i). The assembly of this system is compared to that of a system made of gold nanoparticles functionalized with the same DNA sequences. In particular, the melting curves of DOPG liposomes (ii) and gold NPs (iii) are reported in the presence or not of a spacer made of 9 adenine nucleobases (A9). It can be noticed that the assembly temperature of liposomes is not influenced by the presence of the spacer, in contrast with what was reported for gold NPs. *Credits* Figure **a** adapted from [56] copyright (2010) by the American Physical Society. Figure **b** adapted from [191] and Figure **c** adapted from [48]

vesicles: SNAs maintain their size upon heating, show lower degradation over time and, moreover, facilitate cellular internalization and gene regulation in SKOV-3 cells, a human ovarian cancer cell line with epithelial-like morphology, Fig. 8c(ii) and (iii). Liposomal cores are advantageous because those are biocompatible and maintain many of the specificities of gold nanoparticle-based SNAs.

5 Selective Targeting Using DNA Aptamers

Targeting using DNA conjugated nanovectors is not restricted to surfaces presenting the complementary DNA strands but can be extended to other cell surface receptors (like membrane bound proteins) thanks to the use of DNA aptamers. DNA aptamers are single-stranded DNA sequences (usually made of up to 40 nucleotides) featuring selective targeting proprieties resulting from peculiar secondary structures. DNA aptamers, often called "nucleic acids antibodies" [121], can be isolated and exploited to target specific receptors expressed by the surface of different biological entities [184]. These DNA constructs can be obtained through repeated rounds of in vitro selection and amplification using a SELEX, systematic evolution of ligands by exponential enrichment, procedure. This process starts with the synthesis of a wide oligonucleotide library made of randomly generated sequences of fixed length (of approximately 100 nucleotides) with conserved sequences at the 5′ and 3′ terminals that serve as primers. These randomly generated sequences are then selected on the base of their ability or not to bind to the target. The selected sequences are then amplified by polymerase chain reaction (PCR) and further screened in subsequent rounds of selection with less favourable binding conditions. Affinity chromatography

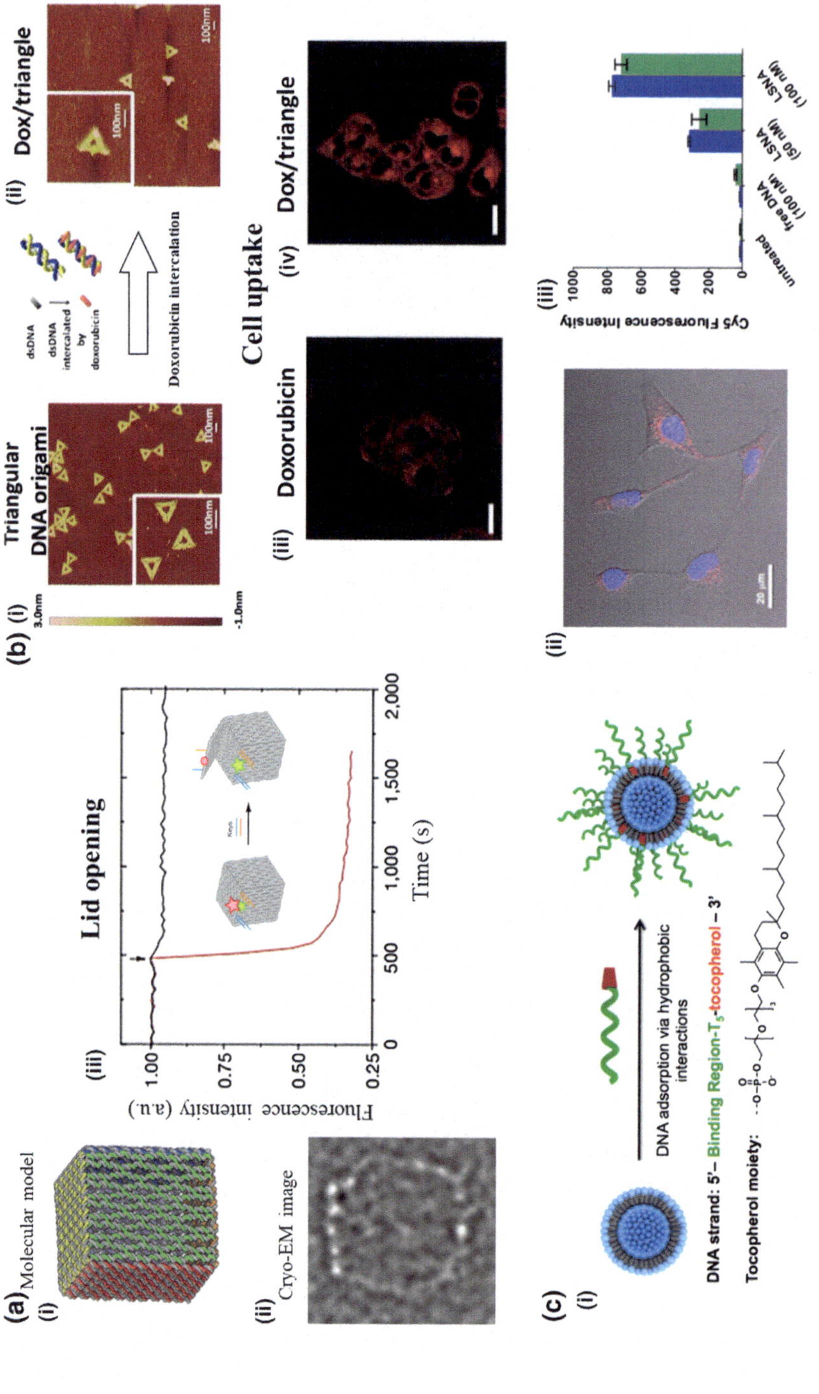
(a)
(i) Molecular model
(ii) Cryo-EM image
(iii) Lid opening
Fluorescence intensity (a.u.)
Time (s)
(b) (i) Triangular DNA origami
(ii) Dox/triangle
Doxorubicin intercalation
dsDNA
dsDNA intercalated by doxorubicin
Cell uptake
(iii) Doxorubicin
(iv) Dox/triangle
(c) (i)
DNA adsorption via hydrophobic interactions
DNA strand: 5'– Binding Region-T5-tocopherol – 3'
Tocopherol moiety:
(ii)
(iii)
Cy5 Fluorescence Intensity
untreated
free DNA (100 nM)
LSNA (50 nM)
LSNA (100 nM)

◂**Fig. 8** Drug carriers made of DNA. **a** Cubic DNA origami with a lid: (i) molecular model and (ii) cryo-EM image. (iii) Förster Resonance Energy Transfer (FRET) experiments confirm the opening of the lid driven by two ssDNA keys (highlighted in blue and orange). The black arrow indicates the time at which either the keys (red curve) or an unrelated oligonucleotide (black curve) are added to the solution. Opening of the lid is detected by a loss of emission from the fluorophore Cy5 due to the increase of the distance between fluorophores. **b** DNA origami loaded with Doxorubicin. After the folding of a single-stranded 7 kilobases genomic DNA into triangular shaped structures (i), the origami is loaded with Doxorubicin which intercalates into the double helix, as shown in the sketch. The origami structure is not modified by the presence of Doxorubicin (ii). Confocal images showing the degree of drug internalization by MCF 7 cells in vitro when delivered with (iv) and without (iii) the origami nanostructure. Scale bar of confocal images is 20 μm. **c** Spherical nucleic acid nanostructure (SNAs) presented in Banga et al. [16]. The SNA is formed by DOPC liposomes functionalized with DNA strands modified with a tocopherol tail immersed into the phospholipid layer (i). Confocal experiments show a high degree of internalization of SNA by SKOV-3 cells (ii). The fluorescence intensity after 1 h (blue bars) and 36 h (green bars) of incubation for untreated cells, cells treated with free DNA and with two different concentrations of loaded SNAs is reported in histogram (iii). *Credits* Figure **a** adapted from [6], Fig. **b** adapted from [88] and Figure **c** adapted from [16], further permissions should be directed to the ACS

is usually used to identify the optimal sequences [63, 186]. From its formulation in 1990 [186], SELEX was applied to find aptamer binding several biological entities, such as small molecules [15, 24, 30, 84], proteins [43, 57, 76], and even entire cells [33, 35, 36, 47, 78, 91, 165]. Although the molecular origin of aptamer-target binding is still not completely understood, electrostatic interactions should play an important role. Moreover, a key advantage of aptamers is the richness of secondary and tertiary structures that ensure isolating selective aptamers for a large variety of target elements [63, 72, 186]. Aptamers are largely used for in vivo applications because this class of ligands is non-immunogenic and stable against wide changes in pH (4–9), temperature, and in different organic solvents without loss of activity [65]. Furthermore, their synthesis does not rely on biological systems and is an entirely synthetic process, which decreases batch-to-batch variability when production is scaled up, an issue that is usually associated to the production of antibodies [37, 40]. Finally, the large variability of aptamer structures provides an almost infinite repertoire of docking molecules. The increasing interest in this targeting technology [29, 75, 99, 106, 160, 180], its proved reliability in vivo [2, 53, 64, 80, 102, 125] and in therapeutic applications [95, 108, 148, 181, 215] let us think that aptamers represent a promising strategy to achieve selective and, when conjugated to nanomaterials, super selective delivery in biological systems.

For example, Farokhzad et al. [64] have developed a Docetaxel (Dtxl) carrier made of biocompatible and biodegradable polymers (PLGA and PEG) functionalized with A10 2′-fluoropyrimidine RNA aptamers. This aptamer selectively adheres to the extracellular domain of the prostate-specific membrane antigen (PSMA), a well-characterized antigen expressed on the surface of prostate cancer cells (PCa) [10, 150]. The authors postulated that controlled release of Dtxl targeted to PCa cells would result in enhanced cytotoxicity and antitumor efficacy, making it a potential therapeutic method for the treatment of prostate cancer. The system of Farokhzard

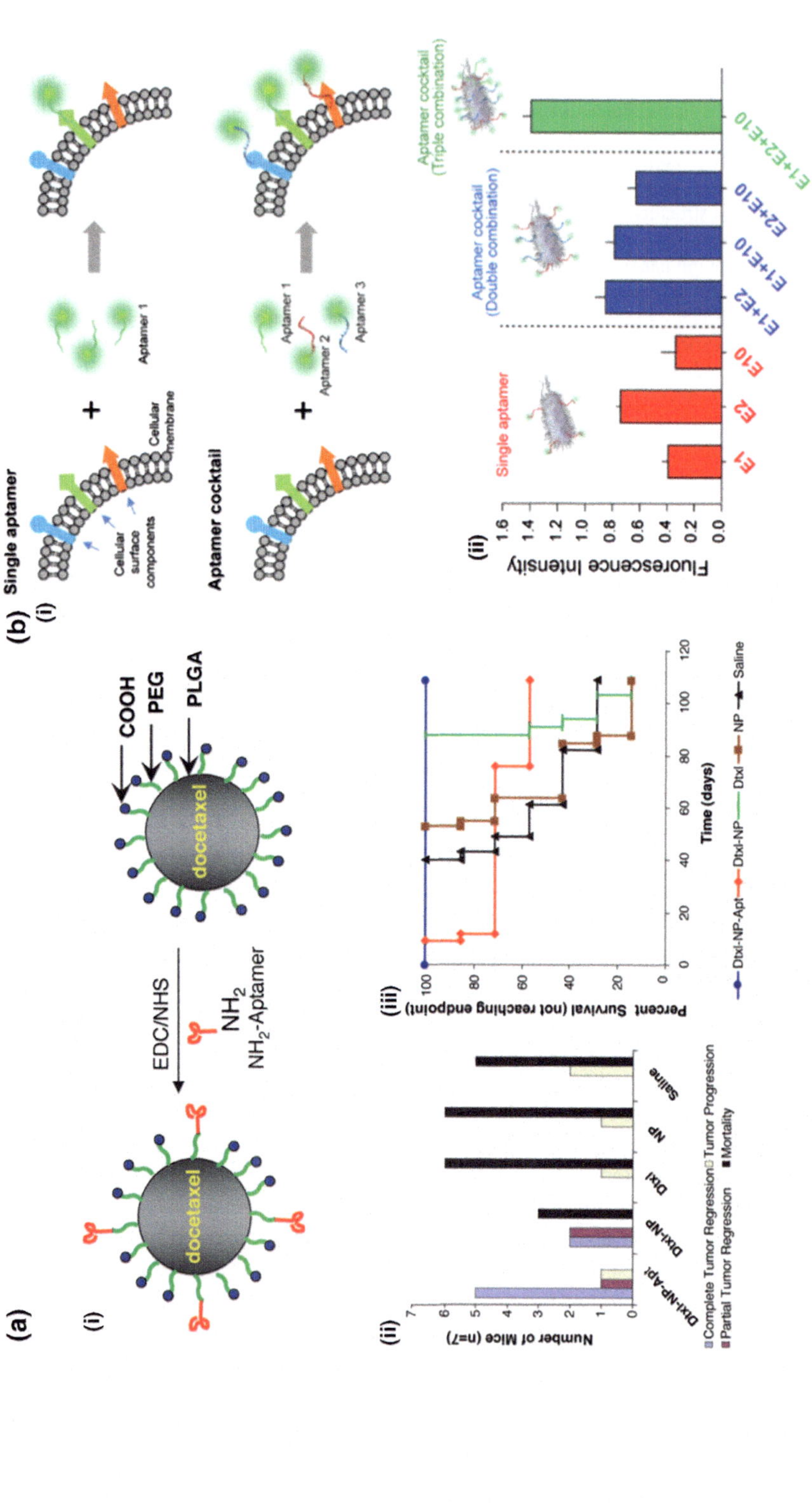
(a)
(i)
docetaxel
EDC/NHS
NH2
NH2-Aptamer
COOH
PEG
PLGA
(ii)
Number of Mice (n=7)
Dtxl-NP-Apt
Dtxl-NP
Dtxl
NP
Saline
Complete Tumor Regression
Tumor Progression
Partial Tumor Regression
Mortality
(iii)
Percent Survival (not reaching endpoint)
Time (days)
Dtxl-NP-Apt
Dtxl-NP
Dtxl
NP
Saline
(b)
(i)
Single aptamer
Cellular surface components
Cellular membrane
Aptamer 1
Aptamer cocktail
Aptamer 2
Aptamer 3
(ii)
Fluorescence Intensity
Single aptamer
Aptamer cocktail (Double combination)
Aptamer cocktail (Triple combination)
E1
E2
E10
E1+E2
E1+E10
E2+E10
E1+E2+E10

◀**Fig. 9** Aptamers targeting biological entities. **a** (i) Dtxl-encapsulated nanoparticles (NPs) functionalised with a preselected aptamer (Apt), as described in Farokhzad et al. [64]. The performance of this carrier (as compared to a control system, left) is studied in histogram (ii), reporting the outcomes of experiments on mice divided into four categories: complete tumour regression (blue), incomplete tumour regression (red), tumour growth (yellow), and mortality (black). (iii) Kaplan–Meier survival curves demonstrate that 100% of the individuals in the Dtxl-NP-Apt group were alive on day 109. **b** Simultaneous use of a mixture of three different aptamers enhances targeting of *E. coli* (KCTC 2571) [98]. Fluorescent experiment results (ii) show that the targeting efficiency is maximized when using multiple aptamers in the same batch. *Credits* Figure **a** adapted from [64] and Figure **b** adapted from [97]

et al. exhibited remarkable efficacy and reduced toxicity: after only a single intratumoural injection, a complete tumour reduction was observed in five out of seven treated mice, and 100% of these animals survived the 109-day study, Fig. 9a. This example testifies to the high efficiency and reliability of aptamers as a new class of targeting elements.

Vectors functionalized with mixtures of ligands are expected to enhance selectivity [215]. The possibility of targeting pathogens using multiple aptamers has been proven by Kim et al. [97]. The authors used three different combinations of DNA aptamers independently isolated to target *E. coli*. Their results proved that the sensitivity and detection limits were increased compared to the case in which the recognition scheme employed a single type of aptamer, Fig. 9b.

Multiple families of ligands have been used to design systems of functionalized nanoparticles featuring complex phase behaviours [127]. In the future, such knowledge will be useful to develop new selective schemes [44]. It is worth mentioning that, as compared to other molecules, DNA allows engineering systems in which, for instance, inter-particle and intra-particle linkages compete [130, 143]. Successful implementation of these techniques for targeting purposes will boost current selective designs beyond the ability to detect specific receptor densities.

6 Conclusions

The development of smart DDS able to selectively target diseased cells and deliver a controlled amount of one (multiple) drug(s) upon an external stimulus would represent a dramatic development for nanomedicine. The road to this goal is however still paved with many challenges. One of them is the elaboration of selective strategies that would allow targeting selected cells among a large variety of closely related ones. Several theoretical and computational works have demonstrated that superselectivity, the ability to recognise sharply cells featuring different receptor concentrations, can be achieved using multivalent interactions. Indeed, by using a multitude of ligand-receptor interactions to program the adhesion of a nanovector to a biological membrane, it is possible to target cells expressing a specific receptor, as conventional strategies, but only above a certain threshold of surface density.

The ability to sort cells expressing different concentrations of receptors is not the only possible selective scheme. Recently, Curk et al. [44] have theoretically predicted that vectors carrying different types of ligands could be used to achieve selective targeting with respect to relative stoichiometric concentrations of two types of receptors at constant coating density. This methodology provides a promising route to solve drug delivery specificity problems, allowing targeting of cells that do not express dominant markers.

Although we recently witnessed advances in the modelling of multivalent interactions, many challenges are still ahead of us. The experimental validation of superselective models remains challenging, mainly due to the difficulty of controlling the design parameters, such as ligand and receptor densities, the stability of the ligand/receptor complexes or the ligand flexibility. Understanding how non-selective interactions affect superselective schemes is also an important task that deserves future investigations.

DNA-based nanotechnology could offer a solution to the needs of controlling nanomaterials at the molecular level. In particular, DNA can be used to control the properties of engineered nanosystems, allowing the precise programming of their 3D structure, thanks to the strict DNA pairing rules and double helix structural characteristics. In addition to predictable structures and intermolecular interactions, other advantages, as convenient automated chemical synthesis, availability of well-known modifying enzymes and high functional group density, have led to the use of DNA nanotechnology in material sciences and biotechnologies, but also in healthcare sciences [83]. DNA also allows the design of systems featuring competing linkers [8, 130, 143, 155], suggesting that DNA could lead to new selective schemes. Furthermore, SELEX procedures allow isolating aptamers, short nucleic acids sequences presenting a high affinity and selectivity for a specific target, ranging from small molecules to full organisms. We believe that bridging the knowledge acquired in the fields of thermodynamics of ligand/receptor interactions with the latest outreaches in DNA programmed self-assembly and aptamer isolation is a promising path towards the developments of new selective DDS.

Acknowledgements The authors acknowledge The Wiener-Anspach foundation for financial support. BMM was supported by the Fonds de la Recherche Scientifique de Belgique—FNRS under grant n° MIS F.4534.17.

References

1. Abel, G.: Current status and future prospects of point-of-care testing around the globe. Expert Rev. Mol. Diagn. **15**, 853–855 (2015). https://doi.org/10.1586/14737159.2015.1060126
2. Alibolandi, M., Ramezani, M., Abnous, K., et al.: In vitro and in vivo evaluation of therapy targeting epithelial-cell adhesion-molecule aptamers for non-small cell lung cancer. J. Controlled Release **209**, 88–100 (2015). https://doi.org/10.1016/j.jconrel.2015.04.026
3. Alivisatos, A.P., Johnsson, K.P., Peng, X., et al.: Organization of "nanocrystal molecules" using DNA. Nature **382**, 609–611 (1996). https://doi.org/10.1038/382609a0

4. Allen, T.M., Cullis, P.R.: Liposomal drug delivery systems: from concept to clinical applications. Adv. Drug Deliv. Rev. **65**, 36–48 (2013). https://doi.org/10.1016/j.addr.2012.09.037
5. Amjad, O.A., Mognetti, B.M., Cicuta, P., Di Michele, L.: Membrane adhesion through bridging by multimeric ligands. Langmuir **33**, 1139–1146 (2017). https://doi.org/10.1021/acs.langmuir.6b03692
6. Andersen, E.S., Dong, M., Nielsen, M.M., et al.: Self-assembly of a nanoscale DNA box with a controllable lid. Nature **459**, 73–76 (2009). https://doi.org/10.1038/nature07971
7. Angioletti-Uberti, S.: Exploiting receptor competition to enhance nanoparticle binding selectivity. Phys. Rev. Lett. **118**, 1–5 (2017). https://doi.org/10.1103/PhysRevLett.118.068001
8. Angioletti-Uberti, S., Mognetti, B.M., Frenkel, D.: Re-entrant melting as a design principle for DNA-coated colloids. Nat. Mater. **11**, 518–522 (2012). https://doi.org/10.1038/nmat3314
9. Angioletti-Uberti, S., Varilly, P., Mognetti, B.M., et al.: Communication: a simple analytical formula for the free energy of ligand-receptor-mediated interactions. J. Chem. Phys. **138**, 0211021–0211024 (2013). https://doi.org/10.1063/1.4775806
10. Arundhati, G., Heston, W.D.W.: Tumor target prostate specific membrane antigen (PSMA) and its regulation in prostate cancer. J. Cell. Biochem. **91**, 528–539 (2003). https://doi.org/10.1002/jcb.10661
11. Auyeung, E., Li, T.I.N.G., Senesi, A.J., et al.: DNA-mediated nanoparticle crystallization into Wulff polyhedra. Nature **505**, 73–77 (2014). https://doi.org/10.1038/nature12739
12. Baaske, M.D., Foreman, M.R., Vollmer, F.: Single-molecule nucleic acid interactions monitored on a label-free microcavity biosensor platform. Nat. Nanotechnol. **9**, 933–939 (2014). https://doi.org/10.1038/nnano.2014.180
13. Bachmann, S.J., Kotar, J., Parolini, L., et al.: Melting transition in lipid vesicles functionalised by mobile DNA linkers. Soft Matter **12**, 7804–7817 (2016). https://doi.org/10.1039/C6SM01515H
14. Bachmann, S.J., Petitzon, M., Mognetti, B.M.: Bond formation kinetics affects self-assembly directed by ligand–receptor interactions. Soft Matter **12**, 9585–9592 (2016). https://doi.org/10.1039/C6SM02016J
15. Baker, B.R., Lai, R.Y., Wood, M.S., et al.: An electronic, aptamer-based small-molecule sensor for the rapid, label-free detection of cocaine in adulterated samples and biological fluids. J. Am. Chem. Soc. **128**, 3138–3139 (2006). https://doi.org/10.1021/ja056957p
16. Banga, R.J., Chernyak, N., Narayan, S.P., et al.: Liposomal spherical nucleic acids. J. Am. Chem. Soc. **136**(28), 9866–9869 (2014). https://doi.org/10.1021/ja504845f
17. Beales, P.A., Vanderlick, T.K.: Partitioning of membrane-anchored DNA between coexisting lipid phases. J. Phys. Chem. B **113**, 13678–13686 (2009). https://doi.org/10.1021/jp9006735
18. Bell, G.I., Dembo, M., Bongrand, P.: Cell adhesion. Competition between nonspecific repulsion and specific bonding. Biophys. J. **45**, 1051–1064 (1984). https://doi.org/10.1016/S0006-3495(84)84252-6
19. Bell, N.A.W., Keyser, U.F.: Nanopores formed by DNA origami: a review. FEBS Lett. **588**, 3564–3570 (2014). https://doi.org/10.1016/j.febslet.2014.06.013
20. Biffi, S., Cerbino, R., Bomboi, F., et al.: Phase behavior and critical activated dynamics of limited-valence DNA nanostars. Proc. Natl. Acad. Sci. **110**, 15633–15637 (2013). https://doi.org/10.1073/pnas.1304632110
21. Biffi, S., Cerbino, R., Nava, G., et al.: Equilibrium gels of low-valence DNA nanostars: a colloidal model for strong glass formers. Soft Matter **11**, 3132–3138 (2015). https://doi.org/10.1039/C4SM02144D
22. Björnmalm, M., Thurecht, K.J., Michael, M., et al.: Bridging bio-nano science and cancer nanomedicine. ACS Nano **11**, 9594–9613 (2017). https://doi.org/10.1021/acsnano.7b04855
23. Boal, A.K., Ilhan, F., Derouchey, J.E., et al.: Self-assembly of nanoparticles into structured spherical and network aggregates. Nature **404**, 746–748 (2000). https://doi.org/10.1038/35008037
24. Bock, L.C., Griffin, L.C., Latham, J.A., et al.: Selection of single-stranded DNA molecules that bind and inhibit human thrombin. Nature **355**, 564–566 (1992). https://doi.org/10.1038/355564a0

25. Boltasseva, A., Shalaev, V.M.: Fabrication of optical negative-index metamaterials: recent advances and outlook. Metamaterials **2**, 1–17 (2008). https://doi.org/10.1016/j.metmat.2008.03.004
26. Bomboi, F., Biffi, S., Cerbino, R., et al.: Equilibrium gels of trivalent DNA-nanostars: effect of the ionic strength on the dynamics. Eur. Phys. J. E **38**, 64–69 (2015). https://doi.org/10.1140/epje/i2015-15064-9
27. Bomboi, F., Romano, F., Leo, M., et al.: Re-entrant DNA gels. Nat. Commun. **7**, 1–6 (2016). https://doi.org/10.1038/ncomms13191
28. Brady, R.A., Brooks, N.J., Cicuta, P., Di Michele, L.: Crystallization of amphiphilic DNA C-Stars. Nano Lett. **17**(5), 3276–3281 (2017). https://doi.org/10.1021/acs.nanolett.7b00980
29. Bunka, D.H.J., Stockley, P.G.: Aptamers come of age—at last. Nat. Rev. Microbiol. **4**, 588–596 (2006). https://doi.org/10.1038/nrmicro1458
30. Burke, D.H., Gold, L.: RNA aptamers to the adenosine moiety of S-adenosyl methionine: structural inferences from variations on a theme and the reproducibility of SELEX. Nucleic Acids Res. **25**, 2020–2024 (1997)
31. Cangialosi, A., Yoon, C.K., Liu, J., et al.: DNA sequence–directed shape change of photopatterned hydrogels via high-degree swelling. Science **357**, 1126–1130 (2017). https://doi.org/10.1126/science.aan3925
32. Cao, Y.W., Jin, R., Mirkin, C.A.: DNA-modified core–shell Ag/Au nanoparticles. J. Am. Chem. Soc. **123**, 7961–7962 (2001). https://doi.org/10.1021/ja011342n
33. Cao, X., Li, S., Chen, L., et al.: Combining use of a panel of ssDNA aptamers in the detection of Staphylococcus aureus. Nucleic Acids Res. **37**, 4621–4628 (2009). https://doi.org/10.1093/nar/gkp489
34. Castro, C.E., Kilchherr, F., Kim, D.N., et al.: A primer to scaffolded DNA origami. Nat. Methods **8**, 221–229 (2011). https://doi.org/10.1038/nmeth.1570
35. Cerchia, L., de Franciscis, V.: Targeting cancer cells with nucleic acid aptamers. Trends Biotechnol. **28**, 517–525 (2018). https://doi.org/10.1016/j.tibtech.2010.07.005
36. Chang, Y.-C., Yang, C.-Y., Sun, R.-L., et al.: Rapid single cell detection of Staphylococcus aureus by aptamer-conjugated gold nanoparticles. Sci. Rep. **3**, 1863–1870 (2013). https://doi.org/10.1038/srep01863
37. Cheifetz, A., Mayer, L.: Monoclonal antibodies, immunogenicity, and associated infusion reactions. Mt. Sinai J. Med. **72**, 250–256 (2005)
38. Chen, H.Y.: Adhesion-induced phase separation of multiple species of membrane junctions. Phys. Rev. E **67**, 1–10 (2003). https://doi.org/10.1103/PhysRevE.67.031919
39. Chen, Y.J., Groves, B., Muscat, R.A., Seelig, G.: DNA nanotechnology from the test tube to the cell. Nat. Nanotechnol. **10**, 748–760 (2015). https://doi.org/10.1038/nnano.2015.195
40. Chester, K., Pedley, B., Tolner, B., et al.: Engineering antibodies for clinical applications in cancer. Tumor Biol **25**, 91–98 (2004). https://doi.org/10.1159/000077727
41. Collier, C.P., Vossmeyer, T., Heath, J.R.: Nanocrystal superlattices. Annu. Rev. Phys. Chem. **49**, 371–404 (1998). https://doi.org/10.1146/annurev.physchem.49.1.371
42. Coombs, D., Dembo, M., Wofsy, C., Goldstein, B.: Equilibrium thermodynamics of cell-cell adhesion mediated by multiple ligand-receptor pairs. Biophys. J. **86**, 1408–1423 (2004). https://doi.org/10.1016/S0006-3495(04)74211-3
43. Cox, J.C., Hayhurst, A., Hesselberth, J., et al.: Automated selection of aptamers against protein targets translated in vitro: from gene to aptamer. Nucleic Acids Res. **30**, e108–e108 PMC137152 (2002)
44. Curk, T., Dobnikar, J., Frenkel, D.: Optimal multivalent targeting of membranes with many distinct receptors. Proc. Natl. Acad. Sci. **114**, 7210–7215 (2017). https://doi.org/10.1073/pnas.1704226114
45. Cutler, J.I., Auyeung, E., Mirkin, C.A.: Spherical nucleic acids. J. Am. Chem. Soc. **134**, 1376–1391 (2012). https://doi.org/10.1021/ja209351u
46. Daniel, M.-C., Astruc, D.: Gold nanoparticles: assembly, supramolecular chemistry, quantum-size-related properties, and applications toward biology, catalysis, and nanotechnology. Chem. Rev. **104**, 293–346 (2004). https://doi.org/10.1021/cr030698+

47. Daniels, D.A., Chen, H., Hicke, B.J., et al.: A tenascin-C aptamer identified by tumor cell SELEX: systematic evolution of ligands by exponential enrichment. Proc. Natl. Acad. Sci. U. S. A. **100**, 15416–15421 (2003). https://doi.org/10.1073/pnas.2136683100
48. Dave, N., Liu, J.: Programmable assembly of DNA-functionalized liposomes by DNA. ACS Nano **5**, 1304–1312 (2011). https://doi.org/10.1021/nn1030093
49. de Gennes, P.G.: The Physics of Liquid Crystals. Clarendon Press (1974)
50. Decuzzi, P., Ferrari, M.: The receptor-mediated endocytosis of nonspherical particles. Biophys. J. **94**, 3790–3797 (2008). https://doi.org/10.1529/biophysj.107.120238
51. Demetzos, C., Pippa, N.: Advanced drug delivery nanosystems (aDDnSs): a mini-review. Drug Deliv. **21**, 250–257 (2014). https://doi.org/10.3109/10717544.2013.844745
52. Di Michele, L., Jana, P.K., Mognetti, B.M.: Ligand mobility suppresses membrane wrapping in passive endocytosis (2017). ArXiv ID 1708.03733
53. Ding, F., Guo, S., Xie, M., et al.: Diagnostic applications of gastric carcinoma cell aptamers in vitro and in vivo. Talanta **134**, 30–36 (2015). https://doi.org/10.1016/j.talanta.2014.09.036
54. Dirks, R.M., Bois, J.S., Schaeffer, J.M., et al.: Thermodynamic analysis of interacting nucleic acid strands. SIAM Rev **49**, 65–88 (2007). https://doi.org/10.1137/060651100
55. Doyen, M., Bartik, K., Bruylants, G.: UV–Vis and NMR study of the formation of gold nanoparticles by citrate reduction: observation of gold–citrate aggregates. J. Colloid Interface Sci. **399**, 1–5 (2013). https://doi.org/10.1016/j.jcis.2013.02.040
56. Dreyfus, R., Leunissen, M.E., Sha, R., et al.: Aggregation-disaggregation transition of DNA-coated colloids: experiments and theory. Phys. Rev. E Stat. Nonlinear Soft Matter Phys. **81**, 1–10 (2010). https://doi.org/10.1103/PhysRevE.81.041404
57. Dua, P., Kim, S., Lee, D.: Nucleic acid aptamers targeting cell-surface proteins. Methods **54**, 215–225 (2011). https://doi.org/10.1016/j.ymeth.2011.02.002
58. Dubacheva, G.V., Curk, T., Auzély-Velty, R., et al.: Designing multivalent probes for tunable superselective targeting. Proc. Natl. Acad. Sci. **112**, 5579–5584 (2015). https://doi.org/10.1073/pnas.1500622112
59. Dubacheva, G.V., Curk, T., Mognetti, B.M., et al.: Superselective targeting using multivalent polymers. J. Am. Chem. Soc. **136**, 1722–1725 (2014). https://doi.org/10.1021/ja411138s
60. Dubertret, B., Calame, M., Libchaber, A.J.: Single-mismatch detection using gold-quenched fluorescent oligonucleotides (vol. 19, p. 365, 2001). Nat. Biotechnol. **19**, 680–681 (2001). https://doi.org/10.1038/86762
61. Duncan, G.A., Bevan, M.A.: Computational design of nanoparticle drug delivery systems for selective targeting. Nanoscale **7**, 15332–15340 (2015). https://doi.org/10.1039/c5nr03691g
62. Elghanian, R.: Selective colorimetric detection of polynucleotides based on the distance-dependent optical properties of gold nanoparticles. Science **277**, 1078–1081 (1997). https://doi.org/10.1126/science.277.5329.1078
63. Ellington, A.D., Szostak, J.W.: In vitro selection of RNA molecules that bind specific ligands. Nature **346**, 818–822 (1990). https://doi.org/10.1038/346818a0
64. Farokhzad, O.C., Cheng, J., Teply, B.A., et al.: Targeted nanoparticle-aptamer bioconjugates for cancer chemotherapy in vivo. Proc. Natl. Acad. Sci. **103**, 6315–6320 (2006). https://doi.org/10.1073/pnas.0601755103
65. Farokhzad, O.C., Karp, J.M., Langer, R.: Nanoparticle-aptamer bioconjugates for cancer targeting. Expert Opin. Drug Deliv. **3**, 311–324 (2006).
66. Fernandez-Castanon, J., Bomboi, F., Rovigatti, L., et al.: Small-angle neutron scattering and molecular dynamics structural study of gelling DNA nanostars. J. Chem. Phys. **145**(84910), 1–7 (2016). https://doi.org/10.1063/1.4961398
67. Fraccia, T.P., Smith, G.P., Bethge, L., et al.: Liquid crystal ordering and isotropic gelation in solutions of four-base-long DNA oligomers. ACS Nano **10**, 8508–8516 (2016). https://doi.org/10.1021/acsnano.6b03622
68. Fraccia, T.P., Smith, G.P., Zanchetta, G., et al.: Abiotic ligation of DNA oligomers templated by their liquid crystal ordering. Nat. Commun. **6**, 1–7 (2015). https://doi.org/10.1038/ncomms7424

69. Gao, H., Shi, W., Freund, L.B.: Mechanics of receptor-mediated endocytosis. Proc. Natl. Acad. Sci. U. S. A. **102**, 9469–9474 (2005). https://doi.org/10.1073/pnas.0503879102
70. Gehrels, E.W., Rogers, W.B., Manoharan, V.N.: Using DNA strand displacement to control interactions in DNA-grafted colloids. Soft Matter **14**, 969–984 (2018). https://doi.org/10.1039/C7SM01722G
71. Gestwicki, J.E., Cairo, C.W., Strong, L.E., et al.: Influencing receptor-ligand binding mechanisms with multivalent ligand architecture. J. Am. Chem. Soc. **124**, 14922–14933 (2002). https://doi.org/10.1021/ja027184x
72. Gold, L., Janjic, N., Jarvis, T., et al.: Aptamers and the RNA world, past and present. Cold Spring Harb. Perspect. Biol. **4**, 1–11 (2012). https://doi.org/10.1101/cshperspect.a003582
73. Gothelf, K.V.: LEGO-like DNA structures. Science **338**, 1159–1160 (2012). https://doi.org/10.1126/science.1229960
74. Grzybowski, B.A., Huck, W.T.S.: The nanotechnology of life-inspired systems. Nat. Nanotechnol. **11**, 585–592 (2016). https://doi.org/10.1038/nnano.2016.116
75. Günter, M.: The chemical biology of aptamers. Angew. Chem. Int. Ed. **48**, 2672–2689 (2009). https://doi.org/10.1002/anie.200804643
76. Hamaguchi, N., Ellington, A., Stanton, M.: Aptamer beacons for the direct detection of proteins. Anal. Biochem. **294**, 126–131 (2001). https://doi.org/10.1006/abio.2001.5169
77. Han, D., Pal, S., Nangreave, J., et al.: DNA origami with complex curvatures in three-dimensional space. Science **332**, 342–346 (2011). https://doi.org/10.1126/science.1202998
78. Herr, J.K., Smith, J.E., Medley, C.D., et al.: Aptamer-conjugated nanoparticles for selective collection and detection of cancer cells. Anal. Chem. **78**, 2918–2924 (2006). https://doi.org/10.1021/ac052015r
79. Hohenberg, P.C., Halperin, B.I.: Theory of dynamic critical phenomena. Rev. Mod. Phys. **49**, 435 (1977). https://doi.org/10.1103/RevModPhys.49.435
80. Hongguang, S., Youli, Z.: Aptamers and their applications in nanomedicine. Small **11**, 2352–2364 (2015). https://doi.org/10.1002/smll.201403073
81. Houk, N.K., Leach, G.A., Kim, P.S., Zhang, X.: Binding affinities of host-guest, protein-ligand, and protein–transition-state complexes. Angew. Chem. Int. Ed. **42**, 4872–4897 (2003). https://doi.org/10.1002/anie.200200565
82. Hu, J., Lipowsky, R., Weikl, T.R.: Binding constants of membrane-anchored receptors and ligands depend strongly on the nanoscale roughness of membranes. Proc. Natl. Acad. Sci. **110**, 15283–15288 (2013). https://doi.org/10.1073/pnas.1305766110
83. Hu, Q., Li, H., Wang, L., et al.: DNA nanotechnology-enabled drug delivery systems. Chem. Rev. (2018). https://doi.org/10.1021/acs.chemrev.7b00663
84. Huizenga, D.E., Szostak, J.W.: A DNA aptamer that binds adenosine and ATP. Biochemistry **34**, 656–665 (1995). https://doi.org/10.1021/bi00002a033
85. Jabbari, H., Aminpour, M., Montemagno, C.: Computational approaches to nucleic acid origami. ACS Comb Sci **17**, 535–547 (2015). https://doi.org/10.1021/acscombsci.5b00079
86. Jacobs, W.M., Reinhardt, A., Frenkel, D.: Communication: Theoretical prediction of free-energy landscapes for complex self-assembly. J. Chem. Phys. **142**, 1–5 (2015). https://doi.org/10.1063/1.4905670
87. Jacobson, K., Ishihara, A., Inman, R.: Lateral diffusion of proteins in membranes. Annu. Rev. Physiol. **49**, 163–175 (1987). https://doi.org/10.1146/annurev.ph.49.030187.001115
88. Jiang, Q., Song, C., Nangreave, J., et al.: DNA origami as a carrier for circumvention of drug resistance. J. Am. Chem. Soc. **134**, 13396–13403 (2012). https://doi.org/10.1021/ja304263n
89. Jones, M.R., Seeman, N.C., Mirkin, C.A.: Programmable materials and the nature of the DNA bond. Science **347**, 1–11 (2015). https://doi.org/10.1126/science.1260901
90. Joshi, D., Bargteil, D., Caciagli, A., et al.: Kinetic control of the coverage of oil droplets by DNA-functionalized colloids. Sci. Adv. **2**, 1–9 (2016). https://doi.org/10.1126/sciadv.1600881
91. Joshi, R., Janagama, H., Dwivedi, H.P., et al.: Selection, characterization, and application of DNA aptamers for the capture and detection of Salmonella enterica serovars. Mol. Cell. Probes **23**, 20–28 (2009). https://doi.org/10.1016/j.mcp.2008.10.006

92. Ke, P.C., Lin, S., Parak, W.J., et al.: A decade of the protein corona. ACS Nano **11**, 11773–11776 (2017). https://doi.org/10.1021/acsnano.7b08008
93. Ke, Y., Ong, L.L., Shih, W.M., Yin, P.: Three-dimensional structures self-assembled from DNA bricks. Science **338**, 1177–1183 (2012). https://doi.org/10.1126/science.1227268
94. Ke, Y., Ong, L.L., Sun, W., et al.: DNA brick crystals with prescribed depths. Nat. Chem. **6**, 994–1002 (2014). https://doi.org/10.1038/nchem.2083
95. Keefe, A.D., Pai, S., Ellington, A.: Aptamers as therapeutics. Nat. Rev. Drug Discov. **9**, 537–550 (2010). https://doi.org/10.1038/nrd3141
96. Kim, A.J., Biancaniello, P.L., Crocker, J.C.: Engineering DNA-mediated colloidal crystallization. Langmuir **22**, 1991–2001 (2006). https://doi.org/10.1021/la0528955
97. Kim, Y.S., Chung, J., Song, M.Y., et al.: Aptamer cocktails: Enhancement of sensing signals compared to single use of aptamers for detection of bacteria. Biosens. Bioelectron. **54**, 195–198 (2014). https://doi.org/10.1016/j.bios.2013.11.003
98. Kim, Y.S., Song, M.Y., Jurng, J., Kim, B.C.: Isolation and characterization of DNA aptamers against *Escherichia coli* using a bacterial cell-systematic evolution of ligands by exponential enrichment approach. Anal. Biochem. **436**, 22–28 (2013). https://doi.org/10.1016/j.ab.2013.01.014
99. Kimoto, M., Yamashige, R., Matsunaga, K.I., et al.: Generation of high-affinity DNA aptamers using an expanded genetic alphabet. Nat. Biotechnol. **31**, 453–457 (2013). https://doi.org/10.1038/nbt.2556
100. Kitov, P.I., Bundle, D.R.: On the nature of the multivalency effect: a thermodynamic model. J. Am. Chem. Soc. **125**, 16271–16284 (2003). https://doi.org/10.1021/ja038223n
101. Kitov, P.I., Sadowska, J.M., Mulvey, G., et al.: Shiga-like toxins are neutralized by tailored multivalent carbohydrate ligands. Nature **403**, 669–672 (2000). https://doi.org/10.1038/35001095
102. Kryza, D., Debordeaux, F., Azéma, L., et al.: Ex vivo and in vivo imaging and biodistribution of aptamers targeting the human matrix metalloprotease-9 in melanomas. PLoS One **1**, 16. https://doi.org/10.1371/journal.pone.0149387 (February 22)
103. LaVan, D.A., McGuire, T., Langer, R.: Small-scale systems for in vivo drug delivery. Nat. Biotechnol. **21**, 1184–1191 (2003). https://doi.org/10.1038/nbt876
104. Lanfranco, R., Giavazzi, F., Salina, M., et al.: Selective adsorption on fluorinated plastic enables the optical detection of molecular pollutants in water. Phys. Rev. Appl. **5**, 1–15 (2016). https://doi.org/10.1103/PhysRevApplied.5.054012
105. Lanfranco, R., Saez, J., Di Nicolò, E., et al.: Phantom membrane microfluidic cross-flow filtration device for the direct optical detection of water pollutants. Sens. Actuators B Chem. **257**, 924–930 (2018). https://doi.org/10.1016/j.snb.2017.11.024
106. Lee, J.F.: Aptamer database. Nucleic Acids Res. **32**, 95D–100 (2004). https://doi.org/10.1093/nar/gkh094
107. Lee, J.B., Peng, S., Yang, D., et al.: A mechanical metamaterial made from a DNA hydrogel. Nat. Nanotechnol. **7**, 816–820 (2012). https://doi.org/10.1038/nnano.2012.211
108. Lee, J.F., Stovall, G.M., Ellington, A.D.: Aptamer therapeutics advance. Curr. Opin. Chem. Biol. **10**, 282–289 (2006). https://doi.org/10.1016/j.cbpa.2006.03.015
109. Li, D., Qu, L., Zhai, W., et al.: Facile on-site detection of substituted aromatic pollutants in water using thin layer chromatography combined with surface-enhanced Raman spectroscopy. Environ. Sci. Technol. **45**, 4046–4052 (2011). https://doi.org/10.1021/es104155r
110. Licata, N.A., Tkachenko, A.V.: Kinetic limitations of cooperativity-based drug delivery systems. Phys. Rev. Lett. **100**, 1–4 (2008). https://doi.org/10.1103/physrevlett.100.158102
111. Liu, H., He, Y., Ribbe, A.E., Mao, C.: Two-dimensional (2D) DNA crystals assembled from two DNA strands. Biomacromolecules **6**, 2943–2945 (2005). https://doi.org/10.1021/bm050632j
112. Liu, F., Sha, R., Seeman, N.C.: Modifying the surface features of two-dimensional DNA crystals. J. Am. Chem. Soc. **121**, 917–922 (1999). https://doi.org/10.1021/ja982824a
113. Liu, D., Wang, Z., Jiang, X.: Gold nanoparticles for the colorimetric and fluorescent detection of ions and small organic molecules. Nanoscale **3**, 1421–1433 (2011). https://doi.org/10.1039/c0nr00887g

114. Liu, Y., Zhang, X.: Metamaterials: a new frontier of science and technology. Chem. Soc. Rev. **40**, 2494–2507 (2011). https://doi.org/10.1039/C0CS00184H
115. Lo, P.K., Metera, K.L., Sleiman, H.F.: Self-assembly of three-dimensional DNA nanostructures and potential biological applications. Curr. Opin. Chem. Biol. **14**, 597–607 (2010). https://doi.org/10.1016/j.cbpa.2010.08.002
116. Lowe, J.N., Fulton, D.A., Chiu, S.H., et al.: Polyvalent interactions in unnatural recognition processes. J. Org. Chem. **69**, 4390–4402 (2004). https://doi.org/10.1021/jo030283o
117. Loweth, C.J., Caldwell, W.B., Peng, X., et al.: DNA-based assembly of gold nanocrystals. Angew. Chem. Int. Ed. **38**, 1808–1812 (1999). https://doi.org/10.1002/(SICI)1521-3773(19990614)38:12%3c1808:AID-ANIE1808%3e3.0.CO;2-C
118. Macfarlane, R.J., Lee, B., Jones, M.R., et al.: Nanoparticle superlattice engineering with DNA. Science **334**, 204–208 (2011). https://doi.org/10.1126/science.1210493
119. Mammen, M., Choi, S.K., Whitesides, G.M.: Polyvalent interactions in biological systems: implications for design and use of multivalent ligands and inhibitors. Angew. Chem. Int. Ed. **37**, 2754–2794 (1998). https://doi.org/10.1002/(sici)1521-3773(19981102)37:20%3c2754:aid-anie2754%3e3.0.co;2-3
120. Mammen, M., Dahmann, G., Whitesides, G.M.: Effective inhibitors of hemagglutination by influenza virus synthesized from polymers having active ester groups. Insight into mechanism of inhibition. J. Med. Chem. **38**, 4179–4190 (1995). https://doi.org/10.1021/jm00021a007
121. Marangoni, K., Neves, A.F., Rocha, R.M., et al.: Prostate-specific RNA aptamer: promising nucleic acid antibody-like cancer detection. Nat. Publ. Gr. **5**, 1–13 (2015). https://doi.org/10.1038/srep12090
122. Marchi, A.N., Saaem, I., Vogen, B.N., et al.: Toward larger DNA origami. Nano Lett. **14**, 5740–5747 (2014). https://doi.org/10.1021/nl502626s
123. Martinez-Veracoechea, F.J., Frenkel, D.: Designing super selectivity in multivalent nanoparticle binding. Proc. Natl. Acad. Sci. **108**, 10963–10968 (2011). https://doi.org/10.1073/pnas.1105351108
124. Maye, M.M., Nykypanchuk, D., van der Lelie, D., Gang, O.: DNA-regulated micro- and nanoparticle assembly. Small **3**, 1678–1682 (2007). https://doi.org/10.1002/smll.200700357
125. Mi, J., Ray, P., Liu, J., et al.: In vivo selection against human colorectal cancer xenografts identifies an aptamer that targets RNA helicase protein DHX9. Mol. Ther. Nucleic Acids **5**, e315 (2016). https://doi.org/10.1038/mtna.2016.27
126. Di Michele, L., Bachmann, S.J., Parolini, L., Mognetti, B.M.: Communication: free energy of ligand-receptor systems forming multimeric complexes. J. Chem. Phys. **144**, 1–5 (2016). https://doi.org/10.1063/1.4947550
127. Di Michele, L., Eiser, E.: Developments in understanding and controlling self assembly of DNA-functionalized colloids. Phys. Chem. Chem. Phys. **15**, 3115–3129 (2013). https://doi.org/10.1039/C3CP43841D
128. Milam, V.T., Hiddessen, A.L., Crocker, J.C., et al.: DNA-driven assembly of bidisperse, micron-sized colloids. Langmuir **19**, 10317–10323 (2003). https://doi.org/10.1021/la034376c
129. Mirkin, C.A., Letsinger, R.L., Mucic, R.C., Storhoff, J.J.: A DNA-based method for rationally assembling nanoparticles into macroscopic materials. Nature **382**, 607–609 (1996). https://doi.org/10.1038/382607a0
130. Mognetti, B.M., Leunissen, M.E., Frenkel, D.: Controlling the temperature sensitivity of DNA-mediated colloidal interactions through competing linkages. Soft Matter **8**, 2213–2221 (2012). https://doi.org/10.1039/C2SM06635A
131. Mognetti, B.M., Varilly, P., Angioletti-Uberti, S., et al.: Predicting DNA-mediated colloidal pair interactions. Proc. Natl. Acad. Sci. **109**(7), E378:E379 LP-E379 (2012b)
132. Nakata, M., Zanchetta, G., Chapman, B.D., et al.: End-to-end stacking and liquid crystal formation of 6- to 20-base pair DNA duplexes. Science **318**, 1–4 (2009). https://doi.org/10.1126/science.1143826
133. Narasimhan, B., Goodman, J.T., Vela Ramirez, J.E.: Rational design of targeted next-generation carriers for drug and vaccine delivery. Annu. Rev. Biomed. Eng. **18**, 25–49 (2016). https://doi.org/10.1146/annurev-bioeng-082615-030519

134. Niemeyer, C.M.: DNA as a material for nanotechnology. Angew Chemie Int Ed English **36**, 585–587 (2003). https://doi.org/10.1002/anie.199705851
135. Nykypanchuk, D., Maye, M.M., van der Lelie, D., Gang, O.: DNA-guided crystallisation of colloidal nanoparticles. Nature **451**, 549–552 (2008). https://doi.org/10.1038/nature06560
136. Ong, L.L., Hanikel, N., Yaghi, O.K., et al.: Programmable self-assembly of three-dimensional nanostructures from 10,000 unique components. Nature **552**, 72–77 (2017). https://doi.org/10.1038/nature24648
137. Orgel, L.E.: The origin of life on the earth. Sci. Am. **271**, 76–83 (1994)
138. Orgel, L.E.: The origin of life—a review of facts and speculations. Trends Biochem. Sci. **23**, 491–495 (2018). https://doi.org/10.1016/S0968-0004(98)01300-0
139. Ouldridge, T.E., Hoare, R.L., Louis, A.A., et al.: Optimizing DNA nanotechnology through coarse-grained modeling: a two-footed DNA walker. ACS Nano **7**, 2479–2490 (2013). https://doi.org/10.1021/nn3058483
140. Oyarzún, B., Mognetti, B.M.: Efficient sampling of reversible cross-linking polymers: self-assembly of single-chain polymeric nanoparticles. J. Chem. Phys. **148**, 114110 (2018). https://doi.org/10.1063/1.5020158
141. Park, S.Y., Lytton-Jean, A.K.R., Lee, B., et al.: DNA-programmable nanoparticle crystallization. Nature **451**, 553–556 (2008). https://doi.org/10.1038/nature06508
142. Park, N., Um, S.H., Funabashi, H., et al.: A cell-free protein-producing gel. Nat. Mater. **8**, 432–437 (2009). https://doi.org/10.1038/nmat2419
143. Parolini, L., Kotar, J., Di Michele, L., Mognetti, B.M.: Controlling self-assembly kinetics of DNA-functionalized liposomes using toehold exchange mechanism. ACS Nano **10**, 2392–2398 (2016). https://doi.org/10.1021/acsnano.5b07201
144. Paukstelis, P.J.: Three-dimensional DNA crystals as molecular sieves. J. Am. Chem. Soc. **128**, 6794–6795 (2006). https://doi.org/10.1021/ja061322r
145. Pontani, L.-L., Jorjadze, I., Viasnoff, V., Brujic, J.: Biomimetic emulsions reveal the effect of mechanical forces on cell–cell adhesion. Proc. Natl. Acad. Sci. **109**, 9839 LP-9844 (2012) https://doi.org/10.1073/pnas.1201499109
146. Praetorius, F., Kick, B., Behler, K.L., et al.: Biotechnological mass production of DNA origami. Nature **552**, 84–87 (2017). https://doi.org/10.1038/nature24650
147. Prins, L.J., Haag, R.: Multivalency. Wiley Blackwell (2018)
148. Proske, D., Blank, M., Buhmann, R., Resch, A.: Aptamers—basic research, drug development, and clinical applications. Appl. Microbiol. Biotechnol. **69**, 367–374 (2005). https://doi.org/10.1007/s00253-005-0193-5
149. Qi, S.Y., Groves, J.T., Chakraborty, A.K.: Synaptic pattern formation during cellular recognition. Proc. Natl. Acad. Sci. U. S. A. **98**, 6548–6553 (2001). https://doi.org/10.1073/pnas.111536798
150. Rajasekaran, S.A., Anilkumar, G., Oshima, E., et al.: A novel cytoplasmic tail MXXXL motif mediates the internalization of prostate-specific membrane antigen. Mol. Biol. Cell **14**, 4835–4845 (2003). https://doi.org/10.1091/mbc.E02-11-0731
151. Ramakrishnan, N., Tourdot, R.W., Eckmann, D.M., et al.: Biophysically inspired model for functionalized nanocarrier adhesion to cell surface: roles of protein expression and mechanical factors. R. Soc. Open Sci. **3**, 1–21 (2016). https://doi.org/10.1098/rsos.160260
152. Rao, J., Lahiri, J., Isaacs, L., et al.: A trivalent system from vancomycin-D-Ala-D-Ala with higher affinity than avidin-biotin. Science **280**, 708–711 (1998). https://doi.org/10.1126/science.280.5364.708
153. Raychaudhuri, S., Chakraborty, A.K., Kardar, M.: Effective membrane model of the immunological synapse. Phys. Rev. Lett. **91**, 1–4 (2003). https://doi.org/10.1103/PhysRevLett.91.208101
154. Rogers, W.B., Crocker, J.C.: Direct measurements of DNA-mediated colloidal interactions and their quantitative modeling. Proc. Natl. Acad. Sci. **108**, 15687–15692 (2011). https://doi.org/10.1073/pnas.1109853108
155. Rogers, W.B., Manoharan, V.N.: Programming colloidal phase transitions with DNA strand displacement. Science **347**, 639–642 (2015). https://doi.org/10.1126/science.1259762

156. Romano, F., Sciortino, F.: Switching bonds in a DNA gel: an all-DNA vitrimer. Phys. Rev. Lett. **114**(78104), 1–5 (2015). https://doi.org/10.1103/PhysRevLett.114.078104
157. Rothemund, P.W.K.: Folding DNA to create nanoscale shapes and patterns. Nature **440**, 297–302 (2006). https://doi.org/10.1038/nature04586
158. Rovigatti, L., Bomboi, F., Sciortino, F.: Accurate phase diagram of tetravalent DNA nanostars. J. Chem. Phys. **140**(154903), 1–10 (2014). https://doi.org/10.1063/1.4870467
159. Rovigatti, L., Smallenburg, F., Romano, F., Sciortino, F.: Gels of DNA nanostars never crystallize. ACS Nano **8**, 3567–3574 (2014). https://doi.org/10.1021/nn501138w
160. Rozenblum, G.T., Lopez, V.G., Vitullo, A.D., Radrizzani, M.: Aptamers: current challenges and future prospects. Expert Opin. Drug Discov. **11**, 127–135 (2016). https://doi.org/10.1517/17460441.2016.1126244
161. Saccà, B., Niemeyer, C.M.: DNA origami: the art of folding DNA. Angew. Chem. Int. Ed. **51**, 58–66 (2012). https://doi.org/10.1002/anie.201105846
162. SantaLucia, J.: A unified view of polymer, dumbbell, and oligonucleotide DNA nearest-neighbor thermodynamics. Proc. Natl. Acad. Sci. **95**, 1460–1465 (1998). https://doi.org/10.1073/pnas.95.4.1460
163. SantaLucia, J., Hicks, D.: The thermodynamics of DNA structural motifs. Annu. Rev. Biophys. Biomol. Struct. **33**, 415–440 (2004). https://doi.org/10.1146/annurev.biophys.32.110601.141800
164. Sarvestani, A.S.: The effect of substrate rigidity on the assembly of specific bonds at biological interfaces. Soft Matter **9**, 5927–5932 (2013). https://doi.org/10.1039/C3SM00036B
165. Savory, N., Nzakizwanayo, J., Abe, K., et al.: Selection of DNA aptamers against uropathogenic *Escherichia coli* NSM59 by quantitative PCR controlled Cell-SELEX. J. Microbiol. Methods **104**, 94–100 (2014). https://doi.org/10.1016/j.mimet.2014.06.016
166. Schubertová, V., Martinez-Veracoechea, F.J., Vácha, R.: Design of multivalent inhibitors for preventing cellular uptake. Sci. Rep. **7**, 1–7 (2017). https://doi.org/10.1038/s41598-017-11735-7
167. Seeman, N.C.: DNA Nanotechnology. WTEC Workshop Report R&D Status and Trends in Nanoparticles, Nanostructured Materials, and Nanodevices in the United States, vol. 3, pp. 177–180 (1998). https://doi.org/10.1007/978-1-61779-142-0
168. Seifert, U., Lipowsky, R.: Adhesion of vesicles. Phys. Rev. A **42**, 4768–4771 (1990). https://doi.org/10.1103/PhysRevA.42.4768
169. Shelby, R.A., Smith, D.R., Schultz, S.: Experimental verification of a negative index of refraction. Science **292**, 77–79 (2001). https://doi.org/10.1126/science.1058847
170. Shi, J., Kantoff, P.W., Wooster, R., Farokhzad, O.C.: Cancer nanomedicine: progress, challenges and opportunities. Nat. Rev. Cancer **17**, 20–37 (2017). https://doi.org/10.1038/nrc.2016.108
171. Shimobayashi, S.F., Mognetti, B.M., Parolini, L., et al.: Direct measurement of DNA-mediated adhesion between lipid bilayers. Phys. Chem. Chem. Phys. **17**, 15615–15628 (2015). https://doi.org/10.1039/C5CP01340B
172. Sievers, E.L., Senter, P.D.: Antibody-drug conjugates in cancer therapy. Annu. Rev. Med. **64**, 15–29 (2013). https://doi.org/10.1146/annurev-med-050311-201823
173. Sihvola, A.: Metamaterials in electromagnetics. Metamaterials **1**, 2–11 (2007). https://doi.org/10.1016/j.metmat.2007.02.003
174. Snodin, B.E.K., Romano, F., Rovigatti, L., et al.: Direct simulation of the self-assembly of a small DNA origami. ACS Nano **10**, 1724–1737 (2016). https://doi.org/10.1021/acsnano.5b05865
175. So-Jung, P., Anne, A.L., Chad, A.M., et al.: The electrical properties of gold nanoparticle assemblies linked by DNA. Angew. Chem. Int. Ed. **39**, 3845–3848 (2000). https://doi.org/10.1002/1521-3773(20001103)39:21%3c3845:aid-anie3845%3e3.0.co;2-o
176. Song, D., Yang, R., Wang, C., et al.: Reusable nanosilver-coated magnetic particles for ultrasensitive SERS-based detection of malachite green in water samples. Sci. Rep. **6**, 1–9 (2016). https://doi.org/10.1038/srep22870

177. St John, A., Price, C.P.: Existing and emerging technologies for point-of-care testing. Clin. Biochem. Rev. **35**, 155–167 (2014). PMC4204237
178. Storhoff, J.J., Elghanian, R., Mucic, R.C., et al.: One-pot colorimetric differentiation of polynucleotides with single base imperfections using gold nanoparticle probes. J. Am. Chem. Soc. **120**, 1959–1964 (1998). https://doi.org/10.1021/ja972332i
179. Storhoff, J.J., Mirkin, C.A.: Programmed materials synthesis with DNA. Chem. Rev. **99**, 1849–1862 (1999). https://doi.org/10.1021/cr970071p
180. Sullenger, B.A.: Aptamers coming of age at twenty-five. Nucleic Acid Ther. **26**, 119 (2016). https://doi.org/10.1089/nat.2016.29001.sul
181. Sun, H., Zhu, X., Lu, P.Y., et al.: Oligonucleotide aptamers: New tools for targeted cancer therapy. Mol. Ther. Nucleic Acids **3**, 1–14 (2014). https://doi.org/10.1038/mtna.2014.32
182. Tikhomirov, G., Petersen, P., Qian, L.: Fractal assembly of micrometre-scale DNA origami arrays with arbitrary patterns. Nature **552**, 67–71 (2017). https://doi.org/10.1038/nature24655
183. Tkachenko, A.V., Maslov, S.: Spontaneous emergence of autocatalytic information-coding polymers. J. Chem. Phys. **143**(45102), 1–8 (2015). https://doi.org/10.1063/1.4922545
184. Tombelli, S., Minunni, M., Mascini, M.: Aptamers-based assays for diagnostics, environmental and food analysis. Biomol. Eng. **24**, 191–200 (2007). https://doi.org/10.1016/j.bioeng.2007.03.003
185. Troian-Gautier, L., Valkenier, H., Mattiuzzi, A., et al.: Extremely robust and post-functionalizable gold nanoparticles coated with calix[4]arenes via metal-carbon bonds. Chem. Commun. **52**, 10493–10496 (2016). https://doi.org/10.1039/C6CC04534K
186. Tuerk, C., Gold, L.: Systematic evolution of ligands by exponential enrichment: RNA ligands to bacteriophage T4 DNA polymerase. Science **249**, 505–510 (1990). https://doi.org/10.1126/science.2200121
187. Um, S.H., Lee, J.B., Park, N., et al.: Enzyme-catalysed assembly of DNA hydrogel. Nat. Mater. **5**, 797–801 (2006). https://doi.org/10.1038/nmat1741
188. Valignat, M.-P., Theodoly, O., Crocker, J.C., et al.: Reversible self-assembly and directed assembly of DNA-linked micrometer-sized colloids. Proc. Natl. Acad. Sci. **102**, 4225–4229 (2005). https://doi.org/10.1073/pnas.0500507102
189. Varilly, P., Angioletti-Uberti, S., Mognetti, B.M., Frenkel, D.: A general theory of DNA-mediated and other valence-limited colloidal interactions. J. Chem. Phys. **137**, 1–15 (2012). https://doi.org/10.1063/1.4748100
190. Ventola, C.L.: Mobile devices and apps for health care professionals: uses and benefits. Pharm. Ther. **39**, 356–64 (2014). PMC4029126
191. Wang, Y., Breed, D.R., Manoharan, V.N., et al.: Colloids with valence and specific directional bonding. Nature **491**, 51–55 (2012). https://doi.org/10.1038/nature11564
192. Wang, S., Dormidontova, E.E.: Selectivity of ligand-receptor interactions between nanoparticle and cell surfaces. Phys. Rev. Lett. **109**, 1–5 (2012). https://doi.org/10.1103/PhysRevLett.109.238102
193. Wang, Y., Jenkins, I.C., McGinley, J.T., et al.: Colloidal crystals with diamond symmetry at optical lengthscales. Nat. Commun. **8**, 1–8 (2017). https://doi.org/10.1038/ncomms14173
194. Wang, Y., Wang, Y., Zheng, X., et al.: Crystallization of DNA-coated colloids. Nat Commun **6**, 1–8 (2015). https://doi.org/10.1038/ncomms8253
195. Wang, Y., Wang, Y., Zheng, X., et al.: Synthetic strategies toward DNA-coated colloids that crystallize. J. Am. Chem. Soc. **137**, 10760–10766 (2015). https://doi.org/10.1021/jacs.5b06607
196. Watson, J.D., Crick, F.H.: The structure of DNA. Cold Spring Harb. Symp. Quant. Biol. **18**, 123–131 (1953). https://doi.org/10.1101/SQB.1953.018.01.020
197. Wicki, A., Witzigmann, D., Balasubramanian, V., Huwyler, J.: Nanomedicine in cancer therapy: challenges, opportunities, and clinical applications. J. Controlled Release **200**, 138–157 (2015). https://doi.org/10.1016/j.jconrel.2014.12.030
198. Wilhelm, S., Tavares, A.J., Dai, Q., et al.: Analysis of nanoparticle delivery to tumours. Nat. Rev. Mater. **1**, 1–12 (2016). https://doi.org/10.1038/natrevmats.2016.14

199. Winfree, E., Liu, F., Wenzler, L.A., Seeman, N.C.: Design and self-assembly of two-dimensional DNA crystals. Nature **394**, 539–544 (1998). https://doi.org/10.1038/28998
200. Wu, D., Wang, L., Li, W., et al.: DNA nanostructure-based drug delivery nanosystems in cancer therapy. Int. J. Pharm. **533**, 169–178 (2017). https://doi.org/10.1016/j.ijpharm.2017.09.032
201. Xiong, H., Van Der Lelie, D., Gang, O.: Phase behavior of nanoparticles assembled by DNA linkers. Phys. Rev. Lett. **102**(15504), 1–4 (2009). https://doi.org/10.1103/PhysRevLett.102.015504
202. Xu, G.K., Hu, J., Lipowsky, R., Weikl, T.R.: Binding constants of membrane-anchored receptors and ligands: a general theory corroborated by Monte Carlo simulations. J. Chem. Phys. **143**(243136), 1–16 (2015). https://doi.org/10.1063/1.4936134
203. Yih, T.C., Al-Fandi, M.: Engineered nanoparticles as precise drug delivery systems. J. Cell. Biochem. **97**, 1184–1190 (2006). https://doi.org/10.1002/jcb.20796
204. Yoo, J., Aksimentiev, A.: In situ structure and dynamics of DNA origami determined through molecular dynamics simulations. Proc. Natl. Acad. Sci. **110**, 20099–20104 (2013). https://doi.org/10.1073/pnas.1316521110
205. Zadeh, J.N., Steenberg, C.D., Bois, J.S., Wolfe, B.R., Pierce, M.B., Khan, A.R., Dirks, R.M.P.N.: NUPACK: analysis and design of nucleic acid systems. J. Comput. Chem. **32**, 170–173 (2010). https://doi.org/10.1002/jcc.21596
206. Zahid, M., Kim, B., Hussain, R., et al.: DNA nanotechnology: a future perspective. Nanoscale Res. Lett. **8**, 1–13 (2013). https://doi.org/10.1186/1556-276X-8-119
207. Zanchetta, G., Lanfranco, R., Giavazzi, F., et al.: Emerging applications of label-free optical biosensors. Nanophotonics **6**, 627–645 (2017). https://doi.org/10.1515/nanoph-2016-0158
208. Zenk, J., Tuntivate, C., Schulman, R.: Kinetics and thermodynamics of Watson-Crick base pairing driven DNA origami dimerization. J. Am. Chem. Soc. **138**, 3346–3354 (2016). https://doi.org/10.1021/jacs.5b10502
209. Zhang, S., Li, J., Lykotrafitis, G., et al.: Size-dependent endocytosis of nanoparticles. Adv. Mater. **21**, 419–424 (2009). https://doi.org/10.1002/adma.200801393
210. Zhang, Y., Lu, F., Yager, K.G., et al.: A general strategy for the DNA-mediated self-assembly of functional nanoparticles into heterogeneous systems. Nat. Nanotechnol. **8**, 865–872 (2013). https://doi.org/10.1038/nnano.2013.209
211. Zhang, Y., McMullen, A., Pontani, L.L., et al.: Sequential self-assembly of DNA functionalized droplets. Nat. Commun. **8**, 1–7 (2017). https://doi.org/10.1038/s41467-017-00070-0
212. Zhang, F., Nangreave, J., Liu, Y., Yan, H.: Structural DNA nanotechnology: state of the art and future perspective. J. Am. Chem. Soc. **136**, 11198–11211 (2014). https://doi.org/10.1021/ja505101a
213. Zhao, Y.-X., Shaw, A., Zeng, X., et al.: DNA origami delivery system for cancer therapy with tunable release properties. ACS Nano **6**, 8684–8691 (2012). https://doi.org/10.1021/nn3022662
214. Zhdanov V.P.: Multivalent ligand-receptor-mediated interaction of small filled vesicles with a cellular membrane. Phys. Rev. E **96**, 012408 (2017)
215. Zhou, J., Rossi, J.: Aptamers as targeted therapeutics: current potential and challenges. Nat. Rev. Drug Discov. **16**, 181–202 (2017). https://doi.org/10.1038/nrd.2016.199

PAMAM and PPI Dendrimers in Biophysical and Thermodynamic Studies on the Delivery of Therapeutic Nucleotides, Nucleosides and Nucleobase Derivatives for Anticancer Applications

Michał Gorzkiewicz, Adam Buczkowski, Bartłomiej Pałecz and Barbara Klajnert-Maculewicz

Abstract In the recent years a number of reports have emerged on the use of dendrimers as nanocarriers for many therapeutic compounds. These polyvalent, monodisperse, usually globular macromolecules, due to their well-defined, highly-branched structure, provide improved solubility of the drugs, high loading capacity and controllable biodistribution. Numerous studies on drug-dendrimer complexes indicate that these polymers can serve as efficient delivery devices without the necessity of creation of stable chemical bonds with the therapeutics. However, as non-covalent interactions are based mainly on the electrostatic forces, the stability of such complexes may vary depending on the location of binding sites in dendrimer structures as well as the influence of different environmental conditions on the macromolecule architecture, leading to significant reduction of the transport efficacy in vivo. Dendrimers like PAMAM or PPI have been proposed for the delivery of antiviral and anticancer nucleoside analogues (NAs), which therapeutic application is often limited by fast metabolism, unfavorable biodistribution, low solubility or drug resistance. In order to overcome those limitations and to improve cellular uptake of NAs, numerous studies on the dendrimer-based drug delivery systems have been conducted. Current research is focused on the biochemical, biophysical and thermodynamic characterization of the interactions between dendrimers and their useful ligands like nucleotides, nucleosides and nucleobase derivatives. More data are also available for the biological assessment of their in vitro and in vivo activity. In this chapter, we summarize available data concerning the formation and stability of complexes between dendrimer nanocarriers and (i) nucleoside/nucleotide, (ii) purine and pirimidine antimetabolites for their potential application in innovative therapies.

M. Gorzkiewicz · B. Klajnert-Maculewicz (✉)
Department of General Biophysics, Faculty of Biology and Environmental Protection, University of Lodz, 141/143 Pomorska St, 90-236 Lodz, Poland
e-mail: barbara.klajnert@biol.uni.lodz.pl

A. Buczkowski · B. Pałecz
Department of Biophysical Chemistry, Faculty of Chemistry, University of Lodz, 165 Pomorska St, 90-236 Lodz, Poland

B. Klajnert-Maculewicz
Leibniz Institute of Polymer Research Dresden, Hohe Str. 6, 01069 Dresden, Germany

C. Demetzos and N. Pippa (eds.), *Thermodynamics and Biophysics of Biomedical Nanosystems*, Series in BioEngineering, https://doi.org/10.1007/978-981-13-0989-2_7

1 Introduction

Despite the significant progress and numerous achievements of modern medicine, the efficient anticancer therapy continues to be a challenge for researchers and medical doctors. The choice of therapeutic strategy depends on the type and severity of the disease, as well as patient's overall condition. Among current methods of cancer treatment, chemotherapy is the most common, but in many cases does not bring satisfactory results. Its efficacy is usually hampered by low specificity of action of the therapeutics, their disadvantageous biodistribution and low solubility, as well as systemic toxicity leading to detrimental side effects. Tumor's changeable microenvironment and cellular resistance may decrease the drug effectiveness even further.

One of the promising alternatives to traditional chemotherapy involves the application of drug delivery systems, which may improve the efficiency of routinely used medications, provide their targeted transport and controlled release, reduce side effects and fight down the drug resistance. The delivery devices must be characterized by the lack of toxicity and immunogenicity, ability to carry an adequate amount of the drug molecules, prolonged blood circulation time, efficient tumor accumulation and enhanced cellular uptake [53]. Such carriers may be manufactured thanks to the latest achievements of nanotechnology, which enable synthesis of chemical compounds with a wide range of applications, from single nanoparticles to more complicated structures and polymers. It is believed that introduction of multi-functional macromolecules with nanometric size may in the nearest future improve methods of diagnosis and therapy, making nanoparticles the basis of modern medicine.

For the past few years several nanoparticles, such as liposomes, carbon nanomaterials or dendrimers have been under investigation for the potential application in drug delivery, with particular emphasis on anticancer therapy [30]. They were reported to protect the therapeutics from degradation, improve their solubility and increase blood half-life, as well as enable targeted delivery and controlled release [35]. Furthermore, nanostructures may passively accumulate in tumors thanks to the retention mechanism, known as Enhanced Permeability and Retention effect (EPR) [5]. All this features allow the nanoparticle-based delivery systems to overcome the limitations of classical anticancer treatments. However, to date only a limited number of liposomes and polymeric systems have been approved for anticancer drug transport [76]. Because the application of liposomes may be hindered by their instability, inability to enter tumor vasculature and rapid clearance from the organism, the attention of scientists turned towards nanopolymers, of which dendrimers draw particular attention.

1.1 Dendrimers–Properties and Application in Drug Delivery

Dendrimers, which belong to a group of highly-branched macromolecules, have been under evaluation for over 30 years. The term "dendrimer" has been proposed by Donald Tomalia, whose team synthesized the first branched macromolecular polymers in the early 1980s [91], inspired by the work of Fritz Vögtle and co. [21]. Dendrimers possess sphere-shaped, symmetrical architecture, which is responsible for a number of unique properties giving the advantage over different forms of polymers. Due to the greatly optimized methods of synthesis, dendrimers are featured with well-defined structure, monodispersity and high purity. These methods give an opportunity to design compounds of specific characteristics through the selection of suitable monomers. By contrast, conventional polymerization techniques usually generate heterogeneous products of various molecular weights, which are difficult to characterize and require additional purification methods, lowering the efficiency of the process [9, 67].

Within the structure of a dendrimer, some critical elements may be distinguished, which are responsible for its physicochemical properties. These include a central core molecule and branched "dendrons" radially spreading from the core, composed of repeated monomeric sequences. The type of the core and monomers used for the synthesis defines shape and size of macromolecule. The latter is traditionally described by the dendrimer's generation, which reflects the number of subsequent layers of branches attached to the central molecule (Fig. 1). The structure of lower generation dendrimers is usually more asymmetric and open compared to higher generation macromolecules—with the increase of the generation, dendrimers adopt globular conformation, characterized by low density of the core region that increases regularly towards the surface [1, 64].

The external branches are terminated with chemical moieties, the number of which depends on the chemical character of core molecule and monomers. A large number of surface groups is responsible for high solubility and biopermeability of dendrimer macromolecules. Moreover, terminal moieties are usually highly reactive and may be further modified in order to change the physicochemical features of the dendrimer, or to generate desirable activity, e.g. therapeutic or catalytic [92]. However, their character and charge greatly affect the biocompatibility and cytotoxicity of discussed compounds. Cationic [like the most commonly used poly(amido amine) (PAMAM) and poly(propylene imine) (PPI)] dendrimers exhibit high cytotoxic and hemolytic activity, mainly because of their non-specific affinity to negatively charged cellular membranes [56, 103]. The toxic effects are generation-dependent and increase with the number of positively charged terminal moieties [33, 79]. On the other hand, anionic and neutral dendrimers do not show cytotoxicity [48], but their ability to penetrate biological membranes is greatly reduced.

PAMAM (also referred to by the trade name "Starburst") and PPI dendrimers (Fig. 2a, b, respectively) were among the first macromolecules of this class of polymers. PAMAM dendrimers are built with polyamide branches with tertiary amines as focal points, primary amino groups on the surface and ethylenediamine core. The

Fig. 1 The structure of dendrimer molecule. Reprinted from Gorzkiewicz and Klajnert-Maculewicz [44] with permission from Elsevier

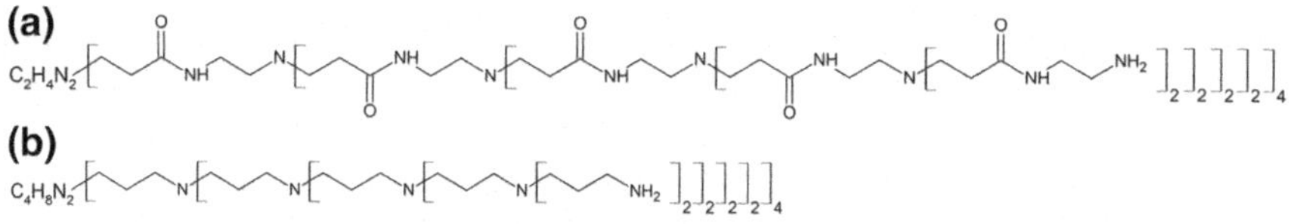

Fig. 2 The structure of PAMAM (**a**) and PPI (**b**) dendrimers of 4th generation

core structure of PPI dendrimers is based on diaminobutane (DAB) with primary amines as end groups and tertiary propylene amines inside the scaffold. It is important to note some discrepancies in scientific literature in naming of PPI dendrimers. According to Tomalia and Rookmaker [93], the uniform generation nomenclature should be used for PAMAM and PPI dendrimers (Table 1), therefore for this chapter we adopted the suggested classification, e.g. describing PPI dendrimer with 64 surface residues as the 4th generation.

Both types of polymers have found a wide range of applications in medical sciences. However, amino-terminated PAMAM and PPI dendrimers behave similarly with regard to cytotoxic and hemolytic effects, including the generation-dependent increase of both [85].

To enable medical application of positively-charged dendrimers, it is crucial to reduce their cytotoxic activity. This can be achieved by the chemical modification of surface moieties, resulting in partial or complete elimination of their cationic character. The most common modifications involve PEGylation (the attachment of poly(ethylene glycol) (PEG) chains) [96], glycosylation [50] and conjugation of biologically inert chemical particles [49]. The attachment of PEG and sugar residues have been additionally reported to prolong blood circulation time of dendrimers

Table 1 The nomenclature of PAMAM and PPI generations

PAMAM		
Generation	Molecular weight (g/mol)	Number of terminal amino groups
1	1429.9	8
2	3256.2	16
3	6908.8	32
4	14214.2	64
5	28824.8	128

PPI			
Generation (old nomenclature)	Generation (new nomenclature)	Molecular weight (g/mol)	Number of terminal amino groups
1	0	316.5	4
2	1	777.3	8
3	2	1686.8	16
4	3	3513.9	32
5	4	7168.1	64

in comparison to their unmodified forms [3, 92]. Surface modifications including conjugation of folate, dextran [54], peptides [68] or antibodies [71] may bring further benefits like targeted transport, receptor-mediated endocytosis and controllable biodistribution patterns.

The three-dimensional architecture of dendrimers with internal cavities and reactive terminal groups makes them the best candidates for drug nanocarriers and gives a number of possibilities for the attachment of biologically active compounds. The drug-dendrimer formulations can take a form of covalent conjugates or complexes, in which the therapeutics may be entrapped inside the scaffold of the dendrimer, or non-covalently bound to the charged surface moieties (Fig. 3). Such formulations are highly stable and capable of efficient transport of the drugs directly to the desired location, thus contributing to elimination of undesirable side effects and circumvention of resistance mechanisms. Application of dendrimers as drug delivery devices improves their solubility, extends blood circulation time and protects against degradation [63, 64].

The use of drug-dendrimer covalent conjugates seems to be the most reasonable approach. Such constructs provide higher stability in comparison to non-covalent complexes due to the permanent chemical bond between the therapeutic and macromolecule. However, their preparation is much more complicated and requires the selection of appropriate linker, which should be subjected to specific, enzymatic or pH-triggered cleavage at the site of action, allowing the release of active compound. Incorrect choice or application of a linker may decrease the therapeutic activity of the drug [97]. Numerous studies on the formation and stability of non-covalent complexes, as well as in vitro and in vivo experiments indicate that dendrimers can serve

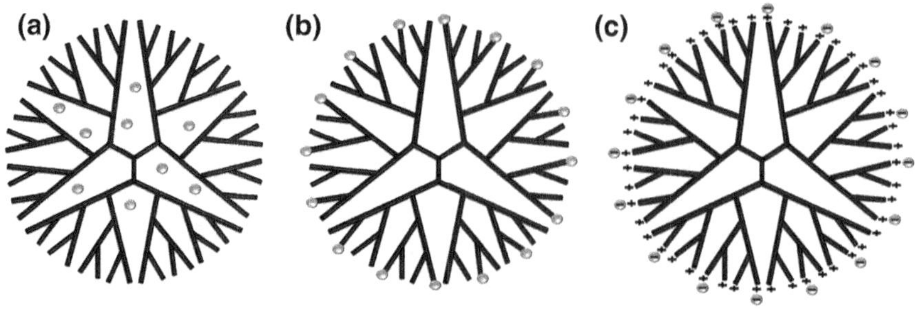

Fig. 3 Possible ways of drug-dendrimer binding: **a** Encapsulation inside the dendritic scaffold; **b** Covalent bonding; **c** Non-covalent interactions with charged functional groups. Reprinted from Gorzkiewicz and Klajnert-Maculewicz [44] with permission from Elsevier

as efficient drug delivery devices without the requirement of creation of stable covalent bonds with the therapeutic molecules. The preparation of complexes is cheaper, less laborious and time-consuming.

Additionally, some types of dendrimers may provide pH-triggered drug release from complexes. PAMAM and PPI dendrimers are characterized by pH-dependent conformations, becoming more open and extended upon lowering the pH due to electrostatic repulsion between internal tertiary amines and surface primary amino groups. At basic pH, so-called "backfolding" phenomenon occurs as a consequence of hydrogen bonding between protonated tertiary amines inside the dendritic scaffold and surface primary amines, resulting in a greater density of the interior. This feature can be utilized to tailor the encapsulation and release properties of dendrimers, especially in drug delivery applications [7].

However, non-covalent drug-dendrimer formulations also have their drawback. As the complexes are created mainly due to the electrostatic and van der Waals interactions, their stability may be significantly affected by the changes of pH, ionic strength and other environmental conditions, leading to the decrease of transport efficiency in vivo. The core size and surface density may dramatically affect the cargo-space of the dendrimers. Also the character of surface modifications may influence the electric charge of macromolecules and hinder the formation of complexes. Moreover, it is evident that size and shape of the guest molecule can significantly affect the maximum loading capacity [95]. It is therefore essential to thoroughly characterize non-covalent complexes for their physicochemical and biological properties before they could enter the therapy.

1.2 Therapeutics Nucleotide and Nucleoside Analogues

Nucleoside analogues (NAs) constitute a class of purine and pyrimidine derivatives featured with several cytotoxic activities. This group of antimetabolites was applied for the first time in antiviral therapy more than 50 years ago. Nowadays nucleoside analogues are used in the treatment of human immunodeficiency virus (HIV) (which involves the drugs like azidothymidine or stavudine) herpes simplex virus (acyclonucleosides) [4], as well as hepatitis B [38] and C [24]. Their activity is primarily based on the inhibition of primer synthesis, activity of polymerases and DNA elongation processes [52], which are crucial for the viral life cycle.

It soon became apparent that NAs exhibit numerous desirable anticancer effects, both against haematological malignancies and solid tumors. Modern antileukaemic therapies taking advantage of this group of compounds utilize cytidine (e.g. decitabine or cytarabine) and adenosine analogues (e.g. fludarabine or cladribine). As for the treatment of solid tumors, gemcitabine and derivatives of uracil (e.g. capecitabine or fluorouracil) are most commonly used [39, 40] (Table 2).

The anticancer potential of nucleoside and nucleotide analogues varies depending on their structure, stability and interactions with cell components, as well as the activity of proteins and enzymes associated with nucleoside metabolic processes [39, 40], and involves:

- the disturbance of metabolism of naturally occurring deoxyribonucleotides (dNTPs). This mechanism includes e.g. inhibition of deoxycytidine kinase (dCK) and ribonucleotide reductase (RR) activity, causing fluctuations in the pool of cellular dNTPs and facilitating incorporation of nucleoside analogues into newly synthesized DNA chain [39, 40].
- inhibition of DNA repair processes and synthesis of DNA and RNA. Incorporation of NAs into DNA and RNA may slow down or terminate chain elongation, which is generally associated with chemical structure and conformation of sugar moieties of analogues [52]. What is more, NAs incorporated into the DNA chain may cause mutations or inhibition of DNA-processing enzymes, e.g. DNA methyltransferases [98]. Nucleoside analogues may also decrease the activity of DNA polymerases, preventing the initiation of DNA synthesis (primase and polymerase α), hampering the DNA repair mechanisms (polymerases β and ε) and inhibiting the synthesis of mitochondrial DNA (polymerase γ) [4, 65, 74].
- activation of apoptotic processes, either through intrinsic mitochondrial pathway or extrinsic, death receptor-associated signaling pathway [77].

All nucleoside analogues share similar metabolic pathways, involving facilitated transport through cellular membrane and subsequent intracellular phosphorylation leading to the formation of active forms of the drugs [52] (Fig. 4). It should be emphasized that in most cases the conversion of NAs to triphosphate derivatives is essential for their cytotoxic activity. Since NAs are hydrophilic in nature, they cannot passively diffuse into the cell interior, and require specialized nucleoside transporters (NTs) enabling the penetration of cellular membrane. These proteins are generally divided into two classes: equilibrative (ENTs) and concentrative (CNTs) nucleoside transporters, with a wide range of specificity towards different nucleosides. It is important to note that human NTs transport only dephosphorylated forms of therapeutics [23, 101], which forces the administration of NAs in the form of prodrugs instead of active triphosphates.

Table 2 Examples of anticancer nucleoside and nucleobase analogues

	Drug	Chemical structure of nucleoside/nucleobase	Main uses
Purine analogues	Fludarabine	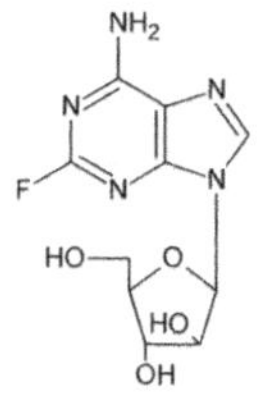	Chronic lymphocytic leukaemia
	Cladribine	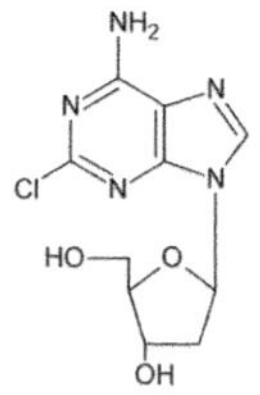	Hairy-cell leukaemia, non-Hodgkin lymphoma
	Mercaptopurine	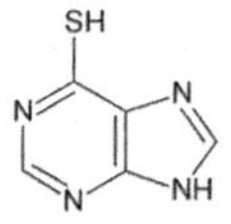	Acute lymphocytic leukaemia, chronic myeloid leukaemia, Crohn's disease, ulcerative colitis
Pyrimidine analogues	Cytarabine	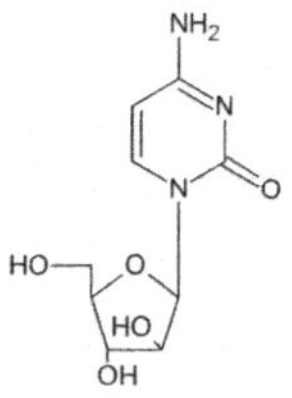	Acute myelogenous and lymphoblastic leukaemias

(continued)

Table 2 (continued)

	Drug	Chemical structure of nucleoside/nucleobase	Main uses
	Gemcitabine	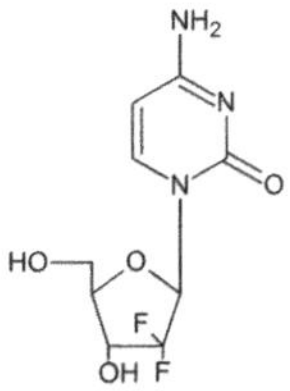	Pancreatic, lung, breast, and bladder cancers
Fluoropyrimidines	Fluorouracil	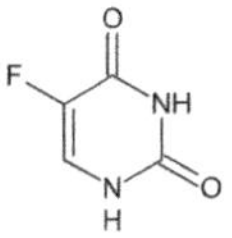	Gastrointestinal, pancreatic, head and neck, renal, skin, prostate and breast cancers
	Capecitabine	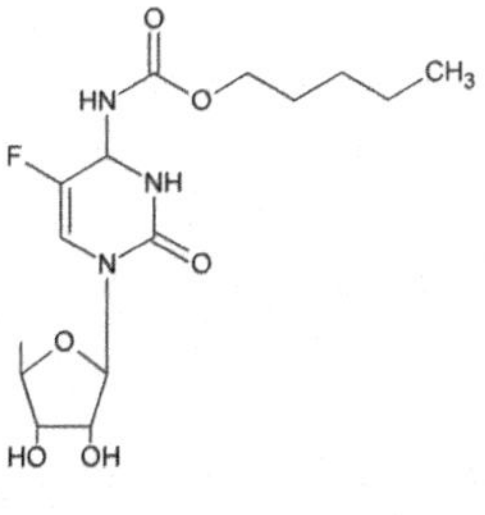	Relapsed breast and colorectal cancers

The application of NAs in antiviral and anticancer therapies may be significantly limited due to their rapid clearance from the organism, unfavorable biodistribution, low solubility and low specificity towards target cells. Further, the complex, multi-stage metabolism of these drugs potentiate the generation of several mechanisms of cellular resistance. The most severe of them involve the decrease of expression of nucleoside transporters and intracellular kinases, as well as the reduction of activity of the latter. These result in ineffective cellular uptake of nucleosides and their insufficient phosphorylation. Resistance might be also caused by alterations of proteins being cellular targets for nucleoside analogues, or involved in the initiation of apoptosis. Finally, nucleoside analogues as antimetabolites may be degraded both in blood and inside the cells [39, 102].

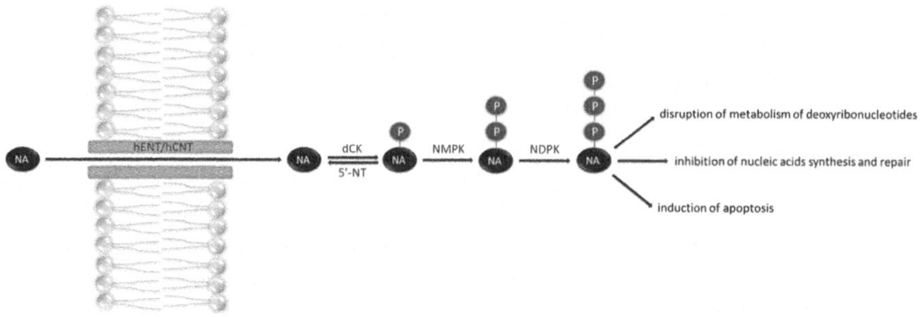

Fig. 4 Metabolism and mechanisms of action of nucleoside analogues. NA: nucleoside analogue; P: phosphate group; hENT/hCNT: human equilibrative/concentrative nucleoside transporter; dCK: deoxycytidine kinase; NMPK: monophosphate kinase; NDPK: diphosphate kinase; 5′-NT: 5′-nucleotidase. Reprinted from Gorzkiewicz and Klajnert-Maculewicz [44] with permission from Elsevier

In order to circumvent the resistance mechanisms and to enhance intracellular accumulation of NAs, an intensive research is being carried out on the application of drug delivery systems. Nanocarriers with their own autonomous ways of intracellular entry and cargo release may help to overcome limitations associated with transmembrane passage of drug molecules, at the same time protecting them from catabolism or degradation. In addition, this approach may enable the transport of active, triphosphate forms of the therapeutics, thereby preventing inadequate phosphorylation of NAs inside the target cells. The use of drug delivery devices could prevent undesirable interactions of nucleoside and nucleotide derivatives with normal cells and diminish toxic side effects of classical chemotherapy. The diversity of nanoparticles proposed as drug carriers for NAs provides several possibilities to improve the efficacy of nucleoside analogue-based therapies [31, 41, 46].

Several research on the use of dendrimers for the delivery of therapeutic nucleoside analogues showed that these macromolecules can form stable, non-covalent complexes with both prodrugs and triphosphates. Currently there is a great emphasis on biophysical and biochemical characterization of such complexes, as well as evaluation of their biological activity in vitro and in vivo, to allow their clinical use in the future.

2 Solubility Measurements

Many anticancer drugs show limited solubility in an aqueous environment, which limits their bioavailability during treatment. Dendrimer macromolecules may be used as nanocarriers of useful small molecule ligands. The transferred drug is protected from the influence of the external environment and should be slowly released near the place of its target action—tumor-altered tissue. The ability of dendrimers to supramolecularly complex anticancer drug molecules: purine and pyrimidine antimetabolites, nucleosides and nucleotides can be utilized to increase the solubility of these ligands in an aqueous environment. They contain delocalized π electrons in their aromatic rings that are capable of absorbing electromagnetic radiation in the UV range. Therefore, UV spectroscopy is a convenient analytical technique enabling quantification of purine and pyrimidine derivatives in aqueous solutions of the tested dendrimers.

In order to determine the effect of dendrimers on the solubility of tested ligands in an aqueous environment, a series of mixtures with increasing concentrations of dendrimer is prepared, usually in the range from several to several tens of μM (μmol/l). The solutions are subsequently added to eppendorf tubes containing weighed amount of the drug. The amount of the drug should be selected so that after saturation of the solution the certain fraction of the drug remains undissolved on the tube bottom. The tubes should be left for a sufficiently long time (after which we no longer observe changes in solubility of the ligand in mixtures) at a fixed temperature in relation to the excess of the crystalline drug.

Prior to determination of the solubility, the samples should be filtered or centrifuged to remove remaining undissolved material. Prepared samples should be diluted before spectrophotometric analysis. The dilution ratio should be chosen so that after dilution the tested substance meets the Lambert-Beer law (in case of the UV-Vis spectrophotometric analysis). Knowing the molar absorption coefficient of the tested ligand and the fold of dilution of the analyzed solution, it is possible to calculate the molar concentration describing the solubility of ligand in the analyzed solution. In order to accurately determine the concentration of the drug in the solutions, the measured absorbance of the drug should be corrected for the absorbance of the background—an aqueous dendrimer solution. Therefore, the corrected absorption of the drug is calculated from the difference of two measurements: the absorbance of the drug in the aqueous mixture with the dendrimer and the residual absorbance of the aqueous dendrimer solution at the same concentration (the same dilution) as the measured sample.

The obtained results can be presented as the dependence of ligand solubility (in units of molar concentration, e.g., μM) on the molar concentration of dendrimer in the mixture (expressed in the same units, e.g. μM). As a result of ligand (drug) binding in a supramolecular complex with the dendrimer, the concentration of free (unbound) drug in the solution decreases. This enables dissolution of another portion of the drug from above the crystalline precipitate (Fig. 5). Thus, the overall solubility of the drug increases (including both the drug dissolved in an uncomplexed form and the drug bound to the soluble complex) and can be determined by UV spectroscopy. Usually, we observe an approximately linear increase of the ligand solubility with the increase of the concentration of dendrimer macromolecules. It indicates that the dendrimer binds the analyzed ligand.

If the determined dependence of the ligand solubility on the molar concentration of the dendrimer in the tested mixture shows a linear character (Fig. 6), then it can be described by the straight line equation ($y = ax + b$) using the least squares method. Therefore, the parameters calculated for the straight line: the direction coefficient a and the free term b, can be assigned with physical sense. The free term b of the linear

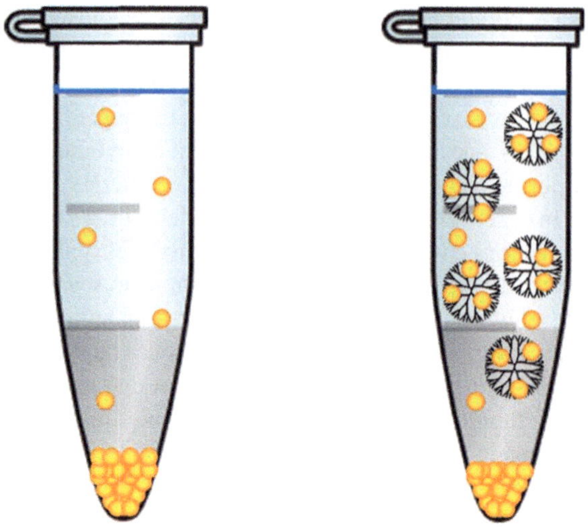

Fig. 5 A diagram of the solubility measurement system: the process of complexation (binding) of the ligand (drug) by the macromolecules of the dendrimer (receptor) results in the ligand solubility increase in a mixture over the precipitate

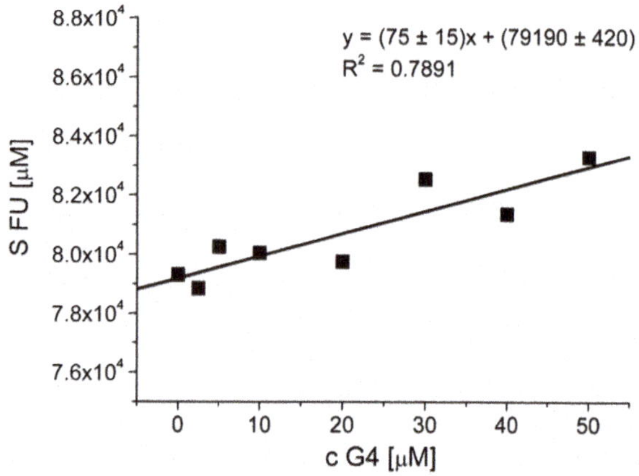

Fig. 6 The dependence of 5-fluorouracil solubility on the concentration of PAMAM-NH_2 G4 dendrimer aqueous solution. Reprinted from Buczkowski et al. [13] with permission from Elsevier

relationship between the drug solubility and the dendrimer concentration describes the solubility of the drug in pure solvent (water), whereas the direction coefficient *a* describes the average number of ligand (drug) molecules bound by the dendrimer macromolecule [15, 16, 18, 100]. The stoichiometric parameter describing the stoichiometry of the complex formation under the conditions of large ligand excess (saturated solution) relative to the macromolecule can be calculated using measurements of the ligand solubility in the dendrimer solutions of increasing concentration.

Determination of the increase in ligand solubility in mixtures containing PAMAM macromolecules were carried out for pyrimidine antimetabolite, 5-fluorouracil (5-FU), with PAMAM-NH_2 G3 [20], G4 [13] (Fig. 6) and G5 dendrimers [11] and with PAMAM-OH dendrimers of the same generations [11, 19, 20]. The determined numbers of 5-fluorouracil molecules bound by macromolecules of dendrimers (Table 3)

Table 3 The number of 5-fluorouracil molecules bound by PAMAM-NH_2 and PAMAM-OH dendrimer macromolecules, as determined by solubility measurements

Dendrimer	G3	G4	G5
PAMAM-NH_2	30 ± 15^a	75 ± 15^b	100 ± 10^d
PAMAM-OH	5 ± 4^a	12 ± 5^c	30 ± 20^d

Note [a]From Buczkowski et al. [20]; [b]From Buczkowski et al. [13]; [c]From Buczkowski et al. [19]; [d]From Buczkowski et al. [11]

indicate that the number of drug molecules bound by the dendrimer increases with the increasing dendrimer generation, due to increase in the number of available active sites (functional groups) in the structure of dendrimers of subsequent generations.

By comparing the amount of 5-fluorouracil molecules bound to the PAMAM dendrimers with amino and hydroxyl terminal groups in the aqueous environment (Table 3), it can be concluded that cationic PAMAM-NH_2 dendrimers bind more molecules of this drug compared to PAMAM-OH dendrimers. This indicates a significant contribution of electrostatic interactions between the protonated amino groups of dendrimer macromolecules and the polarized and anionic 5-fluorouracil molecules to the binding of this ligand with PAMAM dendrimers.

When the tested ligand is very difficult to dissolve in water, it can be assumed that the free term $b \cong 0$. Then the stoichiometry of the complex can be determined by dividing the number of moles of the drug dissolved in a dendrimer solution by the number of moles of the dendrimer. Such an approach was used to describe the results of 6-mercaptopurine solubility analysis in aqueous solutions of PAMAM-OH G4 dendrimer (at a pH adjusted by introducing additional electrolytes), which showed that the macromolecules of this dendrimer bind respectively 38 (pH 7.0), 36 (pH 5.0), or 35 (pH 3.0) drug molecules [69]. On the other hand, calculations of the solubility of 6-mercaptopurine in pure water ($b = 580 \pm 30\,\mu M$) and aqueous dendrimer solutions (without introducing additional electrolytes) indicated that PAMAM-NH_2 G4, PAMAM-COONa G3.5 and PAMAM-OH G4 macromolecules bind 21 ± 1, 14 ± 0.4 and 3 ± 1 6-mercaptopurine molecules, respectively [14].

3 Equilibrium Dialysis

Equilibrium dialysis is an effective method of studying the interactions of macromolecules with ligands in an aqueous environment. Since the results of the content determination of the tested compounds are obtained under equilibrium conditions, the dialysis method enables inferring the interactions in the studied system in equilibrium. An additional advantage of this method is the possibility to study relatively weak interactions between the components of supramolecular complexes. Using the dialysis technique, it is possible to determine:

- the release profile of ligand from supramolecular complex, which allows to characterize the kinetics of the studied system, also under different environmental conditions (temperature and pH)
- the isotherm of ligand binding to the supramolecular receptor, allowing (after mathematical evaluation) calculation of binding parameters—the stoichiometry and the equilibrium constant of the complex.

3.1 Determination of the Ligand Release Profiles from Supramolecular Complexes

In order to determine the release profile of a ligand (drug) bound to a supramolecular complex, the solution of the complex (of known concentration and volume) is enclosed in a dialysis bag (internal phase), which is then immersed in at least several times greater volume (external phase) of the solvent (water or buffer). Samples from the external phase are collected at subsequent intervals. Each time a sample is collected, the volume of the external phase is supplemented with pure solvent. The ligand content is determined (e.g. spectrophotometrically or chromatographically) in each of the successively collected samples. The release profile of the drug from the complex is presented as the percentage of released ligand molecules throughout the dialysis duration. The obtained results depend on the solvent (temperature and pH) and the type of semi-permeable membrane and its permeability limit (molecular weight cut-off, MWCO). Molecules with a molar mass lower than MWCO can penetrate through the semi-permeable membrane, while molecules with molecular weight greater than MWCO are retained by the membrane.

The dialysis technique is widely used in in vitro studies of ligand release from the complex in an aqueous environment. PBS (phosphate-buffered saline) with a composition resembling intraorganic environment, at a pH close to the blood pH (~7.4) or the pH of the extracellular fluid of tumor cells (~5.5), and temperature of 37 °C, is often used as a solvent of choice.

The dialysis technique was used to study the release of 5-fluorouracil bound to PAMAM G4 dendrimer modified with PEG chains and folic acid residues. The aqueous solution of the complex enclosed in the dialysis bag was immersed in water. Samples of an aqueous external solution, in which the drug content was determined spectrophotometrically (UV) were collected at subsequent time intervals. Slow release of 5-fluorouracil from the complex was observed. After 24-h dialysis (MWCO 12–14 kDa) of the solution containing the complex, approximately 50% of this drug was released [81].

5-fluorouracil was slowly and gradually released (sustained release) from PAMAM G4 dendrimer macromolecules (composed of ethylenediamine and methyl methacrylate instead of methyl acrylate, M = 15,932 Da), with 25% of terminal groups substituted with PEG chains. Drug-dendrimer complex was placed in dialysis bag (MWCO 12–14 kDa with pore size of 2.4 nm) and the entrapped drug was

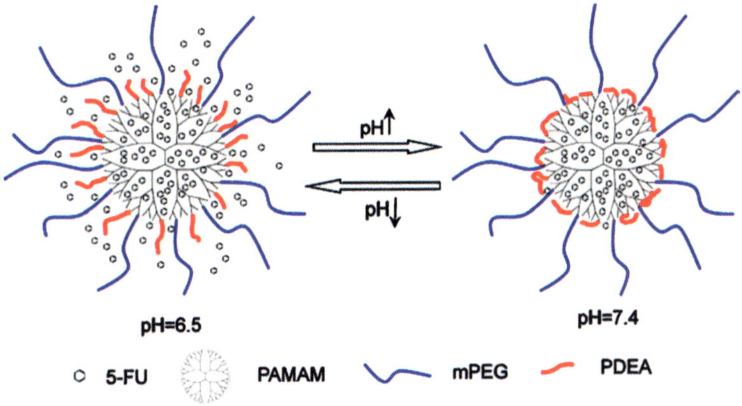

Fig. 7 Illustration of pH effect on the structure of PPD nanocarriers and their encapsulation and release of drugs. Reprinted from Jin et al. [51] with permission from Elsevier

allowed to pass out of the bag against perfect sink condition. In vitro drug release from complex with non-PEGylated PAMAM G4 was finished after 24 h and in the case of PEGylated PAMAM G4 after 144 h [6]. The gradual release of 5-FU from PAMAM G5 dendrimer modified with the antisense micro-RNA 21 (as-miR-21) has also been observed. 30% of 5-fluorouracil was released into the PBS buffer within one hour (MWCO 7 kDa), while 60% of the free (uncomplexed) drug was released during the same time (the control experiment). The release of the entire carried medication occurred after 12 h of dialysis. Interestingly, the presence of the antisense microRNA exerted a minor effect on the observed release profile of 5-fluorouracil, as compared to the PAMAM macromolecules unmodified with the oligonucleotide [62].

The strong effect of pH on the 5-fluorouracil release process was exhibited by PPD composites, i.e. PAMAM G4 macromolecules surface-modified with two types of substituents: poly (2-(N,N-diethylamino) ethylmethacrylate) (PDEA) chains and methoxy-poly (ethylene glycol) (mPEG) chains. The PDEA chains in an acidic environment (pH 6.5 or 4.0) were protonated, resulting in a change in their conformation and increasing the access to the dendrimer macromolecule cavities, while in a slightly basic environment (pH 8.0 or 7.4) they were conformationally closing access to the PAMAM cavities (Fig. 7). The mPEG chains present in the nanocarrier reduced toxicity and prolonged the circulation time of the carrier in the mouse model. The release profiles of 5-fluorouracil from PPD composites in PBS (145 ml, 50 mM, 37 °C, Fig. 8) determined by a dialysis technique (MWCO 3.5 kDa) at two pH values (6.5 or 7.4) confirmed the rapid release of the drug at pH 6.5 (close to the release rate of uncomplexed 5-fluorouracil) and slow release of the drug at pH 7.4 [51].

Similar studies of the release of 5-FU from a composite, in which the drug was encapsulated in PAMAM G5 macromolecules surrounding and stabilizing silver nanoparticles (dendrimer-stabilized silver nanocomposites) indicate that also in this

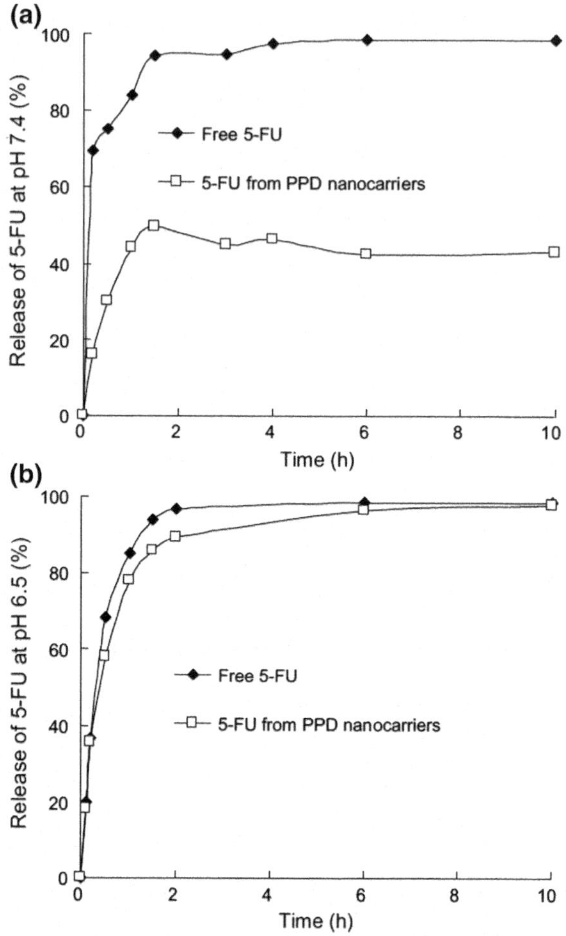

Fig. 8 The release of 5-FU at pH **a** 7.4 and **b** 6.5. Reprinted from Jin et al. [51] with permission from Elsevier

case, the gradual release of 5-fluorouracil molecules from the solution in the dialysis bag (MWCO 10 kDa) to the PBS buffer (pH 7.4, 37 °C) was observed. After 6 h 30% of drug molecules bound by PAMAM G5 macromolecules were released from the silver nanocomposite, while during the same time 90% of the free (uncomplexed) 5-fluorouracil was released in the control experiment [61].

The dialysis technique was also used to assess the stability of PAMAM G4 dendrimers complexes with adenosine-5′-triphosphate (ATP) molecules (ca. 1.6 ATP molecules per dendrimer molecule) in HEPES buffer (pH 7.0, 150 mM NaCl, 4 °C). The complexes were analyzed chromatographically. The obtained results indicate that the ATP-PAMAM G4 complex did not break down even after 48 h of dialysis [86].

The comparison of dialysis (MWCO 12–14 kDa) of gemcitabine (Gem) hydrochloride release profiles in PBS buffer (pH 4.0, 7.4 and 8.0) carried by unsubstituted and maltose-substituted macromolecules of PPI G4 dendrimer indicates that both nanocarriers display a non-linear release profile, which was characterized by a rapid release during the first 0.5–12 h followed by slower and continuous release. The macromolecules of PPI-Mal G4 dendrimer with maltose terminal groups released gemcitabine more slowly, as compared to the unmodified PPI G4 dendrimer (at all three pH values 4.0, 7.4 and 8.0). However, as the pH was lowered (8.0, 7.4 and 4.0), an increase in the release rate of gemcitabine was observed for both studied dendrimers [83]. The release of Gem in the acetic buffer at pH 4.2 was also more rapid for PAMAM-COOH G5.5 dendrimer-coated magnetic nanoparticles (DcMNPs), than at pH 5.2. After 15 h, the magnetic composite released all bound gemcitabine molecules at pH 4.2, while at pH 5.2 only 75% of bound drug molecules were released [75]. The release of gemcitabine from a copolymer based on the PAMAM-COOH G2.5 dendrimer macromolecule (PEG as core, surface modified with PEG 2000 chains and antibodies) in PBS buffer (pH 5.5 or 7.4, 37 °C) was also investigated by dialysis technique (MWCO 2 kDa). The drug was released from the copolymer at a slower rate, comparing to the free (uncomplexed) gemcitabine. After 24-h dialysis in PBS buffer at pH 7.4, the copolymer released 65% of the drug, while in PBS buffer at pH 5.5, 71% of the drug was released [72].

3.2 Determination of the Binding Isotherm of the Ligand to the Supramolecular Receptor

In order to determine the binding isotherm of the ligand to the supramolecular receptor, the dialysis process can be performed in a two-chamber dialyzer, in which equal volumes of the tested solutions remain separated by a semi-permeable membrane. The solution containing both the macromolecule acting as a supramolecular receptor (dendrimer) and its ligand (drug) is placed in one chamber of the dialyzer, while the solution of the ligand itself in the same solvent (usually water) is placed in the other chamber. The initial concentration of the ligand in both cells is identical.

During dialysis, small ligand molecules can freely penetrate through the pores of the semi-permeable membrane, as opposed to receptor molecules being too large. The concentration of the receptor remains (virtually) constant throughout the course of the dialysis. The osmotic penetration of solvent is severely restricted because the tested solutions tightly fill both dialyzer chambers. However, small changes in the volume of dialyzed solutions may occur due to the flexibility of the membrane.

If the receptor (dendrimer) is able to bind the ligand in a supramolecular complex, then the concentration of free (unbound) ligand in the solution containing the macromolecule is lower than on the opposite side. The decrease in free ligand concentration in the receptor solution is accompanied by a decrease in its chemical potential, which drives the diffusion transfer of a number of free ligand molecules through the membrane, into the macromolecule space. The transport of ligand molecules through the membrane continues until its chemical potential is equalized in both dialyzer cham-

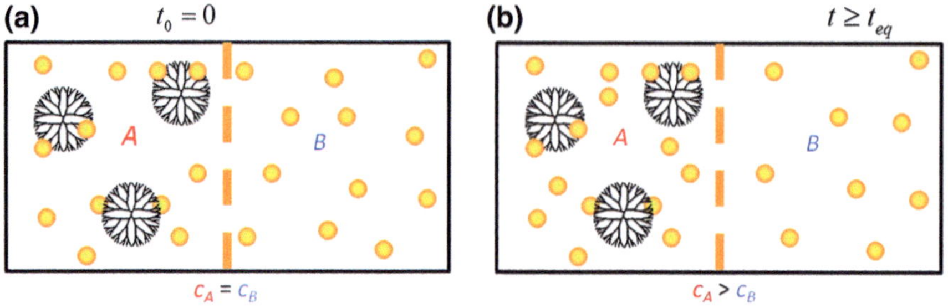

Fig. 9 Schematic representation of the course of dialysis: **a** at the beginning of the process, when the analytical ligand concentration is identical in both solutions, **b** after reaching the state of equilibrium

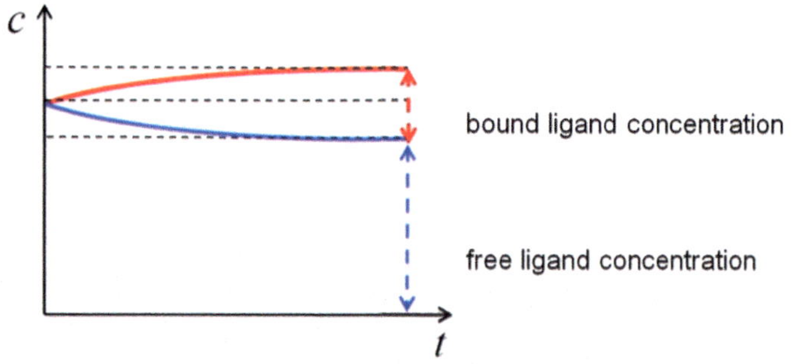

Fig. 10 Changes in analytical concentrations of ligand during dialysis of a macromolecule solution with ligand and ligand solution. The volumes of both dialyzed solutions are equal. The initial concentration of ligand in both solutions is identical

bers separated with a dialytic membrane. The equality of both chemical potentials is achieved under equilibrium conditions, when (in the case of diluted solutions) the concentration of free ligand on both sides of the dialysis membrane will be identical.

In order to equalize the concentrations of free ligand in both dialysis solutions, some of the ligand molecules will be transferred to the receptor solution. The concentration of ligand in the solution without the macromolecule decreases during dialysis. At the same time, due to the equilibration of free ligand concentration in both dialysis solutions, the analytical concentration of the ligand (defined as the sum of concentrations of bound and free ligand) in the macromolecule solution increases (Fig. 9). No further concentration changes will occur under equilibrium conditions (Fig. 10), until the receptor or ligand brakes down, disrupting the established equilibrium.

After reaching the state of equilibrium, the dialysis ends, the samples are collected from both solutions separated by dialysis membrane, and the analytical ligand content is determined (e.g. spectrophotometrically). The assay performed in a one two-chamber dialyzer yields two values: the concentration of bound ligand c (from the difference in concentrations of both dialyzed solutions) and the concentration of free ligand f (in the solution without the macromolecule). A data set in the form of pairs of such values is subjected to further mathematical analysis, in order to determine the binding parameters [28].

In a solution containing a macromolecule (receptor) M, at a concentration r, its ligand usually occurs in two forms: free (f) and reversibly bound (c) to the macromolecule. Reversible ligand binding by n active sites of the macromolecule may be described by the equilibrium constant K, defined as [82]:

$$K = \frac{c}{f(rn - c)} \tag{1}$$

Hence after the transformation:

$$b \equiv \frac{c}{r} = \frac{nKf}{1 + Kf} \tag{2}$$

where b—number of moles of bound ligand per one mole of receptor.

Equation (2) can be generalized to describe the case of reversible ligand binding by a macromolecule having N types (subsets) of active sites. Each subset includes n_i of the same ligand binding sites with a K_i constant. The isotherm of ligand binding by a macromolecule with N types of active sites is described by the following relationship [82]:

$$b \equiv \frac{c}{r} = \sum_{i=1}^{i=N} \frac{n_i K_i f}{1 + K_i f} \tag{3}$$

Thus, in a general case of N types of active sites, the description of the binding isotherm requires $2N$ parameters $n_i = n_1, n_2, \ldots, n_N$ and $K_i = K_1, K_2, \ldots, K_N$. Then, in order to calculate the values of the n_i and K_i set of parameters, based on the knowledge of a set of value pairs b and f, a non-linear multi-parameter regression is used, in which the squared deviations between the experimental data set (f, b) and the set of theoretical values (f_{th}, b_{th}) calculated for the fitted parameters n_i and K_i are minimized.

Equation (2) is satisfied at $N = 1$, which means that the macromolecules of the tested dendrimer have one type of active sites, including n of the same active sites binding the ligand with the equilibrium constant K. Then, the dependence of the number of bound ligand moles per one mole of the receptor on the free ligand concentration is defined by Eq. (2), is hyperbolic and tends to zero at $f \to 0$:

$$\lim_{f \to 0} b(f) = 0 \tag{4}$$

which means that in the absence of a ligand, the active sites of a macromolecule are not saturated with it and the dependence has a horizontal asymptote at $f \to +\infty$, equal to the number of active sites n:

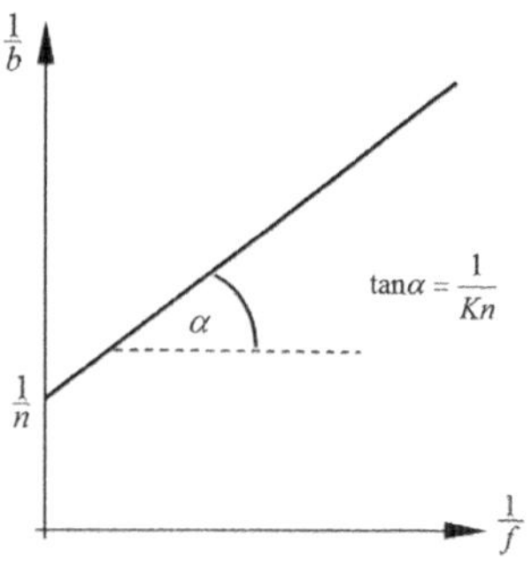

Fig. 11 A Scatchard-Klotz' plot for the reversible ligand binding by a macromolecule with one type of active site

$$\lim_{f \to \infty} b(f) = n \tag{5}$$

which means that all n of active sites of a macromolecule will formally undergo saturation only with an infinite excess of a ligand.

The linearization of the hyperbolic Eq. (2) in a doubly inversely proportional system is called Scatchard-Klotz' equation [8, 78]:

$$\frac{1}{b} = \frac{1}{Kn} \times \frac{1}{f} + \frac{1}{n} \tag{6}$$

The Scatchard-Klotz' Eq. (6) allows to graphically determine binding parameters n and K in case of a ligand binding by a macromolecule to a single type of independent binding sites.

The graph of the Scatchard-Klotz' Eq. (6) in the $\frac{1}{b}\left(\frac{1}{f}\right)$ coordinate system is a straight line with the free term $\frac{1}{\mathrm{n}}$ and direction coefficient $\tan\alpha = \frac{1}{\mathrm{Kn}}$ (Fig. 11).

The number of active sites of a macromolecule can be calculated by knowing the free term A of a linear Scatchard-Klotz' dependence (6):

$$n = \frac{1}{A} \tag{7}$$

The equilibrium constant of a ligand binding to the active site of a macromolecule can be determined by knowing the number of active sites n and the slope coefficient $\tan\alpha$ of a linear Scatchard-Klotz' dependence (6):

$$K = \frac{1}{n\tan\alpha} \tag{8}$$

Due to its simplicity and graphical clarity, the Scatchard-Klotz' Eq. (6) is still used to estimate the number of active sites of macromolecules and the binding constant, especially in biochemical and biophysical studies, despite the fact that it significantly amplifies the measurement errors [66]. More accurate results can be obtained by fitting a set of result pairs (f, b) by a non-linear regression using the hyperbolic Eq. (2).

When macromolecule has two types (subsets) of active sites, the selectable parameters n_1, K_1, n_2 and K_2 can be determined (through the convergence procedure) by describing the set of experimental values (f, b) with Eq. (3), which in this case $(N = 2)$ takes the form [8, 82]:

$$b = \frac{n_1 K_1 f}{1 + K_1 f} + \frac{n_2 K_2 f}{1 + K_2 f} \quad (9)$$

Since the products of $n_1 K_1$ and $n_2 K_2$ are present in the numerators, while the parameters K_1 and K_2 in the denominators of the expression (9), the practical calculation of four parameters n_1, K_1, n_2 and K_2 in such way is very difficult, because the fit error often exceeds the value of fitted parameters for one (out of two) population of sites.

The problem of describing the results of equilibrium dialysis with a regression method in case of a macromolecule with two types of active sites can be solved in two stages: (i) by calculating (estimating) initial values of binding parameters, (ii) by refining these estimates using a coupled two-parameter regression method [12]. The transformation of Eq. (9) in the doubly inversely proportional system is called Scatchard-Klotz' equation for two sets of binding sites [8]:

$$\frac{1}{b} = \frac{\left(\frac{1}{f}\right)^2 + (K_1 + K_2)\left(\frac{1}{f}\right) + K_1 K_2}{(n_1 K_1 + n_2 K_2)\left(\frac{1}{f}\right) + K_1 K_2 (n_1 + n_2)} \quad (10)$$

The relationship described by the Eq. (10) is non-linear [8]. The graph of this non-linear dependence (Fig. 12) enables distinguishing two roughly rectilinear fragments in the areas of low and high ligand concentrations, respectively.

Thus, it is possible to independently provide rough description of two rectilinear fragments of the nonlinear Scatchard-Klotz' dependence ($y = S_1 x + I_1$ and $y = S_2 x + I_2$, Fig. 12) with a Single Set of Identical Sites model [8, 13, 55, 78], using Eq. (6). The solutions obtained in this way (in step i) are subject to uncertainties resulting primarily from treating rectilinear fragments of Scatchard-Klotz' dependencies as linear boundaries to which this dependence tends at $1/f \rightarrow 0$ and $1/f \rightarrow \infty$ respectively. Therefore, this solution (from step i) should be refined (in step ii) using the non-linear two-parameter regression method, according to Eqs. (9) and (10).

The course of the function defined by Eq. (9) is particularly sensitive to the change of the ligand binding parameters by the lower affinity active site, e.g. n_2, K_2, while the course of the function defined by Eq. (10) is particularly sensitive to the change of the ligand binding parameters by the higher affinity active site, e.g. n_1, K_1 [12].

In a coupled two-parameter regression method, the ligand binding parameters with low affinity sites (n_2, K_2) are calculated using a double hyperbolic Eq. (9) by treating the parameters of binding with higher affinity sites (n_1, K_1) as constant. Similarly, the ligand binding parameters with the higher affinity sites (n_1, K_1) are calculated

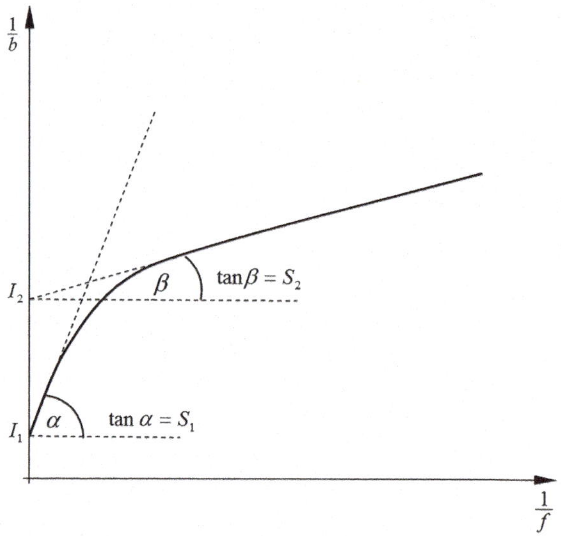

Fig. 12 A Scatchard-Klotz' plot for the reversible ligand binding by a macromolecule with two types of active sites

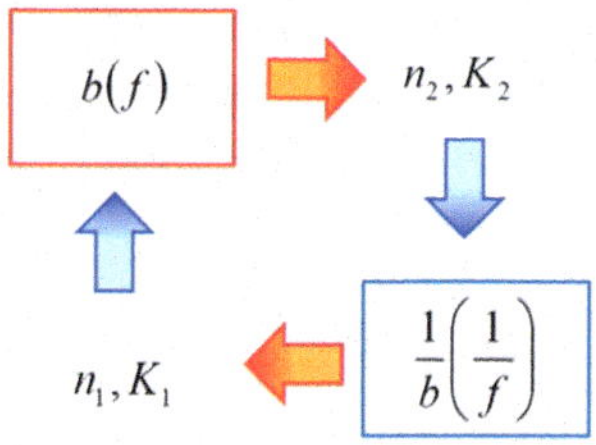

Fig. 13 A flow diagram of input data (fixed parameters) and output data (refined parameters) in a cycle of coupled two-parameter regressions. Reprinted from Buczkowski et al. [12] with permission from Elsevier

using the Scatchard-Klotz' Eq. (10) by treating the binding parameters with lower affinity sites (n_2, K_2) as constant. By alternating both of the described regressions respectively with Eqs. (9) and (10) after just a few iterative cycles (Fig. 13) the converged (relative to both partial regressions) parameter values can be obtained [12].

Using the equilibrium dialysis, we studied the interactions of 5-fluorouracil in aqueous mixtures containing macromolecules of cationic PAMAM-NH_2 G3 [20], G4 [12, 13] and G5 dendrimers [11] and PAMAM-OH G3 [20], G4 [19] and G5 dendrimers [11]. Solutions with a constant (for a given series of measurements) dendrimer concentration and increasing drug concentration were dialyzed. Once the dialysis was completed, 5-fluorouracil was determined spectrophotometrically in the UV range. The resulting binding isotherms of 5-fluorouracil with cationic (Fig. 14) and hydroxyl (Fig. 15) PAMAM G3-G5 dendrimers have been plotted in the $b(f/r)$ system, where r—the concentration of the receptor (dendrimer). The use of this

Fig. 14 The isotherms of 5-fluorouracil binding by amino-terminated PAMAM G3-G5 dendrimers at 20 °C, as determined by dialysis technique. The data for the charts were taken from the works [11, 12, 13, 20]

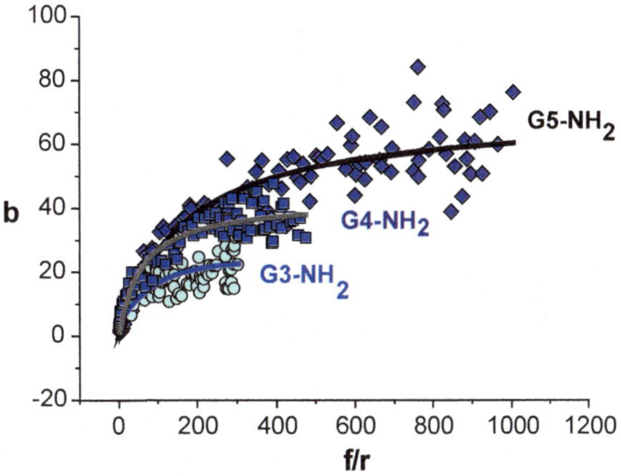

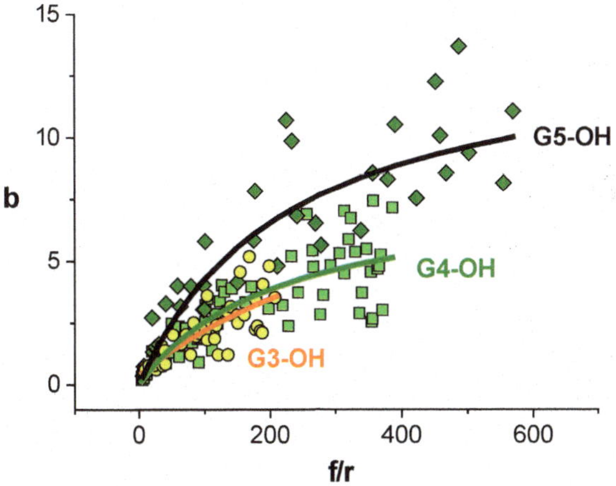

Fig. 15 The isotherms of 5-fluorouracil binding by hydroxyl-terminated PAMAM G3-G5 dendrimers at 20 °C, as determined by dialysis technique. The data for the charts were taken from the works [11, 19, 20]

coordinate system allows to present several data series with differing dendrimer concentrations on one graph.

Using a Single Set of Identical Sites model, the non-linear multiparameter regression method (Eq. 2) was applied to describe the binding isotherms of 5-fluorouracil by amino- or hydroxyl-terminated PAMAM G3-G5 macromolecules obtained during equilibrium dialysis, and to calculate the number of active sites n and the drug-macromolecule active site binding constant K (Table 4).

The obtained results indicate that both PAMAM-NH_2 and PAMAM-OH dendrimers spontaneously bind ($K > 1$) molecules of an anticancer drug, 5-fluorouracil in an aqueous medium at 20 °C. For PAMAM dendrimers, the number of func-

Table 4 The parameters of 5-fluorouracil binding to PAMAM-NH_2 and PAMAM-OH dendrimers in an aqueous environment at 20 °C, as determined by dialysis technique, according to a single set of identical sites model

Dendrimer type	PAMAM-NH_2		PAMAM-OH	
Generation/parameter	n	K	n	K
G3	26 ± 2^a	250 ± 60^a	6.0 ± 1.6^a	20 ± 10^a
G4	43 ± 2^a	420 ± 70^a	8 ± 1^b	60 ± 20^b
G5	70 ± 2^c	310 ± 30^c	14 ± 2^c	110 ± 30^c

Note [a]From Buczkowski et al. [20]; [b]From Buczkowski et al. [19]; [c]From Buczkowski et al. [11]

tional groups (internal—amide and tertiary amine and external—terminal) increases approximately two-fold, with the increase of the dendrimer's generation. With increasing generation of PAMAM-NH_2 macromolecules, the number of bound drug molecules increases more slowly than the increase (approximately two-fold) in the number of functional groups (potential active sites) in the dendrimer structure: $n(\text{G4}) : n(\text{G3}) = 1.7 \pm 0.2$, $n(\text{G5}) : n(\text{G4}) = 1.6 \pm 0.1$. Similarly, the observed increase in the number of 5-fluorouracil molecules bound by subsequent generations of PAMAM-OH dendrimers is smaller compared to the increase in the number of functional groups: $n(\text{G4}) : n(\text{G3}) = 1.3 \pm 0.6$. In both cases, lower increase in the number of bound drug molecules most probably reflects the occurrence of steric hindrance in the proximity of the active sites of the successive generations of dendrimer macromolecules.

The determined binding parameters of 5-fluorouracil by PAMAM-NH_2 and PAMAM-OH dendrimers (Table 4) indicate that the cationic PAMAM-NH_2 dendrimer macromolecules bind more drug molecules and with a higher binding constant, as compared to macromolecules of hydroxyl PAMAM-OH dendrimers of the same generation. The macromolecules of PAMAM dendrimers terminated with the same (amino or hydroxyl) terminal groups show similar values of binding constants, although a tendency for stronger binding of the drug by PAMAM-NH_2 G4 macromolecules can be observed.

The results of the equilibrium dialysis of 5-fluorouracil with PAMAM-NH_2 and PAMAM-OH dendrimers of the G3-G5 generations can also be represented in a doubly inversely proportional coordinate system $\frac{1}{b}\left(\frac{1}{f}\right)$. In two cases (PAMAM-$NH_2$ G3 and G4, Fig. 16) this relationship follows a non-linear course [12, 20], which allows to demonstrate the diversity of active sites binding 5-fluorouracil in these dendrimers. Analysis of this non-linearity enables calculation of binding parameters (Table 5) using a Two Set of Identical Sites model with a coupled two-parameter regression method.

The obtained parameters (Table 5) indicate that the cationic PAMAM-NH_2 dendrimers of the 3rd and 4th generations have two populations of active binding sites for 5-FU. In these macromolecules there is a smaller subset of sites n_1 showing high affinity (equilibrium constant K_1) to 5-fluorouracil and a larger subset of sites n_2 with lower affinity, K_2 to this drug. These parameters confirm the tendency already

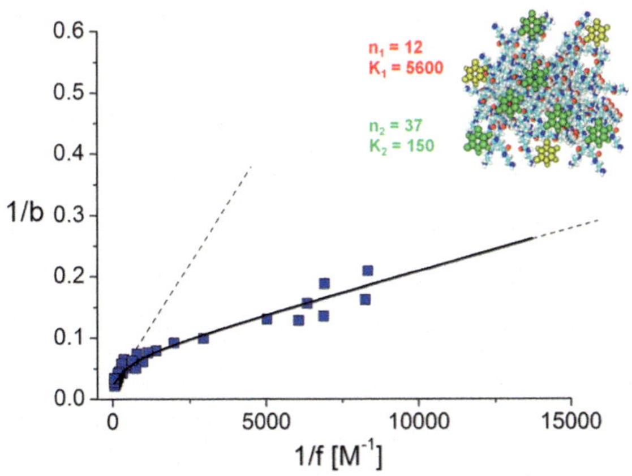

Fig. 16 A non-linear course of the Scatchard-Klotz isotherm describing the process of 5-fluorouracil binding by two types of active sites of the PAMAM-NH_2 G4 dendrimer macromolecule in water at 20 °C. Reprinted from Buczkowski et al. [12] with permission from Elsevier

Table 5 Parameters of binding 5-fluorouracil to the active sites of selected PAMAM-NH_2 dendrimers determined by equilibrium dialysis. The model of Two Sets of Independent Sites were used in each case

PAMAM generation	n_1	K_1	n_2	K_2
G3	5 ± 1^a	3890 ± 930^a	24 ± 3^a	110 ± 30^a
G4	12 ± 1^b	5600 ± 600^b	37 ± 3^b	154 ± 37^b

Note [a]From Buczkowski et al. [20]; [b]From Buczkowski et al. [12]

Table 6 Parameters of binding of 5-fluorouracil to the active sites of PAMAM-NH_2 G4 dendrimer determined by the model of Two Sets of Independent Sites in selected temperatures [12]

Temperature (°C)	n_1	K_1	n_2	K_2
20	12 ± 1	5600 ± 600	37 ± 3	154 ± 37
45	13 ± 2	1820 ± 320	32 ± 5	70 ± 20

observed for the results obtained in accordance with a Single Set of Identical Sites model.

The physicochemical characteristics of binding interactions in the supramolecular dendrimer-ligand system can be broadened by studying the behavior of the system at different temperatures. The thermodynamic studies of interactions of 5-fluorouracil with PAMAM-NH_2 G4 dendrimer were performed using equilibrium dialysis at 20 and 45 °C. The binding parameters for this system at 20 and 45 °C (Table 6), calculated according to a Two Set of Identical Sites model by a coupled non-linear two-parameter regression indicate that with increasing temperature:

(a) the number of 5-fluorouracil binding sites practically does not change within the limits of measurement uncertainties,

(b) the values of 5-fluorouracil binding constant with both types of active sites clearly decrease, which probably reflects an increase in thermal movements weakening the coherence forces of the formed complex.

4 ^{1}H NMR Titration

The ^{1}H NMR proton nuclear magnetic resonance spectroscopy technique can be used to analyze the interactions of dendrimers with their ligands (anticancer drugs) in a heavy water environment. As a result of a ligand-macromolecule binding there is a change in the electron density around hydrogen nuclei located near the functional groups of interacting molecules involved in the binding process, which in turn induces a change in the peak position $\Delta\delta$ on the spectrum for the mixture containing the complex δ_{obs}, in relation to the peak position on the spectrum for the pure components δ_f. For a system in which the exchange equilibrium is fast, the observed chemical shift δ_{obs} can be expressed as a weighted average of chemical shifts over the molar fractions of occupied and vacant active sites [37, 47]. The observed change in a chemical shift $\Delta\delta = \delta_{obs} - \delta_f$ is directly proportional to the fraction of occupied active sites P_b:

$$\Delta\delta = P_b \Delta\delta_{max} \tag{11}$$

where $\Delta\delta_{max}$ is the maximum difference in the chemical shift of the considered proton after a complete saturation of the macromolecule with the ligand.

In the simplest case, when the dendrimer macromolecule contains n active sites of the same type, a fraction of the occupied sites P_b in the macromolecule can be expressed as a function of a composition of the mixture of both components (with dendrimer concentration of r and ligand concentration of l) and K as the binding constant of a ligand with the active site [37, 47].

$$P_b = \frac{1}{2}\left[\left(1 + \frac{l}{nr} + \frac{1}{Knr}\right) - \sqrt{\left(1 + \frac{l}{nr} + \frac{1}{Knr}\right)^2 - \frac{4l}{nr}}\right] \tag{12}$$

By substituting Eq. (12) into (11), we obtain a relationship that can describe the experimental results (measured values $\Delta\delta$ as a function of the composition of the titrated mixture) by a non-linear multi-parameter regression [37, 47]:

$$\Delta\delta = \frac{\Delta\delta_{max}}{2}\left[\left(1 + \frac{l}{nr} + \frac{1}{Knr}\right) - \sqrt{\left(1 + \frac{l}{nr} + \frac{1}{Knr}\right)^2 - \frac{4l}{nr}}\right] \tag{13}$$

Using the results of the NMR titration, it is possible to calculate the number of active sites n in the macromolecule and the ligand binding constant K with the

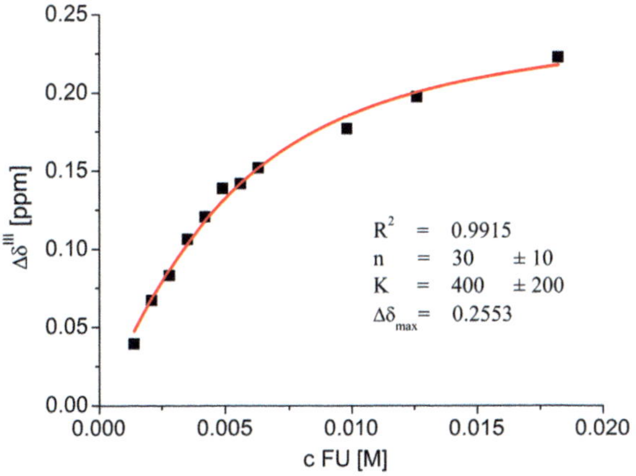

Fig. 17 Dependence of the change in chemical shift on the 5-fluorouracil concentration for methylene protons (band 2.8 ppm) of PAMAM-NH_2 G4. Reprinted from Buczkowski et al. [17] with permission from Elsevier

Table 7 The parameters of 5-fluorouracil binding to PAMAM-NH_2 and PAMAM-OH dendrimers in an aqueous environment at 20 °C, as determined by ^{1}H NMR titration, according to a Single Set of Identical Sites model

Dendrimer type	PAMAM-NH_2		PAMAM-OH	
Generation/ parameter	n	K	n	K
G3	30 ± 13[a]	330 ± 140[a]	(6)[a]	65 ± 10[a]
G4	30 ± 10[b]	400 ± 200[b]	(8)[c]	70 ± 10[c]
G5	(70)[d]	240 ± 50[d]	(14)[d]	110 ± 40[d]

Note n values in parenthesis were determined by dialysis. [a]From Buczkowski et al. [20]; [b]From Buczkowski et al. [17]; [c]From Buczkowski et al. [19]; [d]From Buczkowski et al. [11]

active site of the macromolecule. For this purpose, spectra of mixtures with a constant concentration of the receptor (dendrimer) *r* and increasing concentration of the ligand *l* should be recorded. The excess of ligand in the last solution measured should be large enough to cause (virtually) complete saturation of macromolecules with the drug.

Using a ^{1}H NMR titration, we tested a series of solutions with a constant (in a given series) concentration of PAMAM dendrimer and an increasing concentration of 5-fluorouracil in a heavy water environment. The change of the position of proton signals of PAMAM-NH_2 G3 [20], G4 [17] (Fig. 17) and G5 dendrimers [11] and PAMAM-OH G3 [20], G4 [19] and G5 dendrimers [11] as a function of the mixture composition (increasing concentration of 5-fluorouracil) is described by Eq. (13) using a non-linear multi-parameter regression (Table 7). In case of weakly interacting PAMAM-NH_2 G5 dendrimer and PAMAM-OH G3-G5 dendrimers, the number of active sites determined by the dialysis technique was assumed during calculations.

These results (Table 7) confirm that the macromolecules of cationic PAMAM-NH_2 dendrimers bind 5-fluorouracil molecules with a higher binding constant, as compared to macromolecules of hydroxyl PAMAM-OH dendrimers of the same generation. Analysis of the observed shift changes allows to draw conclusions regarding the site of substitution of drug molecules in the structures of tested dendrimers. The active sites of PAMAM-NH_2 cationic dendrimers that bind 5-fluorouracil molecules are the terminal amino groups and internal amide groups [20, 61], whereas for the PAMAM-OH dendrimers the drug is bound via internal tertiary amino groups [20].

^{1}H NMR titration was used to study the formation of the PAMAM G5 dendrimer complex with guanosine-5′-monophosphate (GMP) in heavy water. The results indicate that both compounds interact forming ion pairs between the protonated amino groups of the dendrimer and the anionic phosphate residues of the ligand. The parameters of complex formation were calculated by a non-linear multi-parameter regression method with Eq. (13). According to the obtained solution, the PAMAM G5 dendrimer has approximately 107 sites binding guanosine monophosphate with a 17,400 M^{-1} binding constant. The binding process occurs on the surface rather than inside the dendrimer macromolecule [47].

5 ITC Titration

The isothermal titration calorimetry (ITC) technique enables measurement of the thermal effects accompanying interaction of the components of the tested systems as a function of composition of the titrated solution. In a typical experiment, a solution of the ligand (drug) is injected in portions to the solution of dendrimer in the same solvent, which is placed in the calorimeter cell. The calorimeter records the thermal power exchanged during subsequent injections as a function of time. This dependence is integrated by the software included with the calorimeter and presented as an enthalpogram as a function of the molar ratio of the ligand to the macromolecule.

The calorimeter measures the global thermal effect of the titration of the dendrimer solution with the ligand solution that, in addition to the effect of direct interactions of the titrated system components, includes also the effects of dilution of the macromolecule and its ligand. Thermal effects of the direct macromolecule-ligand interaction are determined by subtracting the corresponding, independently determined thermal effects of diluting the ligand solution with pure solvent and diluting the macromolecule solution with pure solvent, from the global thermal effect of mixing macromolecule and ligand solutions. Because in commercial calorimeters of this type the volume of the measuring cell is several times larger than the volume of the syringe, the thermal effects associated with diluting the solution of the receptor contained within the cell with pure solvent from the syringe are often negligible. Therefore, the effects of direct dendrimer-ligand interactions can be calculated by correcting the global effects of the titration only for the independently determined dilution effects of ligand solution. The effects of direct interactions of dendrimer macromolecules with ligand molecules as a function of the titrated solution composition and corrected for the dilution effects, can be described by a Single Set of Identical Sites model by non-linear multiparameter regression, using the software included with the calorimeter.

Since the ITC titration is a sensitive and highly-responsive experimental technique, it is important that the solutions used for titration are degassed and freshly prepared (not more than few days old). In case of preparation of a dendrimer solution from a highly concentrated stock solution, it is recommended to leave the solution to be titrated for a few up to several hours under low temperature conditions (e.g. in a refrigerator) in order to provide the time necessary for the disaggregation of macromolecules.

The calorimetric titration is a basic technique enabling full thermodynamic characteristics of the system, which enables the determination of stoichiometry and equilibrium constant, as well as standard thermodynamic functions: enthalpy ΔH and entropy ΔS of the ligand binding with the active site of the macromolecule. Due to the dynamic nature of ITC measurements, the equilibrium constants K obtained by this technique may have higher value and the number of detected active sites may be lower comparing to results obtained with static techniques (such as solubility measurements, equilibrium dialysis or ^{1}H NMR titration). This is mainly due to the relatively short time (about dozen minutes) between consecutive injections of a titrant solution and the complex structure of the titrated dendrimer macromolecules that hinders access to active sites with lower affinity and/or located deeper within the structure. As a result, the tested system does not fully reach the thermodynamic equilibrium. It would often take several hundred hours to reach a full equilibrium (with 30–50 injections). The dynamic technique of calorimetric titration, however, provides a direct insight into the thermodynamics of interactions accompanying the saturation of the active sites of dendrimer macromolecules with ligand molecules.

The interactions of basic (unmodified) PAMAM-NH_2 dendrimers (Fig. 18) and PAMAM-OH with 5-fluorouracil [11, 17, 19, 20] and gemcitabine [73] were studied using a calorimetric titration technique under isothermal conditions. This technique was also used to study interactions of PPI dendrimers in unmodified and maltose-modified forms (in 30%—Open Shell, OS or 100%—Dense Shell, DS) with cytidine-5′-triphosphate (CTP) [89], which structural analogues have established antitumor activity (cytarabine (Ara-C) shows an opposite orientation of the hydroxyl group at the 2′ position of the sugar ring). The values of binding parameters and thermodynamic functions obtained for these systems in aqueous environment by non-linear regression were summarized in Table 8.

The obtained parameters (Table 8) indicate that binding of 5-fluorouracil to cationic PAMAM-NH_2 G3-G5 dendrimers is a thermodynamically spontaneous ($K > 1$) process that occurs with a favorable (exothermic) enthalpy change ($\Delta H < 0$). In case of PAMAM-NH_2 G4 dendrimer, the binding constant of 5-fluorouracil by active sites is higher than for the G3 and G5 generations, which is a consequence of the favorable binding entropy ($\Delta S > 0$) [11, 17, 20]. The entropy of drug binding by G3 and G5 generation PAMAM-NH_2 dendrimers is unfavorable ($\Delta S < 0$), which indicates an increase in the system ordering degree associated with the ligand binding to the open G3 structure (which can significantly adapt its confor-

Table 8 The binding parameters of selected ligands with active sites of PAMAM and PPI dendrimer macromolecules in the aqueous environment, as determined by ITC

Dendrimer	Ligand	n	K (M^{-1})	ΔH (kcal mol^{-1})	ΔS (cal K^{-1} mol^{-1})	References
PAMAM-NH_2 G3	5-FU	13 ± 2	1020 ± 160	−10.0 ± 1.9	−19.9 ± 6.7	[20]
PAMAM-NH_2 G4		25 ± 3	2800 ± 300	−1.5 ± 0.2	10.7 ± 0.4	[17]
PAMAM-NH_2 G5		22 ± 8	370 ± 60	−20.2 ± 8.3	−57 ± 29	[11]
PAMAM-OH G3	5-FU	2.5 ± 0.3	3900 ± 600	−0.65 ± 0.09	14.2 ± 0.6	[20]
PAMAM-OH G4		5.3 ± 0.6	3180 ± 530	−0.28 ± 0.04	15.0 ± 1.0	[19]
PAMAM-OH G5		5.0 ± 0.5	6700 ± 5900	−0.64 ± 0.58	15.3 ± 1.0	[11]
PAMAM-NH_2 G4	Gem	25 ± 8	1050 ± 380	−1.65 ± 0.75	8.3	[73]
PAMAM-OH G4		20 ± 9	670 ± 450	−1.20 ± 0.68	8.9	
PAMAM-COONa G4		44 ± 6	12,200 ± 1000	−2.80 ± 0.40	9.3	
PPI G3[a]	CTP	10	19,500	−9.8	−12.4	[89]
PPI G4[a]		32	13,500	−12.4	−22.8	
PPI-Mal OS G4[a]	CTP	15	13,900	−7.2	−5.1	[89]
PPI-Mal DS G4[a]		12	14,400	−4.8	3.0	

Note [a]The generation number is expressed in the new nomenclature (see introduction section)

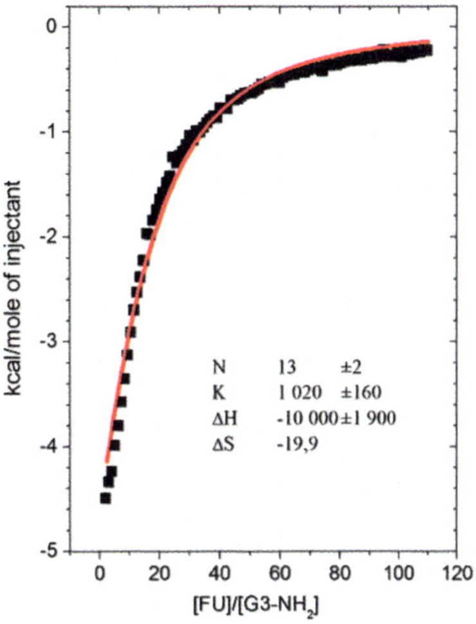

Fig. 18 Thermal effect of the interaction between 60 μM PAMAM-NH_2 G3 dendrimer and 30 mM 5-fluorouracil corrected with the dilution effects and calculated per one mole of the drug. Reprinted from Buczkowski et al. [20] with permission from Elsevier

mation to the ligand) and the highly packed surface of G5 macromolecules (where the significant surface rigidity precludes the structural adjustment of the macromolecule to the structure of the binding ligand).

The calculated binding parameters of 5-fluorouracil with PAMAM-OH G3-G5 hydroxyl dendrimers (Table 8) indicate that the increase in dendrimer generations is accompanied by the increase in the number of bound 5-fluorouracil molecules, but the thermodynamic functions of binding are similar to each other within uncertainty limits. Binding of 5-FU by PAMAM-OH G3-G5 macromolecules is thermodynamically spontaneous ($K > 1$) and occurs with favorable changes in enthalpy ($\Delta H < 0$) and entropy ($\Delta S > 0$) [11, 19, 20]. The PAMAM-OH G3-G5 dendrimers bind smaller amount of drug molecules, as compared to their cationic PAMAM-NH_2 G3-G5 counterparts.

The results of calorimetric titrations of aqueous solutions of PAMAM dendrimers terminated with 64 terminal groups: amino (PAMAM-NH_2 G4), hydroxyl (PAMAM-OH G4) or sodium carboxylate (PAMAM-COONa G3.5) with gemcitabine hydrochloride in aqueous solution (Table 8) indicate that all three dendrimers bind the drug spontaneously ($K > 1$), which was accompanied by favorable changes in enthalpy ($\Delta H < 0$) and entropy ($\Delta S > 0$) of binding. The enthalpy effects associated with gemcitabine binding by the dendrimers become more exothermic in the following order: PAMAM-OH G4 < PAMAM-NH_2 G4 < PAMAM-COONa G3.5. The number of active sites and the calculated values of the equilibrium constants increase in the same order. This order most probably reflects an increase in the contribution of electrostatic interactions between Gem and the tested dendrimers, with a neutral (PAMAM-OH), cationic (PAMAM-NH_2) or anionic (PAMAM-COONa) surface [73].

The calorimetric titration technique was also used to study the interactions of G3 and G4 PPI dendrimers in an unmodified and maltose-modified form (in 30%—Open

Table 9 The binding parameters of CTP with active sites of PPI dendrimer macromolecules in the phosphate-buffered saline (PBS: 10 mM, 137 mM NaCl, 2.7 mM KCl, pH 7.4) environment, as determined by ITC

Dendrimer	Ligand	n	K (M^{-1})	ΔH (kcal mol^{-1})	ΔS (cal K^{-1} mol^{-1})	Reference
PPI G4	CTP	(32)[a]	2400	−10.8	−20.06	[89]
PPI-Mal OS G4		(32)[a]	1500	−11.0	−22.3	
PPI-Mal DS G4		(32)[a]	930	−6.2	−7.2	

Note [a]Fixed parameter value

Shell, OS or 100%—Dense Shell, DS) with CTP. The maltose-modified PPI dendrimers show lower hemolytic activity compared to their unsubstituted counterparts. The obtained results (Table 8) indicate that: (i) the number of active binding sites for CTP increases with an increase of the unsubstituted PPI dendrimer generation, (ii) the number of binding sites decreases, as the degree of substitution of PPI G4 macromolecule terminal groups with maltose increases. The binding of CTP molecules by unsubstituted and maltose-modified dendrimer macromolecules is thermodynamically spontaneous ($K > 1$) and is accompanied by favorable (exothermic) changes in enthalpy. The complete substitution of terminal PPI dendrimer groups with maltose residues (PPI-Mal DS G4) reduces the exothermicity of interactions and occurs with favorable changes in entropy ($\Delta S > 0$) [89].

Calorimetric titrations of solutions of unmodified, as well as maltose-substituted PPI dendrimers of 4th generation (PPI G4, PPI-Mal OS G4 and PPI-Mal DS G4) with CTP solutions were also performed in the phosphate-buffered saline environment (PBS: 10 mM, 137 mM NaCl, 2.7 mM KCl, pH 7.4). The obtained results (Table 9) indicate that the binding constants of tested dendrimers with CTP in PBS buffer are lower by an order of magnitude than in water, which suggests a weakening influence of the buffer counter ions on the interaction. The interaction energetics of the maltose-substituted (at 30%) PPI-Mal G4 dendrimer is similar to this of the unsubstituted PPI G4 dendrimer [89].

6 DSC Study

DSC (differential scanning calorimetry) technique enables measuring the difference in thermal flux generated between the test and the reference samples during the programmed temperature change, which makes it possible to measure thermal capacities and phase transition temperatures. This technique was used to confirm the interaction between gemcitabine and the encapsulating copolymer built around the PAMAM-COOH G2.5 dendrimer macromolecule (PEG as core, surface modified with PEG

2000 chains and antibodies). The sharp peak corresponding to the melting of free (uncomplexed) gemcitabine (287.5 °C), as well as the multiple peaks associated with changes in the PAMAM macromolecule (40–60 °C) were lost following the formation of the drug-copolymer complex, which confirmed the interaction of the drug and its carrier [72].

7 Zeta Potential and Zeta Size Measurements

The surface charge of a particle in solution affects the distribution of ions in the surrounding area. This leads to the increase in concentration of counter ions close to the surface and the development of electrical double layer around each particle. The layer consists of an inner region, called the Stern layer, where the ions strongly interact with the molecule, and an outer, diffuse region. Within the latter the boundary called slipping plane separates ions interacting with the surface of the molecule from those freely moving in solution (Fig. 19).

The electrostatic potential at the electrical double layer surrounding a particle in solution is called the zeta potential. It may be also described as the difference in the electrostatic potential between the solvent and the stationary layer of a particle. Zeta potential is usually measured by determination of the electrophoretic mobility with application of Henry equation:

$$U_E = \frac{2\varepsilon z f(Ka)}{3\eta} \tag{14}$$

where U_E—electrophoretic mobility, ε—dielectric constant, z—zeta potential, $f(Ka)$—Henry's function (usually 1.0 or 1.5), η—viscosity.

The electrophoretic mobility may be evaluated through electrophoresis experiment on the sample and measurement of the velocity of particles using Laser Doppler Velocimetry (LDV). This method determines the speed of the particle movements in a liquid after application of electrical field. Using the velocity and the electrical field values, the zeta potential may be calculated using two additional characteristics of the sample—viscosity and dielectric constant. The boundary between stable and unstable suspensions is usually taken at either +30 or −30 mV. Particles with zeta potentials lower than −30 mV or higher than +30 mV are usually considered stable [25].

The most important factors affecting zeta potential involve pH, electrolyte concentration and ionic strength [45]. Thus, during complexation studies an appropriate choice of solvent is required in order to perform the experiment correctly and to estimate the parameters of ligand binding to the dendrimer. All liquids used to dilute the sample should be filtered to avoid contamination. Also, the bubbles in the sample should be avoided. The measurements should be conducted under controlled temperature conditions.

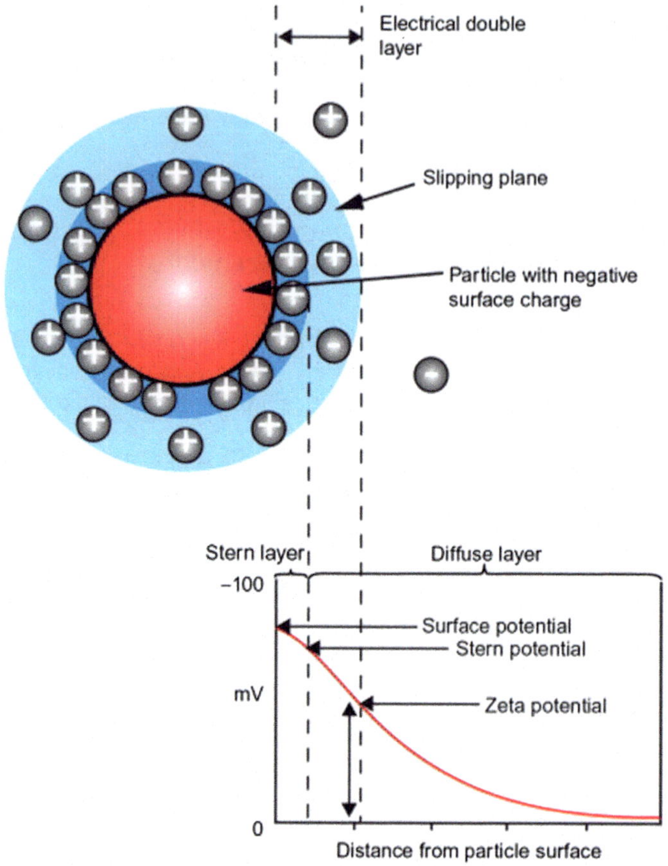

Fig. 19 Ionic concentration and potential difference as a function of distance from the charged surface of a particle in a dispersion medium. Reprinted from Elizondo et al. [34] with permission from Elsevier

The physical properties of sample such as particle size and sample concentration must be considered before performing an experiment. Each type of compound has its own ideal range of sample concentration for optimal measurements. If the sample concentration is too low, there may not be enough light scattered to make a measurement. On the other hand, if the sample is too concentrated, a phenomenon called the multiple scattering may occur, when the light scattered by one particle will be also scattered by another. The upper limit of concentration is additionally governed by the point at which the sample cannot freely diffuse due to the interactions between particles.

Molecules in suspension exhibit Brownian motion consisting of random movements depending on particle size, temperature and viscosity of the solution. Dynamic light scattering (DLS) provides measurements of the time-dependent fluctuations in scattering intensity, in order to determine the translational diffusion coefficient from which the hydrodynamic diameter can be calculated using the Stokes-Einstein equation:

$$D_h = \frac{k_B\,T}{3\,\pi\,\eta\,D_t} \tag{15}$$

where D_h—hydrodynamic diameter, D_t—translational diffusion coefficient, k_B—Boltzmann's constant, T—thermodynamic temperature, η—dynamic viscosity.

DLS measurement allows the estimation of hydrodynamic diameter of dendriplex.

Since drug-dendrimer interactions may significantly affect the zeta potencial values and size of nanocarriers, the application of zeta potential and DLS measurements appears to be a good choice for the experiments involving the determination of stability of complexes and their behavior under different environmental conditions.

To date, zeta potential and zeta size analysis were applied for the characterization of PAMAM-based nanocarriers loaded with 5-fluorouracil and gemcitabine. For instance, zeta potential of PAMAM G5 equaled +49.6 mV and decreased to +44 mV during the formation of dendrimer stabilized silver nanoparticles (DsAgNPs). This slight reduction may be ascribed to the involvement of a several terminal amino groups of the dendrimer in stabilization of the nanoparticle surface. Interestingly, zeta potential increased to +45.6 mV upon 5-FU encapsulation, indicating higher stability of 5-FU-DsAgNPs complexes than that of DsAg nanoparticles [61].

Another example involves a derivative of PAMAM G4, poly(2-(N,N-diethylamino)ethyl methacrylate) (PDEA) with methoxy-poly(ethylene glycol)-poly(amido amine) (PPD). As mentioned before, PDEA is a pH-sensitive polymer—under weakly acidic conditions, PDEA chains of PPD are hydrophilic and extended, allowing 5-fluorouracil to enter or exit the nanostructure. In neutral or weakly alkaline environment, PDEA chains are hydrophobic and contracted, enabling efficient encapsulation of the drug. DLS and zeta potential experiments were used to characterize the drug-loaded nanocarriers at different pH values. The diameter of PPD equaled 43.0 nm at pH 4.0, 41.6 nm at pH 6.5, and 11.6 nm at 7.4, which was most probably associated with the expansion and contraction of PDEA chains at different pH values, affecting the size of nanocarriers. The zeta potential values of PPD nanocarriers varied in different pH media, and equaled 12.4, 9.7 and 7.1 mV at pH 4.0, 6.5 and 7.4, respectively. Since most of primary surface NH_2 moieties of the dendrimer were modified or covered by PEG and PDEA chains, the positive potential may be related to the presence of tertiary amines in PDEA [51].

For the specific delivery of gemcitabine to pancreatic cancer cells, PEG-cored PAMAM G3.5 dendrimers modified with PEG chains (PEGcPAMAM-PEG) were conjugated with Flt-1 antibody targeting a receptor for vascular endothelial growth factors (VEGF). Mean particle diameter was analyzed with DLS, and zeta potential of complexes was measured. The experiments were performed in deionized water.

Drug-dendrimer complexes were prepared using 200, 400 and 600 μM gemcitabine concentrations per 100 μM PEGcPAMAM-PEG dendrimers, which resulted in maximum efficiency of encapsulation of 80, 125 and 132 μM gemcitabine, respectively. Gemcitabine loading increased the size of both PEGcPAMAM-PEG dendrimers and their antibody-conjugated derivatives (PEGcPAMAM-PEG-Flt1) compared to that of unmodified, drug-free dendrimer (174.13 ± 11.3 nm). The particle size values of all gemcitabine-dendrimer complexes were found in the range of approximately 200–300 nm, and were not dependent on the amount of the encapsulated drug. Relatively high particle sizes may be due to aggregation of dendrimers.

Encapsulation of gemcitabine by PEGcPAMAM-PEG led to the decrease in zeta potential values, which correlated with the increase of the amount of gemcitabine loaded, and equaled −16.56 ± 1.3, −8.9 ± 2.25 and −5.34 ± 2.5 mV for PEGcPAMAM-PEG/Gem80, PEGcPAMAM-PEG/Gem125 and PEGcPAMAM-PEG/Gem132, respectively, with the value of −21.9 ± 0.3 mV for free PEGcPAMAM-PEG dendrimer. The interaction between dendrimers and drug molecules could be the result of entrapment of gemcitabine within the dendritic scaffold (involving electrostatic, hydrophobic and hydrogen bonding) and may additionally involve interactions with the surface of dendrimer (e.g. electrostatic and covalent bonding) [72].

PAMAM G4.5-G7.5 dendrimer-coated magnetic nanoparticles (DcMNPs) were synthesized and conjugated with gemcitabine. The conjugation was supposed to occur as result of an electrostatic interaction of the drug with terminal moieties of the dendrimer. The zeta potential values of DcMNPs were measured at pH 7.2. In aqueous solutions, full generation DcMNPs were positively charged with a surface potential greater than +19 mV due to the presence of protonated amino groups of the dendrimer. On the other hand, G5.5 DcMNPs had negative zeta potential (−10 mV) provided by terminal carboxyl groups. The zeta potential of Gem-loaded G5.5 DcMNPs was also negative (−4.85 mV). For this type of nanocarrier, the highest drug loading was observed, while PAMAM G4 and G5 did not effectively bind gemcitabine. These results supported the hypothesis that gemcitabine binding was achieved on the surface of the nanoparticles with negatively charged carboxylic groups. In this study, DLS analysis were used to assess the average diameter of bare MNPs, which equaled 55 ± 15 nm [75].

8 Chromatography

Chromatography is a simple biophysical technique enabling separation, identification and purification of compounds in a mixture, taking advantage of the differences in their affinity for various materials. Basically, the mixture is dissolved in a liquid called "mobile phase", which passes through or over immobilized "stationary phase". The separation is based on differential partitioning of compounds between the stationary and mobile phases. Chromatographic techniques are usually divided into two categories: preparative or analytical. The purpose of preparative chromatography is to separate and purify the components of a mixture for subsequent use. Analytical chromatography is usually carried out with smaller amounts of samples and is meant to characterize the relative proportions of analytes in a solution. General classifica-

Table 10 Classification of chromatographic methods

Type of mobile phase	Gas chromatography	
	Liquid chromatography	
Type of stationary phase	Gas-liquid chromatography (GLC)	
	Gas-solid chromatography (GSC)	
	Liquid-liquid chromatography (LLC)	
	Liquid-solid chromatography (LSC)	
Type of contact between mobile and stationary phase	Planar chromatography	Paper chromatography (PC)
		Thin layer chromatography (TLC)
	Column chromatography	

tion of chromatographic methods in terms of the type of mobile and stationary phase is presented in Table 10.

The type of interaction between stationary phase, mobile phase and components of the mixture is the decisive factor during the separation. Chromatography methods based on partition are very effective for separation and identification of small molecules such as amino acids, carbohydrates or fatty acids. In case of macromolecules such as nucleic acids or proteins, affinity chromatographies (e.g. ion-exchange or size-exclusion chromatography) are more effective [29].

Chromatographic techniques are not standard procedures for characterization of nucleotide-dendrimer complexes. However, Szulc et al. [86] developed interesting application of fast protein/performance liquid chromatography (FPLC) to study formation of complexes between maltose-modified PPI dendrimers and ATP. FPLC has been developed in 1982 by Swedish company "Pharmacia", and called "fast performance liquid chromatography" to differentiate it from high performance liquid chromatography (HPLC). Since it is most often used to analyze or purify mixtures of proteins, the common name found in the scientific literature is "fast protein liquid chromatography". However, because of the wide range of stationary and mobile phases, it has numerous applications for different kinds of biological samples providing ion exchange, gel filtration, hydrophobic interactions and affinity chromatography. In contrast to HPLC, the pressure of the liquid is moderately low (less than 5 bars), with the flow rate being relatively high (1–5 ml/min). Since FPLC does not require high pressures to execute the separation process, the risk of destruction of drug-dendrimer complexes by shear forces is reduced. FPLC also reduces the contamination of the sample and destruction of active compounds.

Due to the lack of pressure control requirement, it is possible to work with transparent and biocompatible columns, which enable monitoring of the process. The flow rate is controlled by a system of positive-displacement pumps and kept constant, while the composition of the buffer can be changed by drawing fluids in different proportions from external reservoirs. The stationary phase is composed of beads (usu-

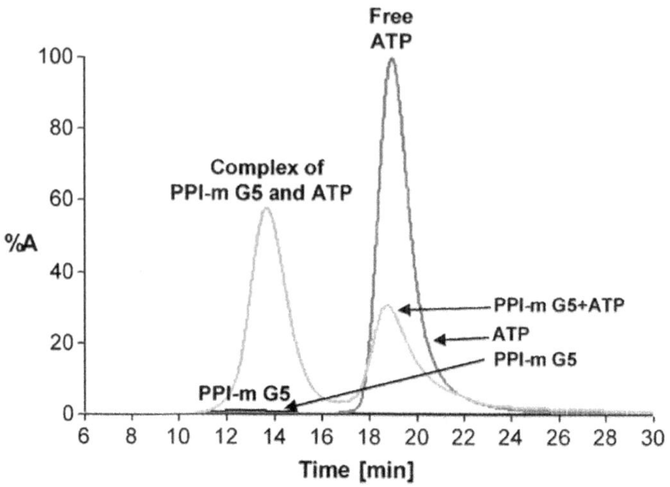

Fig. 20 Chromatograms of separating of PPI-Mal G4 (according to the old nomenclature—G5) dendrimers, ATP and ATP-PPI-Mal G4 dendrimer complex. Reprinted from Szulc et al. [86] with permission from the Centre National de la Recherche Scientifique (CNRS) and the Royal Society of Chemistry

ally of cross-linked agarose) which are available in a different sizes and composition depending on the application [60].

Szulc et al. studied the complexation of maltose-modified G3 and G4 PPI dendrimers with physiological nucleotides (adenosine-5′-monophosphate (AMP), adenosine-5′-diphosphate (ADP) and ATP). Dendrimers were dissolved in 20 mM HEPES buffer (pH 7.0) with an addition of NaCl. A nucleotide (ATP, ADP or AMP) was added to the dendrimer solution at a 25:1 molar ratio. Mixtures were stirred for 1 h at 4 °C and transferred to FPLC HR 10/30 column (Ø 10 mm, 30 cm length) packed with Sephacryl 100-HR. The experiment was performed at room temperature and 1 ml/min flow rate. The absorbance was measured at 254 nm. Sample measurement result is shown in the Fig. 20.

Retention times of the PPI-Mal dendrimer, the ATP-PPI-Mal dendrimer complex and free ATP were 13, 14 and 19 min, respectively. Similar retention times for the pure PPI-Mal dendrimer and the PPI-Mal dendrimer complexed with ATP are a consequence of comparable sizes of the dendrimer and the complex. Surprisingly, PPI-Mal G3 dendrimer did not form complexes with ATP (approximately 1 molecule of ATP was complexed per 10 molecules of PPI-Mal G3 dendrimer). By contrast, about 17 ATP molecules attached to 10 molecules of PPI-Mal G4 dendrimer. This effect was most probably associated with the availability of surface amino groups, responsible for the interactions with nucleotide. The hypothesis was further proved by the fact that ATP was the only nucleotide bound by PPI-Mal G4. Experiments with ADP showed a very weak binding (approximately 1 molecule of ADP was bound per 10 molecules of PPI-Mal G4), and AMP did not create complexes, showing that the triphosphate form is essential to complexation. Moreover, the complexes between PPI-Mal G4 and ATP were formed immediately upon mixing and were stable for at least 48 h.

Additional studies indicated that complex formation depends on concentration of NaCl in the solution and pH of solution. Na^+ and Cl^- ions compete with the nucleotide for binding sites in the dendritic scaffold and in the oligosaccharide shell of PPI-Mal dendrimers, while the pH of the buffer influences protonation of terminal NH_2 moieties, favoring the binding of nucleotide in acidic environment.

Most importantly, these dendrimers protected nucleotides from the enzymatic degradation by alkaline phosphatase, which separates phosphate groups from nucleotides. Complexes of PPI-Mal G4 and ATP were incubated for one hour with the enzyme (5 units/ml) in 20 mM HEPES buffer (pH 8.0, 100 mM NaCl) at 37 °C. Samples were subsequently analyzed by the FPLC method. It was found that when complexes were subjected to the activity of alkaline phosphatase, the peak characteristic for the complex was still visible, although it was a bit smaller due to the fact that 26% of ATP was detached and degraded to adenosine. There was no peak characteristic for unbound ATP, and a new peak with an average retention time equal to 29 min appeared that was attributed to adenosine (Szulc et al. [86]. This outcome indicates that nucleotide forms of therapeutics NAs could be delivered to the cell by dendrimers in an active, phosphorylated form.

For the in vivo evaluation of biodistribution of NAs-dendrimer complexes, HPLC is commonly applied. Although it does not enable direct assessment of the action of complexes, it provides comparison of the concentration of the drug both in free and complexed form in various organs of laboratory animals, thus proving the improved transport of the therapeutic by nanoparticles. HPLC relies on the system of pumps to pass mobile phase through columns under the pressure of 50–350 bars, while ordinary liquid chromatography usually depends on the force of gravity. Typical HPLC columns are made of pressure-resistant stainless steel, 2.1–4.6 mm in diameter and 30–250 mm in length. They are filled with solid phase composed of a granular material of 2–50 μm solid particles with a large resistance to high pressures (e.g. silica), providing superior resolving power. The liquid phase is normally a mixture of solvents (e.g. water, acetonitrile and/or methanol). Its composition and temperature play a major role in the separation process by influencing the physical interactions between sample components and adsorbent.

HPLC offers a quick, reproducible and accurate identification of sample components. Due to the use of pumps and high pressure, HPLC is extremely efficient in comparison to other chromatographic techniques. Due to its high automation, basic HPLC protocols are relatively easy to perform. However, despite the advantages, HPLC can be expensive, requiring large quantities of expensive compounds. Moreover, HPLC has low sensitivity for certain substances, and some of them (mainly volatile ones) cannot be detected [57].

In vivo studies of the activity and biodistribution of drug-dendrimer complexes have been presented in the last section of this chapter.

9 Fluorometric Analysis

Fluorescence is a term used to describe the luminescent emission of light by a substance absorbing electromagnetic radiation. Fluorescence is a feature of certain compounds (usually heterocyclic or polyaromatic), known as fluorophores. These compounds absorb light of a certain wavelength and emit light of a longer wavelength, which is shifted from the absorbed light toward the red end of the spectrum. The difference between the maxima of the wavelengths of absorption and emission is called the Stokes shift. This phenomenon is essential for the sensitivity of fluorescence techniques, as it enables the detection of emission against the background of excitation photons. Since the wavelengths of the excitation and emission light are different, they can be separated using optical filters. Each fluorophore has its own specific absorption and emission spectra, depending on the structure of the molecule and its environment. In biological studies, particular fluorophores called fluorescent probes are used. These compounds are designed to respond to a specific stimuli or to localize within a destined region of a sample [10].

Fluorescence is an effect of three-step process. First, a photon of energy from an external source is absorbed by the fluorophore, exciting it from the ground state to the electronic singlet state. The amount of energy required for this transition differs between fluorophores. This stage differentiate fluorescence from chemiluminescence, in which the excited state is achieved as a result of a chemical reaction. The excited state lasts for a limited time (typically up to 10 ns), when the fluorophore molecule undergoes conformational changes, which may affect its interactions with surrounding environment. The length of the decay time allows to discriminate between fluorescence and phosphorescence, since the latter lasts from milliseconds to seconds. By contrast, fluorescent substances cease to glow immediately when the exciting energy is stopped.

The energy of an excited singlet state is partially dissipated, yielding a relaxed singlet state from which fluorescence emission comes from. In the end, the fluorophore returns to its ground state, emitting a photon of energy. Due to the energy dissipation during the duration of excited state, the energy of emitted photon is lower (resulting in the longer wavelength of emission light) than the excitation photon. This cycle can be repeated many times for a single fluorophore, which enables amplification of the signal. Fluorescence intensity is defined by the Beer–Lambert law, similarly to absorbance [42].

Fluorescence detection sensitivity may be hampered by background signals, originating from autofluorescence of the sample or from nonspecific binding of fluorescent probes. Detection of autofluorescence of biological samples can be minimized either by application of appropriate filters or by selection of probes that can be excited at longer wavelengths (>500 nm). At longer wavelengths, light scattering by dense media such as tissues is significantly reduced, enabling improved penetration of the excitation light.

Fluorescence instruments can be generally divided into four types, basing on information they provide:

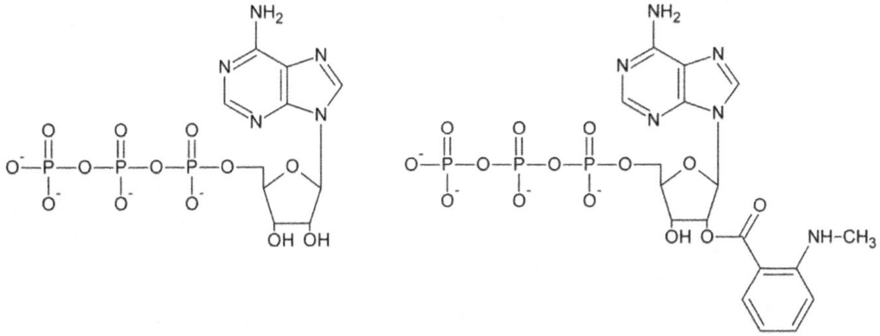

Fig. 21 Chemical structures of ATP and Mant-ATP

- spectrofluorometers and microplate readers, measuring the general fluorescent properties of the sample,
- fluorescence microscopes resolving fluorescence as a function of spatial coordinates for microscopic samples,
- fluorescence scanners resolving fluorescence as a function of spatial coordinates for objects such as chromatograms, blots or electrophoresis gels,
- flow cytometers measuring fluorescence of a single cell in a flowing stream of fluid.

Fluorescence measurements as a part of flow cytometry and microscopy analysis are presented in the last section of this chapter. Here, we indicate the possible application of fluorescence titration for the experiments involving nucleotide-dendrimer interactions. Most dendrimers (except for fluorescent and labelled macromolecules) are able to absorb light but do not exhibit fluorescence, which enables fluorescence experiments with the use of suitable ligands.

Fluorescence titration is a method for determining the properties of a sample by subsequent addition of portions of fluorescent titrant. The titration is followed by measurement of the intensity of fluorescence emitted by the sample. In case of drug-dendrimer interactions, the binding process is determined on the basis of the concentration of the free and complexed fluorescent ligand. The interactions of ligand with the macromolecule is examined by monitoring changes in the emission intensity accompanying complex formation [22].

Szulc et al. [87] applied double fluorimetric titration method to estimate the binding constant and the number of binding centers of dendrimer using 2′-/3′-O-(N′-methylanthraniloyl)-ATP (Mant-ATP), a spectrofluorimetric dye the spectrum of which strongly depends on the features of surrounding microenvironment (Fig. 21). Mant-ATP in aqueous solution emits weak fluorescent signal in the range of 380–540 nm with a maximum at 440 nm. The fluorescence intensity increases significantly in nonpolar environment or upon protein binding. Similar effect was observed for maltose-modified PPI dendrimers.

In the first step, the dendrimers were added in increasing concentrations to a solution of Mant-ATP, and the fluorescence intensity of the dye at the maximum emission wavelength (440 nm) was analyzed. The step when the fluorescent signal reached maximum was considered the final stage of complexation, with all molecules of the ligand in the solution bound to the (added) dendrimer. In the subsequent experiment the situation was reversed: the ligand was added at increasing concentrations to a solution of the dendrimer. In both cases, the fluorescence intensity was recalculated for the concentration of ligands bound to the macromolecule, and the concentrations of bound versus unbound molecules.

It was observed that the intensity of the fluorescent signal increased significantly upon the addition of dendrimers to the solution of Mant-ATP. This change was accompanied by a blue-shift of the emission maximum. The strongest effect was observed for PPI dendrimers partially coated with maltose, weaker for unmodified PPI dendrimers in an acidic solutions, and weakest for unmodified PPI dendrimers at neutral pH. Those experiments confirmed that complexation of PPI dendrimers with nucleotides is generation- and pH-dependent, enabling stronger interactions in acidic environment.

Additionally, this method allowed to determine whether nucleotide is bound more in the interior or on the exterior of dendritic macromolecule. Interestingly, in case of unmodified PPI dendrimers at neutral pH, the number of ligand molecules bound to the surface was not dependent on the generation. By contrast, about three times more Mant-ATP molecules were complexed inside the scaffold of the dendrimer of higher generation. Further, the dye was efficiently encapsulated only in case of PPI-Mal G3 macromolecule, but no Mant-ATP was detected in the interior of PPI-Mal G4 dendrimers, probably due to the dense surface sugar layer. Moreover, incorporation into the interior was not observed for naked PPI dendrimers under acidic conditions, indicating once again the electrostatic interactions between terminal protonated amino groups of cationic dendrimers and negatively charged Mant-ATP molecules [87].

10 Drug-Dendrimer Complexes in Biological Systems—In Vitro and in Vivo Studies

Probably the most important research stage concerning the application of positively charged dendrimers for the delivery of nucleoside derivatives involves evaluation of their activity and impact on living cells and organisms. This phase can be generally divided into two parts:

- basic experiments on in vitro cultured cell lines, in order to verify the activity of complexes in comparison to free drugs, and to select the appropriate concentrations for in vivo testing; those experiments primarily involve evaluation of mechanisms of action/toxicity and cellular uptake of drug-dendrimer formulations;
- animal studies enabling confirmation of cell culture experiments and overall safety of drug-dendrimer formulations before proceeding to clinical trials; these are based

on the assessment of pharmacokinetics, pharmacodynamics and biodistribution patterns.

In this section we summarize techniques and methods most commonly used for the evaluation of activity of drug-dendrimer complexes in biological systems.

It is important to note that despite the confirmed stability of drug-dendrimer non-covalent complexes, it is difficult to determine their relative durability during prolonged periods of storage. Several studies revealed that dendritic polymers in solutions may undergo spontaneous degradation [59], which can greatly affect the reproducibility of biological experiments. Therefore, the complexes should be freshly prepared and used immediately upon preparation. Moreover, it is important to use buffering solvents that do not hamper the complexation process and may be applied in in vivo studies.

10.1 Cell Viability Assays

Cytotoxic activity is the basic mechanism of action of chemicals used in anticancer therapies. For the in vitro studies, techniques enabling the measurement of changes in the basic physiological processes occurring in the living cell are most commonly used. These are mainly based on the evaluation of enzymatic and metabolic activity of cells (e.g. tetrazolium reduction assays (MTT, MTS, XTT, and WST), resazurin reduction assay or measurement of ATP content), integrity of cellular membranes (e.g. lactate dehydrogenase (LDH) activity assay, neutral red uptake assay or propidium iodide (PI)-based assay), proliferation rate (clonogenic assay) or DNA and protein synthesis. The choice of method depends on the type of cell line (adherent or suspension culture) and its metabolic activity, as well as the activity and specificity of the drug.

Since nucleoside analogues are classified as antimetabolites, metabolic activity assays seem to be the most appropriate selection. However, it is possible that dendrimers and/or drugs may interact with compounds used in the particular assays or impact their intracellular metabolism, significantly reducing they reliability. It is crucial to take it into consideration during the evaluation of results.

One of the most common colorimetric methods involves the reduction of tetrazolium dye 3-(4,5-dimethylthiazol-2-yl)-2,5-diphenyltetrazolium bromide (MTT) to insoluble formazan by NADH or NADPH-dependent cellular oxidoreductases [84]. Yellow tetrazolium salt is endocytosed by living cells and reduced to insoluble purple formazan crystals that accumulate inside the cell, being incapable of penetrating the cell membrane. Those crystals are soluble in organic solvents such as DMSO or isopropanol, giving colored solution. Formazan is formed only in undamaged, metabolically active cells, and hence the intensity of the spectrophotometrically measured absorbance of colored solution is directly proportional to the number of living cells. The absorbance can be quantified at a certain wavelength, usually between 500 and 600 nm.

Sample MTT assay protocol

1. *Seed the cells at optimum density into 96-well plates.*
2. *Add tested compounds.*
3. *Incubate for desired period of exposure.*
4. *After the incubation, gently pour the medium from the wells, dry the plates on a paper towel.*
5. *Wash the cells with PBS.*
6. *Add 50 μl of MTT solution (0.5 mg/ml) to each well.*
7. *Incubate the plate for 3 h at 37 °C*
8. *Remove MTT solution, dry the plates on a paper towel.*
9. *Gently dissolve formazan crystals, e.g. with DMSO.*
10. *Measure the absorbance at 570 nm.*

Water-soluble tetrazolium salts, prepared by introduction of charged moieties to the phenyl ring, were proposed as alternatives to MTT. Among them, XTT (2,3-bis-(2-methoxy-4-nitro-5-sulfophenyl)-2H-tetrazolium-5-carboxanilide) exhibits high sensitivity and dynamic range. MTS (3-(4,5-dimethylthiazol-2-yl)-5-(3-carboxymethoxyphenyl)-2-(4-sulfophenyl)-2H-tetrazolium) in the presence of phenazine methosulfate produces a formazan product with an absorbance maximum at 490 nm in PBS. The MTS assay offers the advantage of adding the reagent directly to the cell culture, contrary to the MTT assay. However, this makes the MTS assay susceptible to colorimetric interference as the subsequent steps in the MTT assay remove residues of colored compounds. It is suggested that MTS results should be confirmed with qualitative observation under a microscope. WSTs (Water-soluble tetrazolium salts) are other water-soluble dyes for MTT assays, featured with different absorption spectra of formazan products. WST-1 and WST-8 are advantageous over MTT being reduced outside the cells. WST assays do not require solubilization step, provide stronger signal than MTT and are less toxic to the cells.

Since the reduction of tetrazolium salts depends on the activity of oxidoreductase enzymes, it is also dependent on the NADH/NADPH pool. Cells with slow metabolism such as splenocytes or thymocytes may reduce MTT ineffectively. By contrast, rapidly dividing cells show quick MTT reduction. It is also crucial to note that reaction conditions may change metabolic activity and thus tetrazolium dye reduction without altering cell viability. The mechanism of reduction of tetrazolium dyes, i.e. extracellular (WST-1) vs. intracellular (MTT, MTS) may also influence the amount of product.

Another test based on enzymatic activity is the resazurine reduction assay [70]. Resazurin (also known under the commercial trademark Alamar Blue) is a cell permeable redox indicator that can be used to assess cell viability with protocols similar to those utilizing tetrazolium compounds. It can be dissolved in physiological buffers and added directly to cell culture. Blue non-fluorescent resazurin penetrates to the cytoplasm where it is reduced to pink fluorescent resorufin. Reduction of this compound is most possibly performed by reductase called diaforase, present both in mitochondria and cytoplasm. The amount of produced resorufin is proportional to the number of viable cells which can be quantified using a microplate fluorometer with a 530 nm excitation/590 nm emission filters. The quantity of resorufin may be also evaluated by the measurement of absorbance, however it is not recommended since measuring absorbance is significantly less sensitive than fluorescence measurement.

Resazurin assays exhibit good correlation with MTT, while being more sensitive and much easier. The incubation time required to acquire an adequate fluorescent signal is usually 1 to 4 h and depends on the metabolic activity of the cells, as well as cell density, type of culture medium and other assay conditions. The incubation period should be optimized and provide the right balance between reagent toxicity and assay sensitivity. The drawbacks of resazurin include the possibility of fluorescent interference of compounds under evaluation and direct toxic effects on the cells during prolonged incubations.

Sample resazurin assay protocol

1. *Seed the cells at optimum density into 96-well plates.*
2. *Add tested compounds.*
3. *Incubate for desired period of exposure.*
4. *After the incubation, add resazurin to the culture medium to a final concentration of 10 μg/ml.*
5. *Incubate the plates at 37 °C in the darkness for 2 h to allow conversion of resazurin to resorufin.*
6. *Measure the fluorescence of metabolized resazurin at 530 nm excitation and 590 nm emission wavelengths.*

MTT assay is the most commonly used for evaluation of cytotoxicity of complexes between positively charged dendrimers and nucleoside derivatives. In most cases, the complexes exhibited superior cytotoxic activity compared to free drug. PAMAM G5 dendrimers stabilizing silver nanoparticle surface (DsAgNPs) were used to encapsulate 5-FU, and such constructs were found to elicit an antiproliferative effect in A549 (human lung cancer) and MCF-7 (human breast cancer) cells with enhanced IC50 compared to free drug [61]. 5-FU-PAMAM G5 complexes combined with antisense micro-RNA 21 (as-miR-21) were also tested for growth suppression of breast cancer cells, giving positive results. The codelivery of as-miR-21 significantly improved the cytotoxicity and chemosensitivity of 5-FU, increased the apoptotic percentage of the MCF-7 cells and decreased their migration ability [62].

PEG-cored PAMAM dendrimers with surface PEGylation were loaded with gemcitabine and examined on CFPAC-1 pancreatic adenocarcinoma cells for 24 or 48 h. When gemcitabine was loaded into dendrimers, its anticancer effect was improved but free dendrimers also contributed to the reduction of cell viability. The conjugation of the antibody specific for Flt-1 (surface protein expressed by CFPAC-1 pancreatic adenocarcinoma cell line) intensified both these effects [72]. Further, the MTT studies showed that gemcitabine in complex with mannose-modified PPI dendrimers had significantly higher cytotoxicity than Gem-PPI formulations and free drug, probably due to the enhanced endocytosis of sugar-dendrimer conjugates [83].

Gemcitabine-loaded PAMAM dendrimer-coated magnetic nanoparticles (DcMNPs) were able to increase the cytotoxicity of the drug against pancreatic cancer cell lines (SU86.86, T3M4, Panc-1) and primary human pancreatic stellate cells (PSCs) [99]. Similar results were obtained by XTT assay on SKBR-3 and MCF-7 breast cancer cell lines [75].

Resazurin-based test was used to evaluate the potential of maltose-modified PPI G4 dendrimers to transport active triphosphate form of Ara-C (Ara-CTP) and facilitate its delivery to the cancer cells. The cytotoxicity assay showed enhanced activity of Ara-CTP complexed with PPI-Mal OS G4 in comparison to free Ara-C and Ara-CTP against 1301 (acute lymphoblastic T-cell leukaemia) cell line. What is more, an enhanced uptake and cytotoxic activity of drug-dendrimer complexes in 1301 cells with blocked human equilibrative nucleoside transporter (hENT1) have been observed, indicating that such a system could overcome drug resistance associated with decreased expression of membrane transporters [88]. This phenomenon has been confirmed with similar approach for another anticancer nucleoside analogue—PPI G4 dendrimers partially coated with maltose were able to deliver fludarabine triphosphate (Ara-FATP) to CCRF (acute lymphoblastic T-cell leukaemia, subline 1301), THP-1 (acute monocytic leukaemia) and U937 (histiocytic lymphoma) cells with blocked hENT1 [43].

10.2 Flow Cytometry

Flow cytometry is a widely used biophysical method for analysis and characterization of different cell types in a heterogeneous culture. This technique enables the measurement of optical and fluorescent characteristics of a single cell in a fluid stream passing through a laser beam. The cells are differentiated on the basis of their size, granularity and fluorescent features. Flow cytometry analysis takes advantage of light scattering and fluorescence emission, which occur as light from the excitation source strikes the moving cells. The obtained data provide valuable information about biochemical, biophysical and molecular characteristics of cells. Light scattering is directly related to structural and morphological properties of the cell, while fluorescence emission is proportional to the amount of fluorescent probe bound to the cellular components [2].

The procedure involves preparation of a cell suspension from culture or tissue samples, incubation in tubes or plates with unlabelled or fluorochrome-labelled antibodies and analysis on the flow cytometer. Cells are usually suspended at a density of 10^5–10^7 cells per milliliter to allow their undisturbed flow through the measuring apparatus. The concentration also influences the flow sorting, which typically progresses at 2000–20,000 cells per second. Higher sorting speeds can lower the efficiency of measurement. PBS is commonly used as suspension buffer.

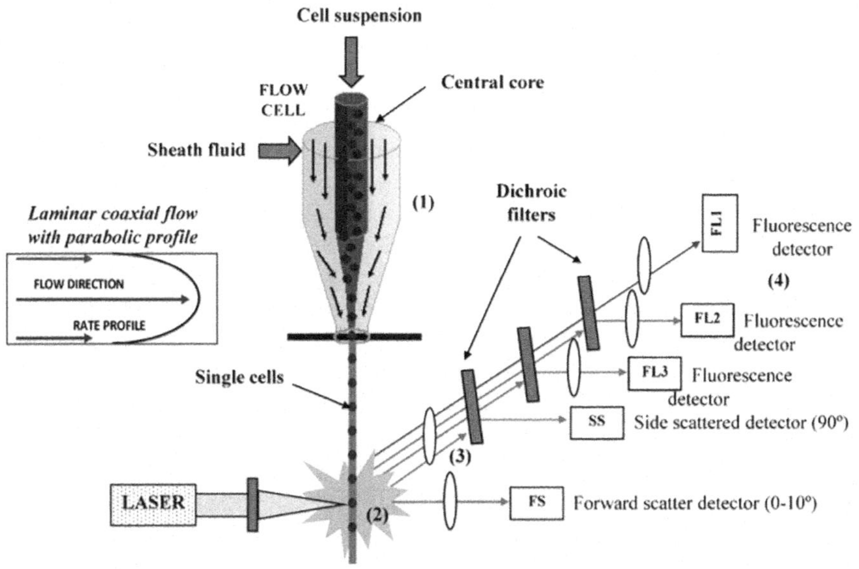

Fig. 22 Scheme of flow cytometer. (1) Formation of a single stream of cells due to the hydrodynamic focusing. (2) Signal emission by cells passing through a laser beam. (3) Separation of scattered and fluorescence signals of each cell according to certain wavelengths. (4) Signal collection by detection system. Reprinted from Diaz et al. [32] with permission from Elsevier

The most basic samples for flow cytometry include cells from suspension cultures, however it is possible to analyze adherent cells or those from solid tissues, which requires mechanical or enzymatic disaggregation. To study intracellular components by flow cytometry, plasma membrane must be permeabilized to allow dyes or antibodies through while retaining the overall integrity of cells. Low concentrations (e.g. 0.1%) of nonionic detergents like Triton X-100 are most appropriate.

When a cell suspension is run through the cytometer, each cell passes separately through one or more laser beams. Detection and processing of light signals generated by passing cells are performed by photodetectors. Two types of detectors, photodiodes (PDs) and photomultiplier tubes (PMTs) can be chosen based on their sensitivity. Light measured in the forward direction is collected by photodetectors in the forward scatter (FSC) channel and correlates with the cell size. Light scattered at an approximately 90° angle to the excitation line is called side scatter (SSC), and provides information about granularity of a cell and its internal structures (Fig. 22). Both FSC and SSC are unique for each cell, and their combination may be used to differentiate cell types in a heterogeneous sample. However, as the outcome is dependent on the sample type and quality, additional fluorescent labelling is recommended. Fluorescence measurements at different wavelengths can provide qualitative and quantitative information on fluorophore-labelled cell surface receptors or intracellular components such as DNA.

Any time a cell passes through the stream of light, a voltage pulse is generated and detected by PMTs or PDs. These pulses, known as "events" can be mapped by plotting signal as a function of time. Analog signals are generated after the conversion of initial light signal into an electrical current by the detectors, and subsequently converted into digital signal for computer processing. Digital data may be displayed as plots or histograms.

Flow cytometry analysis is primarily based on the principle of gating. "Gates" are drawn on scatter histograms and plots to select the populations of interest. The first step in gating aims at the differentiation of cells due to their light scatter properties. For example, subcellular debris can be distinguished from living cells on the basis of size, estimated by forward scatter. Moreover, dead cells are usually characterized by higher side scatter and lower forward scatter than living cells.

Flow cytometry additionally enables the electrostatic cell sorting for separation of the cells with preferred features. Once the cells of interest are collected, they can be used for further analysis such as microscopic, biochemical and functional studies. Most commonly used fluorescence-activated cell sorting (FACS) provides the sorting of heterogeneous mixture of fluorescently labelled cells based on the fluorescent characteristics.

In case of the analysis of toxicity of NAs-dendrimer complexes, flow cytometry has been used for the determination of number of apoptosis-involved cells. Apoptotic cells exhibit low FSC values due to the reduction in size, with initially high SSC values corresponding to chromatin condensation and DNA fragmentation, which increases the degree of light reflection and refraction. In the late stages of apoptosis, both parameters decrease considerably. These studies may additionally involve the application of propidium iodide (PI), a fluorescent, DNA intercalating agent that can be used to evaluate cell viability or DNA content in cell cycle analysis. PI cannot cross the membrane of living cells, which makes it useful to differentiate necrotic, apoptotic and healthy cells.

A common signal of apoptosis involves rapid transfer of phosphatidylserine (PS) from the inner to the outer side of plasma membrane. The externalized PS enables the recognition of apoptotic cells by the interactions with annexin V, anionic lipid-binding protein. Annexin V is widely applied for the detection of apoptotic cells in vitro and shows the potential for determination of treatment efficacy in patients with cancer [58]. During flow cytometry analysis, annexin V-positive and PI-negative cells may be defined as early apoptotic cells with an intact cell membrane with externalized phosphatidylserine, while both annexin V- and PI-positive cells may be considered as dead cells. This method was used by Mei et al. [62] to investigate the role of as-miR-21 in 5-FU-induced apoptosis in MCF-7 cells. The decrease of miR-21 expression increased the apoptotic rate of MCF-7 cells (19%) compared with control (4%) or that induced by 5-FU-PAMAM G5 complex as a single agent (9%). In these studies, flow cytometry was also used to evaluate the cellular uptake of as-miR-21, which has been shown to increase from 3.58 to 54.54% upon binding with the dendrimer. This outcome was additionally confirmed by fluorescence microscopy observation.

Sample protocol for apoptosis assay by flow cytometry

1. *Seed cells (1 × 10^6 cells) in culture flasks for experiments and controls (unstained, Annexin V-stained and PI-stained).*
2. *After desired period of exposure, collect the medium and trypsinize the adherent cells from each flask.*
3. *Wash the collected cells twice with PBS and centrifuge (400× g, 5 min, RT).*
4. *Re-suspend each pellet in Flow Cytometry Staining Buffer (400 μl).*
5. *For experimental samples, add 100 μl of Flow Cytometry Staining Buffer containing 2 μl of Annexin V [1 mg/ml] and 2 μl of PI [1 mg/ml].*
6. *For control samples:*
   - *Unstained control—add 100 μl of Flow Cytometry Staining Buffer*
   - *Annexin V-stained control—add 100 μl of Flow Cytometry Staining Buffer with 2 μl of Annexin V (1 mg/ml)*
   - *PI-stained control—add 100 μl of Flow Cytometry Staining Buffer with 2 μl of PI (1 mg/ml)*
7. *Incubate samples on ice in the darkness for 20 min.*
8. *Analyze the cells using flow cytometer. Cells that are PI-negative and Annexin V-negative are considered healthy. PI-negative and Annexin V-positive cells are considered apoptotic, and cells that are positive to both PI and Annexin V are considered necrotic.*

Similar approach involving the use of PI and YO-PRO-1 dyes was applied by Szulc et al. [88] to evaluate the toxicity of Ara-C, Ara-CTP and Ara-CTP-PPI-Mal OS G4 complexes. The determined signal of cell death was the loss of membrane integrity. The membrane of apoptotic cells is permeable only to YO-PRO-1, whereas that of dead cells is permeable to both PI and YO-PRO-1. The obtained results for percentages of apoptotic and dead cells after incubation with tested compounds corroborated the preliminary conclusions drawn from viability assay.

As mentioned before, flow cytometry enables the evaluation of cellular uptake of nanoparticles. For instance, to determine intracellular accumulation PEG-cored PAMAM dendrimers modified with PEG prepared for the delivery of gemcitabine, CFPAC-1 cells were treated with rhodamine 123-loaded macromolecules. Rhodamine 123 (Rho 123) is a cationic, green-fluorescent dye readily sequestered by active mitochondria without cytotoxic effects. It is also frequently used as a tracer dye. It was shown that coating with anti-Flt-1 antibody increased the cellular uptake of dendrimers under evaluation. Even at low dendrimer concentrations (5 and 10 μM) Flt-1-expressing cells exhibited higher uptake level. This subpopulation was also visualized under the fluorescent microscope [72].

Evaluation of intracellular reactive oxygen species production is also possible with flow cytometry. This method was used for the studies on 5-FU complexed with DsAgNPs. MCF-7 and A549 cells were treated with appropriate concentrations of compounds, and ROS generation was determined with 2,7-dichlorofluorescin diacetate (DCFH-DA) dye. DCFH-DA is a cell permeant fluorogenic probe. Upon entering the cell, it is converted to highly fluorescent 2,7-dichlorofluorescein (DCF) during oxidation. After treatment, samples were analyzed for DCF fluorescence using flow cytometer. The results showed an increase in the number of cells with elevated DCF fluorescence with an increase in 5-FU-DsAgNC concentration for both cell lines, suggesting that complexes may induce oxidative stress. These findings corroborated the MTT results [61].

10.3 Microscopy

Microscopy is a technique commonly used for observation of samples that are beyond the resolution range of the human eye and cannot be evaluated without proper magnification. It can be generally classified into three groups: optical (including light and fluorescence), electron and scanning probe microscopy. Optical and electron microscopy are based on the diffraction, reflection or refraction of laser/electron beams interacting with the sample. The collection of generated signal is necessary to create an image for further evaluation. This may be performed by wide-field irradiation of the specimen or by scanning a fine beam over it. Scanning probe microscopy creates images of surfaces using a physical probe scanning the sample.

Microscopy techniques may be a valuable tool for the characterization of architecture and activity of complexes of cationic dendrimers or their derivatives and nucleoside analogues. The selection of appropriate method may provide information on both the structure of drug-dendrimer formulations and their action in living cells, e.g. intracellular accumulation or mode of inducing cell death.

10.3.1 Light Microscopy

The optical/light microscopy is one of the simplest techniques which uses visible light and a system of lenses to create magnified images of small samples. Light is transmitted from a source on the opposite side of the sample to the objective lens. Usually, the light passes through a condenser to focus it on the specimen to get maximal illumination. The received image can be detected by the direct observation, as well as imaged or saved digitally. Limitations of this technique include low apparent resolution and inadequate contrast of most biological samples. However, the simplicity of sample preparation is a significant advantage of this method [90].

Inverted optical microscope was used to confirm the internalization of dendrimer-coated iron oxide nanoparticles (DcMNPs) into cancer (SU86.86, T3M4, Panc-1) and PSC lines. The microscopy images demonstrated that most of the DcMNPs (also those loaded with gemcitabine) were internalized at 37 °C after 4 h of treatment, while the bare MNPs were not taken up by the cells [99]. Application of light microscopy also enabled the assessment of cell invasiveness upon treatment with 5-FU-PAMAM complexes. In vitro invasion assays were performed with modified Boyden chambers consisting of Transwell membrane filter inserts. In brief, in this method cells are put in an upper chamber to migrate, whereas the migration stimuli are placed in the bottom chambers or in an intermediate filter. The total number of migrating cells that reached the bottom chamber may be subsequently counted under the microscope. This approach allowed to demonstrate that the application of 5-FU complexed with the dendrimer significantly reduced the cell invasiveness compared to free drug, and the addition of as-miR-21 intensified the effect [62].

10.3.2 Fluorescence Microscopy

Fluorescence microscopes allow direct visualization of intracellular structures and processes in a living cell or tissue [26, 27]. The aim of fluorescence microscopy is to examine samples prepared with the use of fluorophores. A specimen labelled with fluorescent dye is illuminated with filtered light of the absorbing wavelength. The illumination light is separated from the much weaker emitted fluorescence due to the use of a spectral emission filter. The emission light is observed through the microscope or by a camera followed by computer digitization. Additional experimental information can be derived from the combination of optical and biochemical responses exhibited by the fluorescent probe. Multi-color images of several types of fluorophores may be composed by combining a number of single-color images.

Typical components of a fluorescence microscope include a light source (mercury-vapor lamp or xenon arc lamp; more advanced microscopes are equipped with high-power LEDs and lasers), excitation filter, dichroic mirror/beamsplitter and the emission filters. The filters and the dichroic beamsplitter are chosen to match the spectral excitation and emission characteristics of the fluorophore for the experiment.

The main techniques of preparation of fluorescent sample involve labeling with fluorescent stains. Some of them are small molecules which are intrinsically fluorescent and bind biological particles of interest. Most of them are nucleic acid stains (like DAPI and Hoechst) which bind to the minor groove of DNA, thus labeling the nuclei of cells. Other examples include drugs or toxins with high affinity to specific cellular structures, which have been derivatized with a fluorescent moiety. The main example of this class of fluorescent stains is phalloidin which is used to stain actin fibres in mammalian cells. Further, there are many fluorescent molecules such as fluorescein, Alexa Fluors or DyLight 488, which can be chemically linked to a different molecule which binds the desired target within the sample. The experiments on living cells also provide a possibility of expression of a fluorescent protein. Moreover, in some cases, autofluorescence (the intrinsic fluorescence of a sample) of the specimen can be used.

Unlike light microscopy techniques, fluorescence microscopy enables observation of only the specific structures which have been labelled for fluorescence. Moreover, fluorophores lose their properties as they are illuminated in a process called photobleaching, caused by the accumulation of chemical changes resulting from the damage by electrons excited during fluorescence. Photobleaching can severely limit the time of observation. Several techniques exist to reduce photobleaching such as the use of more robust fluorophores, by minimizing illumination, or by using photoprotective scavenger chemicals. It is also important to note that cells are susceptible to phototoxicity, particularly during exposure to short light wavelengths.

Confocal microscopy, or confocal laser scanning microscopy (CLSM) aims to overcome several limitations of traditional fluorescence microscopes. Confocal microscopes use a spatial pinhole to block out-of-focus light during image formation, and optical sectioning to get improved resolution of the image. Capturing numerous two-dimensional images at various depths in a sample allows to reconstruct the three-dimensional structures within an object. The software of confocal microscopes allows to perform complex 3D (z-stack), 4D (z-stack over time), or even 5D (z-stack over time including spectral imaging) experiments. This software facilitates data acquisition for complex methods such as Fluorescence Resonance Energy Transfer (FRET) or Fluorescence Recovery After Photobleaching (FRAP).

Fluorescence microscopy has been used to characterize mechanism of cellular death after treatment with 5-FU-DsAgNCs complexes. A549 and MCF-7 cells were stained with acridine orange/ethidium bromide (AO/EB) DNA intercalating fluorescent dyes. The AO dye can penetrate the nuclear envelope and emit green fluorescence upon binding with double stranded DNA, while EB enters the compromised nucleus to emit orange-red fluorescence. Those experiments allowed to conclude about apoptotic mechanism of cell death in both cell lines. This outcome was confirmed by additional staining with Hoechst 33342 (DNA labeling) and rhodamine B (labeling of mitochondria and cytoplasmic compartments) after 6, 12 and 24 h of treatment, which enabled determination of nuclear chromatin compaction and alteration in cytoskeleton [61].

Fluorescence microscopy also enabled the analysis of uptake of DcMNPs by MCF-7 cells, detected by Prussian blue staining method. Prussian blue reaction involves the treatment of cells with acid solutions of ferrocyanides. Ferric ions present inside the cells combine with the ferrocyanide and form a light blue pigment called Prussian blue, or ferric ferrocyanide. Those experiments indicated the accumulation of DcMNPs around the nucleus. Interestingly, the agglomerates formed by DcMNPs on the cell surface were observed even after the intense washing of MCF-7 cells [75].

In order to evaluate cellular uptake of PEG-cored PAMAM dendrimers with surface modified with PEG, CFPAC-1 cells were treated with the dendrimers loaded with rhodamine 123 and monitored under a fluorescence microscope. The results confirmed the outcome of flow cytometry experiments [72].

10.3.3 Transmission Electron Microscopy

Transmission electron microscopy (TEM) provides a possibility to generate an image of the internal structure of a thin sample using a beam of electrons. The image is created due to the interaction of electrons with the sample as the beam is transmitted through it. This technique enables crystallographic and chemical characterization of materials with high spatial resolution, reaching 0.1 nm. Transmission electron microscopes are featured with significantly higher resolution than light microscopes, due to the smaller de Broglie electron wavelengths. This allows to generate images presenting precise details of the sample, even as small as a single rows of atoms. Different variants of TEM give the possibility to observe modulations in chemical, crystal and electronic structure, electron phase shift, as well as the absorption-based imaging. Since the images are formed by the variation of the intensity of electrons transmitted through the specimen, samples must be permeable to electrons. Moreover, the sample must be thin enough to minimize beam broadening and to keep the spatial resolution close to that of the electron beam. The specimen is most often an ultrathin section of less than 100 nm, or a suspension of investigated particles placed on a grid [36].

The local density of dendrimers is sufficient to create fine images providing information on size, shape, or 3D architecture of macromolecules. In addition, the visualization of drug-dendrimer complexes may be helpful to confirm their structure and morphology, and to plan further biological experiments.

TEM at 1,80,000× was used to characterize surface and shape of PEGylated PAMAM dendrimers for the delivery of 5-FU, after drying on beryllium grid. It has been shown that non-PEGylated dendrimers are featured with relatively open outer surface layer. PEGylation was found to increase the size of dendrimer molecules [6]. Moreover, TEM enabled analysis of larger formulations containing dendrimers. For example, the suspension of PPD nanocarriers was dropped onto carbon-coated copper nets, air-dried at room temperature and moved to TEM for observation. The nanocarriers were found to be spherical with the dynamic sizes depended on pH. The diameter of PPD equaled 43.0 nm at pH 4.0, 41.6 nm at pH 6.5, and 11.6 nm at 7.4. The dependence of particle size on pH was consistent with the expansion and contraction of PDEA chains (Fig. 7) [51]. The physical properties, morphology and drug release behavior of 5-FU-PAMAM complexes were also investigated with TEM. One drop of the 1 mg/ml 5-FU-PAMAM sample was added to a copper-supported grid, the excess solution was removed with filter paper and then sample was dried at room temperature. The morphology of nanoparticles was found to be spherical or elliptical [62]. Transmission electron microscopy analysis was done to characterize DsAgNPs and their complexes with 5-FU. Aqueous solution of dendrimer-stabilized AgNPs (1 mg/ml) was dropped onto a carbon coated copper grid and subsequently air-dried before measurements. DsAgNPs were found to be spherical in shape with an average diameter of 8.32 nm, greater than this of the PAMAM G5 dendrimer (5.4 nm) indicating the formation of dendrimer-stabilized AgNPs. Negative staining enabled the evaluation of the homogeneous distribution of DsAgNPs loaded with 5-FU [61]. TEM images also proved nanoscale size (10–14 nm) of DcMNP particles. There were no significant differences between the full and half generations of dendrimeric nanoparticles in terms of sizes and shapes. The experiments with gemcitabine complexed with G5.5 DcMNPs showed almost spherical morphology and uniform size distribution. The particle size of DcMNPs did not change upon interaction with gemcitabine [75].

10.4 In Vivo Studies

In vivo research involves experiments in which the effects of biologically active compounds are tested on living organisms. These include animal studies and clinical trials. In vivo testing is often preferred over in vitro experiments as its outcome may give a direct answer on the activity of the tested compounds and their possible side effects in the biological systems, or enable subsequent clinical trials on humans.

In vivo studies on dendrimer-based drug delivery systems primarily involve the evaluation of their biodistribution, as well as pharmacokinetics and pharmacodynamics. All these parameters complement each other and allow to estimate the overall effects of drug-dendrimer complexes in living organisms:

- biodistribution techniques allow to track the tested compound and determine its location in an animal subject,
- pharmacokinetics enable the characterization of fate and metabolism of compound from the administration to the moment when it is completely removed from the organism,
- pharmacodynamics involve the study of biochemical, physiologic and molecular effects of substances on the organism.

Suffice to say, pharmacodynamics is the study of "how a drug affects an organism", whereas pharmacokinetics is the study of "how the organism affects the drug".

In vivo experiments can bring results that are considerably divergent from the outcome of tests performed on in vitro cultured cell lines. In case of cationic dendrimers, this is particularly evident at the level of cytotoxic activity, which significantly differs in vitro and in vivo. Virtually all cationic dendrimers of low and average generations are non-toxic and non-immunogenic in vivo, despite showing counter effects in vitro. As for the positively charged macromolecules of higher generations, undesirable effects in vivo tend to diminish during prolonged periods of administration or after no-treatment period. Further, introduction of surface modifications meant to reduce the cationic charge of those dendrimers eliminates their toxic activity [80].

Unmodified cationic dendrimers are very quickly removed from blood circulation. They accumulate mainly in kidney, spleen and liver, without causing permanent damage to these organs. The increase in blood-to-organ ratio correlates with increasing size and molecular weight of the dendrimers. The biodistribution and pharmacokinetics of cationic dendrimers can be modulated by their surface modification. For instance, PEG coating may significantly prolong blood half-life of PAMAM dendrimers, and glycosylation of PPI macromolecules may increase their concentration in liver due to the lectin-carbohydrate interactions [80].

Most frequently conducted in vivo research on NAs-dendrimer complexes involve derivatives of PEG-modified PAMAM macromolecules. PEGylation of PAMAM G4 was found to increase their drug-loading capacity, at the same time reducing the drug release rate. In the course of these experiments, male albino rats were given 1000 μg of 5-FU intravenously, as well as PEGylated and non-PEGylated dendrimer-drug formulations containing 5-FU equivalent to 1000 μg. The drug level in blood was determined using spectroscopic method, to determine the release rate and occurrence of sustained release. The samples were also analyzed for hemoglobin and blood cells level (RBC, WBC, monocytes, lymphocytes and neutrophils). Drug-dendrimer formulations were found to follow sustained release characteristics for 5-fluorouracil. The release rate was increased in vivo as compared to in vitro data, possibly due to the metabolism by the enzymes and hydrolysis in the body. These studies showed that PEGylated dendrimers are suitable for prolonged delivery of 5-FU, without the induction of haematological disorders, although it had an effect on the amount of blood cells [6].

CFPAC-1 tumor-bearing male CD-1 nude mice were administered with PEG-cored PAMAM dendrimers coated with PEG chains and their derivatives conjugated with anti-Flt-1 antibody, loaded with Rho123. After 12 h, mice were sacrificed and dissected, and cell suspensions were prepared from the tissues for evaluation of in vivo cellular uptake. These cells were read on flow cytometer and fluorescent microscope as mentioned above. It was shown that antibody-conjugated formulations were concentrated in larger intratumoral areas compared to PEG-cored PAMAM-PEG dendrimers, which indicated that addition of anti-Flt-1 antibody improved the accumulation of dendrimers in the pancreatic tumors. Moreover, the conjugation of antibody allowed to decrease the size of tumors by 30–50% upon administration of dendrimers loaded with gemcitabine, whereas the effect of formulations without the antibody was similar to the activity of free drug [72].

PAMAM dendrimers up to the 4th generation were modified with folate and folate-PEG-NHS (N-hydroxysuccinimide) conjugates. Such constructs were tested for 5-FU delivery potential in vivo. Female BALB/c mice were injected with human carcinoma KB cells overexpressing folate receptors. These mice subsequently developed palpable solid tumors. After 21 days after injection, dendrimers and free drug were introduced through tail vein of animals at day 0 and day 7. Following administration, blood samples were collected from tail vein at 0.1, 0.5, 1.0, 2.5, 5, 8, 12, and 24-h intervals. "Plasma concentration over time" studies revealed prolonged retention of PEGylated dendritic formulation in systemic circulation. Pharmacodynamic studies showed that folate- and PEG-conjugated PAMAM G4 displayed a significant inhibition of the tumor growth as compared to the drug entrapped in free dendrimer. The difference in tumor growth was also observed between macromolecules modified only with folic acid and with both folate and PEG moieties. This may be attributed to the fact that the PEGylated system may provide a relatively slow release compared to folate-conjugated and unmodified PAMAM dendrimers. Modification of dendrimers via PEG-folic acid prolonged drug release and increased accumulation in the tumor. Biodistribution studies were conducted to evaluate drug delivery at various sites of interest, especially tumor. For this purpose, mice were sacrificed at 2, 8, and 24 h. Organs and tumor tissue were subsequently removed, homogenized and analyzed for 5-FU concentration using HPLC. Calibration curves of 5-FU in serum and tissues of various organs were also prepared. All tested macromolecules were found to accumulate in kidneys and liver. This observation suggests that kidney may be the main clearance organ for dendritic systems. Moreover, kidneys and liver are known to express high levels of folate receptors, leading to longer retention of folate-modified dendrimers. The drug levels observed in organs from mice treated with dendrimer formulations were higher as compared to plain drug [81].

For the evaluation of 5-FU-loaded PPD nanocarriers, H22 cancer cells were subcutaneously injected into mice. Free drug and 5-FU-loaded nanocarriers were administrated through intravenous injection. The 5-FU concentration was determined in blood and tissue homogenates using HPLC. Additionally, tumors were dissected and weighed to calculate the tumor inhibitory rate. It was shown that 5-FU was rapidly distributed in the circulation after intravenous administration of drug-loaded nanocarriers to healthy or tumor-bearing mice. Tissue distribution further indicated the tumor-targeting effect of PPD nanocarriers. The biodistribution of 5-FU in tumor-bearing mice was significantly different from that in healthy mice. The amounts of 5-FU in the blood, liver, spleen and lungs of tumor-bearing mice were lower than in healthy mice, while the concentrations in kidneys and heart were similar. By contrast, drug accumulation in tumors was higher after administration of nanocarriers with 5-FU. Interestingly, the concentration of 5-fluorouracil in tumors was much higher than that in the liver. Tumor-specific targeting and pH-responsive rapid drug release were the major reasons for this enhancement [51].

In vivo studies in albino rats were also performed for PAMAM dendrimers coated with fatty acid, carrying 5-fluorouracil. Bioavailability and pharmacokinetic parameters were evaluated on the basis of the plasma profile of 5-FU. The phospholipid-coated dendrimer formulation was found to be more effective upon oral administration than free drug. The lymphatic uptake was also improved showing absorption of the formulation through the lymphatic route [94].

Albino rats of Sprague–Dawley strain were used for in vivo evaluation of gemcitabine-loaded mannose-modified PPI dendrimers up to the 4th generation. The pharmacokinetic study was carried out following intravenous administration of developed formulations to study the release profile and retention of dendrimers in blood circulation. Biodistribution studies were performed to determine the lung targeting efficiency of developed formulations. For this purpose, various organs including spleen, kidney, liver, lungs, and brain were carefully removed, wiped, and weighed. Weighed tissue samples were immediately homogenized, and gemcitabine concentration was assessed via HPLC. Drug-dendrimer formulations exhibited blood concentration comparative to intravenous administration of free gemcitabine. However, the blood half-life of the drug was prolonged by PPI glycodendrimers, which additionally increased the accumulation of gemcitabine in lung tissue. This indicated that the metabolism and elimination of the drug may be minimized significantly by entrapping it in the dendrimer [83].

11 Conclusions

Comprehensive scientific literature on the formation and stability of complexes between cationic dendrimers and anticancer nucleoside and nucleotide analogues, as well as evaluation of their activity in vitro and in vivo indicates that these macromolecules may serve as efficient nanocarriers, improving the therapeutic activity of NAs. This type of drug-dendrimer non-covalent formulations have been widely studied over the last few years, bringing promising results and hope for the future clinical use. Further in vivo studies are crucial to evaluate the mechanisms of action and the impact of complexes on living organisms, to exclude potential side effects and to fully exploit their biomedical applications.

References

1. Abbasi, E., Aval, S.F., Akbarzadeh, A., et al.: Dendrimers: synthesis, applications, and properties. Nanoscale Res. Lett. **9**(1), 247 (2014)
2. Adan, A., Alizada, G., Kiraz, Y., et al.: Flow cytometry: basic principles and applications. Crit. Rev. Biotechnol. **37**(2), 163–176 (2017)
3. Appelhans, D., Klajnert-Maculewicz, B., Janaszewska, A., et al.: Dendritic glycopolymers based on dendritic polyamine scaffolds: view on their synthetic approaches, characteristics and potential for biomedical applications. Chem. Soc. Rev. **44**(12), 3968–3996 (2015)
4. Berdis, A.J.: DNA polymerases as therapeutic targets. Biochemistry **47**(32), 8253–8260 (2008)
5. Bertrand, N., Wu, J., Xu, X., et al.: Cancer nanotechnology: the impact of passive and active targeting in the era of modern cancer biology. Adv. Drug Deliv. Rev. **66**, 2–25 (2014)
6. Bhadra, D., Bhadra, S., Jain, S., et al.: A PEGylated dendritic nanoparticulate carrier of fluorouracil. Int. J. Pharm. **257**(1–2), 111–124 (2003)
7. Boas, U., Christensen, J.B., Heegaard, P.M.H.: Dendrimers: design, synthesis and chemical properties. J. Mater. Chem. **16**, 3785–3798 (2006)
8. Bobrovnik, S.A.: Ligand–receptor interaction. Klotz-Hunston problem for two classes of binding sites and its solution. J. Biochem. Biophys. Methods **52**, 135–143 (2002)
9. Bosman, A.W., Janssen, H.M., Meijer, E.W.: About dendrimers: structure, physical properties, and applications. Chem. Rev. **99**(7), 1665–1688 (1999)
10. Brand, L., Johnson, M.L. (eds): Fluorescence Spectroscopy (Methods in Enzymology, vol. 450). Academic Press, Cambridge, MA (2008)
11. Buczkowski, A., Olesinski, T., Zbicinska, E., et al.: Spectroscopic and calorimetric studies of formation of the supramolecular complexes of PAMAM G5-NH_2 and G5-OH dendrimers with 5-fluorouracil in aqueous solution. Int. J. Pharm. **490**, 102–111 (2015)
12. Buczkowski, A., Piekarski, H., Palecz, B.: Stoichiometry and equilibrium constant of the complex of PAMAM-NH_2 G4 and 5-fluorouracil. J. Mol. Liq. **173**, 8–12 (2012)
13. Buczkowski, A., Sekowski, S., Grala, A., et al.: Interaction between PAMAM-NH_2 G4 dendrimer and 5-fluorouracil in aqueous solution. Int. J. Pharm. **408**, 266–270 (2011)
14. Buczkowski, A., Stępniak, A., Pałecz, B. et al.: Spectroscopic studies of PAMAM-NH_2 G4, PAMAM-OH G4 and PAMAM-COONa G3.5 dendrimers interactions with 6-mercaptopurine in aqueous solutions (in Polish). In: Nyćkowiak, J., Leśny, J. (eds.) Monography on Chemistry: Studies and development of Young Researchers in Poland. Poznań, Młodzi Naukowcy (2015b)
15. Buczkowski, A., Stepniak, A., Urbaniak, P., et al.: Spectroscopic studies of interactions of PAMAM G5-NH_2 and PAMAM G5-OH dendrimers with flutamide and cyclocreatine in aqueous solution. Pharm. Chem. J. **3**, 23–33 (2016)
16. Buczkowski, A., Urbaniak, P., Belica, S., et al.: Formation of complexes between PAMAM-NH_2 G4 dendrimer and L-α-tryptophan and L-α-tyrosine in water. Spectrochimica Acta Part A **128**, 647–652 (2014)
17. Buczkowski, A., Urbaniak, P., Palecz, B.: Thermochemical and spectroscopic studies on the supramolecular complex of PAMAM-NH_2 G4 dendrimer and 5-fluorouracil in aqueous solution. Int. J. Pharm. **428**, 178–182 (2012)

18. Buczkowski, A., Urbaniak, P., Palecz, B.: Interaction between PAMAM-NH_2 G4 dendrimer and paracetamol in aqueous solution. J. Mol. Liq. **186**, 70–75 (2013)
19. Buczkowski, A., Urbaniak, P., Piekarski, H., et al.: Spectroscopic and calorimetric studies on the interaction between PAMAM G4-OH and 5-fluorouracil in aqueous solutions. Spectrochim. Acta A Mol. Biomol. Spectrosc. **171**, 401–405 (2017)
20. Buczkowski, A., Waliszewski, D., Urbaniak, P., et al.: Study of the interactions of PAMAM G3-NH2 and G3-OH dendrimers with 5-fluorouracil in aqueous solutions. Int. J. Pharm. **505**, 1–13 (2016)
21. Buhleier, E., Wehner, W., Vögtle, F.: "Cascade"- and "nonskid-chain-like" synthesis of molecular cavity topologies. Synthesis **9**(2), 155–158 (1978)
22. Bujalowski, W., Jezewska, M.J.: Macromolecular competition titration method accessing thermodynamics of the unmodified macromolecule-ligand interactions through spectroscopic titrations of fluorescent analogs. Methods Enzymol. **488**, 17–57 (2011)
23. Cano-Soldado, P., Pastor-Anglada, M.: Transporters that translocate nucleosides and structural similar drugs: structural requirements for substrate recognition. Med. Res. Rev. **32**, 428–457 (2012)
24. Carroll, S.S., Olsen, D.B.: Nucleoside analog inhibitors of hepatitis C virus replication. Infect. Disord. Drug Targets **6**(1), 17–29 (2006)
25. Clogston, J.D., Patri, A.K.: Zeta potential measurement. Methods Mol. Biol. **697**, 63–70 (2011)
26. Coling, D., Kachar, B.: Theory and application of fluorescence microscopy (Chap. 2:Unit 2.1). Curr. Protoc. Neurosci. (2001)
27. Combs, C.A., Shroff, H.: Fluorescence microscopy: a concise guide to current imaging methods. Curr. Protoc. Neurosci. **79**, 2.1.1–2.1.25 (2017)
28. Cooper, A.: Biophysical Chemistry. The Royal Society of Chemistry, Cambridge (2004)
29. Coskun, O.: Separation techniques: chromatography. North Clin. Istanb. **3**(2), 156–160 (2016)
30. De Jong, W.H., Borm, P.J.: Drug delivery and nanoparticles: applications and hazards. Int. J. Nanomed. **3**(2), 133–149 (2008)
31. Diab, R., Degobert, G., Hamoudeh, M., et al.: Nucleoside analogue delivery systems in cancer therapy. Expert Opin. Drug Deliv. **4**(5), 513–531 (2007)
32. Diaz, M., Herrero, M., Garcia, L.A., et al.: Application of flow cytometry to industrial microbial bioprocesses. Biochem. Eng. J. **48**(3), 385–407 (2010)
33. Duncan, R., Izzo, L.: Dendrimer biocompatibility and toxicity. Adv. Drug Deliv. Rev. **57**(15), 2215–2237 (2005)
34. Elizondo, E., Moreno, E., Cabrera, I., et al.: Liposomes and other vesicular systems: structural characteristics, methods of preparation, and use in nanomedicine. Prog. Mol. Biol. Transl. Sci. **104**, 1–52 (2011)
35. Emeje, M.O., Obidike, I.C., Akpabio, E.I., et al.: Nanotechnology in drug delivery. In: Sezer, A.D. (ed.) Recent Advances in Novel Drug Carrier Systems. InTech, Rijeka (2012)
36. Fernando, J.F.S., Zhang, C., Firestein, K.L., et al.: Optical and optoelectronic property analysis of nanomaterials inside transmission electron microscope. Small **13**(45) (2017)
37. Fielding, L.: NMR methods for the determination of protein–ligand dissociation constants. Prog. Nucl. Magn. Reson. Spectrosc. **51**, 219–242 (2007)
38. Fung, J., Lai, C.L., Seto, W.K., et al.: Nucleoside/nucleotide analogues in the treatment of chronic hepatitis B. J. Antimicrob. Chemother. **66**(12), 2715–2725 (2011)
39. Galmarini, C.M., Mackey, J.R., Dumontet, C.: Nucleoside analogues: mechanisms of drug resistance and reversal strategies. Leukemia **15**(6), 875–890 (2001)
40. Galmarini, C.M., Mackey, J.R., Dumontet, C.: Nucleoside analogues and nucleobases in cancer treatment. Lancet Oncol. **3**(7), 415–424 (2002)
41. Galmarini, C.M., Warren, G., Kohli, E., et al.: Polymeric nanogels containing the triphosphate form of cytotoxic nucleoside analogues show antitumor activity against breast and colorectal cancer cell lines. Mol. Cancer Ther. **7**(10), 3373–3380 (2008)
42. Goldys, E.M. (ed.): Fluorescence Applications in Biotechnology and Life Sciences. Wiley-Blackwell, Hoboken (2009)

43. Gorzkiewicz, M., Jatczak-Pawlik, I., Studzian, M., et al.: Glycodendrimer nanocarriers for direct delivery of fludarabine triphosphate to leukaemic cells: improved pharmacokinetics and pharmacodynamics of fludarabine. Biomacromol **19**(2), 531–543 (2018)
44. Gorzkiewicz, M., Klajnert-Maculewicz, B.: Dendrimers as nanocarriers for nucleoside analogues. Eur. J. Pharm. Biopharm. **114**, 43–56 (2017)
45. Grządka, E., Chibowski, S.: Influence of a kind of electrolyte and its ionic strength on the adsorption and zeta potential of the system: polyacrylic acid/MnO_2/electrolyte solution. Physicochem. Prob. Min. Process. **43**, 31–42 (2009)
46. Hillaireau, H., Le Doan, T., Couvreur, P.: Polymer-based nanoparticles for the delivery of nucleoside analogues. J. Nanosci. Nanotechnol. **6**(9–10), 2608–2617 (2006)
47. Hu, J., Fang, M., Cheng, Y., et al.: Host-guest chemistry of dendrimer-drug complexes. 4. An in-depth look into the binding/encapsulation of guanosine monophosphate by dendrimers. J. Phys. Chem. B. **114**, 7148–7157 (2010)
48. Jain, K., Kesharwani, P., Gupta, U., et al.: Dendrimer toxicity: let's meet the challenge. Int. J. Pharm. **394**(1–2), 122–142 (2010)
49. Janaszewska, A., Ciolkowski, M., Wróbel, D., et al.: Modified PAMAM dendrimer with 4-carbomethoxypyrrolidone surface groups reveals negligible toxicity against three rodent cell-lines. Nanomedicine **9**(4), 461–464 (2013)
50. Janaszewska, A., Mączyńska, B., Matuszko, G., et al.: Cytotoxicity of PAMAM, PPI and maltose modified PPI dendrimers in Chinese hamster ovary (CHO) and human ovarian carcinoma (SKOV3) cells. New J. Chem. **36**, 428–437 (2012)
51. Jin, Y., Ren, X., Wang, W., et al.: A 5-fluorouracil-loaded pH-responsive dendrimer nanocarrier for tumor targeting. Int. J. Pharm. **420**(2), 378–384 (2011)
52. Jordheim, L.P., Durantel, D., Zoulim, F., et al.: Advances in the development of nucleoside and nucleotide analogues for cancer and viral diseases. Nat. Rev. Drug Discov. **12**(6), 447–464 (2013)
53. Kakde, D., Jain, D., Shrivastava, V., et al.: Cancer therapeutics-opportunities, challenges and advances in drug delivery. JAPS **01**(09), 01–10 (2011)
54. Kesharwani, P., Tekade, R.K., Gajbhiye, V., et al.: Cancer targeting potential of some ligand-anchored poly(propylene imine) dendrimers: a comparison. Nanomedicine **7**(3), 295–304 (2011)
55. Klajnert, B., Pastucha, A., Shcharbin, D., et al.: Binding properties of polyamidoamine dendrimers. J. Appl. Polym. Sci. **103**, 2036–2040 (2007)
56. Klajnert, B., Pikala, S., Bryszewska, M.: Haemolytic activity of polyamidoamine dendrimers and the protective role of human serum albumin. Proc. R. Soc. Lond. A: Math. Phys. Eng. Sci. **466**(2117), 1527–1534 (2010)
57. Kromidas, S. (ed.): The HPLC Expert: Possibilities and Limitations of Modern High Performance Liquid Chromatography. Wiley-VCH, Weinheim (2016)
58. Lee, S.H., Meng, X.W., Flatten, K.S., et al.: Phosphatidylserine exposure during apoptosis reflects bidirectional trafficking between plasma membrane and cytoplasm. Cell Death Differ. **20**(1), 64–76 (2013)
59. Lloyd, J.R., Jayasekara, P.S., Jacobson, K.A.: Characterization of polyamidoamino (PAMAM) dendrimers using in-line reversed phase LC electrospray ionization mass spectrometry. Anal. Methods **8**(2), 263–269 (2016)
60. Madadlou, A., O'Sullivan, S., Sheehan, D.: Fast protein liquid chromatography. Methods Mol. Biol. **681**, 439–447 (2011)
61. Matai, I., Sachdeva, A., Gopinath, P.: Multicomponent 5-fluorouracil loaded PAMAM stabilized-silver nanocomposites synergistically induce apoptosis in human cancer cells. Biomater. Sci. **3**, 457–468 (2015)
62. Mei, M., Ren, Y., Zhou, X., et al.: Suppression of breast cancer cells in vitro by polyamidoamine-dendrimer-mediated 5-fluorouracil chemotherapy combined with antisense micro-RNA 21 gene therapy. J. Appl. Polym. Sci. **114**, 3760–3766 (2009)
63. Menjoge, A.R., Kannan, R.M., Tomalia, D.A.: Dendrimer-based drug and imaging conjugates: design considerations for nanomedical applications. Drug Discov. Today. **15**(5–6), 171–185 (2010)

64. Mintzer, M.A., Grinstaff, M.W.: Biomedical applications of dendrimers: a tutorial. Chem. Soc. Rev. **40**(1), 173–190 (2011)
65. Mislak, A.C., Anderson, K.S.: Insights into the molecular mechanism of polymerization and nucleoside reverse transcriptase inhibitor incorporation by human PrimPol. Antimicrob. Agents Chemother. **60**(1), 561–569 (2015)
66. Munson, P.J., Rodbard, D.: Number of receptor sites from Scatchard and Klotz graphs: a constructive critique. Science **220**, 979–981 (1983)
67. Nanjwade, B.K., Bechra, H.M., Derkar, G.K., et al.: Dendrimers: emerging polymers for drug-delivery systems. Eur. J. Pharm. Sci. **38**(3), 185–196 (2009)
68. Narsireddy, A., Vijayashree, K., Adimoolam, M.G., et al.: Photosensitizer and peptide-conjugated PAMAM dendrimer for targeted in vivo photodynamic therapy. Int. J. Nanomedicine **10**, 6865–7688 (2015)
69. Neerman, M.F.: The efficiency of a PAMAM dendrimer toward the encapsulation of the antileukemic drug 6-mercaptopurine. Anticancer Drugs **18**, 839–842 (2007)
70. O' Brien, J., Wilson, I., Orton, T., et al.: Investigation of the Alamar Blue (resazurin) fluorescent dye for the assessment of mammalian cell cytotoxicity. Eur. J. Biochem. **267**(17), 5421–5426 (2000)
71. Otis, J.B., Zong, H., Kotylar, A., et al.: Dendrimer antibody conjugate to target and image HER-2 overexpressing cancer cells. Oncotarget **7**(24), 36002–36013 (2016)
72. Öztürk, K., Esendağlı, G., Gürbüz, M.U., et al.: Effective targeting of gemcitabine to pancreatic cancer through PEG-cored Flt-1 antibody-conjugated dendrimers. Int. J. Pharm. **517**(1–2), 157–167 (2017)
73. Palecz, B., Buczkowski, A., Piekarski, H.: Thermodynamic interaction between PAMAM G4-NH2, G4-OH, G3.5-COONa dendrimers and gemcitabine in water sotutions. Int. J. Second. Metabolite **3**, 21–26 (2016)
74. Parker, W.B.: Enzymology of purine and pyrimidine antimetabolites used in the treatment of cancer. Chem. Rev. **109**(7), 2880–2893 (2009)
75. Parsian, M., Mutlu, P., Yalcin, S., et al.: Half generations magnetic PAMAM dendrimers as an effective system for targeted gemcitabine delivery. Int. J. Pharm. **515**(1–2), 104–113 (2016)
76. Piktel, E., Niemirowicz, K., Wątek, M., et al.: Recent insights in nanotechnology-based drugs and formulations designed for effective anti-cancer therapy. J. Nanobiotechnology **14**(1), 39 (2016)
77. Sampath, D., Rao, V.A., Plunkett, W.: Mechanisms of apoptosis induction by nucleoside analogs. Oncogene **22**(56), 9063–9074 (2003)
78. Scatchard, G.: The attractions of proteins for small molecules and ions. Ann. N. Y. Acad. Sci. **51**, 660–672 (1949)
79. Shao, N., Su, Y., Hu, J., et al.: Comparison of generation 3 polyamidoamine dendrimer and generation 4 polypropylenimine dendrimer on drug loading, complex structure, release behavior, and cytotoxicity. Int. J. Nanomedicine **6**, 3361–3372 (2011)
80. Shcharbin, D., Janaszewska, A., Klajnert-Maculewicz, B., et al.: How to study dendrimers and dendriplexes III. Biodistribution, pharmacokinetics and toxicity in vivo. J. Control Release **181**, 40–52 (2014)
81. Singh, P., Gupta, U., Asthana, A., et al.: Folate and folate-PEG-PAMAM dendrimers: synthesis, characterization, and targeted anticancer drug delivery potential in tumor bearing mice. Bioconjug. Chem. **19**(11), 2239–2252 (2008)
82. Soltes, L., Mach, M.: Estimation of drug-protein binding parameters on assuming the validity of thermodynamic equilibrium. J. Chromatogr. B **768**, 113–119 (2002)
83. Soni, N., Jain, K., Gupta, U., et al.: Controlled delivery of Gemcitabine Hydrochloride using mannosylated poly(propyleneimine) dendrimers. J. Nanopart. Res. **17**, 458 (2015)
84. Stoddart, M.J.: Cell viability assays: introduction. Methods Mol. Biol. **740**, 1–6 (2011)
85. Svenson, S., Tomalia, D.A.: Dendrimers in biomedical applications-reflections on the field. Adv. Drug Deliv. Rev. **57**(15), 2106–2129 (2005)
86. Szulc, A., Appelhans, D., Voit, B., et al.: Characteristics of complexes between poly(propylene imine) dendrimers and nucleotides. New J. Chem. **36**, 1610–1615 (2012)

87. Szulc, A., Appelhans, D., Voit, B., et al.: Studying complexes between PPI dendrimers and Mant-ATP. J. Fluoresc. **23**(2), 349–356 (2013)
88. Szulc, A., Pulaski, L., Appelhans, D., et al.: Sugar-modified poly(propylene imine) dendrimers as drug delivery agents for cytarabine to overcome drug resistance. Int. J. Pharm. **513**(1–2), 572–583 (2016)
89. Szulc, A., Signorelli, M., Schiraldi, A., et al.: Maltose modified poly(propylene imine) dendrimers as potential carriers of nucleoside analog 5'-triphosphates. Int. J. Pharm. **495**, 940–947 (2015)
90. Thorn, K.: A quick guide to light microscopy in cell biology. Mol. Biol. Cell **27**(2), 219–222 (2016)
91. Tomalia, D.A., Baker, H., Dewald, J., et al.: A new class of polymers: starburst-dendritic macromolecules. Polym. J. **17**, 117–132 (1985)
92. Tomalia, D.A., Christensen, J.B., Boas, U.: Dendrimers, Dendrons, and Dendritic Polymers. Discovery, Applications, and the Future. Cambridge University Press, Cambridge (2012)
93. Tomalia, D.A., Rookmaker, M.: Polymer Data Handbook. Oxford University Press, New York (2009)
94. Tripathi, P.K., Khopade, A.J., Nagaich, S., et al.: Dendrimer grafts for delivery of 5-fluorouracil. Pharmazie **57**(4), 261–264 (2002)
95. Tripp, S., Appelhans, D., Striegler, C., et al.: Oligosaccharide shells as decisive factor for moderate and strong ionic interactions of dendritic poly(ethylene imine) scaffold under shear forces. Chem.—Euro. J. **20**, 8314–8319 (2014)
96. Wang, W., Xiong, W., Wan, J., et al.: The decrease of PAMAM dendrimer-induced cytotoxicity by PEGylation via attenuation of oxidative stress. Nanotechnology **20**(10), 105103 (2009)
97. Wong, P.T., Choi, S.K.: Mechanisms of drug release in nanotherapeutic delivery systems. Chem. Rev. **115**(9), 3388–3432 (2015)
98. Wyczechowska, D., Fabianowska-Majewska, K.: The effects of cladribine and fludarabine on DNA methylation in K562 cells. Biochem. Pharmacol. **65**(2), 219–225 (2003)
99. Yalçin, S., Erkan, M., Ünsoy, G., et al.: Effect of gemcitabine and retinoic acid loaded PAMAM dendrimer-coated magnetic nanoparticles on pancreatic cancer and stellate cell lines. Biomed. Pharmacother. **68**(6), 737–743 (2014)
100. Yang, W., Li, Y., Cheng, Y., et al.: Evaluation of phenylbutazone and poly(amidoamine) dendrimers interactions by a combination of solubility, 2D-NOESY NMR, and isothermal titration calorimetry studies. J. Pharm. Sci. **98**, 1075–1085 (2009)
101. Young, J.D., Yao, S.Y., Baldwin, J.M., et al.: The human concentrative and equilibrative nucleoside transporter families, SLC28 and SLC29. Mol. Aspects Med. **34**(2–3), 529–547 (2013)
102. Zhang, J., Visser, F., King, K.M., et al.: The role of nucleoside transporters in cancer chemotherapy with nucleoside drugs. Cancer Metastasis Rev. **26**(1), 85–110 (2007)
103. Ziemba, B., Halets, I., Shcharbin, D., et al.: Influence of fourth generation poly(propyleneimine) dendrimers on blood cells. J Biomed Mater Res A. **100**(11), 2870–2880 (2012)

Thermal Analysis of Glass-Ceramics and Composites in Biomedical and Dental Sciences

George S. Theodorou, Dimitra Patsiaoura, Eleana Kontonasaki and Konstantinos Chrissafis

1 Basic Principles and Brief Description of Thermal Analysis Techniques Applied in Biomedical Sciences

1.1 Introduction

The terms "*thermal analysis*" and "*calorimetry*" denote the conduction of various experimental measurements upon temperature alteration. The physical quantities measured by thermal analysis and calorimetry techniques may involve changes in the state of the specimen (e.g. temperature, mass), or in the properties of the materials (e.g. crystalline structure). Those values are subsequently transformed into electric signals and depicted as experimental curves. The outcome obtained by the abovementioned techniques depend both on specimen parameters (e.g. mass, shape, internal structure) and on operational specifications (e.g. atmosphere, heating rate, pressure). The longest-established measurement techniques that will be briefly discussed below are: Thermogravimetry (TG), Differential Scanning Calorimetry (DSC), Differential Thermal Analysis (DTA) and Dynamic Mechanical Analysis (DMA).

1.2 Thermogravimetry (TG)

TG is a technique in which the mass of a specimen is recorded upon temperature alterations. TG curves exhibit mass loss or percent mass loss plotted against temperature or time, while the temperature program is chosen according to the information needed.

G. S. Theodorou · D. Patsiaoura · K. Chrissafis (✉)
Physics Department, AUTh, Thessaloniki, Greece
e-mail: hrisafis@physics.auth.gr

E. Kontonasaki
Dentistry Department, AUTh, Thessaloniki, Greece

C. Demetzos and N. Pippa (eds.), *Thermodynamics and Biophysics of Biomedical Nanosystems*, Series in BioEngineering, https://doi.org/10.1007/978-981-13-0989-2_8

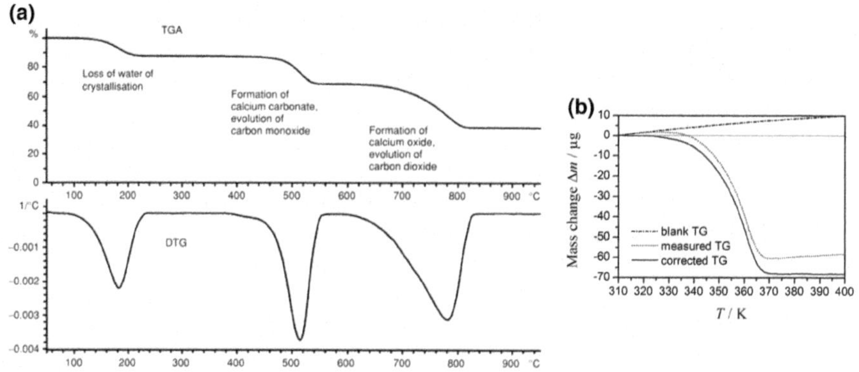

Fig. 1 **a** TG and DTG curves of the stepwise decomposition of calcium oxalate monohydrate [1], **b** TG curves of a blank measurement (empty pan), of calcium oxalate monohydrate sample and the corrected measurement, after subtracting blank TG from the measured one [2]

TG measurements may be carried out either in dynamic mode (constant heating rate) or under isothermal conditions (constant temperature). Moreover, atmosphere used in the experiment may be inert, reactive or oxidizing. Furthermore, first Derivative (DTG) curves are very helpful in examining TG thermographs, since they display the rate of mass change and they exhibit the mass loss steps more distinctively in case of overlapping phenomena. Nowadays, TG instruments are often equipped appropriately in order to further record more signals than the actual thermogravimetric measurements; e.g. simultaneous TG-DTA and TG-DSC for concurrent recording DTA and DSC signals along TG, respectively. The aforementioned techniques, applied simultaneously with TG, will be individually discussed below.

TG curves consist of mass change steps (Fig. 1a), which correspond to losing or even gaining weight processes. Interpretation of most common phenomena involve: H_2O loss or uptake, volatile component evaporation, gas desorption or adsorption, thermal decomposition in inert atmosphere, oxidative decomposition in oxygen or air, metal oxidation, etc. Finally, even quantitative evaluation of TG data can be achieved through the determination of step heights recorded in TG curves.

During measurements, a number of issues must be considered, such as sample preparation, measurement rate, atmosphere, heating rate, etc. Concerning measurement procedure, buoyancy effect must be taken into account. This phenomenon, caused by change of gas density upon temperature alterations, leads to an apparent mass increase, especially at the initial stages of the measurement. In order to eliminate this phenomenon, a blank measurement needs to be performed and then subtracted from the experimental measurement curve of the sample (Fig. 1b).

1.3 Differential Scanning Calorimetry (DSC)

DSC, probably the most widely used thermal technique, was developed by Watson E.S. and O'Neil M.J. of PerkinElmer in 1962 [3]. It constitutes a fast and easy method for the acquisition of various information regarding specimens, covering a plethora of different applications. DSC measures energy changes that occur in samples upon temperature alteration, while DSC thermographs exhibit heat flow signal plotted against temperature or time. More specifically, DSC mainly measures heat flow, which is the energy flow into or out of the specimen as a function of both time and temperature. These energy changes permit the evaluation of transitions, as well as the estimation of temperatures in which they occur. As a consequence, certain events can be monitored, leading to the determination of a wide variety of quantities, with the most common being the melting temperature (T_{m}), the crystallization temperature (T_{c}), the glass transition temperature (T_{g}), enthalpy changes and curing processes.

There are two types of DSC, the power compensation DSC and the heat flux DSC, with their main differences being mainly on their instrumentation design. The power compensation DSC obtains two identical furnaces, one for the reference and one for the sample to study, and when they are both subjected to programmed temperature alterations under specific heating or cooling rate, the power compensations are applied accordingly to the required furnace, in order to maintain this rate stable in both furnaces. On the contrary, in heat flux DSC there is only one furnace, with the reference and the sample being located in the interior. When they are both subjected to programmed temperature alterations under specific heating or cooling rate, the occurring transitions create a temperature difference between reference and sample. This temperature difference creates the heat flow signal.

During measurements, a series of practical issues must be taken under caution, such as encapsulation of samples in the right pan, appropriate temperature range, measurement rate, sample size and purge gas. More specifically, the material of the pan depends on the characteristics of the sample to be studied and on the ultimate temperature that the DSC will reach, while oxidation of samples must be avoided by applying inert atmosphere inside the furnace.

Concerning interpretation of DSC thermographs (Fig. 2), the most common phenomena are: (i) Melting, appearing as a well-defined endothermic peak, since energy must be absorbed by the sample in order to get melted. (ii) Heat of fusion ΔH which is determined by integrating the peak area. (iii) Moisture loss appearing as shallow and broad endotherms. (iv) Crystallization, appearing as a well-defined exothermic process. (v) Cure reactions appearing as shallow and broad exotherms. (vi) Glass transition temperature (T_g) which is the most common measured transition by means of thermal analysis techniques. Glass transition consists a boundary between glassy (hard) and rubbery (soft) state of non-fully crystalline materials and it appears as a single step in heat flow curve. Moreover, T_g is not a single point, but it covers a broad range of temperatures, permitting its calculation with various approaches (Fig. 3). The most widespread manner of T_g calculation is the midpoint, which is defined as the point of the half step height. Other common modes of T_g evaluation are the onset

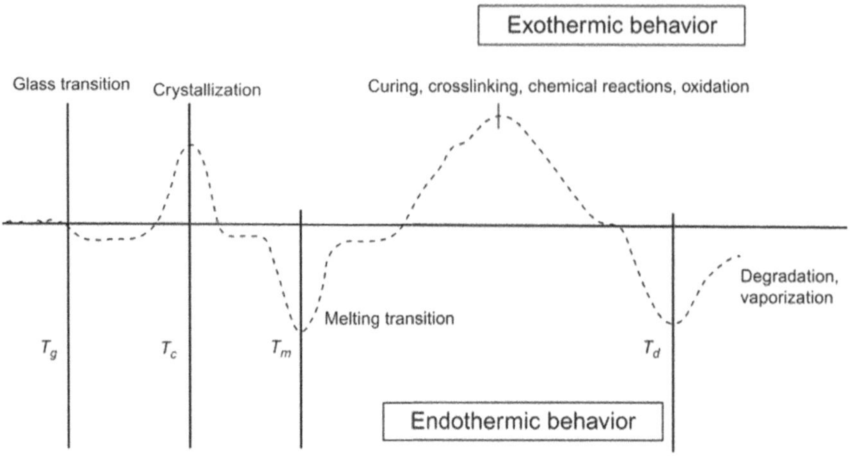

Fig. 2 Schematic illustration of main events detected by DSC [4]

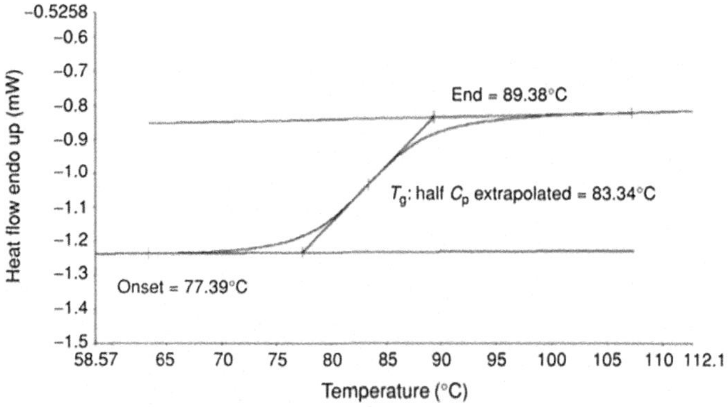

Fig. 3 Different calculations of T_g on a DSC thermograph [1]

and the endpoint, which stand for the intersections of tangents of the heat flow step with initial and final baseline, respectively. Furthermore, least common processes are identified, such as crosslinking, chemical reactions, oxidation, degradation and vaporization.

1.4 Differential Thermal Analysis (DTA)

The instrumental design of DTA is similar to that of the heat flux DSC. More specifically, if the sample undergoes a phase change upon heating, energy is emitted or absorbed and a temperature difference is detected between the reference and the

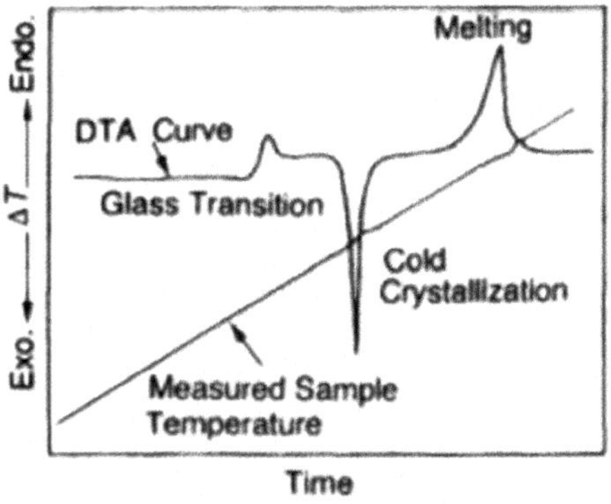

Fig. 4 Schematic illustration of DTA curve as a function of time [5]

sample. The only difference between DSC and DTA lies on the recorded quantity of the y-axis, since the output signal of DTA corresponds to the temperature difference between the sample and the reference (Fig. 4), while DSC displays heat flow values. Furthermore, DTA may record all phenomena that DSC does, as well as it can quantitatively determine any enthalpy changes, as long as its signal gets calibrated using the melting peaks of standard reference materials.

1.5 Dynamic Mechanical Analysis (DMA)

DMA is a simple probing technique which measures mechanical properties of samples upon temperature alterations. During DMA measurements, a force is applied on the samples and the displacement response to this particular force is measured. Upon temperature changes, measured properties markedly change, yielding information of the molecular structure of the samples, as well as of their transitions. Different geometry arrangements which correspond to different stiffness are employed to the measurements, accordingly to the expected magnitude of sample's modulus value as well as to its size. These are the simple shear, tension, clamped bending and three-point bending.

The typical signal outfits of DMA instruments are storage modulus E', loss modulus E'' and tanδ (the ratio E''/E') plotted against temperature (Fig. 5). Storage modulus E' is considered as the measure of the elastic behavior of the sample, indicating its ability to store the energy applied by external forces without the strain deformation to be permanent. The loss modulus E'' value is a good indicator of the material's viscous behavior, while it is sensitive to different types of molecular motions. Tanδ, often called damping coefficient, is the measure of the sample's energy dissipation under load. T_g detection can be carried out by different DMA signals. T_g is usually considered as the maximum of the tanδ curve, although some researchers identify T_g as the large drop in the E' curve.

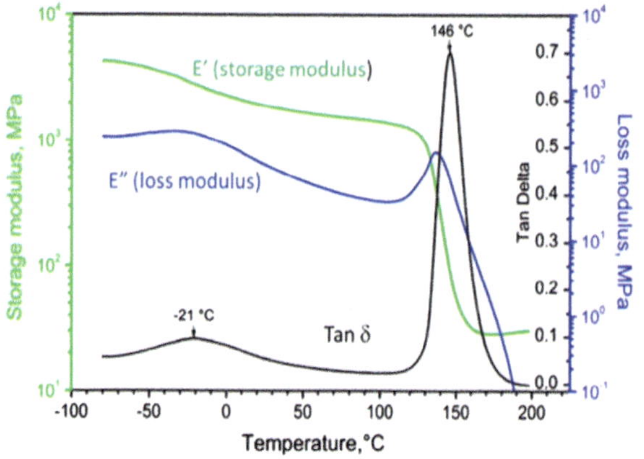

Fig. 5 Schematic illustration of DMA curves [6]

2 Thermal Analysis of Glass-Ceramics and Composites for Dental Prosthetic Restorations

Dental prosthetic restorations are in general fabricated from silicate glass-ceramics containing modifying cations, such as aluminum, potassium and sodium, in appropriate percentages in order to monitor specific properties such as Coefficient of Thermal Expansion (CTE), glass transition temperature, melting point etc. [7]. Dental glass ceramics behave as a viscous liquid above T_g and solidify after cooling. The change from viscous-liquid to solid during cooling is followed by shrinkage which depends on temperature and heating/cooling rates. It has been shown that slow cooling is important to prevent thermal gradients and high-magnitude transient stresses at the interface between the ceramic veneer and the metal or ceramic substructure of a dental prosthetic restoration [8].

Various GCs have been used as veneering coatings on metallic or ceramic cores for the fabrication of dental metal-ceramic or all-ceramic fixed partial dentures or as coatings on implants for bone tissue anchoring. The T_g of these GCs is critical, as the matching of their CTE with the corresponding CTE of the alloy of the metallic or ceramic core should be performed at this temperature area, in order to avoid detrimental cracks in its mass [8, 9]. In this respect, thermal analysis is fundamental in determining crystallization temperatures (T_c), glass transition and melting points and in combination to other complementary characterization techniques such as Fourier Transform Infrared spectroscopy (FTIR), X-Ray Diffraction Analysis (XRD), etc. can unveil significant information for the structure of the GCs. In the following paragraphs a comprehensive thermal analysis of various GCs for dental and biomedical applications will be performed, with the aim of presenting the significant information that can be retrieved by the TG for this group of materials.

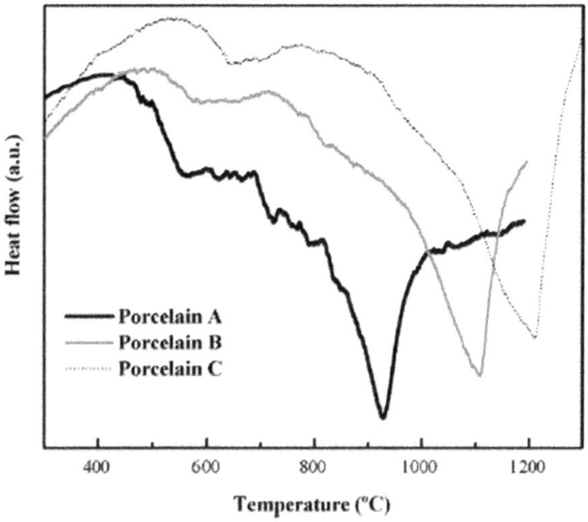

Fig. 6 The heat flow curves of the dental ceramics [10]

2.1 *Silicate Glass Ceramics as Veneers for Metal-Ceramic or All-Ceramic Restorations*

Feldspathic glass ceramics are silicate low-to-moderate leucite-containing feldspathic glasses called also "feldspathic porcelains". They are widely applied in dental restorations either as veneering materials to metallic or ceramic cores or as inlays and laminate veneers. They consist mainly of leucite crystals, while certain compositions may contain fluorapatite or lithium disilicate crystalline phases. Crystallization of glass-ceramics is generally achieved through a two-stage heat treatment consisting of an isothermal nucleation stage followed by an isothermal growth stage. By thermal analysis the heat treatment can be monitored for the production of a glass-ceramic with the desired microstructure and crystallinity. In a study of the thermal behavior of various dental glass ceramics, the thermal properties of ceramic specimens were evaluated with TG with a heating rate of 10 °C/min in argon atmosphere [10]. The Differential scanning calorimetric analysis (DSC) curve of ceramic C (Fig. 6) presents T_g at 607 °C and a melting peak at ~1200 °C. Similar trend in the curves of the rest two commercial ceramics was observed with a shift towards lower T_g and melting temperatures. It was shown by heat flow curves that the glass transition and melting temperatures present a descending order following the firing temperature of the ceramics [10]. This information along with the CTE calculation of the respective glass ceramics give useful information concerning their potential application as veneering layers of metallic substrates.

Recently, new dental feldspathic glass-ceramics were synthesized containing 75 wt% feldspar, 20 wt% quartz and 5 wt% kaolin [11]. Heating was performed up to 1200 °C with a heating rate of 10 °C/min to analyze their thermal behavior after the addition of different B_2O_3 amounts (1, 3 and 5 wt%) which aimed to

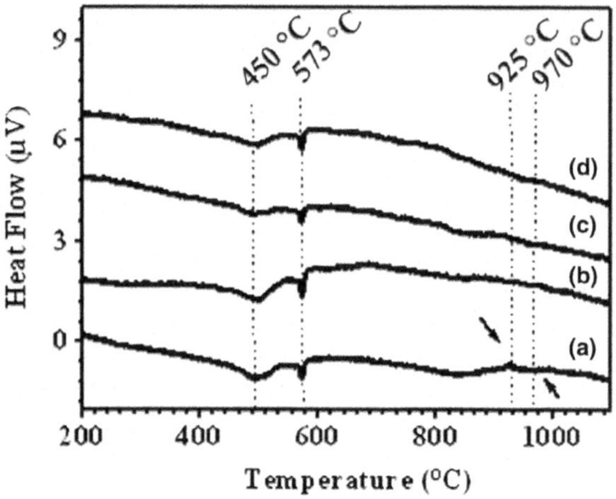

Fig. 7 Heat flow curves of the synthesized dental glass-ceramics **a** 0 wt% B_2O_3, **b** 1 wt% B_2O_3, **c** 3 wt% B_2O_3, **d** 5 wt% B_2O_3 [11]

improve the density and mechanical properties of the sintered specimens. As shown in Fig. 7, B_2O_3 addition does not seem to alter significantly the curve trend compared to control (a). A broad endothermic peak is observed at 450°C corresponding to the burnout of organic residues. The sharp and distinct endothermic peak at 573 has been correlated with the transformation of the $\alpha \rightarrow \beta$ quartz [12]. The small exothermic peak at 925 °C is disappeared in the heat flow curves of the B_2O_3 compositions and the authors justified this phenomenon by the decrease of the crystallization temperature (delay of crystallization) of leucite in the presence of B_2O_3. Furthermore, the sintering temperature was decreased about 25 and 50 °C for 3 and 5 wt% B_2O_3 additions, respectively.

Lithium disilicate glass-ceramics are unique dental materials and they currently represent the materials of choice for aesthetic dental restorations. Recently, novel niobate-containing lithium disilicate glass-ceramics as veneers for zirconia based dental restorations were developed in optimum compositional ranges in mol%: SiO_2 (53–69%)—Al_2O_3 (0.5–3.5%)—B_2O_3 (0.5–4.0%)—Li_2O (19–38%)—Na_2O (2–6%)—Nb_2O_5 (2.0–4.5%)—ZrO_2 (0.2–2.0%) [13]. Dispersions of the synthesized glasses were spray-coated on zirconium substrates and a two-step thermal treatment was applied: the first was a step of 5 min at 820 °C which was followed by a second step of 5 min at 920 °C. According to the heat flow curve in Fig. 8, the small endothermic peak at 480 °C corresponds to T_g while the two crystallization peaks at 700 and 740 °C, are attributed to the formation of Li_2SiO_3 and $Li_2Si_2O_5$ respectively, as confirmed by XRD analysis. The melting temperature was determined at 925 °C, as evidenced by the strong endothermic peak at that temperature. Variations among Li_2O, SiO_2 and Nb_2O_5 and thermal treatments, resulted in varying heat flow curves and to partially crystallized specimens. The authors investigated various Li_2O/SiO_2 ratios and concluded that the most favorable in terms of crystallization phases and the formation of a strong bond to the zirconia substrate was that with the Li_2O/SiO_2

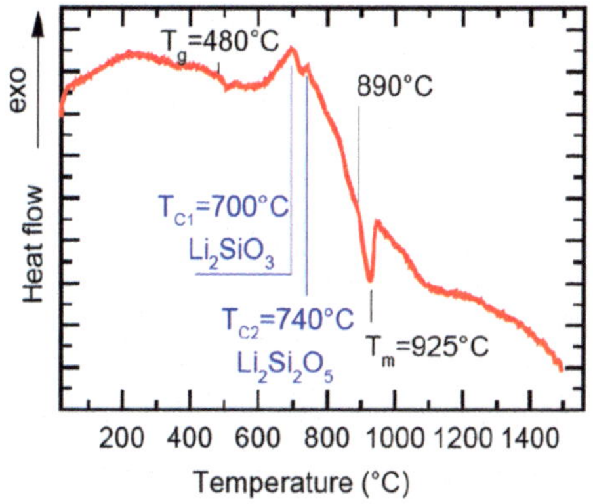

Fig. 8 Heat flow curve of a representative lithium disilicate glass powder [13]

ratio 1/26 and that the formation of $(Na_{(1-x)}Li_x)_3NbO_4$ crystalline phases after the specific heat-treatment resulted in a smooth surface free of flaws and irregularities.

2.2 *Bioactive Glass Ceramic Composites as Veneers for Metal-Ceramic or All-Ceramic Restorations*

Sol-gel method has been recently introduced for the synthesis of GCs. This is a low-temperature chemistry-based synthesis method, where a solution of appropriate chemical precursors is turned to gel under certain conditions near room temperature. Gels are wet inorganic networks of silica that can be converted to glass or glass ceramic after appropriate heating at a temperature range of 400–700 °C.

Sol-gel-derived microporous bioactive GCs have been synthesized with potential application as veneers for fixed partial dentures, by the combination of bioactive glasses and feldspathic ceramics. Compositions of glasses, ceramics, glass-ceramics and composites that have been shown to bond to bone are known as "bioactive ceramics". Bioactive ceramics can bond to hard- and in some cases to soft tissues-through specific physicochemical reactions upon contact with body fluids, which lead to the formation of a hydroxycarbonate apatite layer (HCAp) at the bonding interface [14]. The first bioactive material was a glass of the composition 46.1 mol% SiO_2, 24.4 mol% Na_2O, 26.9 mol% CaO and 2.6 mol% P_2O_5 synthesized by Professor Hench [14], that could form a strong bond to living bone. The key property of this material was that it could release Si in the surrounding medium through network dissolution, and that could rapidly form a silica-gel layer which accelerated the precipitation of amorphous calcium phosphate. This precipitated layer could in turn crystallize to HCAp, which adsorbed proteins, incorporated collagen fibrils and attracted bone progenitor cells that could differentiate to osteoblasts able to produce extracellular matrix. New bone formation occurred after biomineralization of the excreted extracellular matrix. Apart from this physicochemical route, bone forma-

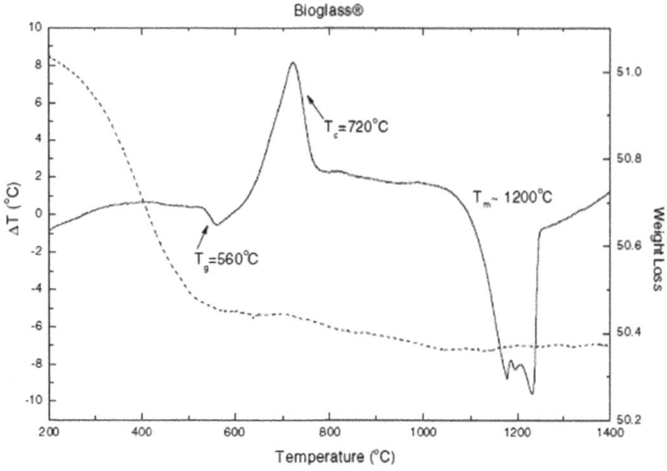

Fig. 9 The TG-DTA curves of the 45S5 Bioglass® [19]

tion can be further controlled by the bioactive glass dissolution products, which cause rapid expression of genes that regulate osteogenesis and the production of associated growth factors [15].

Only certain compositions of these materials can express this phenomenon, called bioactivity that can lead to new tissue formation. Factors that have been correlated to bioactivity are the material composition [16], atomic structure [17], network connectivity [18], sintering temperature and thermal processing [19], activation energy of silica dissolution in the glass [20], surface chemistry and topography etc. From the differential thermal analysis that was performed by Chatzistavrou et al. [19], the T_g, T_c and T_m temperatures of Bioglass® could be recorded. From the DTA curve (Fig. 9, solid line) of Bioglass® the endothermic peak observed at 560 °C can be attributed to the glass transition, while the exothermic event at 720 °C can be related to the formation of crystal phases. Finally, a melting process can be observed in the range 1100–1250 °C. Moreover, the weight loss curve (Fig. 9, dashed line) shows a small reduction in mass of Bioglass® which can be assigned to entrapped water evaporation [21].

The thermal behavior of a melt-derived 45S5 bioactive glass in the form of powdered samples of different grain size ranges was thoroughly investigated using differential scanning calorimetry with a heating rate of 10 °C min^{-1}, up to 1400 °C in argon atmosphere [22]. In the heat flow curve of a powdered sample (Fig. 10a) with particle size <20 μm there are four characteristic temperature ranges: the T_g at 555 °C, two exothermic crystallization peaks (T_{c1} and T_{c2}) at 700 and 795 °C respectively, and the melting process in the range 1100–1250 °C.

In the respective curves (Fig. 10b) of all powdered samples the presence of two crystallization peaks is distinct, however, for particle size <80 μm the two peaks are clearly separated and their behavior is clearly dependent on the particle size. By

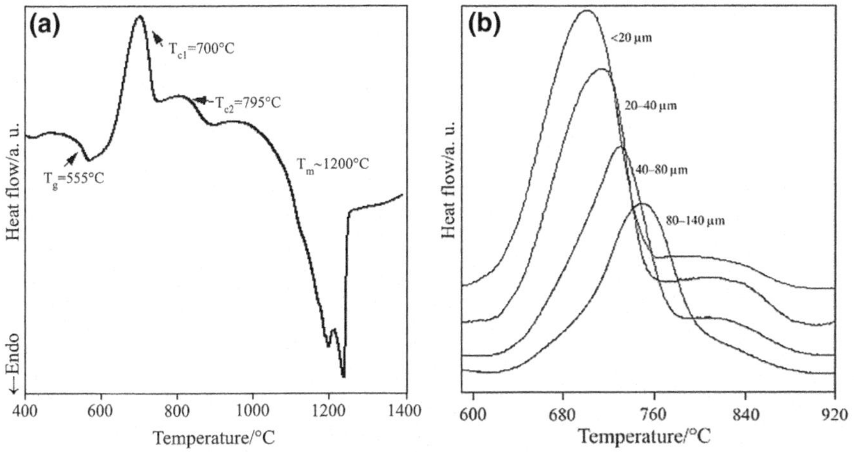

Fig. 10 **a** Heat flow of a melt-derived 45S5 bioactive glass with particle size <20 μm) at heating rate of 10 °C min^{-1}. **b** Heat flow curves as a function of particle size [22]

increasing the particle size, the first peak becomes shorter and broader, resulting in an overlap of the second peak after a critical value of particle size (80–140 mm). The authors performed peak profile analysis and reported that the crystallization peak height (δT_c) of the first peak decreases with increasing particle size, while for the second peak the height remains almost unchanged. This behavior was attributed to different crystallization mechanisms according to the technique for identifying surface and bulk crystallization developed by Ray and Day [23, 24]. The height of the first peak was decreased with increasing particle size, indicating surface crystallization while the height of the second crystallization peak remained unchanged indicating a volume crystallization mechanism. The decrease of the Tc with the decrease of particle size was attributed to faster heat transfer in the center of small particles compared to slow heat transfer in the center of larger particles [25].

Chatzistavrou et al. [26] applied the sol–gel method for the fabrication of a dental glass-ceramic in the system $SiO_2$60%–$P_2O_5$3%–$Al_2O_3$14%–CaO6%–Na_2O7%–K_2O10% (wt%) and a related composite material of this glass-ceramic and the bioactive glass 58S ($SiO_2$58%–$P_2O_5$33%–CaO9%) [27]. Although the synthesized glass ceramic failed to reveal any crystallization peaks (Fig. 11b), due to significant mass loss (4%) at the same temperature range attributed to nitrates, several exothermic peaks were recorded in the heat flow curve of the composite, in the temperature range of 800–1000 °C, where the crystallization of wollastonite occurs (~900 °C) (Fig. 11a). The exothermic peaks observed at lower temperature (~850 °C) were assigned to the crystallization process of the apatite-like phase, as revealed by the XRD analyses. Although these peaks are of lower intensity, they are still distinguishable since the associated mass loss is not significant (~0.65%, total mass loss ~2.5%). This

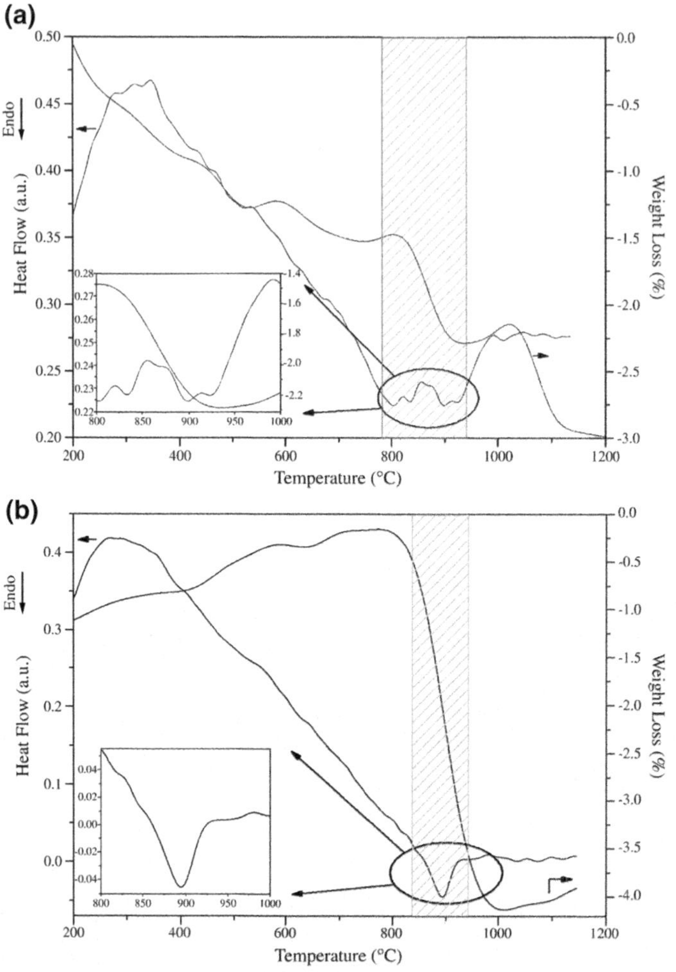

Fig. 11 Weight loss-Heat flow curves of a dental glass-ceramic in the system $SiO_2$60%–$P_2O_5$3%–$Al_2O_3$14%–CaO6%–Na_2O7%–K_2O10% (wt%) and a composite material with 58S bioactive glass [26]

composite resulted in fast HCAp formation after 3 days immersion in SBF and was suggested as a promising dental glass-ceramic.

Another attempt was performed by Goudouri et al. [28], who synthesized an experimental sol–gel dental ceramic composite (DC80) derived from the combination of a feldspathic dental ceramic (DC) and 58S bioactive glass in low percentage (20%). The thermal properties of the synthesized materials were evaluated with TG-DSC, with heating rate 10 °C min^{-1}—in an argon atmosphere. The heating was performed up to 1400 °C and the T_g, T_c and T_m temperature were determined.

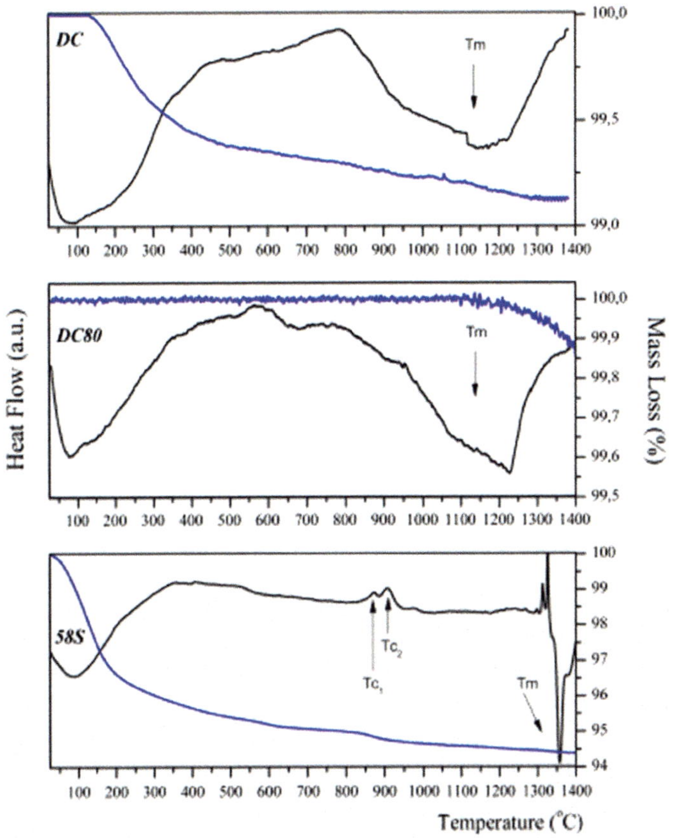

Fig. 12 Mass loss-Heat flow curves of a commercial feldspathic GC, 58S bioactive glass and the composite DC80 [28]

The DC80 composite follows similar thermal behavior with the composite DC80, although its melting point seems to start at 900 °C, at slightly higher temperature compared to the commercial DC (800 °C) (Fig. 12). This is attributed to the presence of 58S bioactive glass, which presents a melting peak at ~1350 °C. The two endothermic peaks at 870 and 905 °C in the curve of 58S are attributed to one or more crystal phase's formation, which according to the respective FTIR spectra and XRD patterns are leucite and various Ca-P phases. The experimental composite DC80 presents no mass loss for a temperature range of 1200 °C (RT–1200 °C) indicating better thermal stability than the starting materials. In the case of the commercial DC, it seems to be stable until 150 °C and it presents a slight mass loss of 1 wt% at 1400 °C. On the other hand, 58S loses the 4 wt% of its mass until 250 °C and then it presents only a slight mass loss (<2 wt%) until 1400 °C.

2.3 *Bioactive Coatings on Dental Implants*

Titanium (Ti) is the most widely used material for dental implants due to its biocompatibility, resistance to corrosion, high mechanical properties, and osseointegration. Currently, there are four commercially pure Ti grades and one alloy that are used for manufacturing of dental implants [29–31].

Surface chemistry and topography are fundamental aspects of an implant. The surface of Ti implants determines how they will interact with the surrounding tissues. Coating of metallic implants through a thin ceramic layer is a widely investigated approach to enhance the formation of new bone at the bone-implant interface. Various glass and nano- ceramic materials have been utilized for the manufacturing of coatings on Ti implants and most of them are bioactive. They mainly include HA [32], bioactive glass [33], and other calcium phosphates. In general, bioceramic coatings are considered osteoconductive. Calcium release from ceramic coatings allows for the deposition of a thin layer of biological apatite, which facilitates cells to adhere, proliferate and differentiate, leading to osseointegration of the implanted material [34]. The main drawbacks of these ceramic coatings are their brittleness and their susceptibility to crack propagation leading to catastrophic failure. Furthermore they usually present thermal expansion mismatch with the titanium substrate, which on heat treatment result with interfacial cracks and delamination of the coating. The most common ceramic nanoparticles utilized as coatings on titanium biomedical implants are presented in the following paragraphs.

2.3.1 Hydroxyapatite Coatings

Nano-HA is probably the most preferable material used as a coating for Ti, Ti alloys and stainless steel implants because it has chemical composition similar to the inorganic component of bone. In vivo experiments with HA nanoparticles have shown enhanced bone-implant contact, bone bonding strength and new bone formation at the early stages of bone healing [35–37]. They can be produced by various synthesis methods such as electrophoretic deposition (EPD), sol-gel deposition, electrostatic spray deposition, electrodeposition, hydrothermal deposition, aerosol deposition, pulse electrodeposition, radiofrequency magnetron sputtering, electrohydrodynamic spray deposition (EHDSD), liquid precursor plasma spraying (LPPS) etc. [38].

In a recent study of Anjaneyulu et al. [39], ethanol-based HAP nanoparticles (NPs) were prepared by the sol–gel method and spin coated on Ti–6Al–4V implants. Their thermal properties were evaluated by TG-DTA at temperatures between 32 and 1000 °C in the presence of N_2 atmosphere at a heating rate of 10 °C/min.

The TG curves presented in Fig. 13 show a sequential weight loss of 53% total at 30–115 °C, 115–300 °C, 300–425 °C and 425–615 °C which according to the authors may be due to the evaporation of ethanol and physically adsorbed water molecules, loss of chemically incorporated water molecules in HAP crystal lattice, removal of nitrates from calcium source used in the synthesis and elimination of carbonates from the raw powder respectively. No significant weight loss was observed from 615 to 1000 °C which confirms the phase formation of HAP and its thermal stability even at higher temperatures. The DTA curve shows an endothermic peak at 85 °C and a broad exothermic peak at 350–620 °C which further confirmed the loss of moisture, ethanol and removal of residual nitrates respectively. Supportive FTIR and XRD analysis were in agreement with the TG-DTA and revealed the formation of pure HAP at the temperature of 600 °C. The pure HA phase was stable even after sintering at 1200 °C where it presented the highest crystallinity, which was gradually developed by the increase of the sintering temperature. Although nanoparticles of pure HA were synthesized by this method, the coatings developed cracks and pores when the sintering temperature increased to 600 °C probably due to mismatch of the thermal expansion coefficient between the coatings and the titanium substrate [40] as well as firing shrinkage during sintering. The authors concluded that the 500 °C heat-treated coatings developed with a spin speed of 3000 rpm were the most stable and uniform and presented highest corrosion resistance and biocompatibility. Similarly, Fathi and Hanifi [41] synthesized sol-gel nano-hydroxyapatite coatings that were deposited by the dip coating method onto commercially pure titanium (cpTi) substrates. Thermogravimetric analysis conducted on as dried gels obtained from solutions revealed three weight losses at 250, 400 and 500 °C which correspond to the evaporation of crystalline water from the $Ca(NO_3)_2.4H_2O$ in the precursor, the removal of $–NO_3$ or –OH groups from the alkoxynitrate species in the precursor and crystallization of HA respectively. No significant loss was observed at temperatures above 550 °C which verifies the stability of HA phase. The authors by applying thermogravimetric analysis in conjunction with XRD proved that an aging time greater than 24 h is required for the production of a uniform nanostructured pure HA thin coating.

Another useful application of TG is presented by Farokhi-Rad et al. who applied EPD for the synthesis of porous nanostructured hydroxyapatite (HA) coatings [42]. In order to introduce porosity in the coatings, carbon black (CB) particles were used as porogen for the sacrificial method. However, the final coating should be free of CB particles, so it's of critical importance to thermally treat the produced coatings in such a way that the minimum amount of CB particles is present. The determination of the temperature range where CB particles burn out completely was performed by TG at the temperature range of 25–800 °C with heating rate 5 °C/min. As shown in Fig. 14, CB particles completely burn out at a temperature range of 480–650 °C depending on their initial amount.

The liquid precursor plasma spraying (LPPS) process has been utilized for the synthesis of nanostructured hydroxyapatite coatings on Ti–6Al–4V alloy substrate [43]. The LPPS process is a plasma spraying technique which uses the liquid precursor as the feedstock for the plasma spraying, which is atomized into mist and injected into the plasma jet, instead of the powder feedstock in the atmospheric plasma spraying (APS) process which is usually applied for the synthesis of HA coatings on orthopedic and dental implants. The HA liquid precursor was prepared by a wet chemical method [43]. The thermal behavior of the HA liquid precursor was

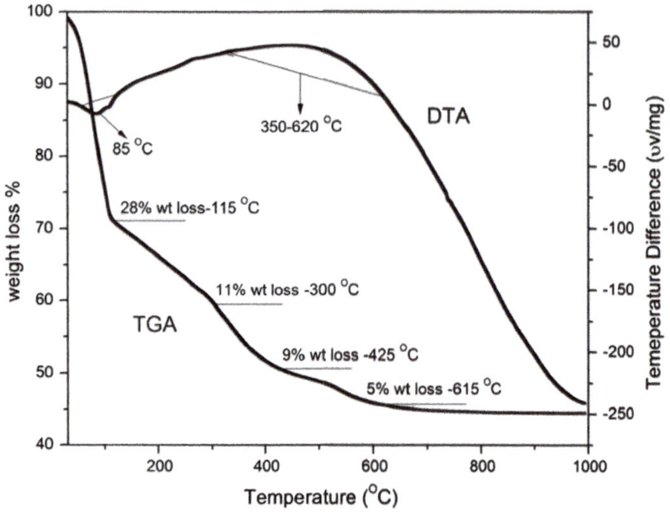

Fig. 13 TG-DTA curves of a nano-HA powder [39]

Fig. 14 Weight loss of the powder scratched from the coatings deposited at 60 V and 1 min from the suspensions with 10 (**a**) and 20 g/L (**b**) of HA nanoparticles and the different concentrations of CB particles [42]

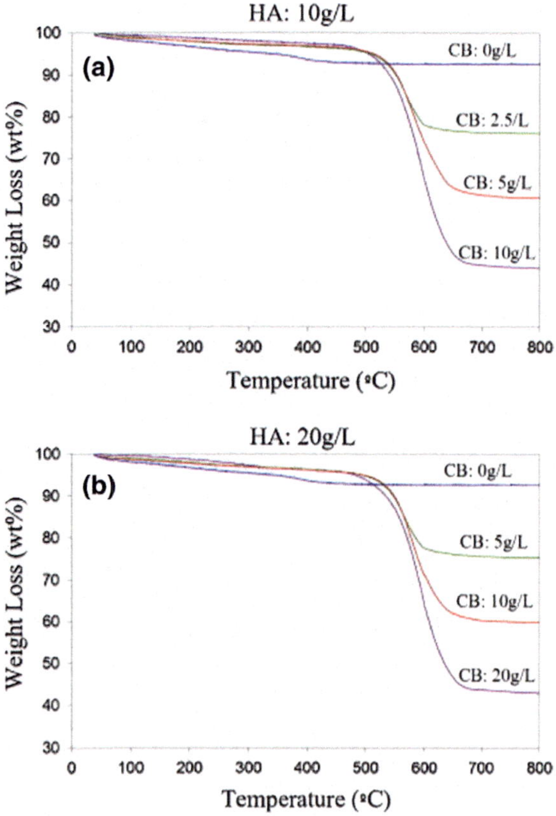

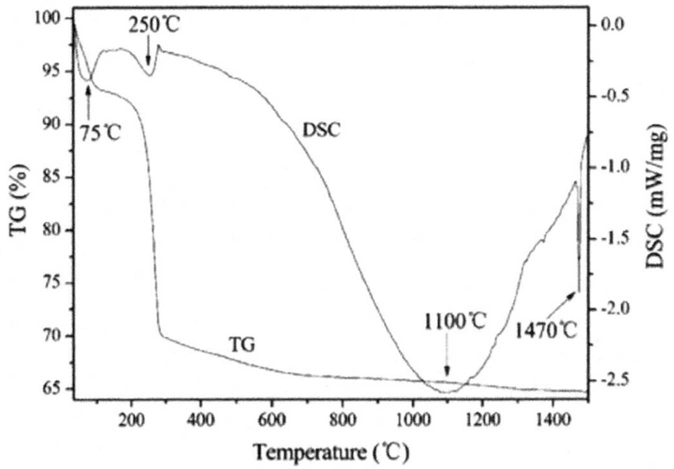

Fig. 15 The TG-DSC curves of the as-dried nano-HA precursor powder [43]

analyzed with TG-DSC heated up to 1500 °C in nitrogen atmosphere at a heating rate of 10 °C min^{-1}.

Figure 15 shows the TG-DSC curves of the as-dried HA precursor powder. Four distinct exothermic peaks are present in the DSC curve. The two peaks at 75 and 250 °C correspond to the loss of water (weight loss 6 wt%) and the decomposition of ammonium nitrate, (weight loss 24%) respectively. At higher temperatures, a broad exothermic peak at 1100 °C is observed that according to the authors corresponds to the transformation from HA to β-TCP and another one at 1470 °C attributed to the transformation of β-TCP to α-TCP. Due to the high temperature of the plasma spray process the decomposition of HA powder in APS results in the formation of ACP, TCP, TTCP and CaO phases, as verified by XRD. The use of the liquid precursor in the LPPS reduced the temperature in such a way that less decomposition took place, leading to the formation of mainly pure HA and a minor fraction of β-TCP. Furthermore, the presence of liquid precursor in LPPS resulted in a lower degree of dehydroxylation with temperature, resulting in coatings with high amount of OH, which could be beneficial to bone bonding.

2.3.2 Bioactive Glass Coatings

Bioactive glasses (BGs) and glass-ceramics have been proposed as coatings for dental implants due to their favorable biological responses in contact with body fluids and fast apatite formation ability. However, their thermal expansion coefficients are usually much larger (~15–17 × 10^{-6} °C^{-1}) than those of Ti oxide and its alloys (~9.2 × 10^{-6} °C^{-1}), and this significant mismatch limits their use to applications bearing low biomechanical loads such as: bone fillers, bone cements and scaffolds

[14]. The first attempt of Tomsia et al. [44, 45] to develop bioactive glass coated implants of titanium alloys revealed that they are very brittle and prone to crack propagation leading to catastrophic failure.

Currently the only melt-quenched BGs accepted by the USA Food and Drug Administration for use in clinical practice are the 45S5 Bioglass® (SiO_2 45 wt%, Na_2O 24.5 wt%, CaO 24.5 wt%, P_2O_5 6 wt%) and S53P4 BonAlive® (wt%: SiO_2 53 wt%, Na_2O 23 wt%, CaO 20 wt%, P_2O_5 4 wt%). However, although various research studies have shown their potential application as coatings on titanium and its alloys [46], none of them has been introduced as implant coating in the dental market yet.

A 45S5 bioactive glass coating on Ti6Al4V alloy was obtained in 2010 by Berbecaru et al. through magnetron sputtering [47]. The authors performed a thermal treatment at 700 °C for 2 h to obtain a glass-ceramic coating. In a recent study [48], EPD was used to coat Ti6Al4V with 45S5 bioactive glass that was synthesized by the sol-gel method. After the hydrolysis of the precursors and the polycondensation, the suspension was transferred to an electrophoretic cell for the deposition of coatings. The authors received a partly densificated coating after thermal treatment at 600 °C, slightly above the 45S5 bioactive glass T_g, while an homogenous coating was received at 700 °C. At higher temperatures coarser and larger crystals were observed and different rich in phosphorous sodium-calcium-silicate phases were detected with XRD.

Fathi and Doostmohammadi [49] attempted to synthesize bioactive glass nanopowder by the sol–gel technique in order to develop bioactive glass coatings on 316L stainless steel. A bioactive glass with composition 57.44% CaO, 35.42% SiO_2 and 7.15% P_2O_5 was synthesized for the improvement of corrosion resistance, biocompatibility and bone bonding. Crack-free amorphous bioactive glass coatings were received (Fig. 16) even after thermal treatment up to 900 °C, however at higher temperatures the crystallization of larnite was identified by XRD, downgrading the mechanical and the bioactive properties of the material.

The necessity for antimicrobial properties of dental implant coatings to prevent peri-implantitis and loss of osseointegration, has led to the vast investigation of silver as doping agent in various bioactive glass formulations [50, 51]. Bioactive glasses containing different percentages of silver with composition (in mol%): 70% of SiO_2 (S), 30% of CaO (C) and x% of Ag_2O (A) where $0.08 \leq x \leq 0.27$ have been synthesized via the sol–gel method and used to coat titanium implants by means of a dip coating technique [51]. Simultaneous FTIR and TG-DTA analysis under inert Ar atmosphere with a constant heating rate of 10 °C min^{-1} from room temperature to 1200 °C were used to characterize the materials. In order to perform kinetic analysis, TG experiments were performed with heating rates of 5, 7, 10 and 15 °C min^{-1} under the same thermal conditions in order to reveal the optimum heating conditions for the treatment of the as-produced gels.

Mass loss follows the same trend in all compositions irrespectively of the silver content and is typically to other calcium silicate ceramics (Fig. 17). The first step up to 160 °C with a mass loss of about 27%, is attributed to dehydroxylation (loss of physically absorbed water). The second step, with a mass loss of 7%, has an

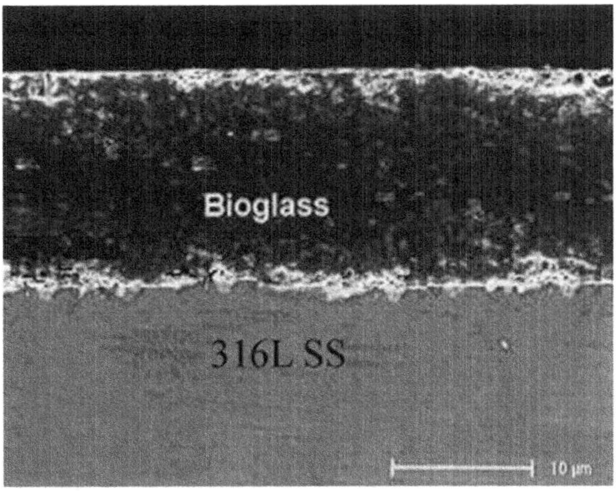

Fig. 16 SEM micrograph of a cross-section view of prepared bioactive glass coating on the 316L SS substrate [49]

exothermic effect, attributed to the thermal desorption of chemically adsorbed water. The third step between 390 and 600 °C with a mass loss of about 12–13% is related to the burning of the nitrates form the sol precursors. An exothermic DTA peak without mass loss is evidenced in the temperature range of 900–1000 °C which is attributed to crystallization of a calcium silicate phase [52, 53]. As the authors investigated the properties of the as-received coatings and based on the performed thermal analysis, future research is needed for the evaluation of proper thermally treated coatings and the investigation of the crystallized calcium silicate phases on the adhesion strength, bioactivity and antibacterial properties of this kind of coatings.

2.3.3 Miscellaneous Bioactive Composite Coatings

Due to the limited feasibility of pure bioactive glasses to form stable and crack-free coatings and/or nanocoatings on dental implants, various composite and nanocomposite coatings have been synthesized incorporating bioactive glasses and/or nano-HA.

Bioactive glass-nano-HA composites

A series of glasses in the SiO_2–Na_2O–ZnO–CaO–MgO–P_2O_5 system were synthesized by the melt-quenching method in order to develop coatings on Ti–6Al–4V alloy [54]. Bioactive glass–nano-HAp composites were prepared with the addition of 10 or 25% nano-HAp. The glass particles (5–20 μm) were milled with nano-HAP (crystallite size 20–60 nm) to achieve a homogeneous microstructure in the resultant glass ceramic composite and applied as coating by spraying or dipping. An analysis of the composites' thermal properties was performed by differential thermal analysis (DTA), derivative differential thermal analysis (DDTA) and thermogravimetry (TG). From DTA and DDTA results, glass softening temperature was found to be below 750 °C, which is critical as at temperatures between 955 and 1010 °C titanium trans-

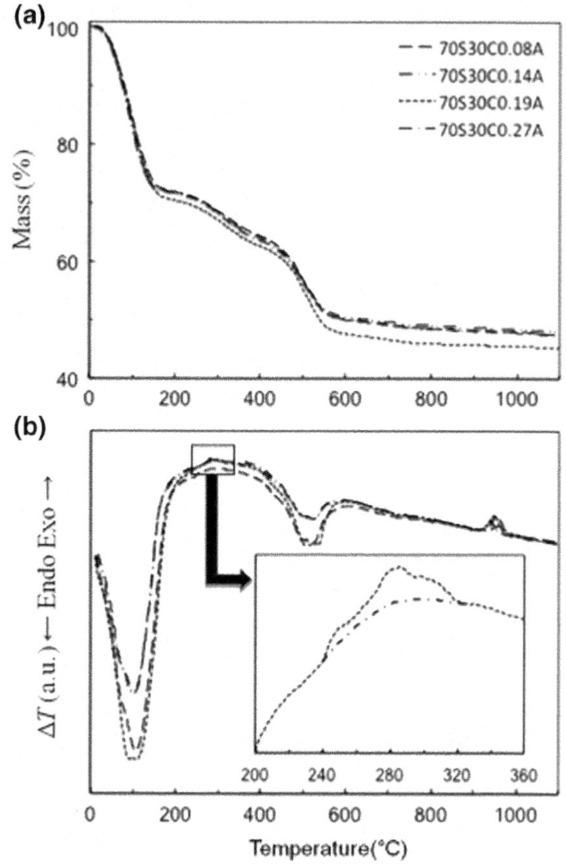

Fig. 17 TG-DTA curves of the four 70S30CxA synthesized gels [51]

forms from its α to β phase, and the coefficient of thermal expansion was higher but comparable with that of the alloy. As shown in Fig. 18 a distinct crystallization peak is absent, indicating a predominantly stable glassy material.

Bioactive glass-chitosan composites

To overcome the issues of bioactive glass brittleness the group of Boccaccini in Erlangen combined 45S5 bioactive glass with chitosan to form a soft composite coating appropriate for coating metallic implants [55]. Bioglass® particles were dispersed in a chitosan solution and EPD was applied for the deposition of coatings. In order to thermally analyze the coatings, they were removed from the metallic substrates and TG-DSC was performed in air at a heating rate of 10 °C/min.

The thermal analysis of the coatings (Fig. 19) reveals subsequent stages of moisture evaporation and combustion of chitosan in the temperature range of 220–600 °C. A small (3%) weight loss due to dehydroxylation is presented in the TG curve of the Bioglass powder (BG) while an increased percentage is recorded for the composites (both loaded and unloaded with gentamycin). Bioglass particles in the composites

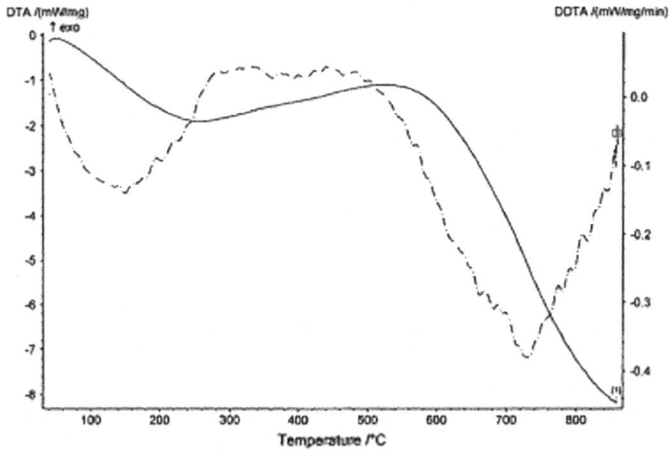

Fig. 18 A typical DTA/DDTA curve of a bioactive glass in the system SiO_2–Na_2O–ZnO–CaO–MgO–P_2O_5 [54]

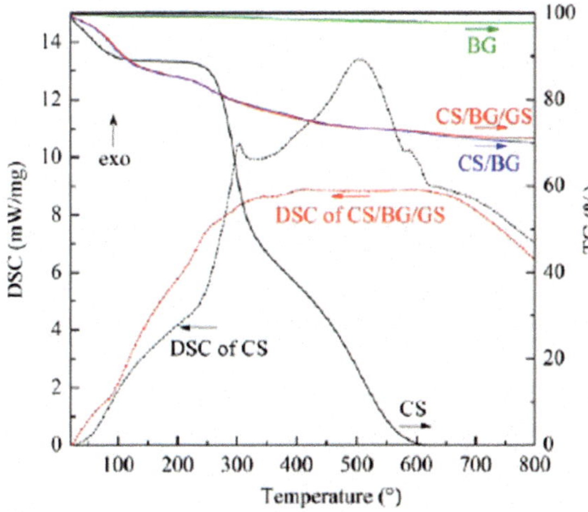

Fig. 19 TG-DSC curves of Bioglass powder (BG), chitosan (CS), CS/BG, and CS/BG/gentamycin (GS) EPD coatings [55]

restricted chitosan burning as shown by the respective TG curves, and significantly modified the DSC curve of pure chitosan (less pronounced exothermic peaks). The preservation of chitosan in the coatings and the restriction of its combustion in the presence of bioglass particles—that was evidenced by the TG—were further evaluated with FTIR, which revealed the presence of hydrogen bonding between chitosan and glass particles. This bonding is responsible for the stabilization of the coating and the co-deposition of the glass and polymer components. Recently the feasibility of these composites as coatings was verified by the same group on magnesium alloy (WE43) substrates [56].

Calcium silicate–reduced graphene oxide and graphene oxide–chitosan–hydroxyapatite composites

In the last 5 years, graphene has attracted enormous interest in biomedical engineering due to the improved mechanical properties of its composites [57, 58]. Recently, calcium silicate (CS)/graphene oxide and graphene oxide reinforced chitosan–hydroxyapatite nanocomposites have been used to improve the biological and mechanical fixation of titanium implants [59, 60]. The CS–rGO composite powders with different GO contents (0, 0.5 and 1 wt%) were produced by a hydrothermal method from the dissolution of appropriate precursors in water, and stable suspensions of CS–rGO powders with concentration of 1 mg/mL were prepared for the EPD coating deposition process [59]. After EPD, the samples were dried in air at room temperature for 12 h and oven dried at 100 °C for 24 h. To achieve appropriate coating adhesion the dried samples were sintered at 900 °C for 1 h in a vacuum furnace (argon atmosphere) with heating and cooling rates <5 °C /min to prevent the appearance of cracks due to CTE differences between the coatings and the substrates. Before sintering, XRD analysis revealed that the samples consisted of xonotlite, which is a calcium silicate-type hydrated phase which was converted to $CaSiO_3$ after sintering.

The TG curve of CS presents a total mass loss of 12.5% due to water removal and loss of water and carbonates up to 570 °C. The respective percentages for the composites were 11.8 and 14.6% (Fig. 20). The presence of rGO did not affect the thermal behavior of the composites. Compared to pure CS coating, the CS–1 wt% rGO coating improved adhesion to Ti by 70%, hardness by 150% and elastic modulus by 240%. The CS–rGO composite coatings exhibited good apatite-forming ability in simulated body fluid (SBF) and biocompatibility. Similarly in the work of Shi et al. [60], who synthesized a ternary graphene oxide–chitosan–hydroxyapatite (GO–C-S–HA) composite coating on Ti substrate by EPD, the associated mass loss for the composites was less compared to pure chitosan or grapheme oxide. It is worth mentioning that the decreased mass loss of the GO–CS–HA composites compared to the respective ones without GO was attributed to higher amount of HA particles incorporation in the composites.

3 Glass-Ceramics for Regenerative Medicine

Glass-ceramic materials have been in usage as biomaterials for many decades. Despite the fact that many ceramic materials are bioinert (e.g. ZrO_2, TiO_2) some other can be designed to promote tissue response. Hence, they have attracted significant attention as they are thought to be suitable candidates, as implants, in regenerative medicine.

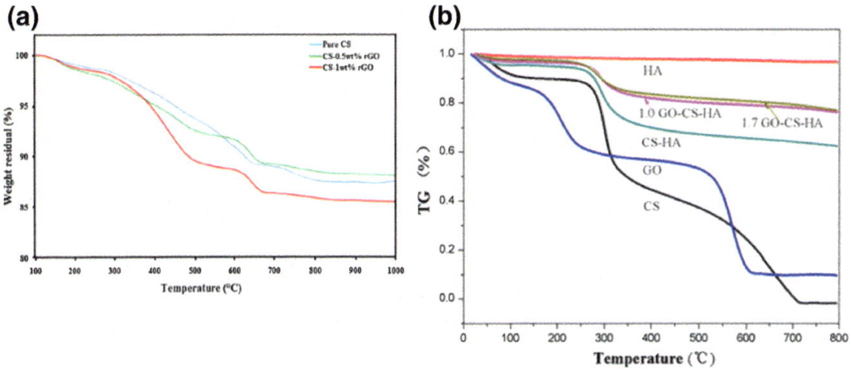

Fig. 20 **a** Weight loss curves of the CS–rGO composite coatings on Ti, **b** TG curves for the pristine materials CS, HA, GO, and the CS–GO, 1.0 GO–CS–HA and 1.7 GO–CS–HA composite coatings [57, 58]

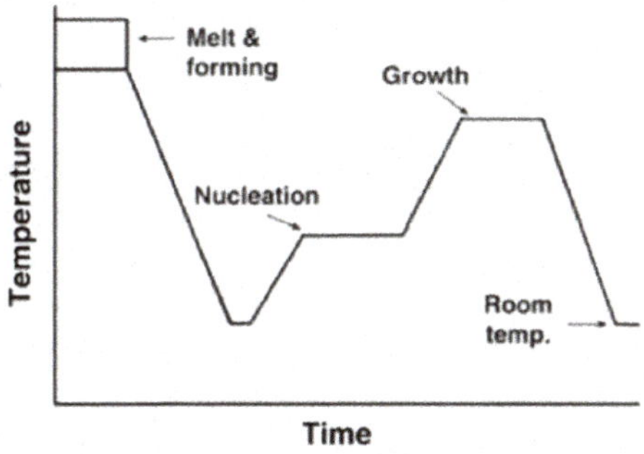

Fig. 21 Temperature-time cycle for a glass ceramic [61]

3.1 *Glass-Ceramic Material Formation*

The most common ways to produce a glass is by cooling a liquid into a rigid position (melt-derived glass) or by water evaporation from a liquid solution (sol-gel glass). Because the crystalline state is more stable than the glass state, in the case of applying temperature (below melting point) the result would be to receive a glass-ceramic or ceramic material. As Fig. 21 illustrates, in order for the glass-ceramic to form, nucleation of microcrystals has to commence in a temperature much below the melting point. By applying even higher temperatures would lead to the growth of the polycrystalline structure.

In the concept of producing bioactive glass-ceramic materials for the purposes of regenerative medicine, the thermal treatment of the final product is highly essential due to the diversity of properties that could be given to the material. Hence the optimum heat treatment conditions for each biomaterial have to be investigated. Usually firing step precedes the sintering of the material, which occurs at relatively low or intermediate temperatures so as possible organic binders are burned out [62,

63]. When the material is ready to undergo a sintering process, the densification window should be taken into account. That is the temperature window in which the material densifies without the nucleation process of the microcrystalline phase. This window is determined by the glass transition temperature (T_g) and the temperature at the onset of crystallization, for each glass. Whereas the driving force for densification is to minimize the surface energy, the existing pores of the material are removed though material movement mechanisms [64, 65]. In general the crystallization and the densification process of a glass should be considered separately.

The densification window of a glass can be affected by a number of factors, such as the particle size, the heating rate and the chemical composition of the starting material. In this regard, materials with finer particle or pore size and rapid heating rate lead to rapid sintering. Moreover, faster heating rates at relatively low temperatures can cause the suspension of the occurrence of crystallization resulting in highly dense materials.

The DTA curves, in Fig. 22, of two different glass-ceramic materials showcase the densification window between T_g and T_{onset}.

What is really interesting in the case of these two materials is that they have similar composition (**45S5**: 24.5Na_2O, 24.5CaO, 45SiO_2, 6P_2O_5; wt%, **13–93**: 6Na_2O, 12K_2O, 5MgO, 20CaO, 53SiO_2, 4P_2O_5; wt%); however they exhibit unlike densification windows. Silicate glass 13–93 exhibit a broader window conversely to 45S5 silicate glass. As a consequence 45S5 is much more difficult to sinter into a dense material without crystallization to occur to some degree. On the contrary 13–93 silicate glass can be densified effectively resulting in materials with higher mechanical strength. The critical factor resulting in such a profound difference is the presence of additional network modifiers (K^+, Mg^{2+}) in the composition of 13–93 [67–69].

Similarly Denry et al. [70] has reported that with increasing the Strontium percentage in a GS (glass-ceramic), from zero to 12, 18 and 24 (mol% SrO), the heat flow curves (Fig. 23) of the glasses presented two exotherms for which the temperature difference followed a linear increment. Despite the glass transition temperature slightly falls from 654 °C for the glass without Sr, to 627 °C (12 mol% SrO), 637 °C (18 mol% SrO) and 643 °C (24 mol% SrO) for the Sr-containing glasses, the presence of the second exotherm at lower temperatures for the Sr-containing glasses has an impact on the densification window of the material, since it is decreased by 20–30 °C. This is another example on how the thermal properties of a material could alter due change in their chemical concentration.

From the same authors came a study on how the heating rate could affect basic properties (sintering behavior, mechanical properties, structure) of a glass and in extension of a glass-ceramic scaffold [71]. The heating rate of the glass, as shown in Table 1, was initially set at 2 °C/min and then increased to 10, 20, 30, 40 and 55 °C/min. The results (Fig. 24) they came to showed that the T_g temperature was increased with increasing the heating rate in a linear manner, whereas the onset of crystallization followed the same trend with the exception of the glass heated at 55 °C/min, which showed a different behavior as a whole.

An increase in the densification window, from 146 to 201 °C, was observed for the glasses heated up to 40 °C/min. Moreover the exotherm peaks of the glass shifted

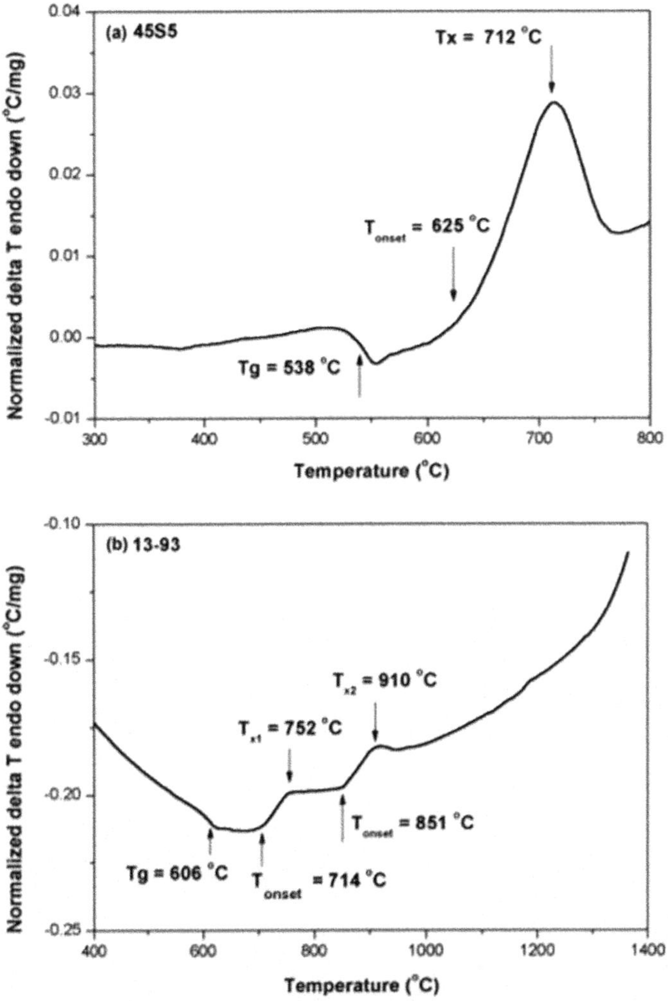

Fig. 22 Differential thermal analysis (DTA) curves for **a** 45S5 glass particles (<150 μm) and **b** 13–93 glass particles (<45 μm) at a heating rate of 10 °C per minute. (T_g: glass transition temperature; T_x: crystallization temperature; T_{onset}: onset of crystallization) [66]

to higher temperatures, whereas the second exotherm peak was not observed at 55 °C/min.

The case studies presented here, give a very good example of how the densification window and the thermal properties of a glass are in general affected by factors such as the chemical composition of the material and the heating rate. Therefore, careful consideration of the heating conditions and the chemical composition are required. Regardless the densification process can be completed; the crystallization of the glass will start if enough heating is present. Therefore, as already said, the conditions of the

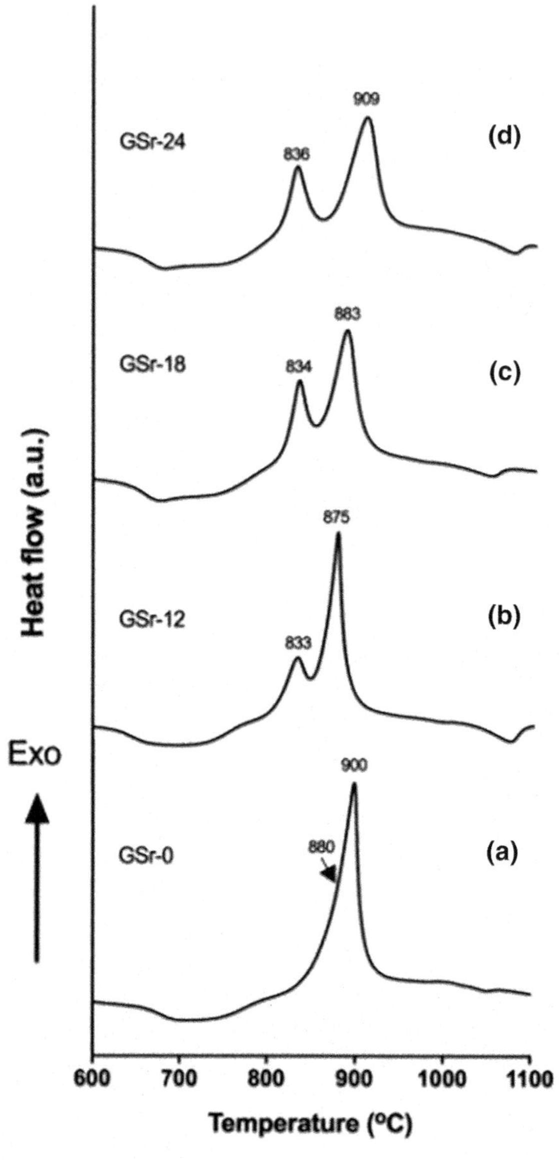

Fig. 23 Heat flow curves for the various glasses [70]

Table 1 Thermal behavior characteristic temperatures as a function of heating rate [71]

Heating rate (°C/min)	Glass transition T_g (°C)	Onset of crystallization (°C)	Processing window (°C)	Exotherm 1 (°C)	Exotherm 2 (°C)
2	646	792	146	818	925
10	659	830	171	848	965
20	665	853	188	873	995
30	676	870	194	892	1015
40	681	882	201	906	1038
55	698	877	179	910	N/A

densification are very important in addition to how large the window is. It is worth mentioning here that in many cases the final product is preferred to be partially crystallized in order to produce a glass-ceramic material with elevated mechanical strength, having both a densified glass structure and a small amount of crystalline structure. Conversely to this fact, it is reported that while increasing the crystalline percentage, the level of tissue reactivity greatly diminishes [21, 72, 73].

3.2 Bioactive Glass-Ceramic Scaffolds for Regenerative Medicine

3.2.1 Properties of Scaffolds

Scaffolds are key components for tissue engineering applications; functioning, in conjunction to other factors such as cells and growth factors, as a host for the development of new tissue, hard and soft alike. While scaffolds possess a number of advantageous properties, rendering them suitable for a number of applications, they do have some limitations [74–76]. In regard of their desired properties first and foremost, a scaffold for biomedical use in tissue regeneration needs to possess the appropriate architecture in order to facilitate 3-D organization of the tissue to be replaced, therefore it has to be tailored to match the defective areas' size and shape [77, 78]. Scaffolds should be highly porous; usually well over 60% should suffice, despite that the ideal scaffold should have approximately 90% porosity [79]. Moreover, the pores should be interconnected ensuring ease of cell infiltration, evenly cell distribution and survival in order to proliferate, with optimum size range between 100 and 500 μm so as to promote tissue regrowth [80, 81]. Another significant property of scaffolds is their mechanical response, ensuring mechanical stability in the defective area while giving way to newly formed tissue [82]. Furthermore, its biocompatibility, biodegradability and ability for osteoconduction should be examined thoroughly. Thus the scaffold implant should be able to host cells promoting,

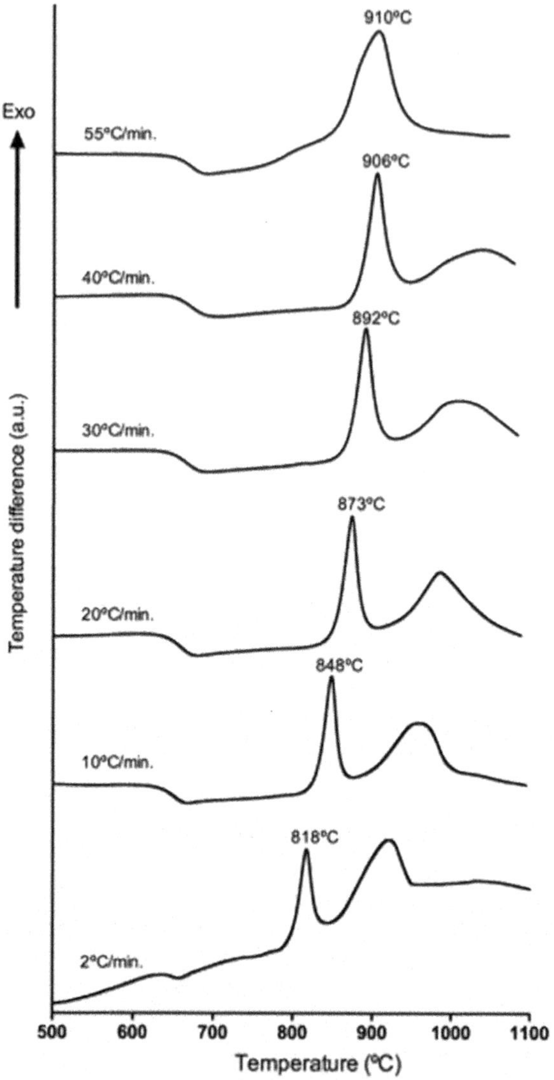

Fig. 24 Differential thermal analysis curves of the base glass as a function of heating rate [71]

proliferation and differentiation without causing toxic or inflammatory responses at the defective site [83–85]; doing so by degrading over time after implantation [86]. However, the degradation rate of a scaffold should be in accordance with the newly formed tissue rate, in a fashion that it can be controllable in order for the material to be used in a range of TE applications [83–85].

3.2.2 Major Scaffold Fabrication Techniques

A wide variety of scaffold fabrication techniques have been utilized in recent years, among them techniques like Rapid prototyping [79, 85] Electrospinning [79, 87], Foam Scaffolding [79, 84, 85], 3D printing and more. Table 2 shows a list of these techniques showcasing the major representatives of each one with a brief comment on the main advantages and disadvantages of each one of them. In collectiveness these techniques aim at developing materials with open pore structures and interconnected porous network but some are focusing in the introduction of a more controlled structure regarding the macro-nanoporosity scales via employment of cutting edge technologies [14, 88]. Each one of the scaffold developing methods exhibit advantages and disadvantages over other techniques, due to different developmental conditions resulting in different structures; meaning pore size and pore network interconnectivity. Thus it is best to consider each of these techniques when aiming at specific tissue types [89]. Another factor influencing scaffold development is the cost of the processing method, where a low cost technique would allow it to be a candidate for large-scale production, whereas, a high cost method would be used selectively.

Table 2 Fabrication techniques for 3D bioceramic scaffolds

Technique	Major representatives	Main advantages	Main disadvantages
Rapid prototyping[a]	SLS, SLA, RC, FDM 3D-Printing	Complex shapes with high accuracy, good mechanical strength	Requires high temperatures during the fabrication process, trapped, powder issue
Electrospinning		Accurate control of scaffold's structure	Limited shape and macro architecture capabilities and
Foam scaffolds[b]	FR, GFCR, SAFA	Versatile shape of final product, does not require additional machining because the solution is being directed in molds	Difficulty achieving high interconnectivity, probable low mechanical stability
Starch consolidation	–	Environment-friendly, low cost	Poor interconnectivity

Major representatives, advantages and disadvantages. Request copyright [79, 85]

[a]**SLS** Selective laser sintering, **SLA** Stereolithography, **RC** Robocasting, **FDM** Fused deposition modeling

[b]**FM** Foam replica, **GFCR** Gas foaming by chemical reaction, **SAFA** Surface active foaming agents

In comparison to other routes dealing with diseased or injured organs and tissues, tissue engineering aims at regenerating the damaged or lost tissue. The problem of limited resources in other routes, such as limited availability of donors and even rejection and surgical complications causing problems, has opened the opportunity for tissue engineering to achieve the regeneration by using other means which are in abundant, such as host cells, bioactive materials or a combination.

Bioactive materials have been widely used as implants in the past decades accommodating an interfacial bonding between the surrounding tissue and the implant [66].

Glass-ceramic materials have been in the centerpiece for some time regarding their use in regenerative medicine in the form of scaffolds. Scaffolds are highly porous 3-Dimensional constructs that are considered to be biodegradable and bioresorbable and are able to support the growth of new tissue [61].

In order to proceed to the scaffold fabrication step, a thorough study of the glass properties is fundamental. For example, two glass–ceramic systems that are intended to be used for scaffold fabrication are presented by Dessou et al. (SiO_2 60, CaO 30, MgO 5, SrO 5 (Sr5) and SiO_2 60, CaO 25, MgO 5, SrO 10 (Sr10) in wt%). In this study the glasses are investigated as powders so as to determine the optimum properties for scaffold fabrication [90]. Figure 25 shows the heat flow curves of the Sr5 and Sr10 glasses. The crystallization temperature of the glasses is in the range of 890–905 °C, while the Sr5 glass exhibits one exotherm and the Sr10 glass exhibits two exotherms. The authors selected to sinter the powders to 892 and 903 °C for Sr5 and Sr10 samples respectively so as to obtain a partially crystallized material in each case. The results indicated that the materials despite the presence of crystalline phases exhibited enhanced apatite forming ability even after sintering, hence making them suitable for the synthesis of Sr-containing bioactive scaffolds for bone tissue regeneration [90].

3.3 *Nanodimensional Ceramics for Regenerative Medicine*

Nanosized materials are considered those that contain particles of grains with less than 100 nm in size. Specific properties of bioactive glasses can be improved and controlled when synthesized in nanometer scale. It is reported that nanodimensional materials exhibit distinct mechanical and optical properties compared to their micro-sized counterparts. It is most distinctive their immensely lager surface area providing with unique surface properties. It is found that nanosized $CaPO_4$ is more efficient in promoting adhesion, differentiation and proliferation of osteoblast cells than the microcrystalline $CaPO_4$ [91, 92]. More specifically, Webster et. al. in their study, demonstrated that osteoblast proliferation, shown in Fig. 26, was greater on nano-HA, nano-TiO_2 and nano-Al_2O_3 than on their conventional counterparts after 1, 3 and 5 days of culture [91].

Additionally, Ostomel et al. have reported that mesoporous glass ceramic nanoparticles of the system $(x)SiO_2$–$(n = 100 - x - z)CaO$-$(z)P_2O_5$ (where $x = 60$ or 80 and $z = 4$; %mol) promoted bioactive behavior in a more efficient manner in regard

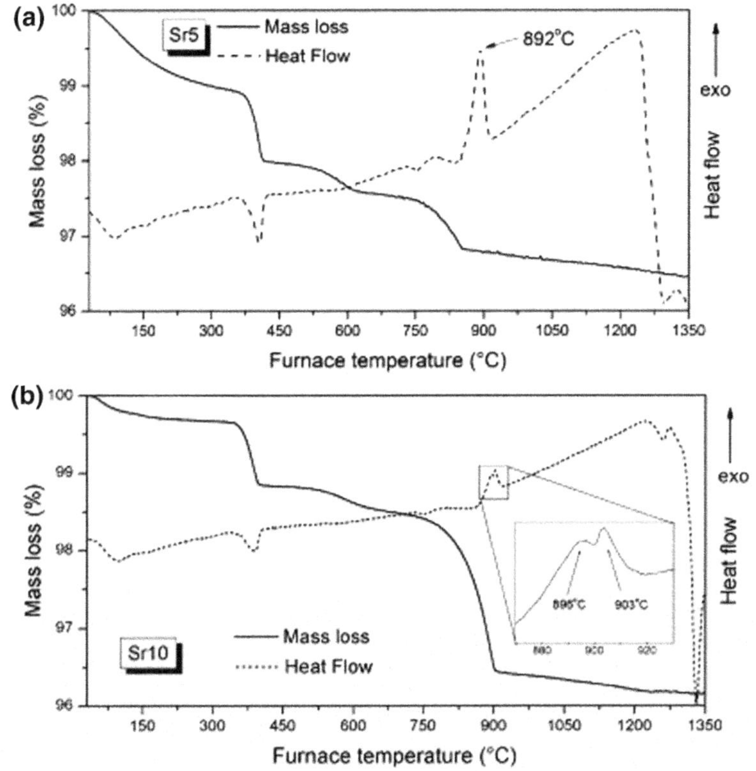

Fig. 25 Heat flow and mass loss curves for **a** Sr5 and **b** Sr10 glass-ceramic materials [90]

to microscaled glass ceramic particles and concluded that this was due to the ability of the nanoscaled glass-ceramics to faster nucleate HCAp on their surface [93]. Figure 27 shows the XRD patterns of the glasses after immersion in SBF solution for 1 h, indicating the significant deposition of HCAp on the surface of the samples.

Moreover, the different particle size affects the solubility of materials; smaller particles tend to have higher solubility which is essential when the material is intended to be bioresorbed. For example, nanodimensional glass-ceramics have been used in biocomposites where the other part of the composite is a biodegradable polymer. The idea is that the organic phase provides elasticity of the construct while the inorganic phase is responsible for mechanical strength and bioactivity, facilitating the formation of new tissue. Tadic et al. demonstrated as shown in Fig. 28 that nanodimensional calcium phosphate had superior solubility than microcrystalline calcium phosphates, second only to amorphous calcium phosphate [94].

Furthermore, a number of nanosized biocomposites are reported to be suitable as drug or growth factors delivery systems; thus increasing the efficiency of the deliverable [95, 96]. More specifically, Govidan et al. produced three different nanocomposite samples (nano-HA, nano-phoshate glass (PG) and PG/HA), loaded with Getam-

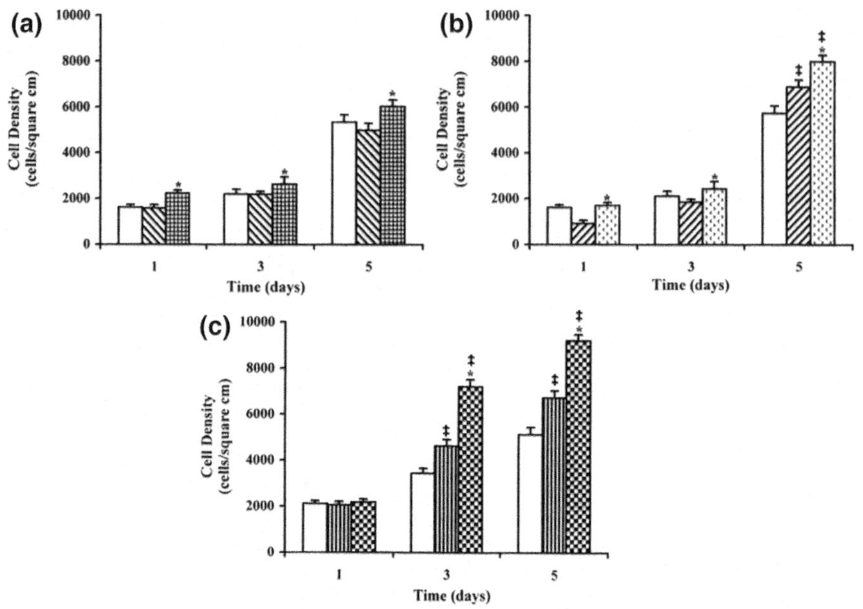

Fig. 26 Osteoblast proliferation on ceramics. Rat calvarial osteoblasts in Dullbecco's modified Eagle medium (suplamented with 10% fetal bovine serum) were seeded (2500 cells/cm^2) on the following substrates: **a** □borosilicate glass (reference material), 167 nm grain size alumina (conventional), and 24 nm grain size alumina (nanophase); **b** □borosilicate glass (reference material), 4520 nm grain size titania)conventional), and 39 nm grain size titania (nanophase); and **c** □borosilicate glass (reference material), 179 nm grain size hydroxyapatite (conventional), and 67 nm grain size hydroxyapatite (nanophase). Osteoblast proliferation under standard cell culture conditions (37 °C, humidified, 5% CO_2/95% air environment) was determined after 1, 3, and 5 days. Values are mean ± SEM; n = 3; *$P < 0.01$ (compared to respective conventional grain size ceramic); ‡ $P < 0.01$ (compared to borosilicate galss) [91]

icin Sulfate (GS) which is used for antibiotic purposes in orthopedic applications. The reported that the PG/HA nanocomposite had superior behavior regarding its degradation, bioactive behavior, biocompatibility and could sustain GS release for longer time periods. Evidence of the GS release over time can be seen in Fig. 29 [95].

Reports form a study, where they produced DNA loaded CaP/PLGA-mPEG hybrid porous nanospheres revealed that the have superior DNA loading capacity from other mesoporous silica vectors reported. Figure 30, shows the mass loss curves of the hybrid porous nanospheres and pure PLGA-mPEG, displaying that from 100 °C up to 425 °C the decomposition of pure PLGA-mPEG occurs. Hence the amount of PLGA-mPEG that can be loaded on the hybrid nanospheres is approximately 18%, which is the mass loss for them [96].

Pouroutzidou [97] has developed nanoparticles of the ternary system SiO_2–CaO–MgO doped with copper ions with intent to be used in a composite material, such as a scaffold, to promote the formation of new tissue. The purpose of

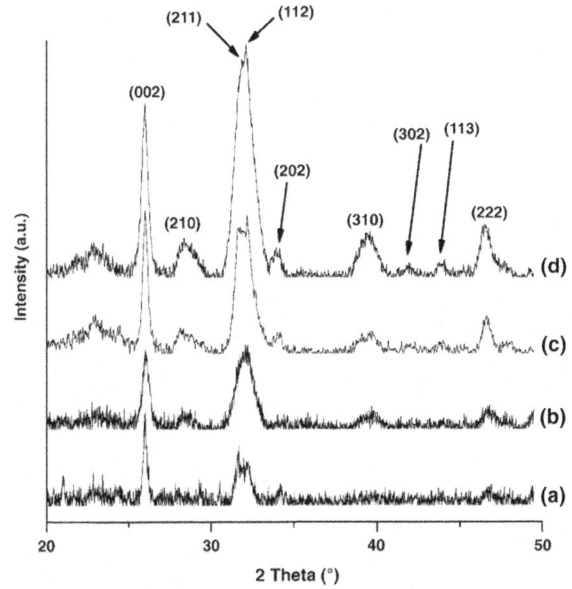

Fig. 27 Wide-angle XRD patterns of **a** MBGM-80 cement, **b** MBGM-60 cement, **c** MBGM-80 cement soaked in SBF for 1 h, **d** MBGM-60 cement soaked in SBF for 1 h [93]

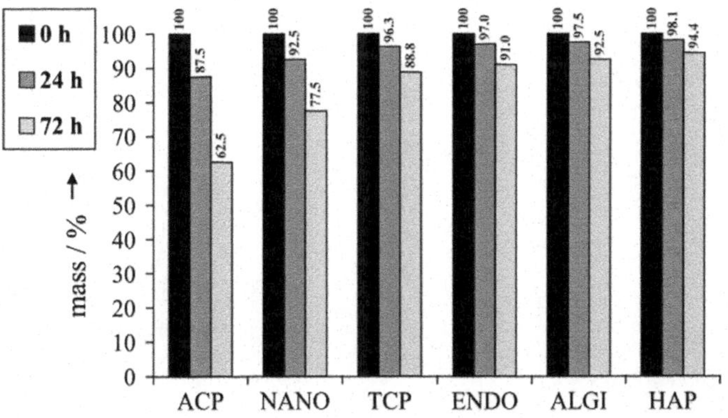

Fig. 28 Solubility of different calcium phosphate phases at constant pH = 4.4 (simulation of osteoblastic resorption). The relative weight after 24 and 72 h is represented, showing the much higher solubility of amorphous carbonated apatite. ACP = amorphous carbonated calcium phosphate; NANO = nanocrystalline calcium phosphate, TCP = β-tricalcium phosphate; ENDO = endobone®; ALGI = algipore®; HAP = sintering crystalline hydroxyapatite [94]

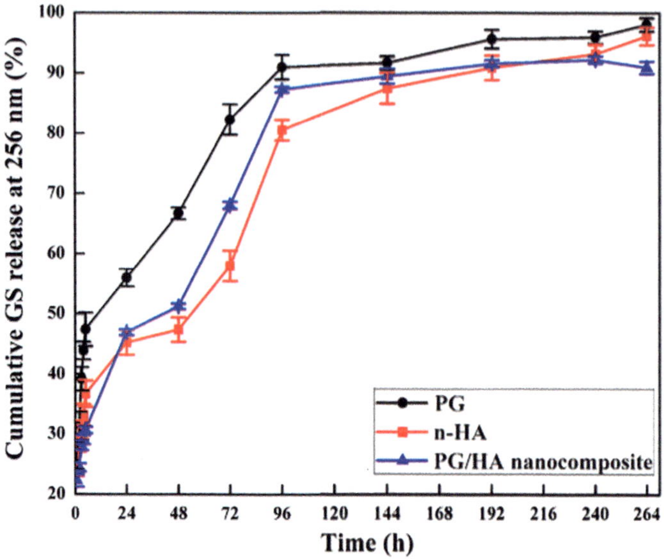

Fig. 29 In vitro cumulative GS release from the PG, n-HA and PG/HA nanocomposite samples [95]

Fig. 30 Mass loss curves of the as-prepared CaP/PLGA-mPEG hybrid porous nanospheres and the pure block copolymer PLGA-mPEG [96]

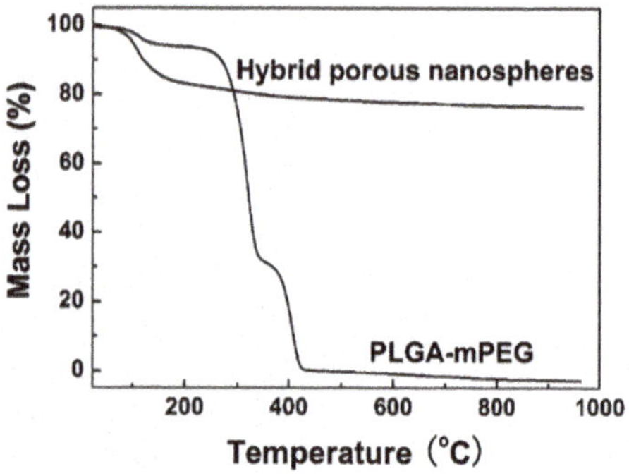

her study was to evaluate the effect of Ethanol to TEOS ratio and the total amount of ammonia (NH_3) during the sol-gel synthesis of nanoparticles of the aforementioned system on the properties of the final product. The Ethanol/TEOS ratio was set to 1, 5 and 10, while the total amount of NH_3 used was 5 and 20 ml. Figure 31 represents the Heat Flow curves of the nanosized glass produced, indicating similar behavior upon heating; presenting exotherms in the same temperature range, 800–900 °C and 1000–1100 °C. The examination of the apatite forming ability of the glasses revealed that all of the samples were able to induce the formation of a Ca/P phase on their surface after 3 days of immersion in SBF, except the one having the larger Ethanol to TEOS ratio accompanied with the higher total amount of NH_3, in which case the formation of HAp was observed on the surface of the sample. Further examination

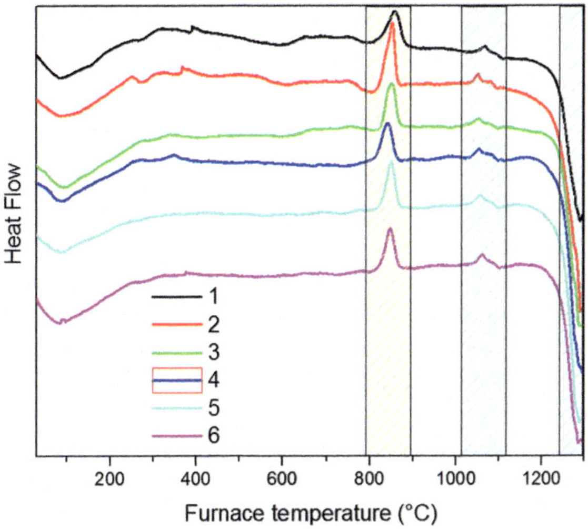

Fig. 31 Heat Flow curves of the nanophase glasses [97]

of the apatite forming ability of the samples revealed that all samples induced the formation of apatite on their surface after 5 days of immersion.

A plethora of studies exist, involving glass-ceramic nanoparticles in regenerative medicine. The examples given in this chapter have a purpose to give a first impression and understanding of the importance of nanosized glass ceramics. For gaining further understanding on nanosized glass-ceramics an article from Dorozhkin [92] in which nanodimensional and nanocrystalline calcium phosphates are discussed.

4 Dental Nanocomposite Resins

4.1 Introduction

During the long-term history of dentistry, various substances were used to fill dental cavities, including alum, honey, ground mastic, gold, lead, tin, amalgam, but it was not until 1871 when the first colored tooth filling material with semi-esthetic properties was produced [98], namely, silicate cement [99]. Before that, Romans are reported to have used gold for dental crowns and bridge restoration applications, while various materials have been applied in restoration dentistry since then, such as vulcanized rubber in denture bases during the 1860s [100]. In 1947 the first synthetic resins came in common use in the field of dentistry, namely methyl methacrylate resins [99], while in 1962 bisphenol A diglycidildimethacrylate (BisGMA) monomer was introduced as an attempt to improve of the properties of acrylic resins [101]. Finally, composite resins were introduced at early 1970s [99], serving a wide range of applications in the dentistry field. They are applied for the restoration of dental lesions (in the manufacture of dentures, crowns, temporary bridges, interim restorations, etc.), as well as for the filling of small and medium sized dental defects, in order to improve performance of dental resins regarding a series of physico-chemical characteristics [102].

Per definition, composite materials are made from different constituent materials with significantly diverse physical and chemical properties, whose combination produces materials which exhibit completely different properties compared to the individual primary ones. It should be mentioned that the particular components remain distinct within the final structure. In the case of dental composites, there are two main categories of constituent materials; the organic resin matrix—acting as binder—and the inorganic filler—acting as reinforcing agent concerning various properties. There are, also, more additives in the dental composites such as coupling agents for the successful bonding of filler particles to the organic matrix, pigments, initiators and stabilizers, whose study is beyond the scope of the present overview [103].

The term "dental composites" refers to a wide range of materials, which can be categorized in various modes, according to final application, composition, mechanical properties, filler type, filler content, filler morphology and polymerization mode [104, 105]. Another common classification approach is based on the particle size and is performed as follows: traditional composites for particles of 8–12 μm size, small particle filled composites for 1–5 μm size, microfilled composites for 0.04–0.4 μm size, hybrid composites for 0.6–1.0 μm size, nanofilled composites (nanocomposite) for fillers in nanometric scale (5–100 nm) and nanohybrid composites [103, 106]. The abovementioned microfilled composites were introduced at around 1977 [107], while nanofilled composite resins were introduced in 2002 in the field of dentistry [106], with the scientists trying to fully exploit the advantages that nanotechnology offers to a wide range of sciences.

4.2 Nanoparticles in Dentistry

Nanoparticles are reported to provide a series of enhanced properties to dental composites, such as high esthetic quality, high strength, increased modulus of elasticity, improved optical properties and favorable wear resistance [103]. A great deal of nanoparticles has been employed in the synthesis of dental nanocomposites, either alone or in combinations, according to the final application. Some examples of those nanoparticles are alumina nanoparticles (Al_2O_3 NPs), titania nanoparticles (TiO_2 NPs), nano-zirconia (ZrO_2), nanohydroxyapatite (HA), nanodiamonds (ND) and nano-silica (SiO_2 NPs) [108, 109].

Safi et al. [110] investigated the effect of Al_2O_3, TiO_2 and SiO_2 nanoparticles on the thermal properties of PMMA, reporting a decrease in the value of coefficient of expansion due to the greater interfacial interaction between nanoparticles and the matrix, as well as a significant increase in T_g and an improved thermal stability. Furthermore, silanized Al_2O_3 nanoparticles added to acrylic resins in 1, 2 and 3% filler loading [111], increased the values of flexural strength at lower contents, while a decrease of the abovementioned property was reported regarding the higher content. Surface hardness values increased proportionally with the increase in nanoparticle loading and there was, also, a significant decrease in water sorption and solubility [111]. Moreover, thermal properties were reported to get enhanced, since thermal conductivity and thermal diffusivity values got increased upon filler loading [111]. According to Sanjay et al. [112], use of nano-alumina also improved the elastic properties of a photo-polymerizable resin based on Bisphenol A glycidyl dimethacrylate monomer.

Kul et al. [113] used nano-HA as a filler in PMMA, achieving increased thermal conductivity and decreased flexural strength. Safarabadi et al. [114] studied the concurrent effect of nano-HA and nano-Al_2O_3 on the mechanical properties and the shrinkage behavior of PMMA, resulting in an increase of flexural strength and impact strength, while hardness was also raised due to the inherent characteristics of Al_2O_3 and the high hardness of HA. Furthermore, longitudinal shrinkage got reduced.

TiO_2 in 1% filler content enhanced impact strength, while in 5% filler content increased microhardness of heat polymerized acrylic resins [115]. On the contrary, flexural strength considerably got decreased upon increasing nanofiller content. Song et al. [116] investigated the effect of nano-TiO_2 combined with nano-SiO_2 on certain properties of a denture base resin, improving the antibacterial property, the tensile strength and the frictional resistance of the neat resin.

Gad et al. [117] incorporated nano-zirconia particles into an autopolymerized resin in 2 and 5 wt% loading, significantly improving the transverse strength of the specimens. Gad et al. [118] also confirmed the increase of flexural strength of PMMA resins upon increase of nano-ZrO_2 content, while impact strength was decreased, especially for high nanofiller loading. Alhavaz et al. [119] reported an increase in the surface hardness of those specific nanocomposite resins. Furthermore, various ZrO_2 nanofiller loadings, namely 1.5, 3, 5 and 7%, incorporated in a heat-cure acrylic resin have been reported to significantly increase hardness, fracture toughness and flexural strength [120]. Furthermore, Chan et al. [121] investigated the effect of zirconia nanoparticles dispersed in a bisphenol A glycol dimethacrylate-based monomer blend, achieving an increase in terms of fracture toughness.

Alnamel et al. [122] investigated the effect of surface treated nano-SiO_2 on various properties of PMMA, improving the impact strength and the transverse strength of the resin, while an enhancement of surface hardness was also reported. Balos et al. [123] improved the mechanical properties of PMMA by incorporating small quantities of nano-SiO_2, namely 0.023–0.91% by volume. It was found that the lowest nanosilica amount achieved the maximum fracture toughness and microhardness values, while T_g remained unchanged. On the contrary, the maximum content of nano-SiO_2 induced a great T_g increase. PMMA-nanosilica composites were also prepared by Hu et al. [124], where thermal properties of the resin got enhanced. More specifically, storage, loss modulus and T_g of the nanocomposites got increased, whereas thermal stability got enhanced.

As it is clear, a variety of properties get altered upon nanoparticle incorporation in resins. Thereafter, amongst others, thermal properties of polymer-based nanocomposites for dental applications need to be investigated, since several characteristics

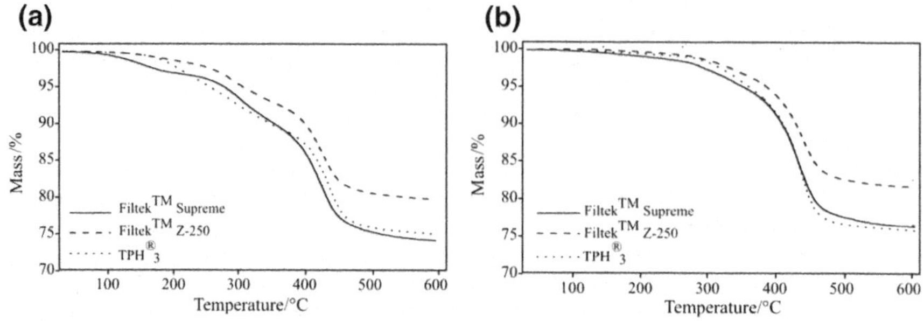

Fig. 32 TG curves of **a** uncured and **b** cured dental composite resins [106]

elicited by thermal experimental measurements are directly connected to the performance of composites, enabling scientists to improve the physico-chemical characteristics of the final products through corrective actions.

4.3 Characterisation of Dental (Nano)Composite Resins by Means of Thermal Analysis Techniques

TG technique mainly studies thermal stability of dental (nano)composites, though it may revel certain structural characteristics and determine the particle volume fraction. Furthermore, degradation kinetic studies may reveal information on the chemical structure of studied materials. DSC measurements are often employed in the evaluation of nanocomposite resins, regarding a series of characteristics, such as glass transition temperature which is directly connected to crosslinking process, the initial degradation stages and the curing state of nanocomposite resins. T_g can also be evaluated by means of DMA, which correlates to changes of polymer chain mobility as composites pass from the glassy to the rubbery state

Bernardi et al. [106] employed TG, DTG and DSC methods in order to investigate degradation of three different light-cured dental composite resins, to study their thermal stability and to identify their T_g. More specifically, they studied composite resins with zirconia/silica and bariumaluminoborosilicate glass/bariumaluminofluorosilicate/highly dispersed silicon dioxide as fillers in great loadings, namely Filtek^{TM} Supreme^{XT}, Filtek^{TM} Z-250 and TPH$_3^{®}$ in cured and uncured state. TG measurements were performed in N_2 atmosphere at a constant heating rate of 10 °C/min (Fig. 32), revealing that in the case of cured resins, there is a 2% enhancement of residual mass percentage, ought to the restricted mobility of the polymer chains in the presence of particles, when resins are cured. Moreover, FiltekTM Z-250, both in its cured and its uncured state, delay thermal degradation in comparison to the rest composite resins, proving its superiority in terms of thermal stability.

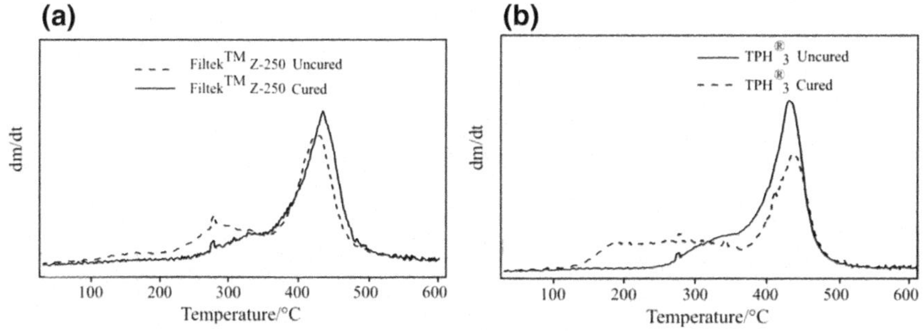

Fig. 33 DTG curves of cured and uncured Filtek™ Z-250 and $TPH_3^{®}$ dental composite resins [106]

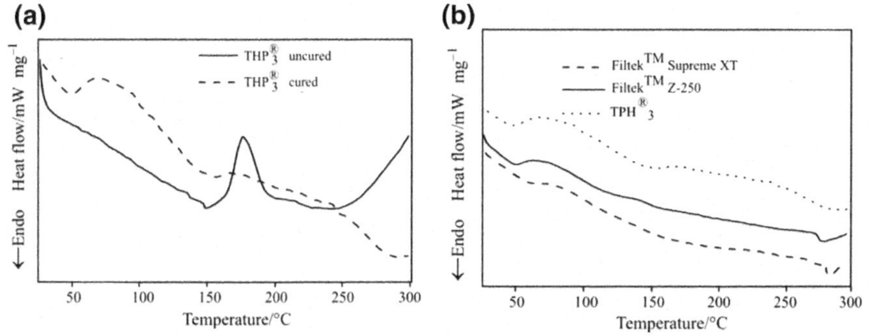

Fig. 34 **a** DSC curves of uncured and cured $TPH_3^{®}$ composite resins, **b** DSC curves of cured dental composite resins [106]

Furthermore, DTG curves of composite resins (Fig. 33) revealed the enhanced thermal stability of cured samples, preventing their premature thermal decomposition compared to uncured ones. That specific behavior, which is clearly seen in Fig. 33 since the first step of thermal degradation disappears for the cured samples, may be attributed to the formed cross-linked network and to the interactions induced between inorganic particles and polymer chains.

Moreover, when dealing with cured composites, curing efficiency must be high. The cure reaction involves a chemical process of polymerization, where monomers are joined together with chemical bonds towards the formation of large-chain molecules. The level of monomer to polymer conversion during the polymerization process is indicative of the final physical and mechanical properties of cured composites [125]. Inadequate polymerization exhibits a low T_g, negatively affecting various properties of the materials. According to Fig. 34a, uncured samples are easily distinguished from the cured ones, since they present an exothermic peak.

As mentioned earlier, T_g is an important parameter to be taken into account when dealing with dental composites, since in order for the materials to preserve their physical and mechanical properties, T_g must be higher than the maximum temperature in the oral cavity [126]. In case that intraoral temperature exceeds T_g, dental composites soften, leading to failure of clinical procedures. These temperature alterations owe their origin to the food and liquid consumption [127]. T_g is defined in DSC curves as a change in heat capacity, when composites pass from the glassy state to the rubbery one. This is a second order endothermic transition, which requires heat to go through and which appears as a step. Therefore, DSC measurements were, also, performed in the study of Bernardi et al. [106], where T_g of all samples lie in the region 48–58 °C (Fig. 34b), with the Filtek^{TM} Z-250 exhibiting the biggest corresponding value, suggesting increased crosslinking provided by the light source compared to the rest composites that were evaluated. Furthermore, according to Fig. 34b, endothermic curves attributed to decomposition of composites are exhibited in the temperature range 270–300 °C.

On the contrary, Topouzi et al. [128] have defined—amongst others- the T_g of synthesized nanocomposite resins by means of DMA, which is correlated to changes of polymer chain mobility as composites pass from the glassy to the rubbery state [126]. More specifically, they investigated the possible reinforcement of PMMA resin with neat and trietoxyvinylsilane-modified silica nanoparticles in low loadings (0.25, 0.50, 0.75, 1% w.t.) as an efficient interim fixed prostheses. Mastication, functional and para-functional movements induce high loads, while food and liquid consumption leads to intense temperature changes, as said above [127]. Moreover, the masticatory system executes complex movements with mean frequency ~1 Hz [127], thus the study of dynamic load needs to be performed in order to simulate real conditions in oral cavity and to predict material's behavior. Therefore, when visco-elastic materials are studied and when changes in the microstructure need to be elucidate upon temperature alterations, DMA measurements are conducted.

During dynamic testing, the samples were subjected to sinusoidal deformation at 1 Hz frequency over the 30–140 °C temperature range, leading to the evaluation of Storage modulus (E′), loss modulus (E″), tangent delta tanδ and T_g (Fig. 35). According to Fig. 35, the storage modulus E' increased by increasing the filler content for both unmodified (SIL) and modified groups (T-SIL), an indication that fillers made composites more stiff. This is explained by the higher rigidity of nano-silica compared to the corresponding value of the matrix. Moreover, E' increase is also attributed to the covalently formed bonds between nano-silica and polymeric matrix and to the restricted macromolecular mobility and deformation of polymer chains induced by nanoparticles. Regarding loss modulus E″, increased filler content leaded to higher E″ values, except for the case of T-SIL 1 specimen. When temperatures were raised, E″ of all materials increased until reaching their maximum value in the glass transition region, with the exception of the T-SIL 1 specimen, for once more. Thereafter, E″ values decreased with increasing temperatures.

Regarding the evaluation of T_g (as the maximum values of tanδ), a gradual increase of its values for both groups is observed, directly dependent on nano-SiO_2 content. That increase was more intense in the T-SIL group, where T-SIL 1 specimen presented a 25 °C value raise. Several mechanisms may be responsible for the aforementioned results. There has been reported that interactions between nanoparticles and polymer chains lead to strong interface [129], which restrict the mobility of macromolecular chains and consequently increase T_g value [130]. However, it has also been reported

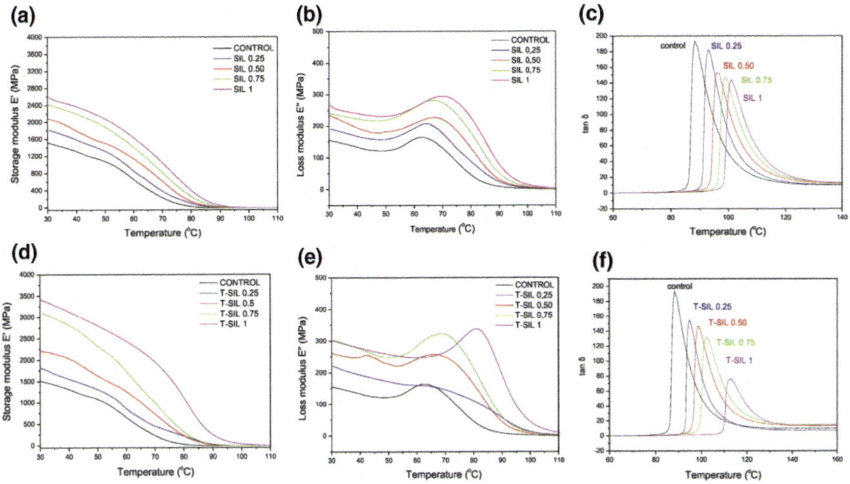

Fig. 35 **a–c** E′, E″ and tanδ of control PMMA and PMMA-SIL composites respectively, **d–f** E′, E″ and tanδ of control PMMA and PMMA—T-SIL composites respectively [128]

that upon increasing filler content, aggregates may be formed which further restrict polymer chain mobility and consequently increase T_g [131].

Furthermore, PMMA was also reinforced with nanodiamonds [132], known for their unique properties such as dopability, hardness and optical transparency over a wide spectral range [133–135]. Synthesized nanocomposite resins were studied regarding their thermomechanical properties, where E′ increased upon filler loading, indicating an additional elastic response compared to neat PMMA. This elastic behavior is desirable to a certain extent, since occlusal loads can be absorbed more effectively from elastically deformed materials and hence the load transmission to the underline dentine is significantly lower. Moreover, nanodiamonds positively are reported to affect the glass transition temperature, since the T_g of nanocomposites increased significantly, greatly influencing the thermal stability and internal stresses within the material developed at high temperatures [136].

Blivi et al. [137] quantified nano-size effects on thermal properties of PMMA-nanosilica composites, prepared at 4% volume fraction with varying particle diameter, namely 15, 25, 60 150 and 500 nm. TG measurements were conducted in nitrogen atmosphere with a heating rate of 10 °C/min, in order to determine the particle volume fraction and the thermal decomposition temperatures. According to TG curves (Fig. 36a), PMMA undergoes a single-step decomposition, whereas nanocomposites exhibit a two-step degradation mechanism. The first step which corresponds to 3–5% weight loss is attributed to the bond breakage between nanoparticles and polymer chains. TG curves of studied samples also revealed that T_d temperature (representing the temperature at 50% of mass loss) of all nanocomposites is higher than that of neat PMMA (Fig. 36b). Consecutively, degradation temperature of nanocomposites is increased with decreasing particle size, since more particles per volume are

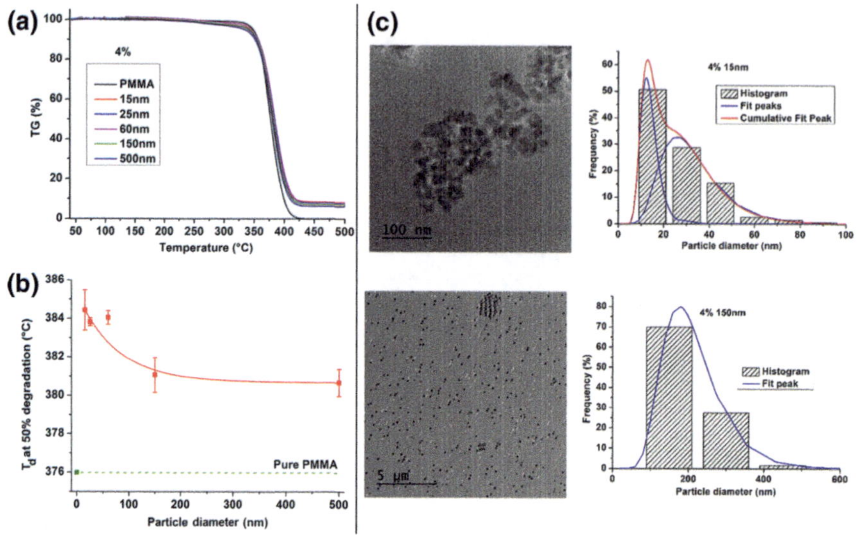

Fig. 36 **a** TG curves of PMMA and PMMA/nanosilica with different particles size, **b** particle diameter effect on degradation temperature of PMMA/silica nanocomposites, **c** TEM graphs of nanosilica dispersed in PMMA matrix for 15 and 150 nm diameter [137]

present, compared to larger nanosilica diameters. This induces more restriction sites for the macromolecular chains, increasing the energy barrier for scission of polymer chains and thus, shifting T_d to higher temperatures.

Moreover, DSC measurements, conducted at a 10 °C/min heating rate, were performed in order to determine the T_g of the samples. Nonetheless, although glass transition temperature was expected to increase upon filler addition, the opposite was observed, probably due to the weak interactions between particles and polymeric matrix [138, 139] as well as due to the formation of aggregates at low loadings [140], as seen in Fig. 36c.

In another study, the effect of various nanosilica particle sizes on certain properties of composites based on Bis-GMA/TEGDMA (50/50 wt/wt) matrix was examined [141]. Various silica nanoparticles were used, with average sizes of 7, 14, 16, 20 and 40 nm, while their surface was treated with the silane coupling agent 3-methacryloxypropyltrimethoxysilane (MPS). TG measurements, performed in nitrogen atmosphere under a 10 °C/min heating rate, were conducted for the determination of the weight percentage of fillers and for the overall study of thermal degradation of the composites. According to Fig. 37a, degradation of the polymeric matrix consists of two steps. The first step is attributed to bond breaking near the cycling points that have been formed in the network during photopolymerization, while the second step corresponds to bond breaking of the main network [142–144]. On the contrary, a three-step degradation is exhibited in the case of nanocomposites, according to Fig. 37b, c, where the first two steps are attributed to the decomposition of Bis-

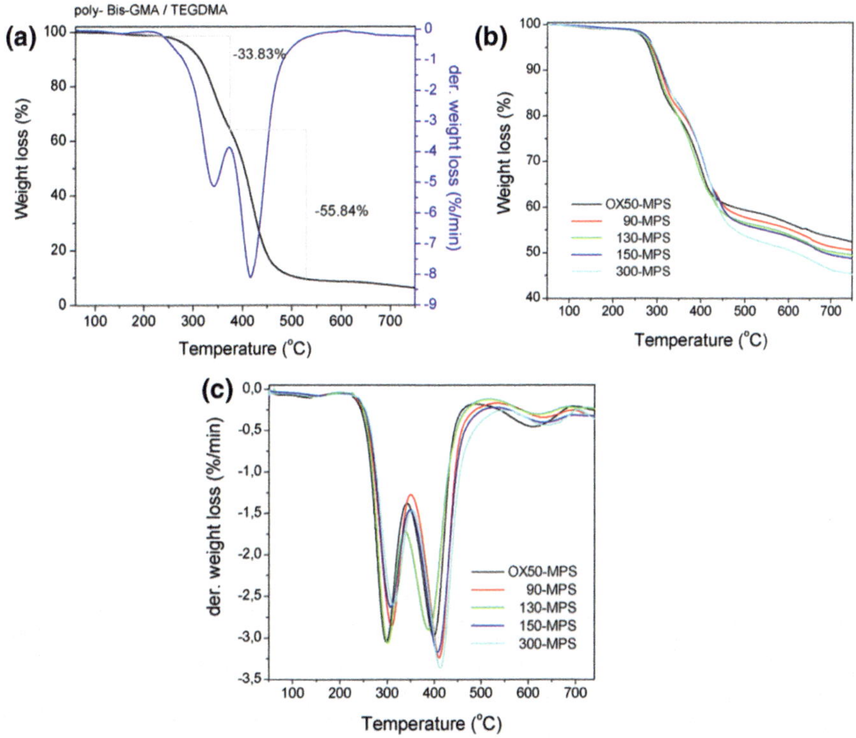

Fig. 37 TG/DTG curves of **a** Bis-GMA/TEGDMA, **b**, **c** nanosilica-based composites [141]

GMA/TEGDMA, while the third step is due to condensation of surface silanols of nanosilica [145].

Vouvoudi et al. [146] have studied thermal degradation of five commercial dental nanocomposites based on a dimethacrylate matrix in an inert (nitrogen) atmosphere, using four different heating rates, namely 2.5, 5, 10 and 20 °C/min, since according to the International Confederation for Thermal Analysis and Calorimetry (ICTAC), the investigation of thermal degradation kinetics requires the application of multiple rates [2]. The aim of that study was to estimate the degradation kinetics in order to gain information on the chemical structure of the nanocomposites. TG curves of all nanocomposites are exhibited in Fig. 38a, while Fig. 38b displays the isoconversional method of Kissinger-Akahira-Sunose (KAS) applied to all nanocomposites. It was found that thermal stability is mainly dependent on the structure and on the percentage of the polymeric matrix. The first two degradation steps are associated to the polymeric matrix, ascribed to the initial decomposition of weak links which are inside the formed macromolecular network followed by the main carbon chain scission. The third observed step is attributed to the inorganic filler. Furthermore, according to the isoconversional analysis, E_α values of degradation start at low val-

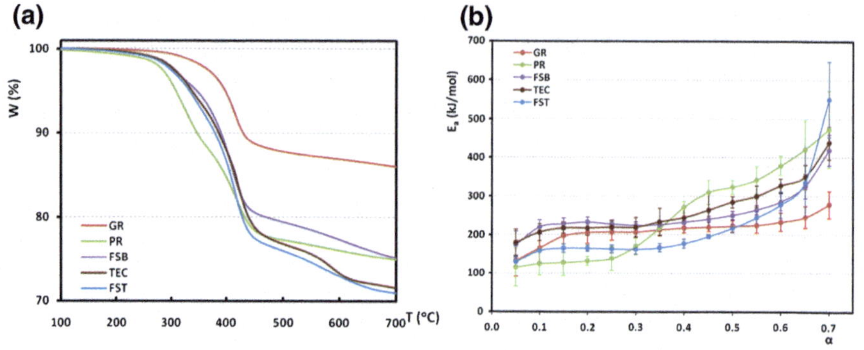

Fig. 38 **a** TG curves of all nanocomposites at a heating rate of 10 °C/min, **b** Variation of the effective activation energy with conversion of all dental composites investigated using the KAS method [146]

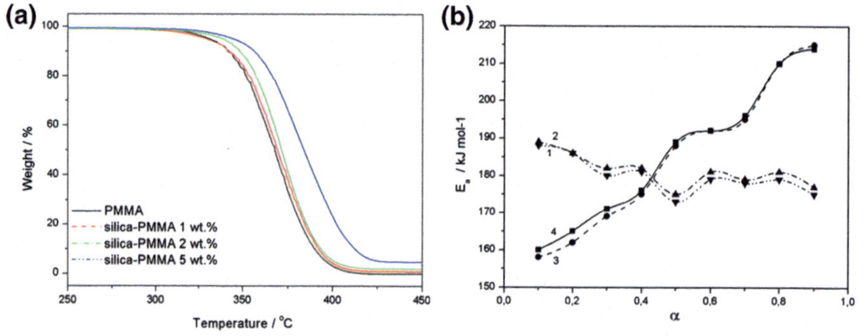

Fig. 39 **a** TG curves of PMMA and its nanosilica-based nanocomposites at a heating rate of 10 °C/min, **b** calculated activation energies by the OFW and KAS methods: (1) PMMA (KAS), (2) PMMA (OFW), (3) silica e PMMA (5 wt%) (KAS), (4) silica e PMMA (5 wt%) (FWO) [147]

ues, though they increase upon degree of conversion (α) raise, leading to a more difficult decomposition process.

Saladino et al. [147] have prepared PMMA-silica nanocomposites, with filler loading in the range 1–5 wt% and—amongst others- they have thoroughly studies their thermal stability. TG measurements were conducted in nitrogen at different heating rates (3, 5, 7 and 9 °C/min). The kinetic analysis was conducted according to Ozawa-Flynn-Wall (OFW) and Kissinger-Akahira-Sunose (KAS) isoconversional methods. TG curves of PMMA and its nanosilica-based nanocomposites at a heating rate of 10 °C/min are exhibited in Fig. 39a, while calculated activation energies (E_α) are presented in Fig. 39b.

According to Fig. 39a, all samples exhibit a single-step degradation process, while curves are shifted to higher temperatures upon increasing filler content. Kinetic study was conducted for neat PMMA and silica-PMMA 5 wt%, which exhibits a significant thermal enhancement of about 10 °C. As it refers to Fig. 7b, the nanocomposite presents lower activation energies of degradation up to ~40% mass loss compared to PMMA, while above that specific mass loss percentage the opposite is occurred (Fig. 39b). In the case of neat PMMA, increase of degree of conversion (α) leads to a decrease in E_α values, probably due to an increase of free radicals which accelerate the degradation process. On the contrary, for small α values, nanoparticles catalyze the reaction, so that less energy is essential for the initiation of the degradation process. However, as the degradation proceeds, the observed E_α values are increased due to the interactions between the escaping monomers and the nanoparticles which trap volatile degradation products.

References

1. Gabbott, P.: Thermogravimetric Analysis. In: Principles and Applications of Thermal Analysis, pp. 87–118 (2008)
2. Vyazovkin, S., Burnham, A.K., Criado, J.M., Pérez-Maqueda, L.A., Popescu, C., Sbirrazzuoli, N.: ICTAC kinetics Committee recommendations for performing kinetic computations on thermal analysis data. Thermochim. Acta **520**(1–2), 1–19 (2011)
3. Watson, E., O'Neil, M.: Differential microcalorimeter. U.S. Patent 3263484 (1962)
4. Tanzi, M.C.: Characterization of thermal properties and crystallinity of polymer biomaterials. In: Characterization of Polymeric Biomaterials, pp. 123–146 (2017)
5. Hatakeyama, T., Quinn, F.X.: Thermal analysis: fundamentals and applications to polymer science (1994)
6. Saba, N., Jawaid, M., Alothman, O.Y., Paridah, M.T.: A review on dynamic mechanical properties of natural fibre reinforced polymer composites. Constr. Build. Mater. **106**, 149–159 (2016)
7. Della Bona, A.: Bonding to ceramics: scientific evidences for clinical dentistry. Nat. Publ. Gr. **209**(2), 99 (2009)
8. Benetti, P., Kelly, J.R., Sanchez, M., Della Bona, A.: Influence of thermal gradients on stress state of veneered restorations. Dent. Mater. **30**(5), 554–563 (2014)
9. Lopez-Esteban, S., Saiz, E., Fujino, S., Oku, T., Suganuma, K., Tomsia, A.P.: Bioactive glass coatings for orthopedic metallic implants. J. Eur. Ceram. Soc. **23**(15), 2921–2930 (2003)
10. Kontonasaki, E., et al.: Microstructural characterization and comparative evaluation of physical, mechanical and biological properties of three ceramics for metal-ceramic restorations. Dent. Mater. **24**(10) (2008)
11. Harabi, A., Guerfa, F., Harabi, E., Benhassine, M.T., Foughali, L., Zaiou, S.: Preparation and characterization of new dental porcelains, using K-feldspar and quartz raw materials. Effect of B_2O_3 additions on sintering and mechanical properties. Mater. Sci. Eng. C **65**, 33–42 (2016)
12. Carbajal, L., Rubio-Marcos, F., Bengochea, M.A., Fernandez, J.F.: Properties related phase evolution in porcelain ceramics. J. Eur. Ceram. Soc. **27**(13–15), 4065–4069 (2007)
13. Möncke, D., Ehrt, R., Palles, D., Efthimiopoulos, I., Kamitsos, E.I., Johannes, M.: A multi technique study of a new lithium disilicate glass-ceramic spray-coated on ZrO_2 substrate for dental restoration. Biomed. Glas. **3**, 41–55 (2017)
14. Jones, J.R.: Reprint of: review of bioactive glass: from Hench to hybrids. Acta Biomater. **23**(S), S53–S82 (2015)
15. Xynos, I.D., Edgar, A.J., Buttery, L.D.K., Hench, L.L., Polak, J.M.: Gene-expression profiling of human osteoblasts following treatment with the ionic products of Bioglass® 45S5 dissolution. J. Biomed. Mater. Res. **55**(2), 151–157 (2001)
16. Vallet-Regí, M., Ragel, C.V., Salinas, A.J.: Glasses with medical applications. Eur. J. Inorg. Chem. **2003**(6), 1029–1042 (2003)

17. Christie, J.K., Ainsworth, R.I., De Leeuw, N.H.: Investigating structural features which control the dissolution of bioactive phosphate glasses: beyond the network connectivity. J. Non. Cryst. Solids **432**, 31–34 (2016)
18. Hill, R.G., Brauer, D.S.: Predicting the bioactivity of glasses using the network connectivity or split network models. J. Non. Cryst. Solids **357**(24), 3884–3887 (2011)
19. Chatzistavrou, X., Zorba, T., Kontonasaki, E., Chrissafis, K., Koidis, P., Paraskevopoulos, K.M.: Following bioactive glass behavior beyond melting temperature by thermal and optical methods. Phys. Status Solidi Appl. Res. **201(5)** (2004)
20. Arcos, D., Greenspan, D.C., Vallet-Regi, M.: A new quantitative method to evaluate the in vitro bioactivity of melt and sol-gel-derived silicate glasses. J. Biomed. Mater. Res. Part A **65A**(3), 344–351 (2003)
21. Clupper, D.C., Hench, L.L.: Crystallization kinetics of tape cast bioactive glass 45S5. J. Non. Cryst. Solids **318**(1–2), 43–48 (2003)
22. Chatzistavrou, X., et al.: Influence of particle size on the crystallization process and the bioactive behavior of a bioactive glass system. J. Therm. Anal. Calorim. **85**(2), 253–259 (2006)
23. Ray, C.S., Day, D.E.: Determining the nucleation rate curve for lithium disilicate glass by differential thermal analysis. J. Am. Ceram. Soc. **73**(2), 439–442 (1990)
24. Ray, C.S., Day, D.E.: Identifying internal and surface crystallization by differential thermal analysis for the glass-to-crystal transformations. Thermochim. Acta **280–281**(SPEC. ISS.), 163–174 (1996)
25. Li, W., Mitchell, B.S.: Nucleation and crystallization in calcium aluminate glasses. J. Non. Cryst. Solids **255**(2), 199–207 (1999)
26. Chatzistavrou, X., Esteve, D., Hatzistavrou, E., Kontonasaki, E., Paraskevopoulos, K.M., Boccaccini, A.R.: Sol-gel based fabrication of novel glass-ceramics and composites for dental applications. Mater. Sci. Eng. C **30**(5) (2010)
27. Zhong, J., Greenspan, D.C.: Processing and properties of sol-gel bioactive glasses. J. Biomed. Mater. Res. **53**(6), 694–701 (2000)
28. Goudouri, O.-M., et al.: Towards the synthesis of an experimental bioactive dental ceramic. Part I: crystallinity characterization and bioactive behavior evaluation. Mater. Chem. Phys. **145**(1–2) (2014)
29. Xuereb, M., Camilleri, J., Attard, N.: Systematic review of current dental implant coating materials and novel coating techniques. Int. J. Prosthodont. **28**(1), 51–59 (2015)
30. Agarwal, R., García, A.J.: Biomaterial strategies for engineering implants for enhanced osseointegration and bone repair. Adv. Drug Deliv. Rev. **94**, 53–62 (2015)
31. McEntire, B.J., Bal, B.S., Rahaman, M.N., Chevalier, J., Pezzotti, G.: Ceramics and ceramic coatings in orthopaedics. J. Eur. Ceram. Soc. **35**(16), 4327–4369 (2015)
32. Bertazzo, S., Zambuzzi, W.F., Campos, D.D.P., Ferreira, C.V., Bertran, C.A.: A simple method for enhancing cell adhesion to hydroxyapatite surface. Clin. Oral Implants Res. **21**(12), 1411–1413 (2010)
33. Mistry, S., Kundu, D., Datta, S., Basu, D.: Comparison of bioactive glass coated and hydroxyapatite coated titanium dental implants in the human jaw bone. Aust. Dent. J. **56**(1), 68–75 (2011)
34. Zhang, B.G.X., Myers, D.E., Wallace, G.G., Brandt, M., Choong, P.F.M.: Bioactive coatings for orthopaedic implants-recent trends in development of implant coatings. Int. J. Mol. Sci. **15**(7), 11878–11921 (2014)
35. Jimbo, R., et al.: Nano hydroxyapatite-coated implants improve bone nanomechanical properties. J. Dent. Res. **91**(12), 1172–1177 (2012)
36. Cheng, Z., Guo, C., Dong, W., He, F.M., Zhao, S.F., Yang, G.L.: Effect of thin nano-hydroxyapatite coating on implant osseointegration in ovariectomized rats. Oral Surg. Oral Med. Oral Pathol. Oral Radiol. **113**(3) (2012)
37. He, F., Yang, G., Wang, X., Zhao, S.: Effect of electrochemically deposited nanohydroxyapatite on bone bonding of sandblasted/dual acid-etched titanium implant. Int. J. Oral Maxillofac. Implants **24**(5), 790–799 (2009)

38. Lewis, G.: Nanostructured hydroxyapatite coating on bioalloy substrates : current status and future directions **2**(1), 65–82 (2017)
39. Anjaneyulu, U., Priyadarshini, B., Arul Xavier Stango, S., Chellappa, M., Geetha, M., Vijayalakshmi, U.: Preparation and characterisation of sol–gel-derived hydroxyapatite nanoparticles and its coatings on medical grade Ti–6Al–4V alloy for biomedical applications. Mater. Technol. **32**(13), 800–814 (2017)
40. Albayrak, O., El-Atwani, O., Altintas, S.: Hydroxyapatite coating on titanium substrate by electrophoretic deposition method: effects of titanium dioxide inner layer on adhesion strength and hydroxyapatite decomposition. Surf. Coat. Technol. **202**(11), 2482–2487 (2008)
41. Fathi, M.H., Hanifi, A.: Sol–gel derived nanostructure hydroxyapatite powder and coating: aging time optimisation. Adv. Appl. Ceram. **108**(6), 363–368 (2009)
42. Farrokhi-Rad, M., Loghmani, S.K., Shahrabi, T., Khanmohammadi, S.: Electrophoretic deposition of hydroxyapatite nanostructured coatings with controlled porosity. J. Eur. Ceram. Soc. **34**(1), 97–106 (2014)
43. Huang, Y., et al.: Characterization and formation mechanism of nano-structured hydroxyapatite coatings deposited by the liquid precursor plasma spraying process. Biomed. Mater. **5**(5) (2010)
44. Pazo, A., Saiz, E., Tomsia, A.P.: Silicate glass coatings on Ti-based implants. Acta Mater. **46**(7), 2551–2558 (1998)
45. Gomez-Vega, J.M., Saiz, E., Tomsia, A.P.: Glass-based coatings for titanium implant alloys. J. Biomed. Mater. Res. **46**(4), 549–559 (1999)
46. Stuart, B., Gimeno-Fabra, M., Segal, J., Ahmed, I., Grant, D.M.: Preferential sputtering in phosphate glass systems for the processing of bioactive coatings. Thin Solid Films **589**, 534–542 (2015)
47. Berbecaru, C., Alexandru, H.V., Stan, G.E., Marcov, D.A., Pasuk, I., Ianculescu, A.: First stages of bioactivity of glass-ceramics thin films prepared by magnetron sputtering technique. Mater. Sci. Eng., B **169**(1–3), 101–105 (2010)
48. Machado López, M.M., Espitia Cabrera, M.I., Faure, J., Contreras García, M.E.: Electrochemical behavior of 45S5 bioactive ceramic coating on Ti6Al4 V alloy for dental applications. IOP Conf. Ser. Mater. Sci. Eng. **123**(1) (2016)
49. Fathi, M.H., Doostmohammadi, A.: Bioactive glass nanopowder and bioglass coating for biocompatibility improvement of metallic implant. J. Mater. Process. Technol. **209**(3), 1385–1391 (2009)
50. Carta, D., Jones, J.R., Lin, S., Poologasundarampillai, G., Newport, R.J., Pickup, D.M.: Neutron diffraction study of antibacterial bioactive calcium silicate sol-gel glasses containing silver. Int. J. Appl. Glas. Sci. **8**(4), 364–371 (2017)
51. Catauro, M., Bollino, F., Papale, F., Vecchio Ciprioti, S.: Investigation on bioactivity, biocompatibility, thermal behavior and antibacterial properties of calcium silicate glass coatings containing Ag. J. Non. Cryst. Solids **422**, 16–22 (2015)
52. Goudouri, O.M., et al.: Development of highly porous scaffolds based on bioactive silicates for dental tissue engineering. Mater. Res. Bull. **49**(1), 399–404 (2014)
53. Saravanapavan, P., Hench, L.L.: Mesoporous calcium silicate glasses. I. Synthesis. J. Non. Cryst. Solids **318**(1–2), 1–13 (2003)
54. Soundrapandian, C., Bharati, S., Basu, D., Datta, S.: Studies on novel bioactive glasses and bioactive glass-nano-HAp composites suitable for coating on metallic implants. Ceram. Int. **37**(3), 759–769 (2011)
55. Pishbin, F., et al.: Electrophoretic deposition of gentamicin-loaded bioactive glass/chitosan composite coatings for orthopaedic implants. ACS Appl. Mater. Interfaces. **6**(11), 8796–8806 (2014)
56. Heise, S., et al.: Electrophoretic deposition and characterization of chitosan/bioactive glass composite coatings on Mg alloy substrates. Electrochim. Acta **232** (2017)
57. Inam, F., Vo, T., Bhat, B.R.: Structural stability studies of graphene in sintered ceramic nanocomposites. Ceram. Int. **40**(PB), 16227–16233 (2014)

58. Baradaran, S., et al.: Characterization of nickel-doped biphasic calcium phosphate/graphene nanoplatelet composites for biomedical application. Mater. Sci. Eng. C **49**, 656–668 (2015)
59. Mehrali, M., et al.: Journal of the European Ceramic Society electrophoretic deposition of calcium silicate—reduced graphene oxide composites on titanium substrate. J. Eur. Ceram. Soc. **36**(2), 319–332 (2016)
60. Shi, Y.Y., et al.: Electrophoretic deposition of graphene oxide reinforced chitosan–hydroxyapatite nanocomposite coatings on Ti substrate. J. Mater. Sci. Mater. Med. **27**(3), 1–13 (2016)
61. Park and Lakes: Biomaterials. Springer, New York (2007)
62. Kokubo, T.: Bioceramics and their Clinical Applications (2008)
63. Narayan, R.: Biomedical materials. Springer, Boston (2009)
64. Antoniac, I.V.: Handbook of bioceramics and biocomposites (2016)
65. Dorozhkin, S.: Calcium orthophosphate-based bioceramics **6**(9) (2013)
66. Hupa, A., Boccaccini, R., Leena, D.S.B.: Bioactive Glasses : Fundamentals, Technology and Applications (2013)
67. Liu, X., Rahaman, M.N., Fu, Q., Tomsia, A.P.: Porous and strong bioactive glass (13–93) scaffolds prepared by unidirectional freezing of camphene-based suspensions. Acta Biomater. **8**(1), 415–423 (2012)
68. Huang, T.S., et al.: Porous and strong bioactive glass (13–93) scaffolds fabricated by freeze extrusion technique. Mater. Sci. Eng. C **31**(7), 1482–1489 (2011)
69. Chen, Q.Z., Thompson, I.D., Boccaccini, A.R.: 45S5 Bioglass®-derived glass-ceramic scaffolds for bone tissue engineering. Biomaterials **27**(11), 2414–2425 (2006)
70. Denry, I., Goudouri, O.M., Harless, J.D., Hubbard, E.M., Holloway, J.A.: Strontium-releasing fluorapatite glass-ceramics: crystallization behavior, microstructure, and solubility. J. Biomed. Mater. Res.—Part B Appl. Biomater. 1–10 (2017)
71. Denry, I., Goudouri, O.M., Harless, J., Holloway, J.A.: Rapid vacuum sintering: a novel technique for fabricating fluorapatite ceramic scaffolds for bone tissue engineering. J. Biomed. Mater. Res.—Part B Appl. Biomater. (2017)
72. Pereira, M.M., Clark, A.E., Hench, L.L.: Calcium phosphate formation on sol-gel-derived bioactive glasses in vitro. J. Biomed. Mater. Res. **28**(6), 693–698 (1994)
73. Li, P., Yang, Q., Zhang, F., Kokubo, T.: The effect of residual glassy phase in a bioactive glass-ceramic on the formation of its surface apatite layer in vitro. J. Mater. Sci. Mater. Med. **3**(6), 452–456 (1992)
74. Griffith, L.G., et al.: Tissue engineering–current challenges and expanding opportunities. Science **295**(5557), 1009–1014 (2002)
75. Burg, K.J., Porter, S., Kellam, J.F.: Biomaterial developments for bone tissue engineering. Biomaterials **21**(23), 2347–2359 (2000)
76. Sheikh, Z., Sima, C., Glogauer, M.: Bone replacement materials and techniques used for achieving vertical alveolar bone augmentation. Materials (Basel) **8**(6), 2953–2993 (2015)
77. Pirhonen, E., Moimas, L., Haapanen, J.: Porous bioactive 3-D glass fiber scaffolds for tissue engineering applications manufactured by sintering technique. Key Eng. Mater. **240–242**, 237–240 (2003)
78. Moimas, L., Biasotto, M., Di Lenarda, R., Olivo, A., Schmid, C.: Rabbit pilot study on the resorbability of three-dimensional bioactive glass fibre scaffolds. Acta Biomater. **2**(2), 191–199 (2006)
79. Vallet-Regí, M.: Bio-Ceramics with Clinical Applications (2014)
80. Karageorgiou, V., Kaplan, D.: Porosity of 3D biomaterial scaffolds and osteogenesis. Biomaterials **26**(27), 5474–5491 (2005)
81. Rezwan, K., Chen, Q.Z., Blaker, J.J., Boccaccini, A.R.: Biodegradable and bioactive porous polymer/inorganic composite scaffolds for bone tissue engineering. Biomaterials **27**(18), 3413–3431 (2006)
82. Peltier, L.F., Bickel, E.Y., Lillo, R., Thein, M.S.: The use of plaster of paris to fill defects in bone *. Ann. Surg. **146**(1), 61–69 (1957)
83. Gleeson, J.P.: *Composite Scaffolds for Orthopaedic Regenerative Medicine* (2011)
84. H. O. Ylänen, *Bioactive glasses: Materials, properties and applications*. 2011

85. Baino, F., Novajra, G., Vitale-Brovarone, C.: Bioceramics and scaffolds: a winning combination for tissue engineering. Front. Bioeng. Biotechnol. **3** (2015)
86. Kaigler, D., Krebsbach, P.H., West, E.R., Horger, K., Huang, Y.C., Mooney, D.J.: Endothelial cell modulation of bone marrow stromal cell osteogenic potential. Faseb J. **19**(6), 665–667
87. Ducheyne, P., Healy, K.E., Grainger, D.W., Hutmacher, D.W., Kirkpatrick, C.J.: Comprehensive biomaterials. **1**(6) (2011)
88. Marchi, J.: Biocompatible glasses : from bone regeneration to cancer treatment, pp. 305–307 (2016)
89. Colombo, P.: Conventional and novel processing methods for cellular ceramics. Philos. Trans. R. Soc. A Math. Phys. Eng. Sci. **364**(1838), 109–124 (2006)
90. Dessou, N.S., et al.: Influence of strontium for calcium substitution on the glass–ceramic network and biomimetic behavior in the ternary system SiO_2–CaO–MgO. J. Mater. Sci. **52**(15), 8871–8885 (2017)
91. Webster, T.J., Ergun, C., Doremus, R.H., Siegel, R.W., Bizios, R.: Enhanced functions of osteoblasts on nanophase ceramics. Biomaterials **21**(17), 1803–1810 (2000)
92. Dorozhkin, S.V.: Nanodimensional and nanocrystalline calcium orthophosphat. Am. J. Biomed. Eng. **2**(3), 48–97 (2012)
93. Ostomel, T.A., Shi, Q., Tsung, C.K., Liang, H., Stucky, G.D.: Spherical bioactive glass with enhanced rates of hydroxyapatite deposition and hemostatic activity. Small **2**(11), 1261–1265 (2006)
94. Tadic, D., Peters, F., Epple, M.: Continuous synthesis of amorphous carbonated apatites. Biomaterials **23**(12), 2553–2559 (2002)
95. Govindan, R., Girija, E.K.: Drug loaded phosphate glass/hydroxyapatite nanocomposite for orthopedic applications. J. Mater. Chem. B **2**(33), 5468–5477 (2014)
96. Wang, K.-W., et al.: Calcium phosphate/PLGA-mPEG hybrid porous nanospheres: a promising vector with ultrahigh gene loading and transfection efficiency. J. Mater. Chem. **20**(6), 1161–1166 (2010)
97. Georgia Pouroutzidou, Bioceramic Nanomaterials: Synthesis and characterization of the ternary system SiO_2–CaO–MgO with Cu ion doping. Aristotle University of Thessaloniki (2017)
98. Greener, E.H., Harcourt, J.K., Brown, D., Lautenschlager, E.P.: Materials science in dentistry. Int. Endod. J. **7**(2), 86–87 (1974)
99. Schulein, T.M.: Significant events in the history of operative dentistry. J. Hist. Dent. **53**(2), 63–72 (2005)
100. Xu, X., He, L., Zhu, B., Li, J., Li, J.: Advances in polymeric materials for dental applications. Polym. Chem. **8**(5), 807–823 (2017)
101. Hervás-García, A., Martínez-Lozano, M., Cabanes-Vila, J., Barjau-Escribano, A., Fos-Galve, P.: Composite resins. A review of the materials and clinical indications. Med. Oral Patol. Oral Cir. Bucal. **11**, 215–220 (2006)
102. Moszner, N., Salz, U.: New developments of polymeric dental composites. Prog. Polym. Sci. **26**(4), 535–576 (2001)
103. Moszner, N., Klapdohr, S.: Nanotechnology for dental composites. Int. J. Nanotechnol. **1**(1/2), 130 (2004)
104. Vallittu, P.: Non-Metallic Biomaterials for Tooth Repair and Replacement. Woodhead Publishing Limited (2013)
105. de Rastelli, S., Teixeira Carreira, E., Bezerra Dias, H., Hamblin, M.R.: Nanobiomaterials in dentistry: applications of nanobiomaterials. In: Grumezescu, A.M. (ed.) Nanobiomaterials in Dentistry: Applications of Nanobiomaterials, vol. 11, pp. 1–25. Matthew Deans (2016)
106. Bernardi, M.I.B., Rojas, S.S., Andreeta, M.R.B., De Rastelli, A.N.S., Hernandes, A.C., Bagnato, V.S.: Thermal analysis and structural investigation of different dental composite resins. J. Therm. Anal. Calorim. **94**(3) (2008)
107. Minguez, N., Ellacuria, J., Soler, J.I., Triana, R., Ibaseta, G.: Advances in the history of composite resins. J. Hist. Dent. **51**(3), 103–105 (2003)

108. Gad, M.M., Fouda, S.M., Al-Harbi, F.A., Näpänkangas, R., Raustia, A.: PMMA denture base material enhancement: a review of fiber, filler, and nanofiller addition. Int. J. Nanomed. **12**, 3801–3812 (2017)
109. Kishen, A. (Ed.) Nanotechnology in Endodontics (2015)
110. Safi, I.N.: Evaluation the effect evaluation the effect of nano-fillers (TiO_2, AL_2O_3, SiO_2) addition on glass transition temperature, E-Moudulus and coefficient of thermal expansion of acrylic denture base material, **26**, 37–41 (2014)
111. Jasim, B.S., Ismail, I.J.: The effect of silanized alumina nano -fillers addition on some physical and mechanical properties of heat cured polymethyl methacrylate denture base material. J Bagh Coll. Dent. **26**(2), 18–23 (2014)
112. Thorat, S., Diaspro, A., Salerno, M.: Effect of alumina reinforcing fillers in BisGMA-based resin composites for dental applications. Adv. Mater. Lett. **4**(1), 15–21 (2013)
113. Kul, E., Aladağ, Lİ., Yesildal, R.: Evaluation of thermal conductivity and flexural strength properties of poly(methyl methacrylate) denture base material reinforced with different fillers. J. Prosthet. Dent. **116**(5), 803–810 (2016)
114. Safarabadi, M., Mehri Khansari, N., Rezaei, A.: An experimental investigation of HA/AL_2O_3 nanoparticles on mechanical properties of restoration materials. Eng. Solid Mech. **2**(3), 173–182 (2014)
115. Ahmed, M.A., El-Shennawy, M., Althomali, Y.M., Omar, A.A.: Effect of Titanium Dioxide nano particles incorporation on mechanical and physical properties on two different types of acrylic resin denture base. World J. Nano Sci. Eng., 111–119 (2016)
116. Song, R., Tao, X.H., Lin, L.: Improvement of mechanical and antimicrobial properties of denture base resin by nano-titanium dioxide and nano-silicon dioxide particles. Pigment Resin Technol. **40**(6), 393–398 (2011)
117. Gad, M., Arrejaie, A.S., Abdel-Halim, M.S., Rahoma, A.: The reinforcement effect of nano-Zirconia on the transverse strength of repaired acrylic denture base. Int. J. Dent. (2016)
118. Gad, M.M., Rahoma, A., Al-Thobity, A.M., ArRejaie, A.S.: Influence of incorporation of ZrO_2 nanoparticles on the repair strength of polymethyl methacrylate denture bases. Int. J. Nanomed. **11**, 5633–5643 (2016)
119. Alhavaz, A., Rezaei Dastjerdi, M., Ghasemi, A., Ghasemi, A., Alizadeh Sahraei, A.: Effect of untreated zirconium oxide nanofiller on the flexural strength and surface hardness of autopolymerized interim fixed restoration resins. J. Esthet. Restor. Dent. **29**(4), 264–269 (2017)
120. Ahmed, M.A., Ebrahim, M.I.: Effect of Zirconium Oxide nano-fillers addition on the flexural strength, fracture toughness, and hardness of heat-polymerized acrylic resin. World J. Nano Sci. Eng. **4**, 50–57 (2014)
121. Chan, K.S., Nicolella, D.P., Furman, B.R., Wellinghoff, S.T., Rawls, H.R., Pratsinis, S.E.: Fracture toughness of zirconia nanoparticle-filled dental composites. J. Mater. Sci. **44**(22), 6117–6124 (2009)
122. Alnamel, H.A., and Mudhaffer, M.: The effect of Silicon di oxide nano-fillers reinforcement on some properties of heat cure polymethyl methacrylate denture base material. J.Bagh Coll. Dent. **26**(1), 32–36 (2014)
123. Balos, S., Pilic, B., Markovic, D., Pavlicevic, J., Luzanin, O.: Poly(methyl-methacrylate) nanocomposites with low silica addition. J. Prosthet. Dent. **111**(4), 327–334 (2014)
124. Hu, Y.H., Chen, C.Y., Wang, C.C.: Viscoelastic properties and thermal degradation kinetics of silica/PMMA nanocomposites. Polym. Degrad. Stabil. **84**, 545–553 (2004)
125. Maffezzoli, A., Pietra, A.D., Rengo, S., Nicolais, L., Valletta, G.: Photopolymerization of dental composite matrices. Biomaterials **15**(15), 1221–1228 (1994)
126. Ukai, S.: Metal, Ceramic and Polymeric Composites for Various Uses, vol. 3 (2011)
127. Miao, X., Li, Y., Zhang, Q., Zhu, M., Wang, H.: Low shrinkage light curable dental nanocomposites using SiO_2 microspheres as fillers. Mater. Sci. Eng. C **32**(7), 2115–2121 (2012)
128. Topouzi, M., Kontonasaki, E., Bikiaris, D., Papadopoulou, L., Paraskevopoulos, K.M., Koidis, P.: Reinforcement of a PMMA resin for interim fixed prostheses with silica nanoparticles. J. Mech. Behav. Biomed. Mater. **69**, 213–222 (2017)

129. Madathingal, R.R., Wunder, S.L.: Effect of particle structure and surface chemistry on PMMA adsorption to silica nanoparticles. Langmuir **26**(7), 5077–5087 (2010)
130. Bansal, A., Yang, H., Li, C., Benicewicz, B.C., Kumar, S.K., Schadler, L.S.: Controlling the thermomechanical properties of polymer nanocomposites by tailoring the polymer–particle interface. J. Polym. Sci. Part B: Polym. Phys. **44**(20), 2944–2950 (2006)
131. Ramanathan, T., et al.: Graphitic nanofillers in PMMA nanocomposites—an investigation of particle size and dispersion and their influence on nanocomposite properties. J. Polym. Sci. Part B: Polym. Phys. **45**(15), 2097–2112 (2007)
132. Protopapa, P., Kontonasaki, E., Bikiaris, D., Paraskevopoulos, K.M., Koidis, P.: Reinforcement of a PMMA resin for fixed interim prostheses with nanodiamonds. Dent. Mater. J. **30**(2), 222–231 (2011)
133. Raty, J.Y., Galli, G.: Optical properties and structure of nanodiamonds. J. Electroanal. Chem. **584**(1), 9–12 (2005)
134. Dolmatov, V.Y.: Polymer—diamond composites based on detonation nanodiamonds. Part 3. J. Superhard Mater. **29**(4), 199–205 (2007)
135. Shenderova, O., M.G, Carbon Nanomaterials, vol. 1(4). CRC Press Taylor & Francis, Philadelphia (2006)
136. Muhtaroğullari, I.Y., Doğan, A., Muhtaroğullari, M., Usanmaz, A.: Thermal and dynamic mechanical properties of microwave and heat-cured poly(methyl methacrylate) used as dental base material. J. Appl. Polym. Sci. **74**(12), 2971–2978 (1999)
137. Blivi, A.S., Benhui, F., Bai, J., Kondo, D., Bédoui, F.: Experimental evidence of size effect in nano-reinforced polymers: Case of silica reinforced PMMA. Polym. Test. **56**, 337–343 (2016)
138. Yang, F., Nelson, G.L.: Polymer/silica nanocomposites prepared via extrusion. Polym. Adv. Technol. **17**(4), 320–326 (2006)
139. Yang, F., Nelson, G.L.: PMMA/silica nanocomposite studies: synthesis and properties. J. Appl. Polym. Sci. **91**(6), 3844–3850 (2004)
140. Qiao, R., Deng, H., Putz, K.W., Brinson, L.C.: Effect of particle agglomeration and interphase on the glass transition temperature of polymer nanocomposites. J. Polym. Sci. Part B: Polym. Phys. **49**(10), 740–748 (2011)
141. Karabela, M.M., Sideridou, I.D.: Synthesis and study of properties of dental resin composites with different nanosilica particles size. Dent. Mater. **27**(8), 825–835 (2011)
142. Elliott, J.E., Lovell, L.G., Bowman, C.N.: Primary cyclization in the polymerization of bis-GMa and TEGDMA: a modeling approach to understanding the cure of dental resins. Dent. Mater. **17**(3), 221–229 (2001)
143. Lovell, L.G., Berchtold, K.A., Elliott, J.E., Lu, H., Bowman, C.N.: Understanding the kinetics and network formation of dimethacrylate dental resins. Polym. Adv. Technol. **12**(6), 335–345 (2001)
144. Achilias, D.S., Karabela, M.M., Sideridou, I.D.: Thermal degradation of light-cured dimethacrylate resins. Part I. Isoconversional kinetic analysis. Thermochim. Acta **472**(1–2), 74–83 (2008)
145. Liu, Q., Ding, J., Chambers, D.E., Debnath, S., Wunder, S.L., Baran, G.R.: Filler-coupling agent-matrix interactions in silica/polymethylmethacrylate composites. J. Biomed. Mater. Res. **57**(3), 384–393 (2001)
146. Vouvoudi, E.C., Achilias, D.S., Sideridou, I.D.: Dental light-cured nanocomposites based on a dimethacrylate matrix: thermal degradation and isoconversional kinetic analysis in N_2 atmosphere. Thermochim. Acta **599**, 63–72 (2015)
147. Saladino, M.L., Motaung, T.E., Luyt, A.S., Spinella, A., Nasillo, G., Caponetti, E.: The effect of silica nanoparticles on the morphology, mechanical properties and thermal degradation kinetics of PMMA. Polym. Degrad. Stab. **97**(3), 452–459 (2012)

Differential Scanning Calorimetry (DSC): An Invaluable Tool for the Thermal Evaluation of Advanced Chimeric Liposomal Drug Delivery Nanosystems

Maria Chountoulesi, Nikolaos Naziris, Natassa Pippa, Stergios Pispas and Costas Demetzos

Abstract Chimeric liposomal systems are classified as advanced drug delivery nanosystems, being composed of different kinds of biomaterials, such as phospholipids and polymers. Chimeric liposomes present many advantages, compared to conventional ones, such as great functionality, stimuli-responsiveness and increased targeting to the pathological tissue. Among the analytical techniques established for the liposomal system characterization, Differential Scanning Calorimetry (DSC) is applied to indicate their thermotropic behavior and thermal stability, providing useful information, in order to optimize the quality and therapeutic efficiency of the liposomal formulations. Until very recently, several researches have aimed at explaining the behavior of carrier lipid forms by the DSC method. Therefore, DSC has been quite frequently applied in pharmaceutical research for scanning the thermal behavior of the samples and to record the difference between the heat flows, while it provides quick and accurate information about the physical and energetic parts of a material. In the present chapter, a variety of different chimeric liposomal systems is presented that were analyzed in terms of their thermal behavior, describing the utilized DSC protocols and highlighting the interpretation of the DSC results. The aim of this chapter is to prove, through different literature examples of chimeric liposomal systems, the utility of the DSC technique upon the characterization of their thermotropic behavior which is strictly correlated with the interactions and cooperativity of the different biomaterials, as well as how it predicts the efficacy of the examined liposomal platforms.

Keywords Liposomes · Thermal analysis · Differential scanning calorimetry · Advanced drug delivery nanosystems · Chimeric

---

M. Chountoulesi · N. Naziris · N. Pippa · C. Demetzos (✉)
Section of Pharmaceutical Technology, Department of Pharmacy, School of Health Sciences, National and Kapodistrian University of Athens, Panepistimioupolis Zografou, 15771 Athens, Greece
e-mail: demetzos@pharm.uoa.gr

N. Pippa · S. Pispas
Theoretical and Physical Chemistry Institute, National Hellenic Research Foundation, 48 Vassileos Constantinou Avenue, 11635 Athens, Greece

C. Demetzos and N. Pippa (eds.), *Thermodynamics and Biophysics of Biomedical Nanosystems*, Series in BioEngineering, https://doi.org/10.1007/978-981-13-0989-2_9

1 Introduction to Liposomes and Classification of Chimeric Liposomal Nanosystems

Drug delivery nanoSystems (DDnSs) are continuously stimulating the research interest, as well as providing smart solutions to critical problems of therapeutic clinical practice. DDnSs, as nano-devices, can introduce a variety of different therapeutic substances in the body and improve their efficacy and safety by controlling the rate, time and place of release of drugs in the body [1].

Demetzos and Pippa proposed in 2013 a classification system of DDnSs, regarding their biomaterial formulation. DDnSs that exhibit a modulatory controlled release profile (MCR) and are referred also as modulatory controlled release nanosystems (MCRnSs), can be classified into conventional (c) and advanced (a) DDnSs, denoted by the acronyms cDDnSs and aDDnSs, and can be composed of a single or more than one biomaterials, respectively. The classification was based on their characteristics such as: surface functionality (f), the nature of biomaterials used and the kind of interactions between biomaterials. The aDDnSs that are developed by mixing different in nature biomaterials and inorganic materials are further classified to chimeric (Chi-), while the aDDnSs that are composed of same in nature biomaterials and inorganic materials are further classified to hybridic (Hy-). The term chimeric is inspired from the ancient Greek mythology, where Chimera was referred as a monstrous fire breathing female creature of Lycia in Asia Minor, composed of parts of three different animals: a lion, a serpent and a goat. The classification of drug delivery systems is illustrated in Fig. 1 [2].

Liposomes are considered to be one of the most well-investigated DDnSs, presenting major advantages, such as biocompatibility, biodegradability and lack of immunogenicity. Liposomes can be loaded with either drugs (hydrophobic, hydrophilic, and amphiphilic), antigens or nucleic acids. Consisting of artificial membranes and simulating the cell membranes, they can also be used as models to study the behavior, the phase transitions and the interactions of cell membranes with the environment and exogenous molecules, such as drugs. There are many approved liposomal and lipid-based products in the market, as well as a great number of novel products in clinical and preclinical development. Moreover, liposomes exhibit great versatility and can be easily surface modified with functional biomaterials, in order to acquire advanced properties, such as escaping the rapid clearance in circulation, due to cellular uptake by the mononuclear phagocyte system (MPS), and presenting increased targeting to pathological tissues [3, 4].

According to the above classification, we can produce advanced chimeric liposomal systems through the modification of liposomes with different in nature biomaterials, such as block copolymers, or through the loading of different nanosystems into liposomes, such as dendrimers or nanoparticles. The interactions between the biomaterials, being presented in the liposomal formulation, affect the liposomal physicochemical properties, the thermotropic behavior of the lipid membrane, the structural polymorphism, the pharmacodynamics and pharmacokinetics of the liposomal content, as well as their biophysical and biological stability. A variety

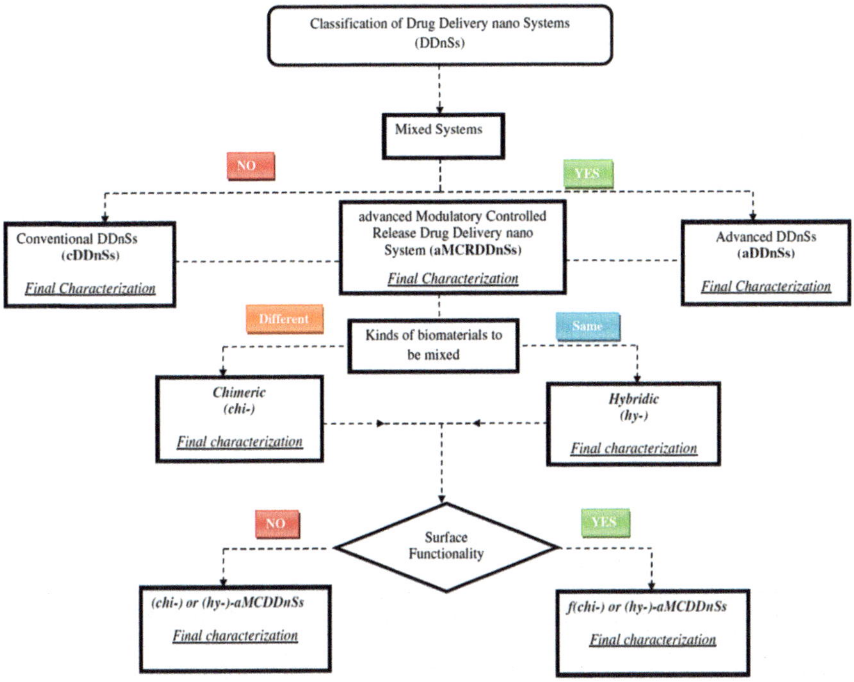

Fig. 1 The classification of drug delivery systems. Adapted from Ref. [2]

of advanced formulations of liposomal platforms are extensively described in the following paragraphs.

2 Introduction to the Differential Scanning Calorimetry (DSC) Technique

Thermal analysis is a commonly used term to describe a gamut of analytical techniques of calorimetry that are applied to investigate the physicochemical properties of materials as functions of temperature or time. During thermal analysis procedure, the material is subjected to a temperature program isothermally or by using different temperature segments, usually at constant pressure. Differential Scanning Calorimetry (DSC) is a typical representative thermal analysis technique, relatively fast, low cost and the most frequent technique used for determining the thermal behavior of a variety of materials that are extensively analyzed below.

DSC scans temperature and measures the difference between the heat flows of a pan containing the sample and a reference pan that are subjected to the same temperature program, under atmospheric pressure, and measures the heat capacity of a material. Both the sample and the reference pan are heated or cooled at the same time and with the same temperature fluctuation rate, so the differences in the heat uptake and heat capacity can be collected. These differences give direct information about the energetics of thermally-induced processes in the sample.

DSC measures the heat flow absorbed into or being released by a material. From that, the heat capacity at constant pressure (C_p) can be calculated. Heat capacity units are JK^{-1}. The amount of heat that is absorbed by an unreacting system in raising its temperature by 1 K at constant pressure is defined as the heat capacity at constant pressure, C_p, which is itself a function of temperature (Eq. 1). Heat capacity is usually normalized by dividing the specimen heat capacity by the number of grams, to get the heat required to raise one gram of specimen by one degree Celsius. This then corresponds to the specific heat capacity C_p. If desired, the heat capacity can be normalized by the number of moles. Heat capacity is defined by:

$$C_p = \left(\frac{\partial q}{\partial T}\right)p, \tag{1}$$

where T is the temperature and q is the heat input.

If the temperature changes from T_0 to T_1, the enthalpy of the reaction (ΔH) is

$$\Delta \mathrm{H} = \int_{T_0}^{T_1} CpdT. \tag{2}$$

Usually, ΔT is small and C_p is independent of temperature between T_0 and T_1. The integral thus reduces to:

$$\Delta H = Cp(T_1 - T_0) = Cp\Delta T. \tag{3}$$

The thermodynamic parameters that are calculated by DSC are described below, and also illustrated in Fig. 2:

T_{onset}: Temperature at which the thermal event begins.

T_{exp}: Extrapolated onset temperature, where the curve extension of the endothermic or exothermic top meet the baseline.

T_m: Main transition temperature at the top of the curve, at which the two phases/states are equal and the system has the maximum heat capacity C_p.

T_{endset}: Temperature at which the thermal event is completed and the curve meets the baseline.

ΔC_p: System thermal capacity change, at constant pressure P, as described by equation (Eq. 1).

ΔH: System enthalpy change, reflecting the amount of energy transferred as heat, under stable pressure conditions, being calculated by integrating the total area under the curve of the transition, as described by equation (Eq. 3).

$\Delta T_{1/2}$: The phase transition range in the middle of the peak [5].

The experimental DSC protocols require careful attention to the sample preparation, the equilibration before heating, the accurate measures of sample mass or

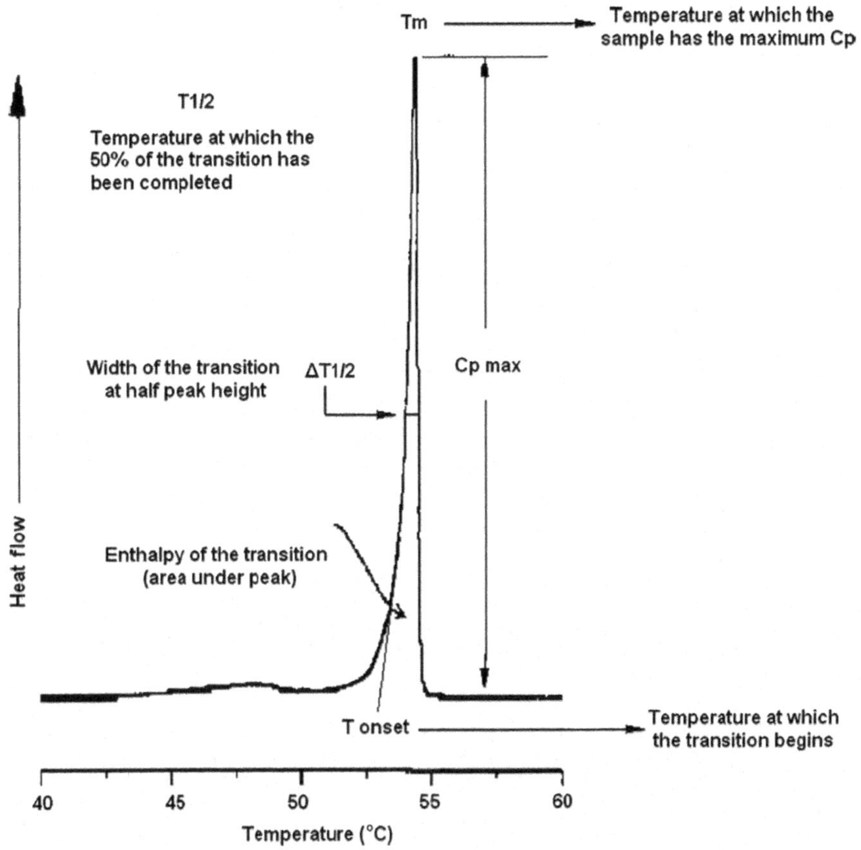

Fig. 2 Typical DSC thermogram, describing the characteristic thermodynamic parameters provided by the DSC technique. Adapted from Ref. [5]

concentration and baseline controls, followed by careful analysis and interpretation of the thermodynamic data [6].

The sample preparation protocol can affect the obtained thermograms. For example, during the study of lipid vesicles, different size vehicles, acquired by different preparation procedures, can alter the thermodynamic parameters. More specifically, small unilamellar vesicles (SUVs) produce a lower resolution peak than multilamellar vesicles (MLVs), as well as a reduction in cooperativity that can be attributed to the smaller radius of SUVs over MLVs, resulting in a less ordered orientation, which increases the free motion of the hydrocarbon chains. The decrease in cooperativity, accompanied by a slightly broader endothermic peak, is also due to SUVs fusion to large unilamellar vesicles (LUVs) upon heating, resulting in size inhomogeneity [7].

The scanning range should certainly include the transition phenomenon. For instance, in the case of phospholipids, it should be used a temperature starting about

20 °C prior to the transition temperature of the phospholipid employed and ending about 20 °C above the transition temperature of the phospholipid employed [8].

While there are many instances of kinetically-determined irreversible process (such as aggregation or chemical degradation at higher temperatures) that are scan rate-dependent, the choice of scan rate is a critical parameter that should be investigated before the experimental procedure. Sometimes, trying different scanning rates is required in order to select the most appropriate one.

The total heat energy uptake by the sample undergoing the transition, reflected in ΔH, depends on the amount of sample present in the active volume of the DSC cell and should be normalized according to the sample size.

"Annealing" (or isothermal heating) the sample one or more times to a temperature prior to the onset of the transition and cooling it can also be useful. This preliminary scan, running at temperatures before the unfolding transition begins, minimizes baseline artifacts that can be induced by the thermal shock involved in the loading of the sample or reference cells. Referring to the number of scan cycles applied, it may be useful to run repeating scans with the same sample, to establish reversibility and reproducibility [6].

2.1 Pharmaceutical and Biological Applications of the DSC Technique

The thermal stress the material experiences during the DSC experimental procedure reveals a great number of phenomena and information about the material behavior. This information may include the recognition of serious impurities, such as the polymorphic forms, highlighting their role in the physical stability of a novel final pharmaceutical product, as well as the kinetics of the material decomposition during accelerated aging experiments, the glass transition of polymers and the level of compatibility of the bioactive molecules with the formulation excipients. All these characteristics are critical for the quality characterization of either the intermediate or the final pharmaceutical forms during the whole drug development and formulation process [5].

2.1.1 Proteins and Nucleic Acids

DSC is also applied at the characterization of pharmaceutical products consisting of biomolecules, such as proteins or enzymes, in order to study the thermally induced, cooperative conformational changes of small proteins, their denaturation, the effect of the environment on their structural relaxation, their kinetic and thermal stability, their aggregation, as well as their interaction with ligands during the binding process with different lipid mixtures. Referring to DNA biomolecules, the base stacking enthalpies and helix-coil enthalpies, calculated by DSC analysis, can be used to determine conformations of DNA, their melting behavior in various duplex or triplex conformations, the denaturation of the conformations, their structural stability, the effect of ions, pH, and temperature, as well as the ligand-DNA interactions. For example, the DSC technique has been used to study the binding of DNA antitumor drugs or the interactions of nonsteroidal anti-inflammatory drugs with DNA. Furthermore, it can be used to investigate methodologies, such as the virus-induced gene silencing [7].

2.1.2 Lipids and Lipid Bilayers

Lipids may be one of the most well-studied biomaterial category by DSC analysis. Membrane lipids can be categorized to glycerol-based lipids (phospholipids), ceramide-based sphingolipids, and cholesterol. Phospholipids are further divided into different groups, depending on their hydrophilic head groups: phosphatidylcholine (PC), phosphatidylethanolamine (PE), and phosphatidylserine (which are largely present in the cellular membrane), accompanied with phosphatidylinositol and cardiolipin that are present in smaller quantities [8].

Phospholipids, being the most common components of mammalian membranes and exhibiting vital structural roles, belong to lyotropic liquid crystals and have extensively been studied regarding their phase transitions (polymorphism behaviour) and especially their gel-to-liquid phase transition [5]. A phospholipid-formed bilayer structure or lamella in aqueous media is selected as a model that is quite close to biological membranes, while the phospholipids phase transitions are of great importance for biological processes occurring on cell membranes, thus their detailed investigation would help us understand many biological phenomena that take place in the cell membranes [9].

The gel-to-liquid crystalline transition, often referred as the melting or main transition, is relatively rapid and highly reversible, represented by high enthalpy DSC peak appearing at the T_m. The ΔH is related to the enthalpic content of lipid bilayers during transition, while the $\Delta T_{1/2}$ indicates the cooperativity of the membrane biomaterials during transition [10]. As described by McElhaney 1982, during the main transition of dipalmitoylphosphatidylcholine (DPPC) lipid, as temperature rises, the all-*trans* conformation of lipid chains in the gel state is transformed to *gauche* conformation that is characteristic in the disordered state, through a reduction of the van der Waals hydrophobic interactions of acyl chains of DPPC. As a result, the liquid crystalline (L_α) phase is also characterized by a large increase in membrane fluidity and disorder. Another intermediate transition that happens prior to the main transition is the pre-transition from the planar gel ($L_{\beta'}$) to the rippled gel phase ($P_{\beta'}$). It has a low enthalpy, is attributed to the interaction between the polar head groups and is sensitive to sample preparation, the scan rate and the presence of impurities. Prior to the pre-transition, some lipid exhibit a sub-transition, from subgel (L_c) to gel phase ($L_{\beta'}$), which is kinetically slow [5, 9, 11, 12].

2.1.3 Liposomes

Except for lipids and lipid bilayers, the DSC technique is considered to be an invaluable method upon the thermal characterization of drug delivery nanosystems consisting of lipids, such as liposomes. The DSC technique provides curious details, such as their material phase transitions that are important during designing and formulation process of such drug delivery systems. These transitions are correlated with their composition, the effects of their environment (hydration, pH, solvent), their physicochemical behavior, their stability over time, as well as the release kinetics of their content, being in strict correlation with the pharmacodynamics, the pharmacokinetics and the bioavailability of the content drugs. A typical example, where the DSC technique can play a critical role, is the recognition of metastable phases (thermodynamic instant nonequilibrium states of a substance or a state of matter, whose properties alter reversibly over time) during the preformulation of nanocarriers, because these metastable phases are proved to govern the drug delivery nanosystems functionality and be correlated to the biophysical properties and behavior of liposomes. The DSC technique and generally the methods associated with the thermodynamics of the systems are useful tools towards the development of innovative nanocarriers, and subsequently, the production and scale-up process of safe and effective nanomedicines, increasing eventually their translational bench-to-clinic efficacy [13].

The regulatory framework for the approval of nanomedicines is a dynamic field that is continually being enriched by new data, while the development of the nanomedicines is constantly upgrading and new innovative nanocarriers are appearing. Moreover, the expiration of the patents of the first generation nanomedicines and the appearance of the follow-up "copies" of patent nanomedicines, referred as nanosimilars, has raised the necessity of redefining the regulatory framework in the aspects of bioequivalence. Owing to the complexity of nanomedicines, showing equivalence is more challenging for the follow-up products of nanomedicines [14]. Not only the bioactive molecules, but also the carrier properties (such as particle size and size distribution, particle surface characteristics, fraction of uncaptured bioactive moiety), which is described as innovative excipient, need to be examined, whether they are similar to the prototype. A pharmacokinetic endpoint based bioequivalence study of the bioactive molecule may not be sufficient to ensure equivalent safety and efficacy. Additional measures such as comparative physicochemical testing (including in vitro release studies) must be undertaken [15]. As the recent literature confirms, the DSC technique can be utilized for further understanding of the composition/structure/function relationships of liposomal products and the statement of comparability between prototypes and generic ones, because DSC technique is capable of detecting slight differences, such as metastable phases, with great impacts on the behavior of the system. Wei et al. [16] used the DSC technique, in order to establish the similarity of Doxil® and Lipodox® (the FDA approved Doxil generic) in terms of thermotropic behavior, satisfying the FDA requirements on thermodynamic characterization of a generic liposomal medicine [16]. In addition, Perinelli et al. [17] carried out a detailed investigation on the thermal behavior of doxorubicin-loaded liposomes, in comparison to the non-loaded liposomal formulation (placebo) [17].

The continuous development and improvement of conventional drug delivery nanosystems to advanced chimeric ones, as classified above, creates new queries about the final nanosystems cooperativity and functionality, the different biomaterials compatibility and whether more complex structures are also better in terms of safety and therapeutic efficiency. The DSC technique can play a crucial role in the characterization of these innovative next generation nanosystems, providing solutions to the above questions. In the following paragraphs, a variety of examples from the literature are presented, referring to different advanced chimeric nanosystems, especially liposomal ones, where the DSC technique has been utilized to characterize their thermotropic behavior.

3 Examples of Chimeric Liposomal Nanosystems Studied by DSC

3.1 Chimeric Inorganic Nanoparticle-Liposomal Nanosystems

3.1.1 Chimeric Magnetic Nanoparticle-Liposomal Nanosystems

Magnetoliposomes (chimeric liposomes containing magnetic nanoparticles) can be classified as multifunctional chimeric liposomal systems, being decorated with magnetic nanoparticles. Magnetoliposomes can supply selective and local heating to the bilayer, so they can be used to control liposomal drug delivery. While lipid phase behavior is sensitive to the presence of exogenous compounds, the DSC technique can highlight the thermotropic alteration that takes place during the incorporation of magnetic nanoparticles into the liposomes.

For example, according to Chen et al. [18] when increasing amounts of SPIO nanoparticles were embedded within DPPC liposomal bilayers, the pretransitions were suppressed and merged with lipid melting at higher temperatures, while the melting regions were broadened and also shifted to higher temperatures. The loaded samples were equilibrated at 15 °C, and sequential heat/cool cycles were performed from 15 to 50 °C at a scan rate of 1 °C min^{-1}. The melting regions were composed of multiple overlapping peaks, a phenomenon that was predominant at the highest nanoparticle loading ratio. From DSC results, authors concluded that the nanoparticles were mixed with lipids in the gel phase rather than the fluid phase. Further results from fluorescence leakage studies, being consistent to DSC results, suggested that the increase of nanoparticle concentration made liposomes more stable and resulted to less spontaneous leakage [18].

On the other hand, Patitsa et al. [19] used the DSC technique, in order to investigate the interactions of iron oxide magnetic nanoparticles coated with polyarabic acid and DPPC lipid bilayers (in this case lipid bilayers were used as lipid model membrane). There were strong interactions of the nanoparticles and especially of the polymeric coating with the polar groups of DPPC phospholipids, leading to a decrease of the pretrasition temperature and causing eventually reorientation of lipid bilayers from rippled to solid-gel liquid crystal phase. DSC results were found to be in agreement with atomistic molecular dynamics (MD) simulations experiments, concluding that magnetic nanoparticles showed excellent cell penetration properties in human breast cancer cells in culture [19].

An attractive strategy of the recent literature combines diagnostic and therapeutic modalities in one single nanocarrier that can be described as "theranostic". Theranostic liposomes can encapsulate theranostic agents enabling simultaneous MRI detection (Magnetic Resonance Imaging) and photodynamic therapy (PDT). Liposomes containing MRI contrast agents (CAs) are attractive carriers because they have been shown to have very good pharmacokinetic properties and higher target specificity [20, 21]. Skupin-Mrugalska et al. [22] proposed several formulations of bifunctional liposomes bearing gadolinium chelate (GdLip), consisting of a lipid derivative of Gd(III) ion, namely (N-diethylenetriaminepentaacetic acid gadolinium salt (GdDTPA1), possessing paramagnetic properties for MRI, in combination with a model photosensitizing agent, namely zinc phthalocyanine (ZnPc) for PDT. Liposomes were mainly composed of 1-palmitoyl-2-oleoyl-sn-glycero-3-phosphocholine (POPC), as well as of additional components, such as phosphatidylglycerol (PG) and 1,2-dioleoyl-3-trimethylammonium-propane chloride (DOTAP), coated with (polyethylene glycol)-2000 (PEG-2000) chains to obtain sterically stabilized liposomes. The DSC technique was used to investigate the effect of liposome membrane composition on the phase behavior of GdLip. DSC measurements were made on liposomal suspensions, over the range from −40 to 60 °C, at a temperature rate of 5 °C/min. Increasing concentrations of GdDTPA1 decreased the main transition temperature and the enthalpy of POPC/PG and GdDTPA1/POPC/PG liposomes containing ZnPc in a concentration dependant manner. ZnPc loaded formulations presented a broadening of the main transition peak with decreasing T_m and ΔH_m in comparison to the non-loaded formulations. In addition, much pronounced peak broadening was noted in the DSC scans of GdLip formulations containing DOTAP, and the transition peak was almost completely diffused at the highest ratio. Referred to the addition of 1,2-dipalmitoyl-sn-glycero-3-phosphoethanolamine-N-[methoxy(polyethyleneglycol)-2000](PEG200-DPPE) component, it led to a slight decrease in T_m peak for all pegylated formulations. The evaluation of the interaction among the materials, reflected by the thermal behavior, was crucial, because the different phospholipids, DOTAP and PEG200-DPPE, caused eventually different oxygen generation and relaxivity, as also confirmed by the results of NMR relaxation experiments [22].

3.1.2 Chimeric Gold Nanoparticle-Liposomal Nanosystems

The incorporation of gold nanoparticles into liposomes is proved to cause significant changes in the thermal release behavior of the liposomal system, depending on the size and the concentration of the particle being incorporated, allowing temperature to be used as a triggered release mechanism.

Preiss et al. [23] tried to incorporate dodecanethiol—stabilized gold nanoparticles (two kinds of nanoparticles of different size) into DPPC liposomes at different molar ratios. The lipid phase behavior was studied by the DSC technique, using a scanning program from 20 to 50 °C at scan rate of 1 °C min^{-1}, while DSC results were associated with fluorescence leakage studies. Results indicated that nanoparticles being smaller than the bilayer thickness affected the bilayer differently than nanoparticles being closer to the bilayer thickness. Analytically, the smaller nanoparticles caused the pretransition and main transition to merge, but the main transition temperature to remain at similar values to pure DPPC liposomes. In contrast to pure DPPC liposomes, which presented the greatest leakage at the pretransition temperature, according to fluorescence leakage studies, smaller gold nanoparticles caused a decrease of spontaneous leakage at lower temperatures and so provided greater bilayer stability. These results are associated with the above alterations of pretransition and main transition, being presented at DSC thermograms. Contrary wise, bigger gold nanoparticles suppressed the pretransition and broadened the main transition, reflecting maximum leakage near the pretransition temperature or at lower temperatures. Authors concluded that liposome leakage and stability can be manipulated by the size of embedded nanoparticles even at low loadings and the effects are also reflected to the thermal behavior of the systems [23]. Thus, the evaluation of the membranes thermal behavior can help at the choice of the proper nanoparticles size that will prevent chimeric nanosystems defects such as premature content leakage.

3.2 *Chimeric Cyclodextrin-Liposomal Nanosystems*

Cyclodextrins (CDs) are cyclic oligosaccharides, formed by glucopyranose units. Their internal cavity is relatively hydrophobic, while their outer surface is hydrophilic. This particular structure confers to CDs the ability to include a large number of hydrophobic guests, and to form inclusion complexes soluble in water, improving drug bioavailability and stability, as well as decreasing toxic side effects [24]. CD/drug inclusion complexes inserted into liposomes were proposed by McCormack and Gregoriadis in 1994, called also as drug-in-CD-in-liposome (DCL) [25]. DCLs can be characterized as chimeric nanosystems and have the ability to modify the release profile of the enclosed drug. Gharib et al. [26] studied recently the effect of the hydroxypropyl-ß-CD (HP-ß-CD) on the fluidity and stability of membranes of different composition at various molar ratios HP-ß-CD: DPPC, by using the DSC technique. The scanning rate employed was 1 °C/min in the temperature range of 20–50 °C. The pre-transition peak was completely abolished in the presence of HP-ß-CD at various DPPC:HP-ß-CD molar ratios and the main transition temperature shifted to higher values in a concentration dependent manner, because the carbohydrate moieties of dimethyl-ß-CD can stabilize bilayers via hydrogen bonding to lipids, involving probably the phosphodiester group of the DPPC head group. Moreover the ΔH_m was altered among the different HP-ß-CD ratios, in contrast to the $\Delta T_{1/2}$, which was increased at all DPPC:HP-ß-CD molar ratios, reflecting disorganization of the lipid packing and decrease of the system cooperativity. DSC results were in accordance with Raman spectroscopy which confirmed that the presence of HP-ß-CD increased the DPPC membrane fluidity in a concentration dependent manner. Authors concluded that such results should be taken into consideration in DCL carrier design, since it may be strictly associated to the release kinetic profile of the enclosed drug [26].

At another recent study, HP-ß-CD was incorporated into DPPC membranes, in order to investigate whether irbesartan (IRB) can be administered alone or into a CD complex and what the impact of the method of incorporation into membranes, such as model DPPC membranes, may be. Liposomal samples were scanned from 10 to 60 °C at least three times until identical thermal scans were obtained using a scanning rate of 2.5 °C/min. The presence of IRB alone resulted in a reduction of the T_m, narrowing of the transition width of the main phase transition and abolishment of the pre-transition, indicating a strong effect on the head-group region of lipid bilayers, while it did not affect ΔH_m considerably. Concerning HP-β-CD incorporation, both the pretransition and the main transition are lowered by a few degrees, assuming that the HP-β-CD incorporation causes small perturbation, decreases the packing density of the lipids, but not so intensively, and sits on the liposomal surface, also confirmed by SAXS and solid state NMR spectroscopy results. IRB encapsulated into the HP-β-CD, exerted significant differences on the membrane internalization and maximizes the fluidization effect of the membranes, as depicted by DSC, in combination with a variety of techniques such as solid state NMR spectroscopy, X-ray diffraction and Molecular Dynamics. Authors concluded that when IRB is encapsulated into the HP-β-CD, IRB is released and perturbs the lipid bilayers. As a result, the DSC technique, combined with other useful methods, can be utilized into the evaluation of the interaction of the drug delivery nanosystems with model membranes as an effort of improving ADME (absorption, distribution, metabolism, and excretion) properties of the enclose drugs [27].

3.3 Chimeric Non-ionic Surfactant-Liposomal Nanosystems

Tween and Span non ionic surfactants are commonly used in pharmaceutical industry due to their low cost, biodegradability and biocompatibility and can be applied in liposomal technology in order to improve the properties of conventional liposomes, such as the drug capturing efficiency. Juárez-Osornio and Gracia-Fadrique [28] managed to produce intermediate nanometric structures (between liposomes and lipid emulsions) prepared of DPPC, Cholesterol (Chol) and a mixture of ethoxylated and non-ethoxylated sorbitan fatty acid esters (Tween 20, Span20, Tween 80 and Span 80), by mixing all of them inside water, near the cloud point temperature (cp) of the ethoxylated surfactant. Calorimetric results, from DSC technique, performed from 10 to 80 °C with scan rate of 1 K/min, of these intermediate structures do not present pre-transition, because non-DPPC molecules are being inserted in the bilayer interrupting the packing of the bilayer and promoting the non-symmetrical lamellar areas, while their main transition is similar to the main transition of pure DPPC, indicating that the prepared nanometric structure is not a typical mixed micelle and that the DPPC molecules are organized as a bilayer, just like in liposomes or in layers in the case of lipid emulsions. Another interesting phenomenon is a difference being observed between the first and third scan, suggesting a spatial reconfiguration of their constituents, due to the heat change, taking place between scans. Authors also suggested that the T_m value decreases as a function of the amounts and the kind of the other hydrophobic compounds inserted in the bilayer. In particular, in the case of Tween 20-Span 20 systems, T_m showed different values, due to the Tween 20 insertion in the DPPC bilayer, while Span 20 was considered to prefer the central core rather than the bilayer, due to its higher hydrophobicity. Concerning the composition, larger amounts of Tween 20 lead to a small peak close to 33.4 °C, while the lower amounts lead to a larger corresponding peak, accompanied with a T_m near to the T_m of the pure DPPC bilayer. Contrariwise at the Tween 80-Span 80 systems, T_m was more similar to the pure DPPC. Authors attributed this observation to the fact that the majority of the Tween 80-Span 80 molecules are not inserted into the bilayer but they form the central mixed micelle of the nanostructure, where the rest of DPPC forms more than one bilayer around the nucleus of the particle. Concerning cholesterol, it is probably distributed non-homogenously into the bilayer, reflected by the shoulders appearing in the calorimetric peaks. In conclusion, the DSC technique, in combination with confocal fluorescence microscopy, SEM and AFM techniques highlighted a bilayer consisting mainly of DPPC that differentiates in the number of bilayers and/or in the kind and amount of the non-lipid compounds inserted (some monomers of Tween), accompanied with a mixed micelle nucleus that is formed by Tween-Span, as well as by some amounts of DPPC [28].

3.4 *Chimeric Carbon Nanotubes-Liposomal Nanosystems*

Carbon nanotubes are considered to be an emerging, well promising category of drug delivery nanosystems, due to their high aspect ratio, chemical stability, robustness, high drug carrier capacity, as well as their ability to penetrate cell membranes. The combination of carbon nanotubes and liposomes can provide an innovative chimeric nanosystem that presents advantages of both individual nanosystems, such as high loading ratio and targeting of carbon nanotubes embedded into biocompatible liposomes, as well as minimizes the potential tissue toxicity of carbon nanotubes.

Pippa et al. [29] achieved to create a platform, made of DPPC and HSPC (hydrogenated soya phosphatidyl choline) liposomes and multi-walled carbon nanotubes (MWCNTs), presenting different kinds of functional chemical groups. The DSC technique, performed by two cooling-heating cycles at 10–60 °C, with 20 °C/min and 2 °C/min scanning rates respectively, was used to describe the thermotropic alteration of lipid bilayers, caused by the incorporation of MWCNTs (Fig. 3). Depending on its chemical groups, each MWCNT had a quite different effect on thermotropic characteristics (ΔH_m, $\Delta T_{1/2}$, T_m), except for the pretransition, which was decreased at all mixed systems. The main transition temperature, which corresponds to the mobility of the acyl chains of phospholipids, remained unaffected for all the mixed lipid bilayers. The transition enthalpy ΔH_m and the system cooperativity (described by $\Delta T_{1/2}$) were altered depending on the different incorporated MWCNT, while there were strong interactions between MWCNT and polar groups of phospholipids. The orientation of the polar groups was also changed. Both HSPC and DPPC lipids presented more or less the same thermotropic profiles, as well as new metastable phases between the gel and the liquid crystalline phases being induced from the incorporated MWCNTs. Authors also concluded that the thermotropic characteristics can also indicate the localization of the MWCNT into the membrane, being oriented in some cases parallel to the membrane and located at the center of lipid bilayers, confirmed also by fluorescence spectroscopy results [29]. Regarding the new metastable phases, which were observed by the DSC technique, are indicative of the thermal stability of the prepared nanosystems and must necessarily be recognized and taken into account during the development of the liposomal nanosystem.

3.5 Chimeric Graphene Oxide-Liposomal Nanosystems

Graphene oxide (GO) is a single layer of graphite with unique thermal, electrical and optical properties. Hence, GO with a large number of surface functional groups on the surface, low production cost and acceptable thermal conductivity, can be considered as a versatile stimuli-triggered candidate for photothermal therapy [30, 31]. Drug can be incorporated into the GO that can also interact with liposomes (GO is usually absorbed onto the liposomal surface), in order to produce advanced chimeric drug delivery systems that present both the advantages of the GO ant the biocompatibility of the liposomes [32, 33]. Recent literature describes the development of a chimeric nanosystem (LBL Lipo-graph), composed of alternating layers of GO and GO conjugated poly-L-lysine (GO-PLL), coating the surface of a thermosensitive cationic liposome from DPPC, Brij 78, DOTAP and cholesterol, containing doxorubicin as a core, using a layer-by-layer assembly method mediated by ionic interaction. DSC measurements were carried out with a heat cycle from 0 to 100 °C, at a rate of 10 °C/min, in combination with TGA application, in order to evaluate the thermotropic behavior of the prepared nanosystems. The main transition temperature T_m was found to be dependent on the materials that were gradually added to the formulation and more specifically, DOTAP and cholesterol shifted the phase transition temperature to higher values, while Brij 78 decreased the phase temperature to lower values. The phase transition temperature slightly shifts to higher values, following layer-by-layer deposition. The increase of the number of layers resulted in main transition peak broadening, exhibiting the deposition of thin layers of complementary materials on the liposome surface. Authors suggested that the second peak observed in the DSC thermogram may be related to the total dissolution of the

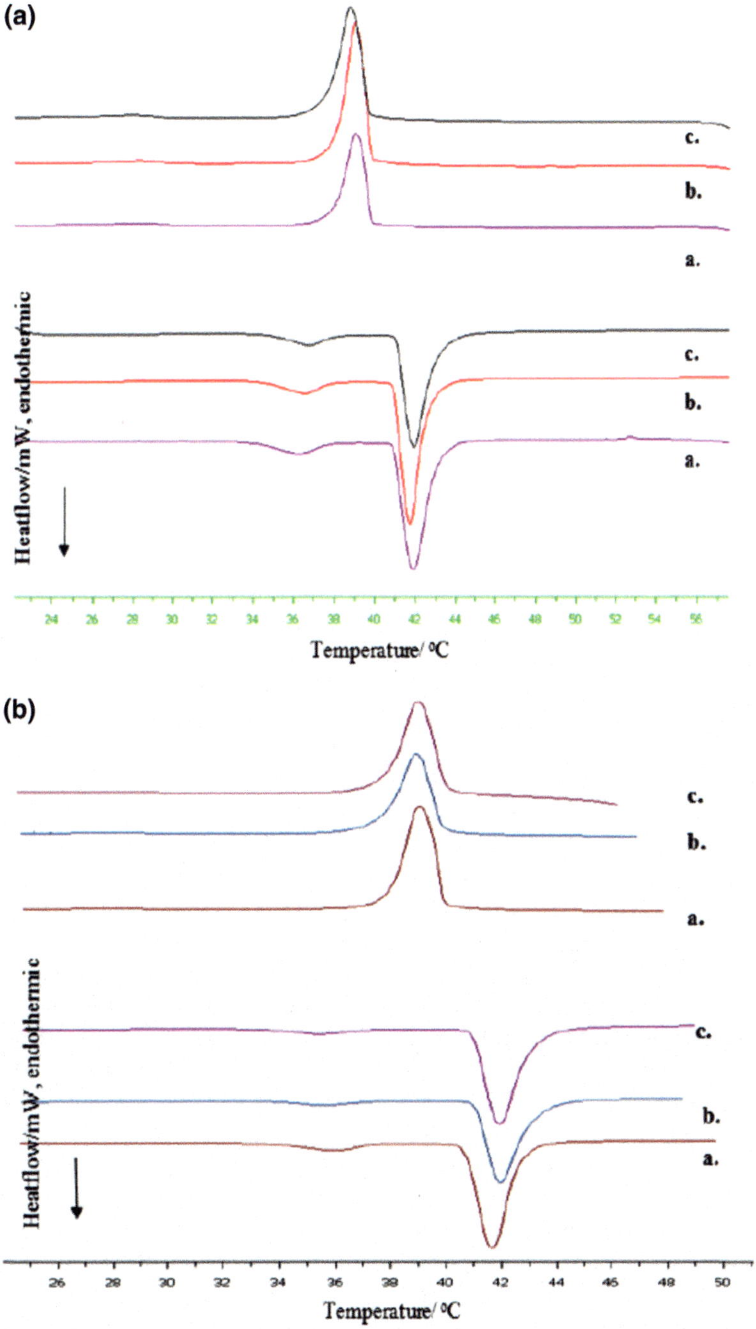

Fig. 3 DSC thermograms. The heating and cooling scans of (**a**). **a**. DPPC, **b**. DPPC/2a and **c** DPPC/2b (2a, 2b are anionic MWCNTs materials) and (**b**). **a**. DPPC, **b**. DPPC/5a and **c**. DPPC/5b lipid hydrated bilayers (5a, 5b are cationic MWCNTs materials). The hydration medium is HPLC-grade water. Adapted from Ref. [29]

liposomal compartment. TGA results described the weight loss upon the temperature increase and so the thermal stability of the nanosystems. One of the main advantages of the prepared nanosystems was the decrease of the initial burst release, which may result in the reduction in systemic toxicity, increased pH responsiveness around the tumor environment and improved NIR light absorption [34].

3.6 Chimeric Dendrimer-Liposomal Nanosystems

Dendrimers belong to a well promising class of highly branched polymeric biomaterials, which were first synthesized in the mid-1980s by Tomalia's group. They present high degree of molecular uniformity, narrow molecular weight distribution, specific size and shape and a highly functionalized terminal surface. Dendrimers are built from a central core molecule, which is surrounded by layers of branched monomers so dendrimers adopt globular shape with a densely packed surface and empty internal cavities [35]. The firstly synthesized, most used and most investigated dendrimers are the poly(amidoamine) (PAMAM) ones. Concerning to their utility, dendrimers have considerable medical applications, such as forming drug delivery systems with high loading capacity, whereas they can be combined with liposomes to provide modulatory liposomal controlled release systems (MLCRSs). Due to their hydrophilicity, dendrimers are localized either on the surface of liposomes, interacting with the head groups or inside the aqueous liposomal core, producing liposomal locked-in dendrimers (LLDs) that can be classified as chimeric advanced Drug Delivery nano Systems (chi-aDDnSs) [36].

A rational design of a novel chimeric system that combines dendrimers and liposomes presupposes some preformulation studies, in order to reveal the different interactions between dendrimers and lipid bilayers, thus the DSC technique has been proven to be a useful tool. Literature describes how the perturbation of dendrimers within the liposomal lipid bilayers depends on the different lipid structure. Briefly, phosphatidylethanolamine (PE) vesicles are disrupted, but phosphatidylcholine (PC) vesicles and even mixed vesicles with a high fraction of PC are protected from disruption, while PAMAM preferentially interact with lipids that present a greater negative curvature tendency. Klajnert and Epand [37] used DSC in order to investigate the interactions of increasing concentrations of PAMAM generation 3 (G3) and DPPC and DMPC (1,2-dimyristoyl-*sn*-glycero-3-phosphocholine) membranes. Samples were heated and cooled repeatedly five times at a rate 30°/h between 15 and 65 °C in the case of DPPC MLVs, and between 5 and 40 °C in the case of DMPC SUVs. There was also a delay of 5 min between sequential scans to allow for thermal equilibration. Analytically, DSC results confirmed that the lowest dendrimer concentration altered neither the main phase transition, nor the pretransition of DPPC and DMPC membranes significantly, but the increased PAMAM dendrimer concentration resulted in a slight broadening of the main transition peak, with a decrease in the pretransition enthalpy. It should be noted that the pretransition enthalpy was significantly lower compared to the dendrimer-free samples. Authors attributed the pretransition decrease to the fact that the dendrimer, inserting as a foreign molecule into the bilayer, increases the spacing between phospholipid molecules and eliminates steric crowding of the PC headgroup, thus eliminating the driving force for the formation of a ripple phase. Moreover, experimental results showed that the method of preparation of dendrimer–liposome systems had a great impact on the final product [37].

More recent literature describes also the interactions of PAMAM generation 4 (G4) and 3.5 (G3.5) and DPPC membranes, using a similar DSC protocol (three to four heating scans from 10 to 60 °C, using a scanning rate of 2 °C/min), combined with Raman spectroscopy. Increased concentrations G4 and G3.5 also eliminate the pre-transition and lower the ΔH, while G4 interacted more intensively than G3.5. The maximum percentage of PAMAM G4 dendrimer that can be incorporated in the DPPC bilayers is 5% and the respective percentage of PAMAM G3.5 is 3%. Authors combined DSC with Raman spectroscopy results and concluded that dendrimers promote endocytosis of the chimeric liposomes through cell membranes and not through disruption, depending on the dendrimer concentration. So the dendrimer not only affects the thermal behavior, but also the biological and the nanosystem-body interaction [36].

Except for dendrimer-lipids interactions, there are also electrostatic interactions between dendrimers and drug molecules. Gardikis et al. [38] tried to investigate the strong interactions being induced by PAMAM G3.5 on DOPC (dioleoylphosphatidylcholine) membranes and improve the pharmacokinetic and pharmacological profile of doxorubicin (DOX) by its encapsulation into DOPC:DPPG liposomes. The electrostatic complex between G3.5 and DOX, which was formed in the containing liposomal medium pH, promoted higher encapsulation efficiency and drug:lipid ratio. Authors attributed this fact to the fluidization effect being caused by low G3.5 concentrations, which enables molecular rearrangements inside the bilayer, and phase separation/displacement of DOX:G3.5 complexes inside the hydrophilic liposomal core. Dendrimer incorporation yielded more controlled drug release and increased therapeutic index in cancer cell lines and as a result, to a more effective DOX formulation [38].

More recently, the interactions between positively charged fifth generation (G5) PAMAM dendrimers and DPPC based liposomes were studied. G5 PAMAM dendrimers affected the pretransition significantly, without changing the main transition temperature, indicating that dendrimers interact with the headgroups of the phospholipids, while the enthalpy decreases, revealing an interaction between the lipid headgroups and the dendrimers. The temperature of the main transition peak did not change, but the $\Delta T_{1/2}$ was increased, causing a loss in cooperativity. However, increasing the ratio of the dendrimers resulted in significant alterations (the above observations were more intensive), thus implies a strong interaction between dendrimers and lipids with a concentration-dependent manner. The used DSC protocol was three scans from 25 to 70 °C, with different scanning rates 1 °C/min at the first time, then 0.5 °C/min during the heating and 1 °C/min during cooling period and the samples were incubated for 30 min at 25 °C [39].

A new Liposomal-Locked in-Dendrimer (LLD) was also developed by Gardikis et al. [40] using DPPC:DPPG (DPPG: Dipalmitoyl-phosphatidyl-glycerol) and PAMAM 3.5, where eventually incorporated Doxorubicin (DOX). DSC analyses were performed on reference multilamellar lipid bilayers (MLBs), on MLBs with G3.5, on conventional liposomes and eventually on LLDs incorporating each one of the three dendrimers. After DOX loading DSC analysis was repeated in all systems, in order to reveal the physicochemical interactions and thermotropic effect induced by the bioactive substance on the system. The results showed that dendrimer introduction caused a slight decrease of T_m, a significant decrease of the transition enthalpy and a reduction of the transition cooperativity with respect to the reference system. Authors concluded that the lipids-dendrimer interaction trigger a phase separation, with dispersion of dendrimer patches within the lipid bilayer that negatively affect the overall cooperativity of the transition and the stability of the system. On the other hand, the presence of DOX in the DPPC:DPPG/PAMAM MLB seemed to recover the destabilization effects of the PAMAM in the DPPC:DPPG MLB, concerning the transition enthalpy and cooperativity, reducing the dispersion of dendrimer patches within the lipid bilayer. Similar to the previous study referred above, there is an electrostatic interaction between DOX and PAMAM, possibly between the protonated amino group of the anthracycline and the charged carboxyl group of PAMAM, thanks to the favorable pH liposomal conditions. In addition, DOX seems to push phase separation to its extreme consequences, namely, the segregation of the dendrimer out of the lipid bilayer, i.e. toward the aqueous side of the interphase, thus allowing the arrangement of lipid molecules in a less disturbed bilayer structure, enhancing eventually membrane stability and transition cooperativity. As a result, these LLD systems can be characterized as a well promising drug delivery system [40].

Another chimeric system was produced from generation 1 and 2 hydroxy-terminated dendrimers (PG1 and PG2) that were locked in DOPC:DPPG liposomes, which were subsequently loaded with DOX. DSC analysis revealed a lowering of the T_m of the membrane due to the incorporated dendrimers that induce a more fluid lipid bilayer, reflecting a higher mixing affinity of the dendrimers in the liquid crystalline phase. Combination of DSC and fluorescence spectroscopy results revealed that the interaction between lipids and dendrimers is of entropic nature, meaning that there is no bond formation between lipids and dendrimers but the interactions are of steric nature and are favored in the interior of the acyl chains. PG2 dendrimer at high concentrations induces a clear phase separation of the lipid bilayer due to its larger size, probably leading to higher encapsulation percentage into the more thermodynamically favorable aqueous interior of a unilamellar liposomal vesicle. Referring to DOX loading, the complex formed by dendrimers and DOX, while crossing the membrane, finds a more thermodynamically stable location inside the interior of the liposome leading to high drug loading and slower release compared to pure liposomes [41].

From an alternative point of view, the DSC technique can be combined with scattering techniques and fractal analysis, in order to fully characterize chimeric nanosystems and highlight their physicochemical, morphological and thermodynamic properties. Pippa et al. [42] managed to evaluate DPPC:PAMAM G4 9:1 liposomes by the above combined techniques and determine their physicochemical and thermodynamic interaction that explains their fractal morphology and colloidal stability. More specifically, DSC results of dendrimer containing chimeric liposomes showed a decrease of the specific enthalpy of the main transition $\Delta H_{m,}$ in comparison to conventional DPPC liposomes, which was correlated to the decreased "effective volume fraction" of lipid nanodomains. Additionally, absence of metastable phases of DPPC pure membranes was observed, confirming alterations of thermodynamic behavior by the dendrimer introduction and implying colloidal stability due to entropic phenomena, while the presence of such metastable phases is considered as "physical impurities" and incompatibilities that lead to aggregation phenomena. The authors concluded that DSC allows a better understanding of the physicochemical parameters, the interaction of drug delivery systems with biological media, as well as their behavior during the formulation process. Therefore, DSC should be included to the regulatory requirements during the design and the development of advanced liposomal drug delivery systems [42].

Another category of dendrimers, namely cationic phosphorus-containing dendrimers (CPDs) can also be introduced into liposomes, providing special surface properties that are strictly associated with the pharmacokinetic profile of the drug delivery system. Ionov et al. [43] used DSC in order to investigate the thermotropic behavior of chimeric liposomes when G3 and G4 CPDs were incorporated into uncharged DMPC and negatively charged DMPC:DPPG membranes. Regarding the DMPC membrane, both CPDs caused concentration-dependent alterations of the pretransition specific enthalpy, while the other calorimetric parameters of the pretransition, such as T_{onset}, T_m and $\Delta T_{1/2}$, as well as the main transition remained unaltered, implying interactions of the positively charged dendrimer end groups with the polar head groups of the phospholipid membranes but not with the lipidic chains. Contrariwise, at negatively charged DMPC:DPPG membranes there was an increase of $\Delta T_{1/2}$, implying a reduction of the cooperativity of the materials due to the reduction of the membrane rigidity. Moreover, $\Delta T_{1/2}$ values alteration of DMPC/DPPG lipid bilayers in the presence of CPD G3 were more expressed than in the case of CPD G4 [43].

Similarly, the influence of G3 and G4 CPDs was also investigated during their incorporation into DMPC and DPPC model membranes. DSC was combined with fluorescence anisotropy experiments in order to extract results about fluidity, anisotropy, and phase transition temperature and evaluate the interactions at both the hydrophobic and the hydrophilic region. DSC revealed changes of transition temperature and enthalpy values and abolishment of pretransition peak. More specifically, both G3 and G4 CPDs penetrated lipid membranes and interacted with both the hydrophobic lipid chains and the hydrophilic polar head groups of model membranes. Fluorescence anisotropy results showed that lipid membranes became more rigid, because dendrimers of both generations decreased membrane fluidity. Authors concluded that there was dendrimer concentration-dependent phenomena, influencing eventually the physical state of the membrane, as well as that dendrimers of higher generations interacted stronger with model membranes [44].

G3 and G4 CPDs were incorporated also into DMPC large unilamellar and multilamellar vesicles (LUVs and MLVs) that were studied through ζ-potential measurements and DSC analysis on lipid bilayers. CPD G3 and CPD G4 influenced DMPC bilayer in a different way (generation-dependent phenomena), as well as in a concentration-dependent manner. The presence of dendrimers enhanced the reduction of cooperativity but no phase separation was noticed even at high CPDs concentration. Both dendrimers shifted main transition enthalpy and phase transition temperature values, while pretransition was abolished, suggesting that dendrimers are localized near the polar head groups of the lipids and interact with the membrane either by hydrogen-bonding or electrostatic attraction. ζ-potential values variation and size values increase, being in agreement with the DSC results in bilayers, determined the formation of complexes between the dendrimers and liposomes, caused by the presence of the positive charged dendrimers in liposomal suspension [45].

Finally, a more recent research presents the influence of maltose-modified poly(propylene imine) (PPI) dendrimers on neutral DMPC and negative DMPC:DMPG liposomes. There were used three kinds of G4 PPI dendrimers, open shell glycodendrimers (OS) and dense shell glycodendrimers (DS), both modified with maltose residues, as well as unmodified PPI. Samples were subjected at two heating/cooling cycles at a range of 5–45 °C, at a 0.5 °C/min scan rate. PPI OS dendrimers increased the main transition temperature, but the change was dependent on the salt concentration of the medium. PPI DS dendrimers increased main transition temperature, but when the environment of interaction was enriched with salt, the transition temperature decreased, as PPI OS also did. DSC results, in combination with spectrofluorimetry, indicated that the DS dendrimer induced higher perturbations in the lipid bilayers than the OS and unmodified dendrimer at the same concentration. The nature of interactions seemed to be mostly entropic and depends on the exclusion volume of the dendrimers. Results from DSC and fluorescence study were also in agreement with DLS results, suggesting that marginal changes in size dimension give first indication on weakly non-covalently driven interactions of neutral dense shell glycodendrimers against neutral and anionic liposomes. Authors concluded that maltose-modified PPI dendrimers may have potential use as a new type of drug delivery system [46].

3.7 Chimeric Polymer Grafted-Liposomal Nanosystems

3.7.1 Stealth Polymer Grafted-Liposomal Nanosystems

PEGylated Liposomes

PEGylation provides to liposomes stealth properties, suppressing the interaction with blood components, decreasing protein adsorption, and preventing the uptake (opsonization) by the macrophages of the reticulo-endothelial system (RES), leading eventually to higher bioavailability and therapeutic efficiency. However, it should be highlighted that liposomes must be completely coated with PEG which is not always achievable during liposomal preparation.

Paolino et al. [47] attempted to prepare super-stealth liposomes with the aim of increasing the localization of PEG moieties on the liposomal surface. They used DSC to study the degree of PEGylation of the PEGylated liposomes and the kinetics of the thermodynamic balance between PEGylation and de-PEGylation. Pre-formed DSPE-PEG_{2000} micelles can self-assemble with non-stealth DSPE/DSPC liposomes, forming long circulating PEGylated liposomes. Kinetic experiments were performed by co-incubating DSPE-PEG_{2000} micelles with non-stealth DSPE/DSPC (1,2-Distearoyl-sn-glycero-3-phosphoethanolamine/ 1,2-Distearoyl-sn-glycero-3-phosphocoline) liposomes at different molar ratios. The analysis was carried out by setting up the scanning program which was run in the range from 4 to 85 °C, an isothermal segment at 85 °C for 1 h and a cooling run from 85 to 4 °C. The DSC kinetic experiments demonstrated that the interaction between DSPE-PEG_{2000} micelles and non-stealth DSPE/DSPC liposomes is a time—(after 4 h incubation) and molar-ratio-dependent process (up to 0.1:1 DSPE-PEG_{2000}: phospholipids), which affects the post-PEGylation of liposomes. More specifically, the DSPE-PEG_{2000}micelles and the DSPE/DSPC liposomes presented no significant alterations of their thermotropic characteristics up to 4 h of incubation. A large peak at 27 °C showed up after 4 h of incubation, thus evidencing the presence of mixed micelles, which were not included in the liposomal bilayer, while the above peak at 27 °C disappeared at a DSPE-PEG_{2000}: DSPE/DSPC molar ratio of 0.1:1. Authors concluded that 0.1:1 ratio is the maximum amount of DSPE-PEG_{2000} that can be dispersed in the liposomal bilayer, without the formation of mixed micelles. The DSPE-$PEG_{2000}$was fully incorporated within the liposomal bilayer below its CMC value. The above proves the utility of the DSC technique at improving the biopharmaceutical profile of stealth liposomes [47].

Pluronic Block Copolymer-Grafted Liposomes

Pluronics are amphiphilic triblock copolymers with a central hydrophobic poly(propylene oxide) (PPO) chain and two flanking hydrophilic poly(ethylene oxide) (PEO) chains on each side. Pluronics can be applied to sterically stabilize liposomes, providing them with stealth properties and prolonged circulation time in the blood, just like PEGylated liposomes, but less expensively. Liposome formation in the presence of Pluronics consists of membrane-spanning conformation of the PPO block, while the PEO blocks are localized at both sides of the bilayer.

Hädicke and Blume [48] used DSC technique to systematically study the interaction of three different kinds of Pluronics (specifically F127, F88, and F87) with DSPC and DMPC liposomes. In all experiments, the heating and cooling rate was 1 °C/min and a time resolution of 4 s/data point was used, while the lipid concentration in the calorimetric cell was always 5 mM and the Pluronic concentration varied from 0.05 up to 5 mM, depending on the desired lipid:Pluronic mixing ratio of 100:1, 10:1, and 1:1. Preformed liposomes composed of DMPC or DSPC were incubated with polymer solution of Pluronics F87, F88, or F127, respectively. The relation between the lipid main transition temperature and the Pluronic micellization temperature, as well as the length of the PEO and PPO blocks and their ratio in the copolymer were found to influence strictly the lipid-Pluronic interaction.

If the main phase transition of the lipid occurs before the micellization of the Pluronic, such as in the case of DMPC lipid, the vesicles are in the liquid crystalline state and the membrane fluidity is high enough to enable the incorporation of single unimers of the Pluronics. As long as the polymer concentration within the chimeric liposomes is concerned, there was very low incorporation of the low amounts of polymer (100:1 and 10:1 mixtures) into the DMPC bilayer. This phenomenon is reflected in the absence of differences in the DSC curves of pure and chimeric bilayers, thus added Pluronics only adsorb on the surface of the vesicles at low concentrations. Contrariwise, equimolar amounts of polymer (1:1 mixture) caused much larger alteration of the lipid vesicles thermotropic behavior. Due to the higher concentration, the interaction of the polymers with the liposomes was increased and changes in the DSC peaks of the DMPC lipid phase transition were obvious. Analytically, the pre-transition occurred at a higher temperature but with lower transition enthalpy, compared to pure vesicles, indicating a destabilization of the ripple gel phase. The broadened main phase transition of the chimeric vesicles reflects PPO block perturbation and possibly a partial phase separation into structures with different Pluronic content. More specifically there were probably three types of aggregates formed: lipid bilayer aggregates with low amount of incorporated Pluronics, mixed polymer micelles with incorporated lipids, and pure Pluronic micelles. The length of the PEO chains has a minor influence because they are facing the bulk water.

Concerning the length of PPO block, Pluronic F127, which consists the most elongated PPO block among the three other Pluronics, destabilized the gel phase of the DMPC lipid and resulted in a slight impurity-induced reduction of the transition temperature regardless of its concentration at the mixture, implying also a greater penetration into the membrane. The large reduction of the transition temperatures at 1:1 mixture and the slight reduction of the transition enthalpy at 100:1 and 10:1 mixtures indicated an enhanced incorporation of the Pluronic F127 compared to F87 and F88.

In contrast to DMPC lipid, the main phase transition of DSPC was unaffected upon addition of the selected Pluronics, while the gel to liquid crystalline transition occurred unchanged with the same high cooperativity, at 54.5 °C, without any peak broadening, even at high polymer content. Authors concluded that this observation takes place, because the micellization of the polymer takes place at a lower temperature compared to the DSPC main transition, and so the polymer prefers the aggregation to micelles instead of incorporation into gel-phase lipid bilayers. Most of the polymer molecules are trapped into micelles, reducing the effective concentration of separated polymer molecules which could interact with liposomes. However, some Pluronic molecules are inserted into the bilayer causing a reduction and broadening of the pretransition at high polymer concentration [48].

3.7.2 Other Block Copolymer-Grafted Liposomes

The incorporation of triblock copolymers into liposomes, such as Pluronic polymers referred above, can provide to the vehicles a steric barrier and longer $T_{1/2}$ in human plasma. Replacing the common used Pluronics with other block copolymers that present amphiphilic character can be a smart idea towards innovative chimeric drug delivery nanosystems. In the next examples of block copolymer grafted liposomes, the performed DSC protocol was two cooling-heating cycles at 10–60 °C, with 20 °C/min and 2 °C/min scanning rates respectively.

Pippa et al. [49] managed to incorporate the block copolymer poly(oligoethylene glycol acrylate)-b-poly(lauryl acrylate) (POEGA-PLA) into DPPC and HSPC liposomes. POEGA-PLA block copolymer provides long hydrophobic side chains in the PLA block which can anchor inside the lipid bilayers, whereas the POEGA block with the short oligoethylene glycol side chains provides a denser PEG corona outside the bilayer, as well as an environment similar to branched polysaccharides at cellular surfaces. DSC was applied to highlight the alterations of lipidic bilayers thermotropic behavior being caused by the polymer incorporation in two aqueous dispersion media. Indeed, copolymer changed the self-assembly/organization, the fluidity and cooperativity within the bilayer. More specifically, the pretransition seems to be sensitive to the presence of POEGA-PLA especially at the highest molar ratio of the polymeric guest. The pretransition peak broadened significantly, especially for the HSPC chimeric lipid bilayers as the POEGA-PLA molar ratio increased, indicating absorption of polymer chains to the head group region of the bilayer. Moreover, the strong alteration of ΔH_m means affection of the mobility of the lipids polar head groups, while the lack of sharp transitions indicates that the POEGA-PLA is not uniformly distributed in the DPPC/HSPC bilayers, leading to polymer-rich and polymer-poor domains, reflecting the kinks being observed on the chimeric membranes by cryo-TEM technique. DSC results were found to be also in agreement with scattering techniques, where scattering intensity indicated incorporation of PLA chains into the lipid bilayers. The authors also concluded that the different amount of the released ibuprofen should be attributed to the different thermal and inherent organizational/morphological properties of the HSPC and DPPC/POEGA-PLA chimeric liposomal membranes [49].

Taking into account other examples, the polymeric guest is responsible for the new metastable phases, which strongly affect the physicochemical properties and the drug release profile. The incorporation of two different gradient copolymers poly(2-methyl-2-oxazoline)-grad-poly(2-phenyl-2-oxazoline) (MPOx) into DPPC liposomes reoriented the lipid bilayers from solid–gel to liquid crystal phase, reflected by an increase of the temperature of main transition (T_m). Increasing quantities of MPOx in DPPC membranes resulted to the additional broadening of the main transition, as indicated by the $\Delta T_{1/2}$ values, which suggests a decreased cooperativity of biomaterials in those systems. The macromolecular architecture of the polymeric guest influences oppositely the cooperativity of the system's components as the mass% hydrophobic component increased, as well as the new metastable phases being produced. The MPOx 2, being more hydrophobic, led to relatively

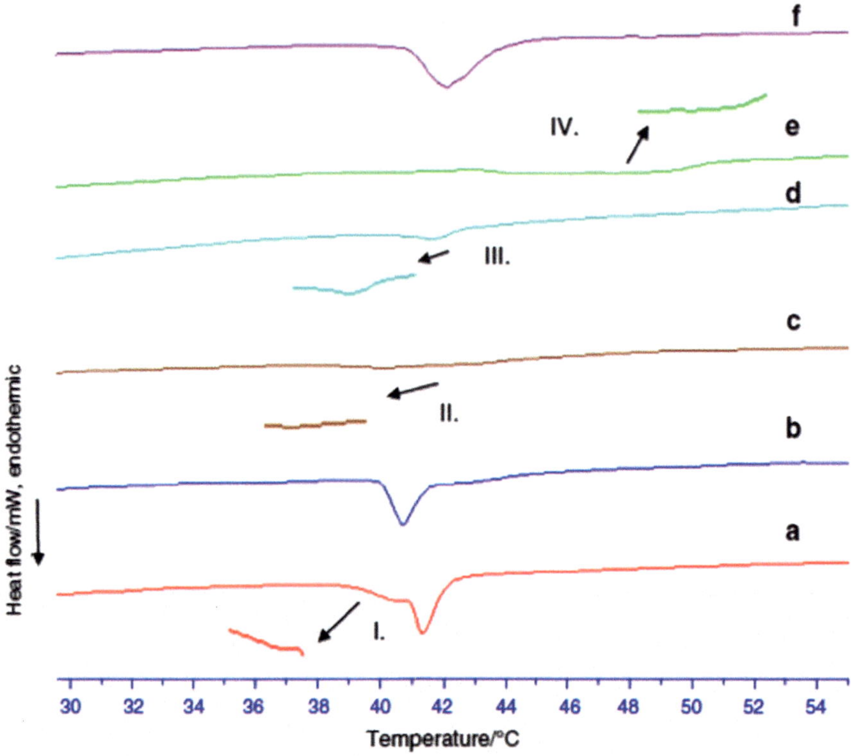

Fig. 4 DSC heating scans of **a** DPPC, **b** DPPC:DPPE-PEG 3000 (9:1 molar ratio), **c** DPPC:PEO-b-PCL 1 (9:1 molar ratio), **d** DPPC:PEO-b- PCL 2 (9:1 molar ratio), **e** DPPC:MPOx-1 (9:1 molar ratio), and **f** DPPC:MPOx-2 (9:1 molar ratio) liposomes. Adapted from Ref. [50]

more decreased cooperativity, while MPOx 1 caused a metastable phase (Fig. 4d III). From the other hand, the new metastable phase (probably an interdigitation that was revealed from the hysteresis being observed between cooling and heating curves) can act as release modulator and more specifically can accelerate the release of amphiphilic molecules such as indomethacin. Moreover, the interdigitation that can also be characterized as "kinetically trapped", yields entropic stabilization of the colloidal dispersion through hydration forces. The above DSC results were associated with fractal studies, which confirmed increased rigidity and curvature due to the new metastable phases.

Similarly, the incorporation of two different block copolymers of poly(ethylene oxide)-block-poly(ε-caprolactone) (PEO-b-PCL) into DPPC liposomes also changed the membrane conformation, fluidity and cooperativity, leading to altered thermodynamic profile. Both PEO-b-PCL block copolymers utilized broadened the main transition peak, which was accompanied by a small reduction of the main transition temperature and a significant reduction of the enthalpy, related to liposomal bilayers fluidization effects. PEO-b-PCL 2 incorporation led to a "solution-like model", while PEO-b-PCL 1 formed "raft-like" nanodomains, which were characterized as metastable phases (Fig. 4c II), "working as a zipper", modulating drug release kinetics. Thus, the release of the contained molecules from liposomal systems can be controlled by modulating the thermotropic characteristics of liposomes, as well as can be predicted by the application of the DSC technique, during preformulation stages of liposomal system development [50].

3.7.3 Stimuli-Responsive Polymer Grafted Liposomes

Stimuli-responsive forms of liposomes have been developed in order to overcome serious disadvantages of the conventional liposomes, such as their rapid recognition and uptake by reticulo-endothelial system (RES) that reduce their plasma half-life, their degradation by lysosomal enzymes, after their endocytosis, as well as their content leakage that sometimes happens before the target tissue. Moreover, the incorporation of stimuli-sensitive biomaterials, such as stimuli-responsive polymers, upgrades liposomes to advanced chimeric nanosystems that respond to different kinds of environmental stimuli by changing their conformation. As a result stimuli-responsive liposomes exploit the environmental conditions of the pathological tissues, presenting increased targeting and decrease of toxic side effects. The stimulus can be endogenous, presenting pathological alterations in comparison to physiological environment, such as pH and temperature alterations, as well as enzyme concentration or exogenous stimuli, which can be applied from external sources, like ultrasounds, local hyperthermia and magnetic field [51–53].

The polymer chain conformational changes that were referred above can be thermodynamically analyzed. In the case of well investigated amphiphilic stimuli responsive polymers, the hydrophilic stimuli responsive segment stretches out of the liposomal membrane and protects liposomes from aggregation (entropic/corona effect), providing physicochemical and biological stability in physiological conditions. When the stimulus is applied, polymer chains shrink and cause destabilization of the liposomal membrane and thus result in release of the content. On the other hand, the hydrophobic polymer segment is usually anchored inside the bilayer, causing perturbation and affecting the cooperativity, the thermal and the physicochemical behavior, as also described in the literature discussed below. The effect of the polymer incorporation either before or after its change in conformation should by investigated and taken into account during the design and development of liposomal systems. Thus DSC has proved to be useful for this strategy as the following literature examples indicate. The DSC protocol that had been used in the next paragraphs is the performance of two cooling–heating cycles from 10–60 °C with 2 or 5 °C/min scanning rate, taking into account the second run for obtaining the calorimetric data. Before each cycle, the samples were subjected to a constant temperature of 20 °C to ensure equilibration.

Several thermoresponsive polymers exhibit lower critical solution temperature (LCST) in aqueous solutions. That indicates water-solubility below the LCST. Increasing temperature through LCST causes coil-to-globule conformational phase transition, so above the LCST the polymers become less hydrophilic, as a result of breakage of hydrogen bonds existing between the polymer segments and water molecules. Conjugation of thermoresponsive polymers onto liposomes provides temperature sensitive properties to the liposomes. Below the LCST, highly hydrated polymer chains stabilize liposomes, protect them partly from blood proteins, improving the circulation time in the blood stream and prevent the association of the liposomes. Above the LCST, polymer chains gradually undergo dehydration and contraction. This results in the augmentation of liposome's surface hydrophobicity and permeability (destabilization of liposomes), as well as release of their contents but only at pathological hyperthermic tissue or alternatively in tissues where the external applied hyperthermia indicates [54–56].

PNIPAM [poly(N-isopropylacrylamide)] is considered to be one of the most popular stimuli responsive polymers that has dual properties of both thermo and pH responsiveness and exhibits LCST at around 32 °C (close to human temperature) that make PNIPAM ideal for biomedical applications, such as the development of drug delivery systems. At temperatures near its LCST, PNIPAM undergoes a spontaneous reversible and endothermic, coil-to-globule transition driven by entropy gain. At temperatures below LCST, the acrylamide groups of the block interact via hydrogen bonding with the water molecules of the solvent, enhancing the solubilization of the polymer chains. Above the LCST, the hydrogen-bonded network is instantaneously disrupted resulting in the release of the water molecules out of the polymer chain, while the intramolecular polymer interactions become more dominant [57, 58].

Pippa et al. [59] used PNIPAM block to develop a functional thermoresponsive amphiphilic homopolymer and more specifically $C_{12}H_{25}$-poly(N-isopropylacrylamide)-COOH ($C_{12}H_{25}$-PNIPAM-COOH), whose hydrophilic segment is the PNIPAM and the lipophilic is the hydrocarbon end group $C_{12}H_{25}$. Two different molecular weight polymers (PNIPAM 1 and 2), (different degree of polymerization of the PNIPAM part) were synthesized by the reversible addition fragmentation chain transfer radical polymerization technique (RAFT). Subsequently, authors managed to create thermoresponsive chimeric liposomes by incorporating the two different forms of the above homopolymer to DPPC liposomes, where they also incorporated the hydrophobic drug indomethacin (IND). Polymer used the $C_{12}H_{25}$ to be anchored into the liposomal membrane, while the end-functional group of –COOH, made it sensitive also to pH alterations.

The DSC technique was used to evaluate the thermotropic behavior of the thermoresponsive nanosystems. More specifically, the incorporation of PNIPAM 1 (20 kDa) in DPPC liposomes affected only the specific enthalpy ΔH_m of the main transition. The specific enthalpy ΔH_m of the main transition was increased for the DPPC:PNIPAM 1 chimeric liposomes at low and medium molar ratios, and was decreased for the DPPC: PNIPAM 1 chimeric liposomes at the two highest molar ratios. The main transition temperature T_m was slightly reduced, while the system cooperativity was reduced at a concentration-dependent manner (severe decrease at the highest polymer ratios). It was very interesting the fact that the incorporation of highest polymer amounts caused the appearance of new peaks (centered at approximately 30 °C), indicating the creation of new metastable phases. The authors concluded that these new metastable phases were the result of lateral phase separation into PNIPAM-1 rich and PNIPAM-1 poor domains. Afterwards, the lack of sharp transitions and the LCST of the polymeric guest indicate that the PNIPAM 1 is not uniformly distributed in the DPPC liposomal membranes. Taking into consideration the amphiphilic character and the molecular structure of PNIPAM 1, the biphasic nature of the transition enthalpy (ΔH_m) at high concentration of the polymeric guest and the hysteresis on cooling (of ~2 °C) that was clearly observed, the authors suggested the formation of an interdigitated phase.

As long as PNIPAM 2 (7 kDa) incorporation was concerned, specific enthalpy ΔH_m of the main transition was increased for the DPPC:PNIPAM 2 chimeric liposomes as the molar ratio of the polymeric guest increased, cooperativity decreased slightly for all systems, except for the highest polymer molar ratio, where it declined dramatically and finally, T_m decreased only for that system too. In contrast to the PNIPAM 1 systems, no new metastable phases were observed for the PNIPAM 2 systems, indicating that polymer characteristics, in respect to hydrophilic/hydrophobic parts constitution/sequence, composition and molecular weight, can play a key role in the thermal behavior of the whole nanosystem.

DSC results helped eventually at the explanation of the IND loading and release results. Indeed, PNIPAM 1 exhibited larger incorporation efficiency, immediate release of IND at 37 °C and no release at 32 °C. This temperature dependant "burst" release was in agreement with PNIPAM LCST. The authors stated that the new metastable phases centered at 30 °C may behave as promoters of the release of IND from the highest ratios of DPPC:PNIPAM 1, due to the inhomogeneous drug distribution inside the chimeric liposomal membrane at 37 °C. On the other hand, DPPC:PNIPAM 2 systems had lower incorporation efficiency and did not release IND at all [59].

Another interesting, widely investigated category of stimuli-responsive liposomes are the pH-responsive ones. The incorporation of different kinds of pH-responsive polymers, such as poly(acrylic acid)s (PAAs), succinylated PEG and N-isopropylacrylamide (NIPAM) containing copolymers, transforms conventional liposomes to pH responsive chimeric ones. In particular, pH-sensitive liposomes remain stable at physiological pH (7.4), but destabilize due to polymer change conformation and acquire fusogenic properties under acidic pH, due to their modulatory materials, containing "ionizable" chemical groups, such as amines, phosphoric acids and carboxylic acids, and changing dramatically their conformation in response to environmental pH alteration. There are many pathological states which are associated with different pH profiles from that of normal tissues, such as ischemia, infection, inflammation and tumors, where pH-responsive liposomes can provide effective and more targeted therapeutic effect.

Ones of the widely used pH-responsive polymers are the block copolymer containing the stimuli responsive block of the weak polyelectrolyte poly(acrylic acid) (PAA). The negative cargo of carboxylic group (COO^-), which PAA blocks exhibit at physiological pH, make liposomes repulse each other, preventing fusion and so destabilization, while its three-dimensional conformation provides steric stabilization. Contrariwise, when the carboxyl groups of PAA block are protonated in acidic pathological tissues, becomes more hydrophobic, shrinkage of polymer chains takes place and leads to destabilization of the membrane and content release [60–63].

Similarly to Pippa et al. [59] research study referred above, Naziris et al. [64] combined PAA homopolymers with hydrophobic $C_{12}H_{25}$ end groups using reversible addition fragmentation chain transfer radical polymerization technique (RAFT), in order to create a pH-responsive amphiphilic homopolymer, the $C_{12}H_{25}$-poly(acrylic acid) ($C_{12}H_{25}$-PAA) one. $C_{12}H_{25}$-PAA has been incorporated into HSPC bilayers and then the resulted liposomes were loaded with IND. The bilayers were fully hydrated with two different dispersion media, HPLC-grade water and PBS. PBS presents pH around 7.4 and ionic strength $I = 0.154$ N, simulating the conditions met within the human body, while HPLC-grade water with $pH = 4.5$ resembles the microenvironment existing within late endosomes and lysosomes. The incorporation of $C_{12}H_{25}$-PAA caused alterations of the thermotropic behavior of HSPC lipid bilayers, affecting extensively the specific enthalpy ΔH_m of the main transition in both dispersion media in a concentration-dependent manner. The pre-transition was eliminated at most of the systems, while cooperativity was decreased especially in HPLC-grade water. The main transition temperature T_m remained unaffected, indicating that the interactions of the PAA polymeric segment with HSPC lipids affect the mobility of the polar head groups of lipids. The temperature at which the thermal event starts ($T_{onset,m}$) changed significantly only for the systems of the highest polymer amount. Concerning the two dispersion media, no significant differences were observed at the calorimetric parameters of the systems. The authors suggested the formation of a new metastable phase, probably of an interdigitated phase, considering the amphiphilic nature of the homopolymer and the hysteresis being observed at the cooling curves. In more detail, the $C_{12}H_{25}$ group of the polymer penetrates into the interior of the bilayers, forming hydrophobic nanoclusters that are heterogeneously distributed and cause membrane disruption and phase separation, especially at the highest molar ratios where the metastable phase has been observed. As the thermal behavior is reflected to the IND release profile, the observed metastable phase probably acted as a drug release "barrier" for one of the chimeric nanosystems (which showed time delayed release of IND), while the in vitro release of IND is faster from the chimeric liposomes with the lower molar ratio of $C_{12}H_{25}$-PAA. These differences in the release rate, as well as at the total percentage of drug release were attributed to the pH-responsiveness of $C_{12}H_{25}$-PAA. DSC results indicate that even small differences in polymer concentration can cause severe alteration to the behavior of the system, due to different structural rearrangement of these systems that are characterized by high grade of complexity, resulting in different release kinetics and thus pharmacokinetic profile [64].

Apart from amphiphilic homopolymers, block copolymers can also be synthesized from stimuli responsive parts. For example, PNIPAM and PAA have been also combined in a dual-responsive diblock copolymer (i.e. PNIPAM-b-PAA, dual responsive to pH and temperature changes) and incorporated into DPPC bilayers. The above chimeric bilayers were thermodynamically investigated, regarding polymer composition, polymeric guest concentration, as well as the nature of the dispersion medium conditions (i.e., pH and ionic strength). Two copolymers having different compositions, PNIPAM-b-PAA1 and PNIPAM-b-PAA 2, with different weight ratio of the polymeric blocks (50 and 30% PAA by mass, respectively) were used in different molar ratios to lipid. Three different dispersion media (i.e., HPLC-grade water pH = 5.5, PBS pH = 7.4, hydrochloric acid solution of pH 4.5) were utilized, in order to investigate the role of dispersion medium on the thermal behavior. In detail, the incorporation of the diblock copolymer PNIPAM-b-PAA 1 caused alterations in the thermodynamic parameters in a concentration-depended manner in all three media. In HPLC-grade water, the pretransition disappeared, probably due to the copolymer penetration into the bilayer, the main transition peak was broadened, reflecting a severe decrease of system cooperativity, while shoulders also appeared at high temperatures and at high molar ratios. It was suggested that the PNIPAM-b-PAA copolymer interacts with the polar headgroups of membrane phospholipids and is heterogeneously distributed into the membrane, creating polymer rich and poor regions. Similar calorimetric shifts were also observed in PBS dispersion medium, but significant differences were observed in acidic medium, due to the PAA block pH responsiveness. The decrease of cooperativity was attributed to the fact that above LCST, PNIPAM block becomes less hydrophilic (coil-to-globule transition) and penetrates to the interior of the bilayers, forming hydrophobic clusters and finally causing membrane disruption (temperature-dependent perturbation). When PNIPAM-b-PAA 2 (with lower PAA content) was incorporated, there were also shifts at the calorimetric profiles. The most important observation for PNIPAM-b-PAA 2 containing systems was the creation of a new metastable phase, probably an interdigitated phase below the LCST temperature, reflected in the biphasic nature of the main transition and the suppression of the pretransition. The authors eventually stated that the study of dual-stimuli-responsive chimeric bilayers can act as a compass for the rational design of respective dual-stimuli-responsive chimeric liposomes that respond to both tissue pH and temperature alterations [65].

Stimuli-responsive block copolymers can also be synthesized from a stimuli-responsive hydrophilic part and a non-responsive hydrophobic part that is however very useful at polymer anchoring inside the liposomal membrane. Hydrophobic block usually perturbs liposomal membrane causing shifts of the calorimetric profiles, as well as changes of the physicochemical behavior. A well investigated example of an amphiphilic pH-responsive block copolymer is described below. PAA block was combined with poly(n-butylacrylate) (PnBA) at two different ratios (70 and 85% content of PAA) using RAFT polymerization to create the PnBA-b-PAA block copolymers. Subsequently, PnBA-b-PAA was incorporated in DPPC (Fig. 5) and HSPC bilayers that were evaluated by the DSC technique. DSC results were further associated with the physicochemical behavior of the respective chimeric liposomes that were prepared. At both kinds of systems, DPPC and HSPC, the presence of the polymeric guest caused the disappearance of the pretransition peak, suggesting that the interaction between PAA blocks and DPPC or HSPC lipids occurs on the membrane surface and that the driving force is the formation of hydrogen bonds between non-ionized polymer carboxyl groups and the lipids polar head groups. Concerning the calorimetric results, the specific enthalpy ΔH_m of the main transition decreased significantly at all molar ratios with a concentration-dependent manner and the temperature at which the thermal events tarts ($T_{onset,m}$) did not change significantly. The main transition temperature, T_m, remained almost unaffected indicating that the interactions of the PAA polymeric segment with DPPC or HSPC lipids affect the mobility of the polar head groups of the lipids. The cooperativity of all the systems decreased significantly, and especially at the highest molar ratio of the polymeric component. In addition, the lack of sharp transitions indicates poor cooperativity, while the presence of shoulders is attributed to the fact that PnBA-b-PAA is not uniformly distributed in the lipid bilayers, suggesting lateral phase separation into PnBA-b-PAA–rich and PnBA-b-PAA-poor domains. The extreme broadening of the main transition peak at the highest molar ratio was also related to bilayers fluidization effects (solution-like model). As long as polymer composition effect is concerned, the increase of the PAA segment (85%) of the copolymer, caused the creation of a new phase, probably an interdigitated phase that is indicated also by the hysteresis being observed at the cooling curves. In regards to the role of the hydrophobic block, the authors suggested that the PnBA block of the block copolymer penetrates into the interior of the bilayers, forming hydrophobic nanoclusters and finally causing membrane disruption and phase separation, especially at the high molar ratios of the polymeric guest. Nanoclusters are heterogeneously distributed to the bilayer lipid interior, forming highly concentrated PnBA regions and are responsible for the significant broadening of the main transition peak. In HSPC bilayers, the study was carried out using both PBS and citrate buffer (pH ~ 4.0), where the decrease of PAA content (70%) yielded bimodal distributions, suggesting the formation of a new phase, while the other calorimetric parameters did not show significant differences. When the block copolymer was incorporated into liposomes, the resultant chimeric vesicles were proved to be physicochemical stable over time due to steric effect of the PAA block, as well as the spacer effect of the PnBA block that can reduce the membrane tension. In conclusion, the systems presented remarkable pH-responsiveness at acidic environment [66, 67].

3.8 Chimeric Polysaccharide-Decorated Liposomal Nanosystems

Flexible liposomes, ethosomes and transethosomes are considered to be suitable carriers for the delivery of therapeutic agents in the deep layers of the skin, while they are capable of breaching the skin barrier, presenting major advantages such as slow, sustained release of the drug, thus allowing reduction of the frequency of administration and so the toxicity. In particular, flexible liposomes present increased bilayer fluidity, due to the addition of an edge activator, usually a surfactant, that promotes a stress dependent adaptability. Franze et al. [68] designed hyaluronic acid polysaccharide decorated flexible liposomes, suitable for enhancing the cutaneous administration of drugs, in order to enhance the skin penetration of the nifedipine drug. At one

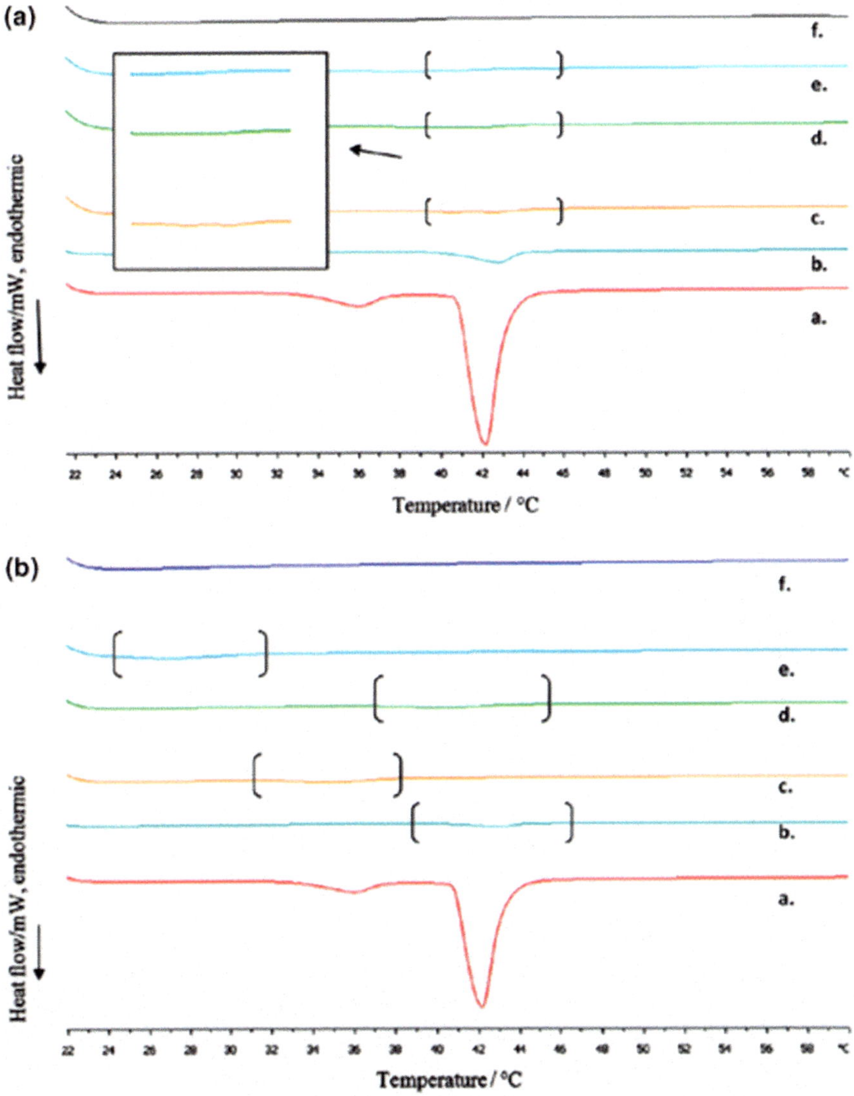

Fig. 5 DSC heating scans of DPPC:PnBA-b-PAA (**a**) 30/70 and (**b**) 15/85 of **a**. 9:0.0, **b**. 9:0.1, **c.** 9:0.5, **d.** 9:1.0, **e.** 9:2.0 and **f**. 9:3.0 molar ratio chimeric bilayers. The limits for the calculation of thermotropic parameters are from 25 to 45 °C. Adapted from Ref. [66]

hand, hyaluronic acid improves the dermal delivery of therapeutic molecules, but at other hand it makes the lipid bilayer stiff, altering the overall flexibility properties of the carriers. Thus, Franze et al. [68] investigated the counterbalance between using hyaluronic acid into the formulation and the stiffening of the bilayer. In particular, they used flexible liposomes, prepared from egg phosphatidylcholine (ePC) and Tween80®, where nifedipine was incorporated. The liposomes were decorated with hyaluronic acid conjugated to DPPE lipid. DSC was applied, using a heating cycle from −20 to 10 °C at the heating rate of 1 K/min, in order to evaluate the hyaluronic acid-induced alteration to the liposomal bilayer and DSC results were correlated with the mechanical properties that reflect the flexibility of the systems. First, the addition of Tween 80® destabilized the bilayer, by reducing its packing order and increasing the fluidity, a phenomenon that was reflected in the acute decrease of ΔH_m, accompanied with the presence of a shoulder that suggests the surfactant caused formation of different lipid domains, as well as by a small transposition of the pretransition. Subsequently, the nifedipine incorporation in the bilayer increased the main transition enthalpy, abolishing the pretransition and the shoulder referred above and reducing the cooperativity. Then, the addition of the hyaluronic acid-conjugate provokes full abolishment of the pretransition, accompanied with a sharper main transition, confirming the increased packing of the lipid domains. The latter observation is in accordance with flexibility studies that suggested a stiffening induced by hyaluronic acid. However, although hyaluronic acid stiffened the bilayers, it was also proved to be essential for an efficient penetration of the liposomes into the skin in a concentration-dependant manner. The authors concluded that the counterbalance between the polysaccharide presence and the bilayer stiffening should be established [68].

Except from skin delivery, hyaluronic acid polysaccharide decoration has been employed to increase the plasma half-life of liposomes, replacing the PEG polymer chains of PEGylated liposomes. Hyaluronic acid also increases liposomal targeting properties towards specific receptors against hyaluronic acid over-expressing them at the cell membrane, (such as the cluster determinant 44 (CD44) receptor, the receptor for hyaluronate-mediated motility (RHAMM), the hyaluronic acid receptor for endocytosis (HARE) and the lymphatic vessel endothelial hyaluronan receptor-1 (LYVE-1)). Cosco et al. [69] used hyaluronan, which is high molecular weight hyaluronic acid (sodium salt, 1500 kDa), either in pure form or conjugated to DPPE lipid, in order to prepare liposomes that can resist the effect of serum proteins. DSC revealed some shifts of the calorimetric patterns that were dose dependant. Samples were subjected to a heat cycle, from 10 to 70 °C at a rate of 5 °C/min. When liposomes were incubated with 1 or 3 mg of pure hyaluran they did not exhibit a change of the T_{onset} of the DPPC transition, neither a drastic change in its enthalpy. Contrariwise the increase of hyaluran amounts to 6 mg shifted the T_m towards higher temperature, with a decreased enthalpy, reflecting a possible adsorption of hyaluran onto DPPC bilayers and increased interactions with polar heads. In the case of hyaluran conjugated to DPPE at different concentrations, it induced a shift of the T_{onset} of DPPC (from 42 to 38 °C) for the highest amount of HA-DPPE (6 mg) with no change of the transition enthalpy of DPPC. Moreover, the DSC peak was irregularly broadened probably as a consequence of both hyaluran adsorption and DPPE insertion. Indeed, DPPE also influences the bilayer in this way, as it was confirmed when pure DPPE was inserted to the DPPC bilayer. The authors concluded that the use of hyaluran decoration altered the physicochemical and thermotropic behavior of liposomes and increased their plasmatic half-life, a feature that can be exploited towards avoiding the appearance of the ABC phenomenon due to multiple administrations of the PEGylated liposomes [69].

3.9 Chimeric Lipoplexes

Lipoplexes are prepared from the combination of cationic liposomes and negatively charged nucleic acids and are used for gene delivery in vitro and in vivo. During the design and the preparation of lipoplexes, the DSC technique has been proven to be very useful in order to evaluate several physicochemical parameters of the lipoplexes, such as the level of hydration at the lipid-DNA interface. For example, a protocol for the water quantification of lipoplexes samples has been presented in the literature by Hirsch-Lerner and Barenholz [70]. Briefly, a weighed amount of water is added, which is always less than the amount of lipids or DNA orlipoplexes. The sample in the sealed crucible is scanned from −30 to 30 °C, with a scanning rate of 2 °C/min. When there is an overlap, the heat enthalpy of the lipid phase transition is subtracted from the ice-water fusion heat enthalpy for the specific hydration calculations of the lipoplexes. The difference between calculated and measured hydration of lipoplexes (based on ice-water fusion heat enthalpy of 325 J/g obtained from a calibration curve) can be used to calculate the change in level of hydration due to lipoplex formation [70].

Apart from physicochemical parameters evaluation, recent literature mentions the DSC technique as a tool to rationalize the different biological behavior of lipoplexes. The incorporation of different non-lipid cationic components in liposomes can provide chimeric lipoplexes that present special structure or transfection activity. In the following, a DSC investigation is described, where the biological features of chimeric lipoplexes, being composed by 1,2-dimyristoyl-sn-glycerophosphocholine (DMPC) and stereomeric cationic gemini surfactants, are rationalized, in order to investigate their DNA condensation and transfection ability, their fusion with biological membranes, as well as the effects of cationic surfactant stereochemistry. Heating scans of micelles (assemblies of pure gemini surfactants), micelle-DNA complexes and lipoplexes and of mixtures were recorded at 1 °C/min and 5 °C/min rate, respectively. The changes in the thermotropic behavior of both lipoplexes and cell membrane models after mixing and incubating them at 45 °C were also investigated by DSC experiments, in order to study the nature and the kinetics of the interaction between cationic lipoplexes and zwitterionic or anionic cell membrane models. The results of DSC experiments confirmed that all lipoplexes interact with cell membrane models (alteration of the peaks relative to the cell membrane models and lipoplexes after gradual increase of incubation), while the kinetics and the mode of interaction depends on the stereochemistry of the gemini surfactant and on the cell membrane model (different ΔH values were observed). Moreover, the aggregation process of gemini surfactants is attributed to the transition from metastable hydrated semicrystalline phases (the so-called coagel phases) to gel phases (reflected by peak complexity), while the diasteromeric interactions of the different surfactant enantiomers with DNA involve differences in the ΔH values (increase of ΔH values, i.e. the associated with the transition of lipoplexes is higher with respect to the ΔH of the corresponding liposomes) of the complexes. The authors also concluded that the similar T_m values reflect a complex compensation of enthalpy and entropy factors and is attributed to the domination of electrostatic interactions over the van der Waals forces between alkyl chains as observed in other systems [71].

3.10 Liposomal Vaccines

Cationic liposomes composed of dimethyldioctadecylammonium bromide and trihalosedibehenate (DDA:TDB) are characterized as efficient adjuvants for TB subunit vaccines, potentiating a strong CD4+ T-cell response [72, 73]. Hamborg et al. [74] prepared a series of cationic liposomes composed of DDA, TDB and the zwitterionic phospholipid distearoylphosphatidylcholine (DSPC) in different ratios. CAF01 was named the cationic liposomal adjuvant consisting of no DSPC, while NAF01 was the neutral adjuvant formulation, in which DDA has been replaced with DSPC. There were also two intermediate ratios consisting of all the three different lipids. The authors used the above formulations in order to investigate their mechanisms of interaction with two equally sized, but oppositely charged highly pure and well-characterized model proteins α-lactalbumin and lysozyme. They used a number of analytical methods, including DSC to elucidate the effects of the interactions on the membrane thermotropic phase behavior, as well as the protein structure. They applied a scanning rate of 0.5 °C/min in the temperature range of 20–60 °C for the DDA liposomes, and from 20 to 75 °C for the DSPC-containing liposomes. While DDA presents one sharp phase transition around 46.7 °C, the formulation CAF01, presented two or more interconnected peaks between approximately 42 and 47 °C, suggesting that the two lipid components (DDA and TDB) are in homogeneously distributed, resulting in the appearance of DDA-rich and TDB-rich microdomains having different phase transitions. At the other hand, NAF01 presented a sharp peak with a T_m of 53 °C. Regarding the proteins effects in the thermotropic behavior of the formulations, the addition of α-lactalbumin yielded a more orchestrated transition at the CAF01, reflected by a narrow transition peak, while α-lactalbumin did not affect the NAF01 significantly. Contrariwise, lysozyme did not influence the thermotropic phase behavior of neither CAF01 nor NAF01. Liposomes consisting of ternary mixtures of DDA, TDB and DSPC presented an interesting phase separation, illustrated by a broadened main transition peak, that was completely separated in two distinct peaks around 42 and 55 °C, by the addition of α-lactalbumin, suggesting a complete separation of the DDA and DSPC domains. The above phenomenon was presented in the first heating scan, but it disappeared in the second one, indicating that DDA-enriched domains, in contrast to DSPC-enriched domains, lost their reversibility between the gel state and the liquid state. The authors, taking into account the experimental results from intrinsic fluorescence and Langmuir monolayer technique about the interfacial properties of the proteins and the membranes, suggested that the above DSC results indicate that α-lactalbumin, upon binding to the cationic head groups of DDA interacts, with the hydrophobic membrane interior of the cationic, DDA liposomes, presumably both in the gel state and certainly in the liquid state when the fluidity of the membrane is increased [74].

4 Conclusions

Through the careful examination of the above characteristic literature examples, we can conclude that the chimeric liposomal systems present a high level of structural and behavioral complexity and variety. Under suitable circumstances, their complexity yields a high level of functionality and subsequently, therapeutic efficacy. These circumstances require careful design of the nanosystems, taking into account all the interfered parameters, such as the types and the amounts of the different biomaterials, as well as the variety of interactions that they exhibit between them and towards the environment.

DSC analysis is able to decode the interactions taking place between the biomaterials inside a chimeric liposomal nanosystem, through characterization of the thermal behavior of the particular system. The thermal behavior may be considered as a projection of the posterior behavior of the system, such as its physicochemical stability, the cooperativity of the combined biomaterials, the release kinetics of its content, as well as the interactions of the nanosystem with plasma proteins and its biological stability. The properties of the chimeric liposomal systems, that are associated with changes of the thermotropic behavior and vice versa, are summarized in the Ishikawa diagram shown (Fig. 6). Regarding the parameters that have a great impact on the thermotropic behavior of chimeric liposomal formulations, as described in the previous paragraphs, those include the concentration and the characteristics of the non-lipid "foreign" molecules, such as their chemical structure, functional chemical groups, ionizable groups, polymerization degrees and sequence of monomers (in the case of polymers) and hydrophilicity/hydrophobicity balance. We should also note that the DSC protocols being performed should be carefully designed. According to the previous examples, a temperature range that includes the phase transitions of the lipids and the alterations inducted by the non-lipid biomaterials is required, as well as slow scanning rates that enable the revelation of the different biomaterial interactions.

As a conclusion, the upgrade of conventional liposomal nanosystems to advanced, chimeric and more functional ones is the means to more effective therapies with fewer side effects. DSC in turn has been proven to provide quick and reliable results on the behavior of these systems, even during the first stages of the preformulation studies.

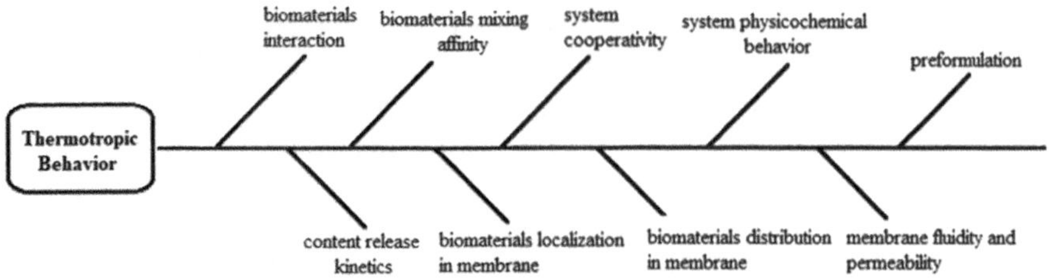

Fig. 6 Ishikawa diagram illustrating the information provided by the investigation of the thermotropic behavior of a liposomal formulation

Thus, DSC techniques can contribute to the rational design, the development and the optimization of the formulation, until reliable chimeric liposomal systems can be produced, in terms of safety and therapeutic efficacy.

References

1. Rowland, M., Noe, C.R., Smith, D.A., Tucker, G.T., Crommelin, D.J., Peck, C.C., Jr, Rocci M.L., Besançon, L., Shah, V.P.: Impact of the pharmaceutical sciences on health care: a reflection over the past 50 years. J. Pharm. Sci. **101**, 4075–4099 (2012). https://doi.org/10.1002/jps.23295
2. Demetzos, C., Pippa, N.: Advanced drug delivery nanosystems (aDDnSs): a mini-review. Drug Deliv. **21**(4), 250–257 (2014). https://doi.org/10.3109/10717544.2013.844745
3. Allen, T.M., Cullis, P.R.: Liposomal drug delivery systems: from concept to clinical applications. Adv. Drug Del. Rev. **65**, 36–48 (2013). https://doi.org/10.1016/j.addr.2012.09.037
4. Akbarzadeh, A., Rezaei-Sadabady, R., Davaran, S., Joo, S.W., Zarghami, N., Hanifehpour, Y., Nejati-Koshki, K.: Liposome: classification, preparation, and applications. Nanoscale Res. Lett. **8**, 102–102 (2013). https://doi.org/10.1186/1556-276X-8-102
5. Demetzos, C.: Differential scanning calorimetry (DSC): a tool to study the thermal behavior of lipid bilayers and liposomal stability. J. Liposome Res. **18**, 159–173 (2008). https://doi.org/10.1080/08982100802310261
6. Cooper, A., Nutley, M.A., Wadood, A.: Differential scanning microcalorimetry. In: Harding, S.E., Chowdhry, B.Z. (eds.) Protein-Ligand Interactions: Hydrodynamics and Calorimetry, pp. 287–318. Oxford University Press, Oxford New York (2000). ISBN: 9780199637461
7. Chiu, M.H., Prenner, E.J.: Differential scanning calorimetry: An invaluable tool for a detailed thermodynamic characterization of macromolecules and their interactions. J. Pharm. Bioallied Sci. **1**(3), 39–59 (2011). https://doi.org/10.4103/0975-7406.76463
8. Sarpietro, M.G., Castelli, F.: Transfer kinetics from colloidal drug carriers and liposomes to biomembrane models: DSC studies. J. Pharm. Bioall. Sci. **1**(3), 77–88 (2011). https://doi.org/10.4103/0975-7406.76472
9. Koynova, R., Caffrey, M.: Phases and phase transitions of the phosphatidylcholines. Biochem. Biophys. Acta **1376**, 91–145 (1998). https://doi.org/10.1016/S0304-4157(98)00006-9
10. Pippa, N., Pispas, S., Demetzos, C.: Physicochemical characterization and basic research principles of advanced Drug Delivery nano Systems (aDDnSs). In: Tiwari, A., Misha, Y.K., Kobayashi, H., Turner, A.P.F. (eds.) Intelligent Nanomaterials, 2nd edn, pp. 107–126. WILEY-Scrivener Publishing LLC, New Jersy-Massachusetts (2016). https://doi.org/10.1002/9781119242628.ch5

11. McElhaney, R.N.: The use of differential scanning calorimetry and differential thermal analysis studies of model and biological membranes. Chem. Phys. Lipid. **30**, 229–259 (1982). https://doi.org/10.1016/0009-3084(82)90053-6
12. Smith, E.A., Dea, P.K.: Differential scanning calorimetry studies of phospholipid membranes: the interdigitated gel phase. In: Elkordy, A.A. (ed.) Applications of Calorimetry in a Wide Context—Differential Scanning Calorimetry, Isothermal Titration Calorimetry and Microcalorimetry, pp. 408–444. InTech (2013). https://doi.org/10.5772/51882
13. Demetzos, C.: Biophysics and thermodynamics: the scientific building blocks of bio-inspired drug delivery nano systems. AAPS Pharm. Sci. Tech. **16**(3), 491–495 (2015). https://doi.org/10.1208/s12249-015-0321-1
14. Astier, A., Pai, A.B., Bissig, M., Crommelin, D.J.A., Flühmann, B., Jean- Hecq, D., Knoeff, J., Lipp, H.P., Morell-Baladron, A., Mühlebach, S.: How to select a nanosimilar. Ann. N.Y. Acad. Sci. **1407**(1), 50–62 (2017). https://doi.org/10.1111/nyas.13382
15. Zheng, N., Jiang, W., Lionberger, R., Yu, L.: Bioequivalence for liposomal drug products. In: Yu, L.X., Li, B.V. (eds.) FDA Bioequivalence Standards, AAPS Advances in the Pharmaceutical Sciences Series 13, pp. 275–296. Springer, New York, ISBN 978-1-4939-1251-3 (2014). https://doi.org/10.1007/978-1-4939-1252-0_11
16. Wei, X., Cohen, R., Barenholz, Y.: Insights into composition/structure/function relationships of Doxil® gained from "high-sensitivity" differential scanning calorimetry. Eur. J. Pharm. Biopharm. **104**, 260–270 (2016). https://doi.org/10.1016/j.ejpb.2016.04.011
17. Perinelli, D.R., Cespi, M., Bonacucina, G., Rendina, F., Palmieri, G.F.: Heating treatments affect the thermal behaviour of doxorubicin loaded in PEGylated liposomes. Int. J. Pharm. **534**, 81–88 (2017). https://doi.org/10.1016/j.ijpharm.2017.09.069
18. Chen, Y., Bose, A., Bothun, G.D.: Controlled release from bilayer-decorated magnetoliposomes via electromagnetic heating. ACS Nano **4**(6), 3215–3221 (2010). https://doi.org/10.1021/nn100274v
19. Patitsa, M., Karathanou, K., Kanaki, Z., Tzioga, L., Pippa, N., Demetzos, C., Verganelakis, D.A., Cournia, Z., Klinakis, A.: Magnetic nanoparticles coated with polyarabic acid demonstrate enhanced drug delivery and imaging properties for cancer theranostic applications. Sci. Rep. **7**(1), 1–8 (2017). https://doi.org/10.1038/s41598-017-00836-y
20. Zeng, L., Luo, L., Pan, Y., Luo, S., Lu, G., Wu, A.: In vivo targeted magnetic resonance imaging and visualized photodynamic therapy in deep-tissue cancers using folic acid-functionalized superparamagnetic-upconversion nanocomposites. Nanoscale **7**, 8946–8954 (2015). https://doi.org/10.1039/c5nr01932j
21. Tan, X., Pang, X., Lei, M., Ma, M., Guo, F., Wang, J., Yu, M., Tan, F., Li, N.: An efficient dual-loaded multifunctional nanocarrier for combined photothermal and photodynamic therapy based on copper sulfide and chlorine. Int. J. Pharm. **503**, 220–228 (2016). https://doi.org/10.1016/j.ijpharm.2016.03.019
22. Skupin-Mrugalska, P., Sobotta, L., Warowicka, A., Wereszczynska, B., Zalewski, T., Gierlich, P., Jarek, M., Nowaczyk, G., Kempka, M., Gapinski, J., Jurga, S., Mielcarek, J.: Theranostic liposomes as a bimodal carrier for magnetic resonance imaging contrast agent and photosensitizer. J. Inorg. Biochem. **180**, 1–14 (2018). https://doi.org/10.1016/j.jinorgbio.2017
23. Preiss, M.R., Hart, A.E., Kitchens, C.L., Bothun, G.D.: Hydrophobic nanoparticles modify the thermal release behavior of liposomes. J. Phys. Chem. B **121**(19), 5040–5047 (2017). https://doi.org/10.1021/acs.jpcb.7b01702
24. Nasir, A., Harikumar, S.L., Kaur, A.: Cyclodextrins: an excipient tool in drug delivery. Int. Res. J. Pharm. **3**(4), 44–50 (2012). ISSN 2230–8407
25. McCormack, B., Gregoriadis, G.: Entrapment of cyclodextrin–drug complexes into liposomes: potential advantages in drug delivery. J. Drug Target **2**(5), 449–454 (1994). https://doi.org/10.3109/10611869408996821
26. Gharib, R., Fourmentin, S., Charcosset, C., Greige-Gerges, H.: Effect of hydroxypropyl-β–cyclodextrin on lipid membrane fluidity, stability and freezedrying of liposomes. J. Drug Deliv. Sci. Technol. **44**, 101–107 (2018). https://doi.org/10.1016/j.jddst.2017.12.009

27. Liossi, A.S., Ntountaniotis, D., Kellici, T.F., Chatziathanasiadou, M.V., Megariotis, G., Mania, M., Becker-Baldus, J., Kriechbaum, M., Krajnc, A., Christodoulou, E., Glaubitz, C., Rappolt, M., Heinz, A., Gregor, M., Theodorou, D.N., Valsami, G., Pitsikalis, M., Iatrou, H., Tzakos, A.G., Mavromoustakos, T.: Exploring the interactions of irbesartan and irbesartan–2-hydroxypropyl-β-cyclodextrin complex with model membranes. Biochim. Biophys. Acta **1859**(6), 1089–1098 (2017). https://doi.org/10.1016/j.bbamem.2017.03.003
28. Juárez-Osornio, C., Gracia-Fadrique, J.: Structures similar to lipid emulsions and liposomes. Structures similar to lipid emulsions and liposomes. J. Liposome Res. **27**(2), 139–150 (2017). https://doi.org/10.1080/08982104.2016.1174944
29. Pippa, N., Chronopoulos, D.D., Stellas, D., Fernández-Pacheco, R., Arenal, R., Demetzos, C., Tagmatarchis, N.: Design and development of multi-walled carbon nanotube-liposome drug delivery platforms. Int. J. Pharm. **528**(1–2), 429–439 (2017). https://doi.org/10.1016/j.ijpharm.2017.06.043
30. Omidi, M., Fathinia, A., Farahani, M., Niknam, Z., Yadegari, A.;, Hashemi, M., Jazayeri, H., Zali, H., Zahedinik, M., Tayebi, L.: Bio-applications of grapheme composites: from bench to clinic. In: Tiwari, A., Syväjärvi, M. (eds.), Advanced 2D Materials, pp. 433–471 (2016). https://doi.org/10.1002/9781119242635.ch11
31. Zhang, L., Xia, J., Zhao, Q., Liu, L., Zhang, Z.: Functional graphene oxide as a nanocarrier for controlled loading and targeted delivery of mixed anticancer drugs. Small **6**(4), 537–544 (2010). https://doi.org/10.1002/smll.200901680
32. Wang, F., Liu, J.: Nanodiamond decorated liposomes as highly biocompatible delivery vehicles and a comparison with carbon nanotubes and grapheme oxide. Nanoscale **5**(24), 12375–12382 (2013). https://doi.org/10.1039/c3nr04143c
33. Wang, F., Liu, B., Ip, A.C.F., Liu, J.: Orthogonal adsorption onto nano-grapheneoxide using different intermolecular forces for multiplexed delivery. Adv. Mater. **25**(30), 4087–4092 (2013). https://doi.org/10.1002/adma.201301183
34. Hashemi, M., Omidi, M., Muralidharan, B., Tayebi, L., Herpin, M.J., Mohagheghi, M.A., Mohammadi, J., Smyth, H.D.C., Milner, T.E.: Layer-by-layer assembly of graphene oxide on thermosensitive liposomes for photo-chemotherapy. Acta Biomater. **65**, 376–392 (2018). https://doi.org/10.1016/j.actbio.2017.10.040
35. Tomalia, D.A., Naylor, A.M., Goddard III, W.A.: Starburst dendrimers: molecular-level control of size, shape, surface chemistry, topology, and flexibility from atoms to macroscopic matter. Angew. Chem. Int. Ed. **29**(2), 138–175 (1990). https://doi.org/10.1002/anie.199001381
36. Gardikis, K., Hatziantoniou, S., Viras, K., Wagner, M., Demetzos, C.: A DSC and Raman spectroscopy study on the effect of PAMAM dendrimer on DPPC model lipid membranes. Int. J. Pharm. **318**(1–2), 118–123 (2006). https://doi.org/10.1016/j.ijpharm.2006.03.023
37. Klajnert, B., Epand, R.M.: PAMAM dendrimers and model membranes: Differential scanning calorimetry studies. Int. J. Pharm. **305**, 154–166 (2005). https://doi.org/10.1016/j.ijpharm.2005.08.015
38. Gardikis, K., Fessas, D., Signorelli, M., Dimas, K., Tsimplouli, C., Ionov, M., Demetzos, C.: A new chimeric drug delivery nano system (chi-aDDnS) composed of PAMAM G 3.5 dendrimer and liposomes as doxorubicin's carrier. In Vitro Pharmacological Studies. J. Nanosci. Nanotech. **11**(5), 3764–3772 (2011). https://doi.org/10.1166/jnn.2011.3847
39. Berényi, S., Mihály, J., Wacha, A., Toke, O., Bóta, A.: A mechanistic view of lipid membrane disrupting effect of PAMAM dendrimers. Colloids Surf., B **118**, 164–171 (2014). https://doi.org/10.1016/j.colsurfb.2014.03.048
40. Gardikis, K., Hatziantoniou, S., Signorelli, M., Pusceddu, M., Micha-Screttas, M., Schiraldi, A., Demetzos, C., Fessas, D.: Thermodynamic and structural characterization of liposomal-locked in-dendrimers as drug carriers. Colloids Surf., B **81**(1), 11–19 (2010). https://doi.org/10.1016/j.colsurfb.2010.06.010
41. Gardikis, K., Hatziantoniou, S., Bucos, M., Fessas, D., Signorelli, M., Felekis, T., Zervou, M., Screttas, C.G., Steele, B.R., Ionov, M., Micha-Screttas, M., Klajnert, B., Bryszewska, M., Demetzos, C.: New drug delivery nanosystem combining liposomal and dendrimeric technology (liposomal locked-in dendrimers) for cancer therapy. J. Pharm. Sci. **99**(8), 3561–3571 (2010). https://doi.org/10.1002/jps.22121

42. Pippa, N., Gardikis, K., Pispas, S., Demetzos, C.: The physicochemical/thermodynamic balance of advanced drug liposomal delivery systems. J. Therm. Anal. Calorim. **116**(1), 99–105 (2014). https://doi.org/10.1007/s10973-013-3406-7
43. Ionov, M., Gardikis, K., Wróbel, D., Hatziantoniou, S., Mourelatou, H., Majoral, J.P., Klajnert, B., Bryszewska, M., Demetzos, C.: Interaction of cationic phosphorus dendrimers (CPD) with charged and neutral lipid membranes. Colloids Surf., B **82**(1), 8–12 (2011). https://doi.org/10.1016/j.colsurfb.2010.07.046
44. Wrobel, D., Ionov, M., Gardikis, K., Demetzos, C., Majoral, J.P., Palecz, B., Klajnert, B., Bryszewska, M.: Interactions of phosphorus-containing dendrimers with liposomes. Biochim. Biophys. Acta **1811**(3), 221–226 (2011). https://doi.org/10.1016/j.bbalip.2010.11.007
45. Ionov, M., Wróbel, D., Gardikis, K., Hatziantoniou, S., Demetzos, C., Majoral, J.P., Klajnert, B., Bryszewska, M.: Effect of phosphorus dendrimers on DMPC lipid membranes. Chem. Phys. Lipids **165**(4), 408–413 (2012). https://doi.org/10.1016/j.chemphyslip.2011.11.014
46. Wrobel, D., Appelhans, D., Signorelli, M., Wiesner, B., Fessas, D., Scheler, U., Voit, B., Maly, J.: Interaction study between maltose-modified PPI dendrimers and lipidic model membranes. Biochem. Biophys. Acta **1848**(4), 1490–1501 (2015). https://doi.org/10.1016/j.bbamem.2015.03.033
47. Paolino, D., Accollac, M.L., Cilurzo, F., Cristiano, M.C., Cosco, D., Castelli, F., Sarpietro, M.G., Fresta, M., Celia, C.: Interaction between PEG lipid and DSPE/DSPC phospholipids: an insight of PEGylation degree and kinetics of de-PEGylation. Colloids Surf., B **155**, 266–275 (2017). https://doi.org/10.1016/j.colsurfb.2017.04.018
48. Hädicke, A., Blume, A.: Interactions of Pluronic block copolymers with lipid vesicles depend on lipid phase and Pluronic aggregation state. Colloid Polym. Sci. **293**(1), 267–276 (2015). https://doi.org/10.1007/s00396-014-3414-6
49. Pippa, N., Stellas, D., Skandalis, A., Pispas, S., Demetzos, C., Libera, M., Marcinkowski, A., Trzebicka, B.: Chimeric lipid/block copolymer nanovesicles: physico-chemical and biocompatibility evaluation. Eur. J. Pharm. Biopharm. **107**, 295–309 (2016). https://doi.org/10.1016/j.ejpb.2016.08.003
50. Pippa, N., Pispas, S., Demetzos, C.: The metastable phases as modulators of biophysical behavior of liposomal membranes. J. Therm. Anal. Calorim. **120**(1), 937–945 (2015). https://doi.org/10.1007/s10973-014-4116-5
51. Eloy, J.O., Souza, M., Petrilli, R., Barcellos, J.P.A., Lee, R.J., Marchetti, J.M.: Liposomes as carriers of hydrophilic small molecule drugs: Strategies to enhance encapsulation and delivery. Colloids Surf., B **123**, 345–363 (2014). https://doi.org/10.1016/j.colsurfb.2014.09.029
52. Honey, P.J., Rijo, J., Anju, A., Anoop, K.R.: Smart polymers for the controlled delivery of drugs—a concise overview. Acta Pharm. Sinica B **4**(2), 120–127 (2014). https://doi.org/10.1016/j.apsb.2014.02.005
53. Naziris, N., Pippa, N., Pispas, S., Demetzos, C.: Stimuli-responsive drug delivery nanosystems: from bench to clinic. Curr. Nanomed. **6**(3), 166–185 (2016). https://doi.org/10.2174/2468187306666160712232449
54. Kono, K.: Thermosensitive polymer-modified liposomes. Adv. Drug Deliv. Rev. **53**(3), 307–319 (2001). https://doi.org/10.1016/S0169-409X(01)00204-6
55. Kono, K., Hayashi, H., Takagishi, T.: Temperature-sensitive liposomes: liposomes bearing poly(N-isopropylacrylamide). J. Control. Release **30**(1), 69–75 (1994). https://doi.org/10.1016/0168-3659(94)90045-0
56. Chountoulesi, M., Kyrili, A., Pippa, N., Meristoudi, A., Pispas, S., Demetzos, C.: The modulation of physicochemical characterization of innovative liposomal platforms: the role of the grafted thermoresponsive polymers. Pharm. Dev. Technol. **22**(3), 330–335 (2017). https://doi.org/10.3109/10837450.2015.1121497
57. Lee, S.M., Nguyen, S.T.: Smart nanoscale drug delivery platforms from stimuli-responsive polymers and liposomes. Macromolecules **46**(23), 9169–9180 (2013). https://doi.org/10.1021/ma401529w
58. Ta, T., Convertine, A.J., Reyes, C.R., Stayton, P.S., Porter, T.M.: Thermosensitive liposomes modified with poly(N-isopropylacrylamide-co-propylacrylic acid) copolymers for triggered

release of doxorubicin. Biomacromolecules **11**(8), 1915–1920 (2010). https://doi.org/10.1021/bm1004993
59. Pippa, N., Meristoudi, A., Pispas, S., Demetzos, C.: Temperature-dependent drug release from DPPC:C12H25-PNIPAM-COOH liposomes: Control of the drug loading/release by modulation of the nanocarriers' components. Int. J. Pharm. **485**(1–2), 374–382 (2015). https://doi.org/10.1016/j.ijpharm.2015.03.014
60. Lin, Y.L., Jiang, G., Birrell, L.K., El-Sayed, M.E.H.: Degradable, pH-sensitive, membrane-destabilizing, comb-like polymers for intracellular delivery of nucleic acids. Biomaterials **31**(27), 7150–7166 (2010). https://doi.org/10.1016/j.biomaterials.2010.05.048
61. Liu, J., Huang, Y., Kumar, A., Tan, A., Jin, S., Mozhi, A., Liang, X.J.: pH-Sensitive nano-systems for drug delivery in cancer therapy. Biotechnol. Adv. **32**(4), 693–710 (2014). https://doi.org/10.1016/j.biotechadv.2013.11.009
62. Felber, A.E., Dufresne, M.H., Leroux, J.C.: pH-sensitive vesicles, polymeric micelles, and nanospheres prepared with polycarboxylates. Adv. Drug Deliv. Rev. **64**(11), 979–992 (2014). https://doi.org/10.1016/j.addr.2011.09.006
63. Wang, L., Geng, D., Su, H.: Safe and efficient pH sensitive tumor targeting modified liposomes with minimal cytotoxicity. Colloids Surf., B **123**, 395–402 (2014). https://doi.org/10.1016/j.colsurfb.2014.09.003
64. Naziris, N., Pippa, N., Meristoudi, A., Pispas, S., Demetzos, C.: Design and development of pH-responsive HSPC:C12H25-PAA chimeric liposomes. J. Liposome Res. **27**(2), 108–117 (2017). https://doi.org/10.3109/08982104.2016.1166512
65. Kolman, I., Pippa, N., Meristoudi, A., Pispas, S., Demetzos, C.: A dual-stimuli-responsive polymer into phospholipid membranes. J. Therm. Anal. Calorim. **123**, 2257–2271 (2016). https://doi.org/10.1007/s10973-015-5080-4
66. Pippa, N., Chountoulesi, M., Kyrili, A., Meristoudi, A., Pispas, S., Demetzos, C.: Calorimetric study on pH-responsive block copolymer grafted lipid bilayers: rational design and development of liposomes. J. Liposome Res. **26**(3), 211–220 (2016). https://doi.org/10.3109/08982104.2015.1076464
67. Kyrili, A., Chountoulesi, M., Pippa, N., Meristoudi, A., Pispas, S., Demetzos, C.: Design and development of pH-sensitive liposomes by evaluating the thermotropic behavior of their chimeric bilayers. J. Therm. Anal. Calorim. **127**(2), 1381–1392 (2017). https://doi.org/10.1007/s10973-016-6069-3
68. Franzé, S., Marengo, A., Stella, B., Minghetti, P., Arpicco, S., Cilurzo, F.: Hyaluronan-decorated liposomes as drug delivery systems for cutaneous administration. Int. J. Pharm. **535**(1–2), 333–339 (2018). https://doi.org/10.1016/j.ijpharm.2017.11.028
69. Cosco, D., Tsapis, N., Nascimento, T.L., Fresta, M., Chapron, D., Taverna, M., Arpicco, S., Fattal, E.: Polysaccharide-coated liposomes by post-insertion of a hyaluronan-lipid conjugate. Colloids Surf., B **158**, 119–126 (2017). https://doi.org/10.1016/j.colsurfb.2017.06.029
70. Hirsch-Lerner, D., Barenholz, Y.: Hydration of lipoplexes commonly used in gene delivery: follow-up by laurdan fluorescence changes and quantification by differential scanning calorimetry. Biochem. Biophys. Acta **1461**(1), 47–57 (1999). https://doi.org/10.1016/S0005-2736(99)00145-5
71. Aleandri, S., Bonicelli, M.G., Giansanti, L., Giuliani, C., Ierino, M., Mancini, G., Martino, A., Scipioni, A.: A DSC investigation on the influence of gemini surfactant stereochemistry on the organization of lipoplexes and on their interaction with model membranes. Chem. Phys. Lipid. **165**(8), 838–844 (2012). https://doi.org/10.1016/j.chemphyslip.2012.11.003
72. Henriksen-Lacey, M., Bramwell, V.W., Christensen, D., Agger, E.M., Andersen, P., Perrie, Y.: Liposomes based on dimethyldioctadecylammonium promote a depot effect and enhance immunogenicity of soluble antigen. J. Control. Release **142**(2), 180–186 (2009). https://doi.org/10.1016/j.jconrel.2009.10.022

73. Perrie, Y., Kastner, E., Kaurm, R., Willkinsonm, A., Inghamm, A.J.: A case-study investigating the physicochemical characteristics that dictate the function of a liposomal adjuvant. Hum. Vaccin. Immunotherapeutics **9**(6), 1374–1381 (2013). https://doi.org/10.4161/hv.24694
74. Hamborg, M., Rose, F., Jorgensen, L., Bjorklund, K., Pedersen, H.B., Christensen, D., Foged, C.: Elucidating the mechanisms of protein antigen adsorption to the CAF/NAF liposomal vaccine adjuvant systems: effect of charge, fluidity and antigen-to-lipid ratio. Biochem. Biophys. Acta **1838**(8), 2001–2010 (2014). https://doi.org/10.1016/j.bbamem.2014.04.013

Drug-Membrane Interactions in the Renin Angiotensin System

Dimitrios Ntountaniotis, Tahsin F. Kellici, Paraskevi Gkeka, Zoe Cournia, Ioannis Galdadas, Gregor Mali, Johanna Becker-Baldus, Clemens Glaubitz, Manfred Kriechbaum, Michael Rappolt, George Liapakis and Thomas Mavromoustakos

Abstract Renin Angiotensin System (RAS) plays a key role in the pathophysiology of the cardiovascular and renal system. Within this system stimulation of the G-protein coupled receptor (GPCR) AT1 by the peptide angiotensin II (AII) has a central role. Molecules have been discovered which either block the formation of AII or compete with the activation of the AT1 receptor. Angiotensin Receptor Blockers (ARBs) or sartans are the first synthetic drugs of the latter class. Up to now the question has not been answered how the drug molecules reach the receptor site. Are these molecules incorporated in the lipid bilayers core

D. Ntountaniotis · T. F. Kellici · T. Mavromoustakos (✉)
Department of Chemistry, National and Kapodistrian University of Athens, 15771 Panepistimiopolis Zografou, Greece
e-mail: tmavrom@chem.uoa.gr

T. F. Kellici
Department of Chemistry, University of Ioannina, 45110 Ioannina, Greece

P. Gkeka · Z. Cournia · I. Galdadas
Biomedical Research Foundation, Academy of Athens, 4 Soranou Ephessiou, 11527 Athens, Greece

P. Gkeka
Sanofi Aventis Group, Paris, France

G. Mali
Department of Inorganic Chemistry and Technology, National Institute of Chemistry, Hajdrihova 19, 1001 Ljubljana, Slovenia

J. Becker-Baldus · C. Glaubitz
Institute of Biophysical Chemistry, Goethe University Frankfurt, Max-von-Laue-Str. 9, 60438 Frankfurt, Germany

M. Kriechbaum
Institute of Inorganic Chemistry, Graz University of Technology, Graz, Austria

M. Rappolt
School of Food Science & Nutrition, University of Leeds, Woodhouse Ln, Leeds LS2 9JT, UK

G. Liapakis
Department of Pharmacology, School of Medicine, University of Crete, 71003 Heraklion, Crete, Greece

C. Demetzos and N. Pippa (eds.), *Thermodynamics and Biophysics of Biomedical Nanosystems*, Series in BioEngineering, https://doi.org/10.1007/978-981-13-0989-2_10

before they freely diffuse to reach the active site or do they enter at the receptor binding site directly from the water phase? Both mechanisms are possible and may have a synergistic action. Various biophysical techniques, such as calorimetric, spectroscopic and X-ray scattering methods can provide valuable information on the dynamic and thermodynamic changes caused when drugs are incorporated into the lipid bilayers. Molecular Dynamics (MD) simulations are a valuable tool to acquire atomic level information on the molecular basis of the interaction between the drugs and the RAS system. The purpose of this chapter is to review the contribution of the various biophysical techniques and MD on drug-membrane interactions on the RAS system. Studies described in the manuscript are mainly referred to drugs acting on the GPCR AT1 receptor, however it is envisaged that the discussed concepts of drug interaction can apply also to other GPCRs.

Keywords Renin angiotensin system (RAS) · SAXS · NMR · DSC · Raman spectroscopy · GPCR · Angiotensin receptor blocker (ARB)

1 Introduction

Overactivation of the Renin Angiotensin System (RAS) plays a vital role in the pathophysiology of the cardiovascular and renal systems as it leads to hypertension, cardiac hypertrophy, heart failure, ischemic heart disease and nephropathy [1]. Thereby the peptidic hormone angiotensin II (AII) stimulates the G-protein coupled receptor (GPCR) Angiotensin II Type 1 (AT1) receptor. There are two ways for interfering in the RAS system. The first way is to block the metabolism of angiotensinogen to angiotensin I (AI) and finally to AII. To achieve this aim renin inhibitors (e.g. aliskiren) and ACE inhibitors (e.g. captopril) have been developed. The second way, which is the main focus of this chapter, is the use of small molecules that act as Trojan horses and block the AII activation on GPCR AT1 receptor. These are named AT1 antagonists or ARBs (Angiotensin Receptor Binding) or sartans as the first synthetic drug of this series was losartan (Fig. 1).

The AT1 receptor consists of seven transmembrane α-helices, an extracellular N-terminus, three intracellular loops, three extracellular loops and an intracellular C-terminus, as revealed by its recently reported crystal structure [2]. Sartans as well as AII interact with residues located both in the transmembrane helices and in the extracellular loops of the AT1 receptor [3–5]. AT1 receptor antagonism of sartans was based on pharmacophore segment similarities with the C-terminal fragment of AII [6]. Details of these interactions are described in a recent review by Balakumar and Jagadesh [7]. Interestingly, while with other GPCRs like cannabinoid and opioid receptors, these interactions are studied in connection with lipid bilayer environments, such studies are lacking for AT1 receptors. The findings with cannabinoid and opioid receptors clearly show that the lipid matrix does not just surround the protein matrix but it also affects the small molecule binding due to conformational changes either to the receptor and/or to the drug [8].

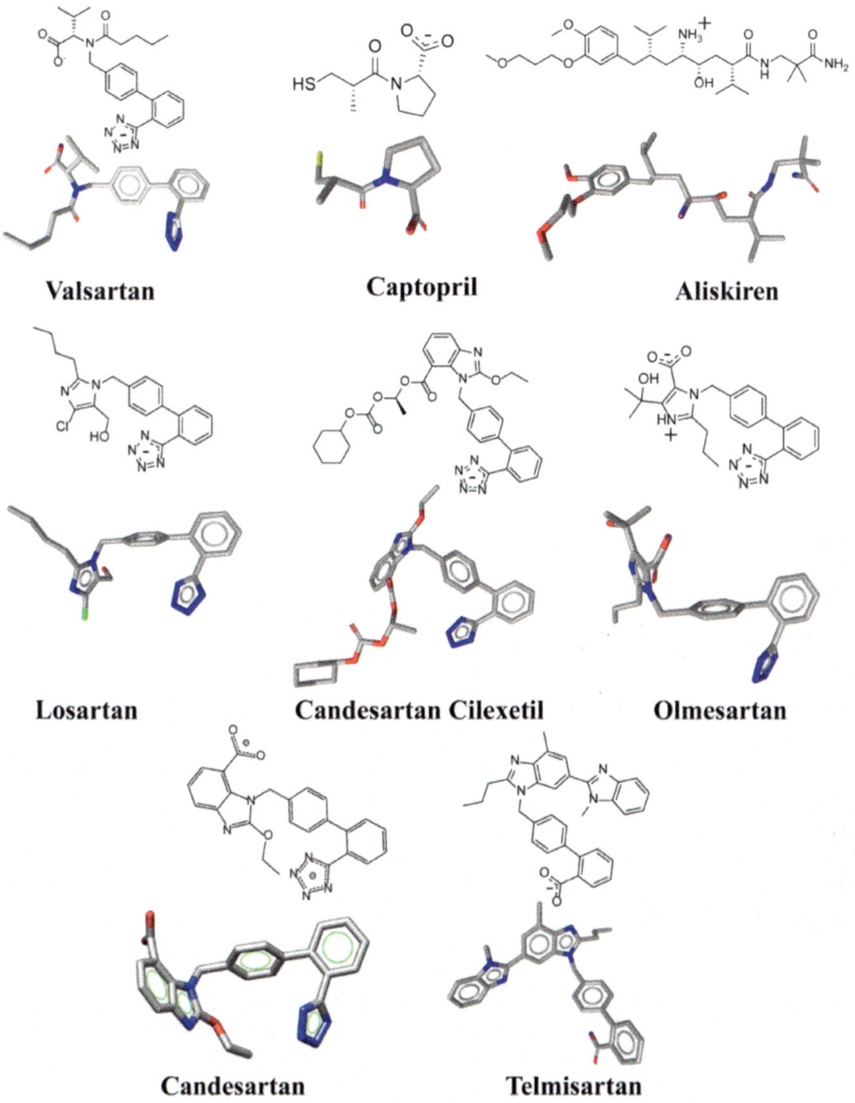

Fig. 1 Most favored ionic forms of drug molecules and the prodrug candesartan cilexetil mentioned in this chapter that act on the RAS system at neutral pH

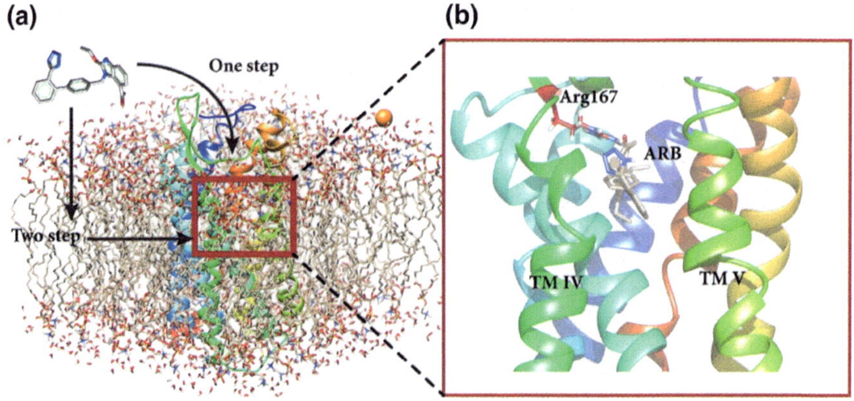

Fig. 2 **a** The mode of insertion for ARBs. In the first step they might be embedded in the lipid matrix and in the second step are laterally diffused to the active site of AT1 receptor. The other mode of insertion is directly through the receptor site. **b** Expansion of the vicinity area of the active site when AT1 antagonist approaches the receptor [12, 15]

Since sartans act in the extracellular loops and in the transmembrane region, a legitimate question that arises is "how do they achieve their binding"? Do sartans bind directly to the receptor, as it is proposed for alprenolol which acts on another GPCR, i.e. the β2 adrenergic receptor [9, 10], or through a two-step mechanism, which involves the partition of the drug in the lipid bilayers and subsequent diffusion towards the active site (Fig. 2) [11–14]? Both mechanisms are plausible, although the second mechanism is more prone to happen for the following reasons: (i) the drug is an amphipathic entity and can easily insert itself in the lipid matrix; (ii) the lipid matrix occupies more space compared to the accessible site of the receptor and therefore drug insertion in the membrane is statistically more favored; (iii) the receptor is a protein that bears ionized aminoacids and groups that can easily interact with corresponding groups of the inserting molecules. Thus, these interactions may prevent incorporation of the drug into the active site; (iv) theoretical calculations for the two step mechanism as it is postulated for dihydropyridine calcium channel antagonists were of three orders of magnitude faster than the aqueous pathway to reach the receptor site [11]. The simultaneous appearances of both mechanisms cannot be excluded. Partial fractions of the drug may follow the receptor pathway and the rest the pathway through the membrane.

Recently, molecular dynamics studies in a lipidic environment and in an environment where the AT1 receptor is embedded in a lipid bilayer have been applied to shed more light on the mechanistic aspects of the drugs during "their journey to Ithaca", thus to reach the active site of the receptor [16–18]. Certainly, the increase in available computational resources has made these studies feasible and provides significant evidence on the mechanism of action of AT1 antagonists [19]. One inherent difficulty with such computational studies was the absence of crystallographic data of the AT1 receptor. Two crystal structures of AT1 receptor with ZD7155 and olmesartan molecules were reported recently [2, 20].

Elegant experiments where the pathway of the drug will be followed towards the active site will definitely aid the understanding of the forces that contribute to these processes. Such experiments will lead to the optimization of the pharmacological profile of novel AT1 antagonists.

In this chapter, we will restrain the description to the efforts made to understand the membrane interactions of AT1 receptor antagonists. We will outline the most important results observed using various biophysical methods and we will conclude with future perspectives from such studies.

2 Experimental Insights on RAS Interacting Drugs

2.1 Model Membranes

DPPC (dipalmitoylphosphatidylcholine) bilayers are almost exclusively used in these studies because: (i) the partition coefficient of the drugs particularly in the fluid state, resembles that of natural plasma membranes of the vasculature [21, 22]; (ii) hydrated DPPC spontaneously forms multilamellar bilayers whose mesomorphic changes occur in a temperature range between 25 and 50 °C; (iii) Phosphatidylcholines (PCs), especially those bearing oleic, linoleic or palmitoyl alkyl chains are the most abundant lipid species found in the plasma membranes and vascular smooth muscle cells [21] and sarcolemma cardiac membranes [22]; (iv) the thermal and dynamic properties of DPPC bilayers are extensively studied.

Cholesterol is a major constituent of biological membranes, which is responsible for the induction of the liquid-ordered phase (L_o) [23, 24] In this phase lipids display ordered chains but still are able to freely diffuse laterally in the membrane monolayer. L_o phase is proposed but not without problems to be connected to membrane rafts and functional platforms that enable the assembly of signaling proteins or transbilayer transport [25].

2.2 Interactions of Losartan with Lipid Bilayers

The effects of losartan on DPPC bilayers and DPPC/cholesterol bilayers were studied using Differential Scanning Calorimetry (DSC), solid-state (ss) ^{13}C Magic Angle Spinning (MAS), cross polarization (CP) ^{31}P NMR experiments, Raman spectroscopy, X-ray scattering, and ESR [26, 27].

The DSC scans show progressive effect of losartan as its concentration is increased. At very low concentration (1 mol%; molar ratio $x = 0.01$) the pretransition broadens and at higher (5 mol%) it vanishes. The use of drug concentrations of 5 and

10 mol% causes broadening of the main phase transition. At 20 mol% the half-width of the phase transition is further increased and a shoulder is observed at the high temperature side of the main phase transition. A similar thermal effect of losartan occurred in dimyristoyl-phosphatidylcholine (DPMC) bilayers at high concentrations (molar ratio DMPC/losartan 80:20 and 60:40) [28]. These thermal profiles are explained by Schneider et al. as (i) an enhanced membrane elasticity which accompanies the lipid state fluctuation on chain melting and (ii) solvent associated interactions that favor a change in membrane curvature [29] or (iii) the formation of domains [30, 31]. Such thermal profiles are very dependent on the thermal history of the sample as we have proved with the anesthetic steroid alphaxalone [32]. The lowering of the main phase transition of DPPC bilayers is also concomitant to the increase of the losartan concentration. ^{13}C MAS experiments were in agreement with DSC results. The peaks attributed to the drugs were higher when their concentration increased showing additional drug incorporation, in accordance with DSC data.

The ^{13}C NMR experiments showed that losartan decreased the chain mobility of phosphatidylcholine bilayers both in gel and liquid phases of membrane bilayers in agreement with ESR measurements [28].

The CP ^{31}P NMR broadline simulations showed that lipid bilayers containing losartan were characterized by an improved packing and losartan caused an upfield shift of the average chemical shift (σ_{iso}) of the ^{31}P signal. Raman spectroscopy showed that losartan affected the peak at 715 cm^{-1} attributed to C-N stretching in the Raman spectra. The combination of the two methods suggests that losartan anchors in the realm of the headgroup. DSC results were in accordance with Raman and CP ^{31}P NMR broadline simulations as losartan caused increase of the enthalpy ΔH, in lipid bilayers.

The profile of the intensity ratio I_{2935}/I_{2880} cm^{-1}, based on the effects of the intermolecular interactions on the Raman spectra, depicts decreased mobility of the hydrophobic alkyl chain of the lipids with the inclusion of losartan, compared to the mobility of the unloaded bilayers. The profile of I_{2935}/I_{2880} cm^{-1} shows increased intermolecular interactions in the liquid crystalline phase between the acyl chains, interpreted as possible tail interdigitation [33, 34].

The effects of losartan on DMPC bilayers and POPC (palmitoyl-oleolyl-phosphatidylcholine) bilayers have been examined using small angle X-ray scattering. Its effects were similar in both lipids and losartan caused a reduction of the bilayer thickness by 3–4%. This is in agreement with all other techniques applied and which point out that losartan causes a partial interdigitation in the fluid state (Fig. 3). When losartan is co-inserted with cholesterol, the effects of losartan on POPC and DMPC bilayers differed strikingly. The condensation effect by cholesterol is decreased by losartan in POPC bilayers in contrary with DMPC bilayers where an increase of the cholesterol concentration expelled losartan towards the aqueous phase. This is indicative that losartan competes the effects of cholesterol and most probably does not like to exert its effects in domains rich in cholesterol.

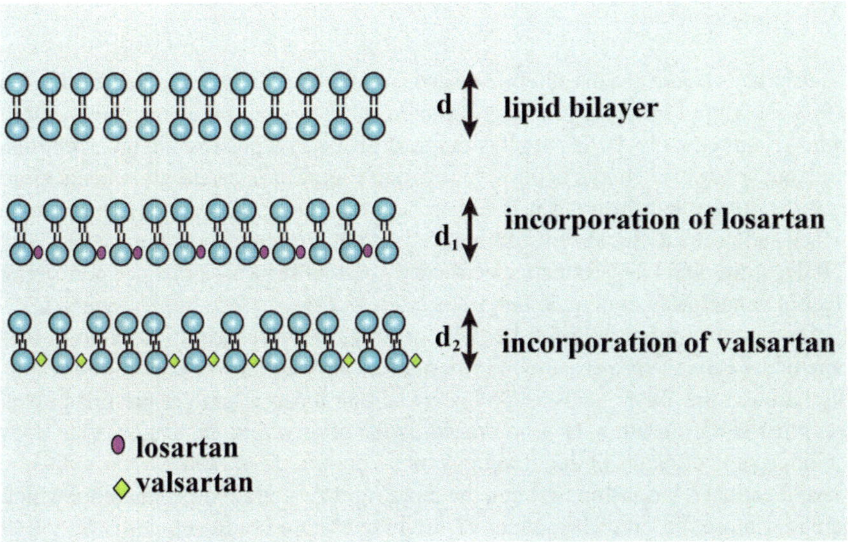

Fig. 3 Interdigitation effect of losartan and valsartan in DPPC bilayers. The bilayer d-spacing follows the order: $d_2 < d_1 < d$

2.3 *Comparison of AT1 Antagonist Losartan with Aliskiren and Captopril*

2.3.1 Valsartan

Valsartan has a relatively low pKa value of 4.9 which indicates that under physiological conditions the majority of valsartan molecules are deprotonated at their tetrazole ring and carboxylate groups as losartan (Fig. 1). Valsartan causes similar thermal effects on DPPC bilayers as losartan. At 20 mol% valsartan induces stronger increase of ΔH in comparison to losartan which indicates a possibly more effective partial interdigitation. Thus valsartan induces the interdigitated $L_{\beta I}$ phase which coexists with valsartan-poor region conserving mainly the $L_{\beta'}$ phase. Valsartan like losartan decreases the thickness of lipid bilayers in the liquid crystalline phase in a higher extent and causes swelling in the gel phase (Fig. 3). In addition, its partial lateral areas are comparable to those of DPPC. Hence, valsartan by partitioning into the membrane may shield the terminal methyl group of the DPPC hydrocarbons in the opposing membrane layer. This suggests that the ΔH increase of the main transition is mainly caused by an increase of the van der Waals energy contribution. Also Raman spectra confirmed that valsartan causes as for losartan an ordering of the alkyl chains.

2.3.2 Candesartan

Candesartan's acidic groups of tetrazole and carboxylate are deprotonized at (pH ~ 7) as it is shown in Fig. (1). Losartan potassium salt bears only the negatively charge acidic group tetrazole. DPPC at physiological pH bears a positive charge at choline group and a negative charge at phosphate group (Fig. 7). As candesartan bears more negative charge than losartan it is anticipated to affect differently the membrane surface and probably localize itself in a distinct topographical position.

Indeed, the DSC and Raman spectroscopy results showed a distinct perturbing effect of candesartan in comparison to losartan and valsartan in lipid bilayers. Candesartan induced more order in the polar head region in respect to losartan as it is depicted by Raman spectroscopy. Furthermore, candesartan was packed more suitably between the alkyl chains as DSC showed that it caused milder thermal effects in comparison to losartan. Of importance, is the observation that candesartan does not induce any domains in the lipid bilayers.

An identical relationship between the more rigid Δ^8-tetrahydrocannabinol which increased more the order parameter of the bilayer core compared with CP-55940 and WIN-55212-2 was reported [35]. The homogeneous broadening parameter (brd) and the observed Raman ΔI changes of the 2935/2880 intensity ratio confirmed that sartans restrict the mobility of the alkyl chains. However, in harmony with DSC thermograms, the different phases are preserved in the presence of candesartan, whereas the pretransition is abolished with losartan.

2.3.3 Olmesartan

Olmesartan interacts strongly with the polar region of the lipid bilayers by abolishing the pre transition and the upper segment of the alkyl chains. However, olmesartan does not exert strong interactions with the hydrophobic segment as compared with losartan and valsartan which have been shown to cause partial interdigitation between the two layers of the lipid bilayers. Candesartan exerts similar thermal profile when is incorporated in lipid bilayers as olmesartan with the only difference that increases ΔH. The fact that olmesartan causes no increase of ΔH is in a harmony with the X-ray data which showed its inability to induce lipid interdigitation [36]. The less disturbation of the chain packing of the lipids by olmesartan and candesartan may related to their shorter alkyl chains, and is in an agreement with Tian et al. results [35].

^{13}C and ^{13}C CP/MAS experiments reveal that olmesartan and cholesterol differ in their dynamic properties when are incorporated in lipid bilayers. This may be due to their rigidification in the bilayer core and different degree of incorporation. DSC data confirm the second explanation as olmesartan does not cause any additional thermal effect upon increase of the concentration [36]. Losartan's aromatic peaks were sharper and more prominent in lipid bilayers than olmesartan's. This is indicative of the different mobility of the biphenyl tetrazole segment of the two molecules. This may reflect also their differential thermal effects in lipid bilayers. Such significant differences in motion that are reflected in their thermal effects have been also observed for the pair of steroids Δ^{16}-alphaxalone and alphaxalone [37] and cannabinoids Δ^8-THC and Me-Δ^8-THC [38–41] and structurally related tricyclic antidepressants desipramine and imipramine [41].

Intramolecular *trans–gauche* conformational changes within the hydrocarbon chain region can be monitored directly with Raman spectroscopy using the intensity ratio I_{1090}/I_{1130} [42] and are in agreement to the results derived from the calorimetric measurements. A strong ΔI increase across the gel to liquid crystalline phase (81.2%) is observed in DPPC bilayers. This is even higher when olmesartan is present (93%). Cholesterol together with olmesartan cause an additional increase during the main phase transition (115.7%). These results signify that both olmesartan alone or in combination with the cholesterol increase entropy changes during the main phase transition. The presence of losartan, valsartan or candesartan lowers the ΔI value [28, 43] in contrast to olmesartan that causes an opposite effect. Thus, the chain mobility decreases by the three AT1 receptor antagonists losartan, valsartan and candesartan and increases for olmesartan.

The stretching mode region 2800–3000 cm^{-1} is commonly used to monitor changes in the lateral packing properties and mobility of the lipid chain in both gel and liquid crystalline bilayer systems [44]. The two bands I_{2850} and I_{2880} are used as an order parameter [45] and the ratio I_{2850}/I_{2880} describes the main change occurring in the hydrocarbon-chain region of the lipids. Thus, it corresponds to intermolecular interactions among aliphatic chains and it is sensitive to subtle changes in conformational order from rotations, kinks, twists and bends of the lipid chains [46]. Alike to the ratio I_{1090}/I_{1130}, the presence of olmesartan or olmesartan/cholesterol causes significant increase also in the ratio I_{2850}/I_{2880}. This means that both, drug alone or in the present of cholesterol cause disorder in the lipid bilayers. The same applies for the ratio I_{2935}/I_{2880}. In conclusion, the three ratios show that olmesartan causes fluidization of the lipid bilayers which is strengthened by the incorporation of cholesterol.

The band at 715 cm^{-1} (represents C-N symmetric stretching vibration of the polar head-group) [47] of DPPC bilayers shows a downshifting of 1 cm^{-1} when olmesartan is inserted in lipid bilayers. This effect is similar to that observed for dipeptide and non-steroidal anti-inflammatory molecules [48, 49] but opposite to candesartan. Last, olmesartan in DPPC/cholesterol bilayers causes an upshift of the 715 cm^{-1} band and the temperature profile of this band behaves like DPPC bilayers alone. The above results show that cholesterol and olmesartan exert antagonistic effect on head-group.

The incorporation of olmesartan into DPPC bilayers causes in the gel-phase the unbinding of membranes like the other sartans. Olmesartan does not induce any unbinding in DPPC/cholesterol bilayers because the affinity of olmesartan for the DPPC/cholesterol bilayers is reduced as compared to pure DPPC bilayers, meaning that the electrostatic repulsion gets reduced in the ternary bilayer system. The lattice spacings (d-values) of the DPPC/olmesartan/cholesterol and DPPC/cholesterol systems show no significant differences. Such behavior to a more significant degree is observed also with losartan.

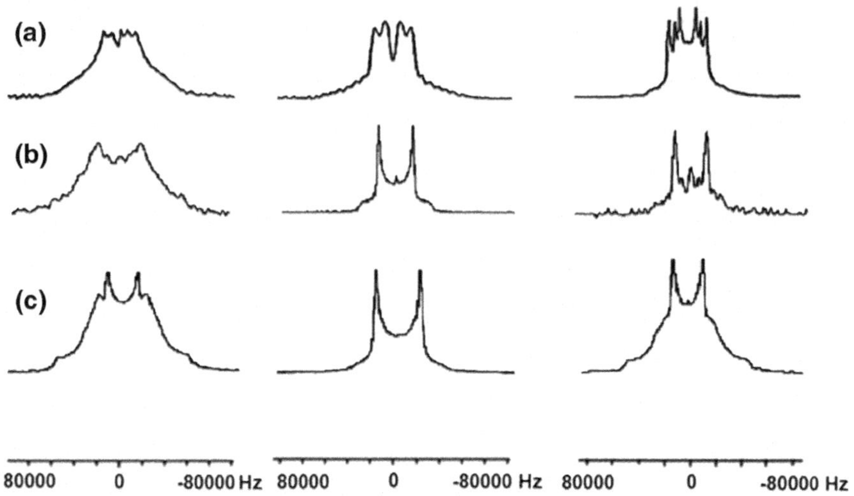

Fig. 4 ^{2}H NMR spectra at a temperature of 35 °C of: **a** (left) 1,2[2,2-$^2H_2$]-DPPC bilayers, (middle) with TCV-116 ($x = 0.20$) and (right) with olmesartan ($x = 0.20$). **b** (left) 1,2[4,4-$^2H_2$]-DPPC bilayers, (middle) with TCV-116 ($x = 0.20$) and (right) with olmesartan ($x = 0.20$). **c** (left) 1,2[7,7-$^2H_2$]-DPPC bilayers, (middle) with TCV-116 ($x = 0.20$) and (right) with olmesartan ($x = 0.20$) [14]

2.3.4 Candesartan Cilexetil (TCV-116)

Candesartan cilexetil is studied for: (a) Comparative physical chemical interactions with candesartan; (b) drug formulations reasons. A prodrug is used for being converted into the active site to the drug. However, candesartan cilexetil if it is transferred (for example via cyclodextrin or calixerene) to the lipid bilayers without being de-esterified might exert even more beneficial effects. Solid state ^{2}H NMR and 2D ^{1}H-^{1}H NOESY experiments were used to compare the localization of TCV-116 and olmesartan [14]. The results were complemented by ^{13}C MAS and ^{13}C CP/MAS studies [14]. As it has been reported, olmesartan has lower lipophilicity in comparison with TCV-116 [50, 51] and fluidizes more the lipid bilayers exerting more pronounced effect at DPPC deuterated sited in the upper segment and middle of the lipid bilayers. Olmesartan causes more significant effect when the deuteration is closer towards the esterified bond and head-group region (Fig. 4) [14].

2.3.5 Telmisartan

Telmisartan and candesartan CV bear condensed aromatic rings, and have higher PISA values and lipophilicity in comparison to losartan and valsartan [52]. Telmisartan and candesartan CV exert lower ionization potential than losartan and valsartan.

The thermal effects of telmisartan resemble those observed by candesartan CV and are differentiated from those of losartan and valsartan.

2.3.6 Irbesartan

In the gel phase the stacking of the DPPC/irbesartan bilayers is less ordered as compared to pure DPPC bilayers as shown by small angle X-ray scattering experiments (SAXS) (Fig. 5: 20–30 °C). The diffraction peaks are clearly broader and only the first three diffraction orders could be observed. While in the ripple phase regime (35–40 °C) almost the same stacking repeat was apparent (Fig. 6), in the liquid crystalline phase the bilayers unbind and only spatially uncorrelated membranes are observed (45–55 °C). Further, irbesartan like various other AT1 antagonists not only decreases the quasi long-range order in the gel, but at the same time suppresses the formation of the ripple phase, i.e. only a lamellar repeat patterns are displayed over the entire gel-phase regime (20–40 °C) [53].

Most strikingly, the DPPC/irbesartan sample above the main transition temperature shows only diffuse scattering arising from bilayer scattering alone. A probable

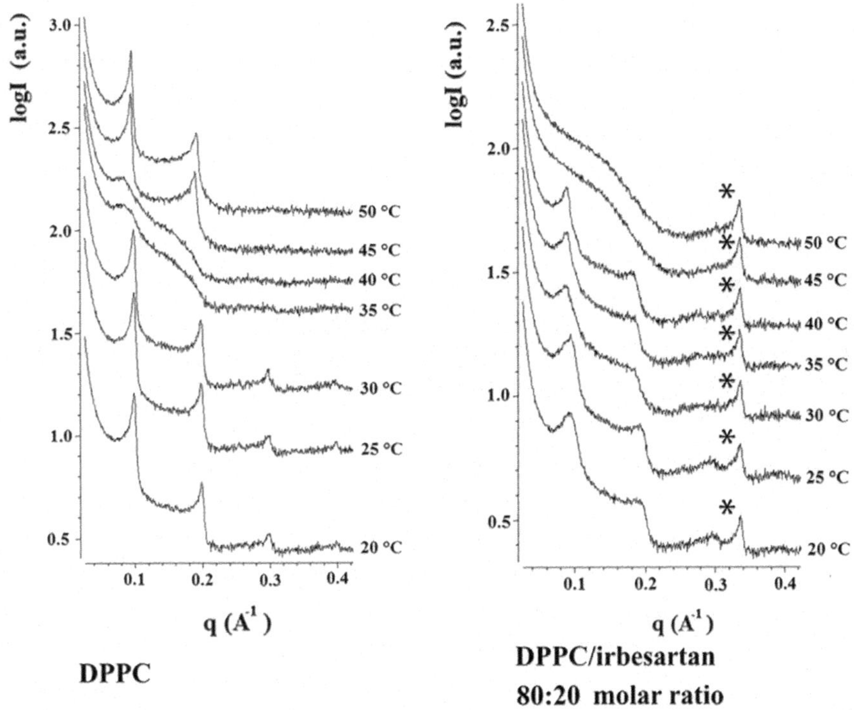

Fig. 5 SAXS experiments for pure DPPC (left) and DPPC bilayers containing irbesartan (right) [53]

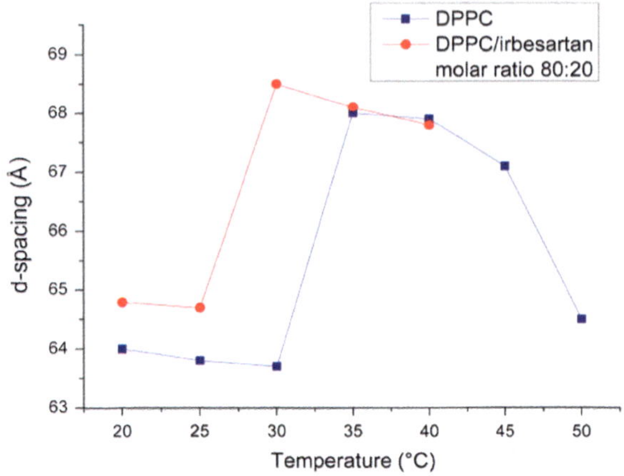

Fig. 6 D-spacing versus temperature for pure DPPC (squares) and DPPC bilayers containing irbesartan (circles) [53]

explanation would be that the incorporation of irbesartan in the liquid crystalline phase leads to an increase in fluidization of the hydrocarbon chains and this in turn causes stronger bilayer undulations, which provokes the unbinding of the membranes. Although irbesartan strongly interacts with DPPC bilayers over the entire studied temperature range, we note that irbesartan is not entirely solvated in the membranes, since a strong diffraction peak at $q = 0.335$ Å-1 is clearly visible throughout the entire temperature range (marked with *). This diffraction peak stems from the crystal form A of irbesartan [54].

^{13}C CP/MAS experiment of irbesartan incorporated in DPPC bilayers at 45 °C are shown in Fig. 7. Additional peaks due to the drug attributed to the flexible segment and rigid segments have been observed. The aromatic region of the drug consists of peaks characterized by increased linewidth which is an indication that this region of the lipid bilayers is squeezed into the lipid bilayers with restrained flexibility. The butyl chain of the drug is clearly seen in all temperature range (25–45 °C). Thus, the butyl chain of irbesartan is embedded in a more flexible hydrophobic core of phospholipid bilayers.

2.4 *Aliskiren and Captopril in Comparison with Other Sartans*

Aliskiren's thermal effects on the pre- and main transition of DPPC bilayers as well as its topographical position in them, show significant similarities to those of AT1 receptor antagonists losartan and valsartan (Fig. 8). Furthermore, when higher concentrations of cholesterol are used, aliskiren gets expelled from the membrane just as it has been demonstrated for losartan. MD simulations and solid state NMR spectroscopy have been used to identify the localization of aliskiren in the lipid bilayers. The NMR data showed drastic changes in the hydrophobic lipid chain which are con-

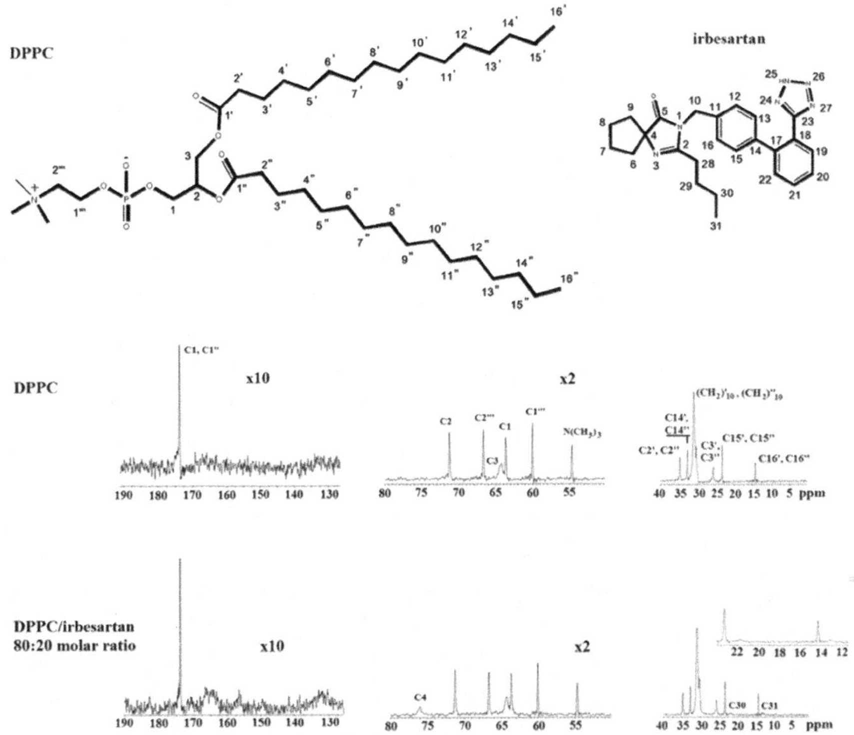

Fig. 7 ^{13}C CP/MAS spectra of DPPC bilayers (upper) and incorporated irbesartan (lower) [53]

firmed by the MD calculations. In addition, the incorporation of aliskiren increases ΔH as losartan and valsartan both with and without the addition of cholesterol at $x = 0.20$ [55]. In contrast, other AT1 receptor antagonists like telmisartan, candesartan CV and olmesartan fail to do so. This is due to the inability of these AT1 receptor antagonists to cause interdigitation effects. Interestingly, captopril (ACE inhibitor) causes increase of ΔH only when cholesterol is present, showing a distinct thermal profile in comparison to other drugs acting on the RAS system (Table 1).

Thus, given the similar drug-membrane behaviour of losartan and aliskiren, AT1 and pro(renin) receptors are blocked when surrounded by cholesterol rich membrane rafts/caveolae (Fig. 9) [14].

3 Molecular Dynamics Studies on RAS Interacting Drugs

Molecular Dynamics (MD) simulations are an indispensable tool for elucidating biophysical phenomena such as protein-membrane or protein-drug interactions at an atomic level of detail [9, 19, 56–61]. The class of GPCRs, that the angiotensin

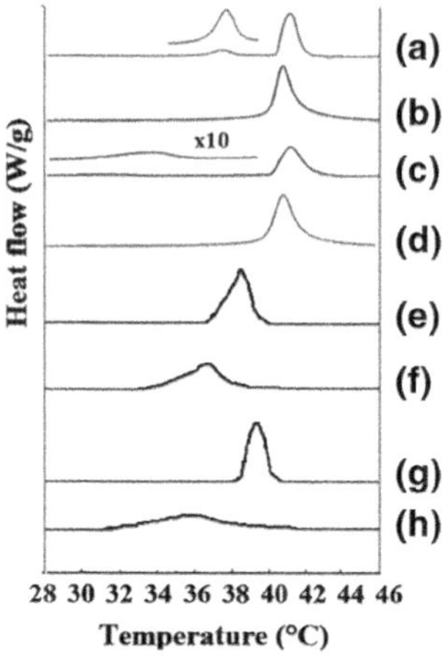

Fig. 8 Thermal scans of **a** DPPC bilayers alone; **b** DPPC bilayers containing x = 0.15 cholesterol; **c** DPPC containing x = 0.20 olmesartan; **d** [DPPC/cholesterol (x = 0.15)] containing x = 0.20 olmesartan; **e** DPPC containing aliskiren hemifumarate x = 0.20; **f** [DPPC/cholesterol (x = 0.15)] containing x = 0.20 aliskiren hemifumarate; **g** DPPC/captopril x = 0.20; **h** [DPPC/cholesterol (x = 0.15)] containing x = 0.20 captopril

Table 1 Mean ΔH values for three experiments (values ranged within 10%) of DPPC with and without drug acting on the RAS system and DPPC/cholesterol (0.85/0.15 molar ratio) with and without drug acting on the RAS system

	Pretransition	Main Phase Transition	ΔH	
Sample	Kcal/mol	J/g	Kcal/mol	J/g
DPPC	1.105	6.3	7.33	41.80
DPCC/cholesterol	–	–	5.12	29.20
DPPC/aliskiren hemifumarate	–	–	10.45	59.59
[DPCC/cholesterol]/aliskiren hemifumarate	–	–	7.33	41.77
DPPC/olmesartan	0.561	3.2	7.49	42.70
[DPCC/cholesterol]/olmesartan)	–	–	5.44	31.00
DPPC/captopril	–	–	7.89	45.00
[DPPC/cholesterol]/captopril	–	–	7.02	39.99

The concentration of the drug used in the two systems was $x = 0.20$

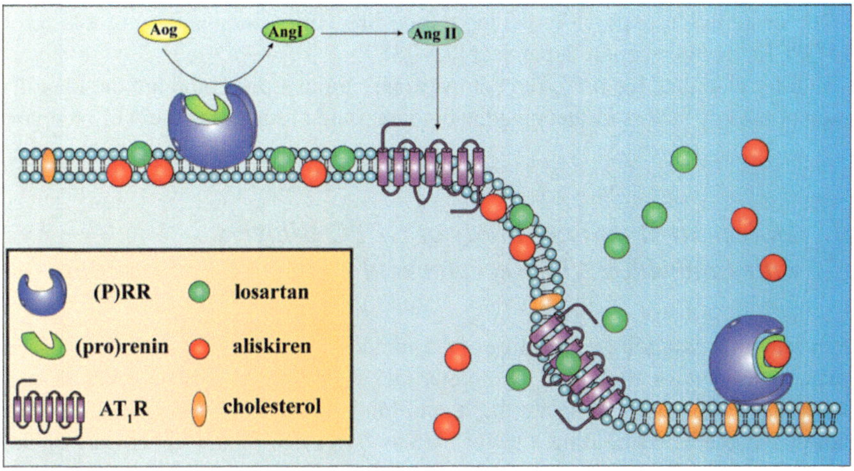

Fig. 9 Schematic representation of cholesterol-induced release of aliskiren and losartan. Left hand side: aliskiren and losartan accumulate in the lipid bilayer. Angiotensinogen (Aog) is converted by (P)RR-bound renin to AI and then by the angiotensin-converting enzyme (ACE) to the active peptide AII. Right hand side: A cholesterol rich caveolae is formed and both, aliskiren and losartan are released from the bilayer to block the receptors [55]

receptors belong to, have been extensively studied using several MD simulations techniques [62, 63]. Filizola et al. performed MD simulations to study the rhodopsin monomer and dimer in an explicit POPC membrane [64]. Biased (steered) and unbiased MD simulations have been also used to study the activation of the serotonin human 5-HT2A receptor, including an agonist and a GRq peptide [65]. Moreover, ligand dissociation from the M3 muscarinic acetylcholine receptor has been studied using all-atom MD simulations where the application of an external force pulled the ligand away from the receptor, but not along any prespecified pathway or direction providing important information regarding the energy barriers that the M3 ligands need to surpass in order for binding to take place [66].

Experimental and molecular modeling studies of losartan in DPPC bilayers indicate that the location of the drug is found at the water-lipid interface, favoring electrostatic interactions between the hydroxyl and tetrazole groups of the drug with the polar headgroups and diffusing water molecules [27, 67]. This observation was also confirmed by a combined experimental and MD study, where losartan was simulated in DPPC bilayers in the presence of the AT1 receptor [19].

Despite the extensive studies that have focused on the binding route of sartans to AT1 receptor [19, 68–70], the exact pathway to the binding site of AT1 has not been elucidated and thus both direct and membrane-mediated routes could underlie in the drug action. However, it was recently shown that losartan does not favor cholesterol-rich membrane domains: thus cholesterol concentration and distribution is another factor that needs to be addressed when investigating the pathway of losartan binding to AT1 receptor.

Recent studies have provided evidence for a direct interaction between cholesterol and GPCRs that is receptor dependent [71–73].

To fully elucidate the preferred pathway, MD simulations could aid the identification of the key factors influencing the binding route of sartans to the AT1 receptor.

3.1 Molecular Dynamics Study of DPPC Bilayers with Embedded AT1 Receptor and Approaching Losartan

Past attempts for homology modeling of the human or the rat AT1 receptors have used structural models of rhodopsin photointermediates and site-directed mutagenesis studies have been used to locate the ligand binding site [74]. According to these studies, the predicted binding site for sartans is located at the upper part of the receptor between the transmembrane helices 3–7 (TM3-7) and the extracellular loop 2 (ECL2); a prediction that was later confirmed by both crystal structures of the AT1 receptor [2, 20]. Moreover, mutational studies have revealed that ligand-specific conformational changes induced to the ECL2 are responsible for the switch between the active and inactive state of the receptor [75]. Earlier MD studies of the losartan-AT1 receptor have been based on this rhodopsin model [19].

Since sartans' pathway to approach the binding site of AT1 receptor has not been elucidated, both direct and membrane-mediated routes could underlie the drug action. Thus, the hypothesis of the lateral diffusion of the antagonist through the membrane in order to access the binding site merits further investigation. Towards this goal, in a recent study we have investigated the (i) the conformational dynamics of losartan (in its neutral form) in a lipid bilayer and at the vicinal region of AT1 receptor; (ii) the positioning of losartan in the bilayer; and (iii) the diffusion pathway of losartan in the bilayer. Using a homology model of the AT1 receptor that was constructed based on the structure of bovine rhodopsin receptor, it was shown that losartan exhibits a strong preference for the membrane environment with respect to the water phase, which is not surprising given the significant hydrophobic component of the drug solvent-accessible surface area and the absence of the negative charge in its neutral form. Within the first few ns all losartan molecules partition spontaneously in the DPPC bilayer from the water phase with the exception of one molecule, which interacts with the extracellular part of the AT1 receptor. Losartan molecules diffuse freely along the membrane plane remaining close to lipid head groups and the water:bilayer interface. Both experimental and simulation results point to an orientation of losartan at the interface between water and the lipid bilayer.

Only one losartan molecule out of eight was found to encounter the receptor from the water phase by interacting with the extracellular loops throughout the ~600 ns of the MD trajectory. The rest are found to diffuse freely within the bilayer though without approaching the receptor. Various reasons may be formulated justifying the inability of the drug to reach the binding site in this time scale either from the extracellular surface or through the bilayer core.

The molecular mechanism of drug-membrane interactions poses an issue of paramount significance for the design and screening of novel drugs. Non-specific binding to the membrane or membrane-embedded receptors could hinder drug access to the binding site; however, diffusion through the membrane pathway could speed up the process of encountering the transmembrane binding site as compared to the random hit of the free ligand at the extracellular part of the receptor [76, 77]. A complex interplay of all these factors will ultimately determine the overall drug efficacy. Lastly, membrane incorporation of drug molecules could possibly aid the ligand to adopt a bioactive conformation in order to interact effectively with a membrane embedded receptor target [76].

The homology model of the AT1 receptor used in Zervou et al. [19], was constructed using the bovine rhodopsin structure as a template, where the extracellular loop 2 sits above the binding site and encloses it. Therefore, the rearrangements needed for our model to adopt the open state of the extracellular loop 2 may occur in the ms time scale, whereas the time scale reachable by the current state-of-the-art all-atom MD simulations is in the order of μs. Moreover, different experimental MD conditions may have to be used (i.e. higher temperature) in order to increase the flexibility of the ligand and receptor and thus avoid the trapping.

The complexity of the membrane in a realistic system (containing cholesterol, rigid ion channels, etc.) hinders the diffusion of the AT1 receptor compared to our model, where the receptor may freely diffuse in the plane of a pure DPPC bilayer. This higher mobility of the AT1 receptor may not favor the encounter with the drug molecules. Lastly, the diffusion of losartan toward the binding site of the AT1 receptor may slow down by its stronger binding to the headgroup and upper lipophilic segment of the lipid bilayers.

Recently-resolved crystal structures of the ligand activated turkey β1- and human β2-adrenergic, the human A2A-adenosine, the human dopamine D3 and the human chemokine CXCR4 GPCR class A receptors with antagonists bound, have identified structural differences of the AT1 receptor with respect to the rhodopsin receptor. Open clefts in the extracellular regions may guide the entrance of the sartans towards the binding site, whereas in rhodopsin the ECL-2 sits above the binding site and encloses it [68, 69]. Based on this structural information, the most recently developed homology model of AT1 receptor is based on the chemokine CXCR4 receptor [70, 78]. This model is characterized by an open ECL2 loop, presumably enabling easier doping of losartan and highlights the higher accessibility of the binding site to the extracellular environment. This AT1 receptor model was thus used in newer MD simulation studies of the AT1-losartan complex, which we report below.

3.2 DPPC Bilayers with Cholesterol and Embedded AT1 Receptor and Approaching Losartan

The crystallographic structure of AT1 receptor complexed with ZD7155 and olmesartan have been reported recently [2, 20]. In the meantime, in order to aid the identification of the key factors influencing the binding route of sartans to the AT1 receptor, we had performed an extensive MD study of the AT1 receptor embedded in a DPPC bilayer and in the presence of losartan using a homology model that was reported in the recent literature [70, 78]. Thus, whatever work is described here would be of interest to be applied for comparative reasons also in the crystal structure of AT1 receptor, substituting olmesartan with losartan.

It was recently shown that losartan tends to avoid cholesterol-rich membrane domains [79], and thus cholesterol concentration and distribution is another factor that needs to be addressed when investigating the pathway of losartan binding to the AT1 receptor. Thus, in our newer simulation study, we used a more realistic representation of the system, which included cholesterol in the membrane model and not only losartan but also its metabolite EXP-3174. Moreover, contrary to our previous studies, losartan was simulated in its anionic form, which is also the oral administration form. We employed MD simulations to assess losartan-membrane and losartan-AT1 receptor interactions in atomic resolution and provide information on the binding route of losartan and its metabolite EXP-3174 to the AT1 receptor. The most recently released model of the AT1 receptor (as described above) [70] was used, embedded in an explicit DMPC membrane with the inclusion of cholesterol was used providing a realistic model of the system. This model also provides the newest information on the losartan binding site. Using MD simulations, we monitored the unprompted insertion, the diffusion pathway and dynamics of losartan and EXP-3174 towards the AT1 receptor binding site. Current knowledge of the losartan pathway to the binding site of AT1 does not allow for a definite conclusion and thus both direct and membrane-mediated routes could underlie the drug action. In the present study, we try to elucidate the mechanism of losartan binding by MD simulations.

All-atom simulations of losartan/EXP-3174 in association with a model DMPC lipid bilayer with 40% cholesterol concentration and AT1 receptor embedded in it were performed. DMPC/cholesterol systems have been widely studied computationally [80, 81], and have been used previously for studies of membrane proteins [73].

Analysis of the positioning of losartan/EXP-3174 relative to the phosphate group of DMPC lipids indicates that the ligands lie close to the hydrophilic lipid heads. This behavior is different to our previous observations where losartan has been shown to lie deeper in the lipid bilayer. This difference is most probably caused by the presence of cholesterol in our membrane models, as cholesterol increases membrane's rigidity and decreases membrane's permeability [82, 83].

Losartan/EXP-3174 preference for DMPC lipid heads is also confirmed by the interaction energies and the hydrogen bonds between the small molecules and the other system components (Figs. 10, 11 and Table 2). Both electrostatic and VDW interactions of losartan and its metabolite with DMPC lipids are significantly stronger

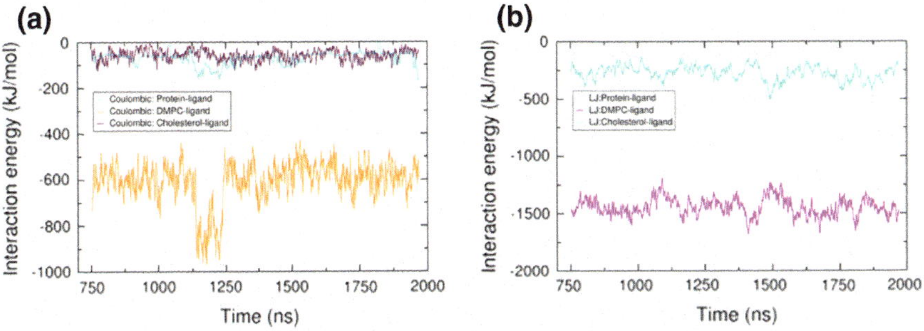

Fig. 10 Coulombic (**a**) and Lennard-Jones (**b**) interaction energies of losartan/EXP-3174 and the system components

compared to the interactions with cholesterol and AT1 receptor. Also, due to the negative charge that the ligands carry, the hydrogen bonds between the nitrogen group of DMPC and tetrazole group of losartan/EXP-3174 are significantly more than those observed in our previous study where losartan did not carry any net charge. The effect of the charge of the losartan/EXP-3174 is more intense in the hydrogen bonds formed between the ligand tetrazole group and water (Table 2). These hydrogen bonds influence the ligands' desolvation that is needed for the interaction with AT1 receptor and/or the membrane. In our simulations we captured a losartan molecule interacting with ECL2 residues Asn 174, Thr 175 and Ile 177 of AT1 receptor and after a few nanoseconds approaching the Angiotensin binding pocket and interacting with Ile172, His183, Ala181, Val179. However, due to the solvation cell around losartan as well as the narrow entrance of the protein losartan stops interacting with residues lining the inner part of the AT1 pore and starts interacting with DMPC lipids. Finally, the specific molecules positions itself at the AT1-membrane interface and interacts with Tyr99 through constant hydrogen bonds and with Leu100 and Ile103 through hydrophobic interactions through its biphenyl and butyl groups respectively. A major issue in modeling GPCRs is selecting appropriate structural model or state, monomeric or interacting with other proteins in order to capture the phenomena of interest, since only a few crystal structures are available [84]. Simulation studies like the ones presented herein provide key information about the systems, potential obstacles, methodology and model limitations and most importantly may give directions for future experimental and computational investigations.

4 Future Perspectives

The studies of the drug-membrane interactions in the RAS system are fruitful and offer more challenges in the future because: (i) the role of lipid bilayers in the drug action needs to be further enlightened and the advance of MD calculations contributed already significantly in the development of this field. It is of paramount importance

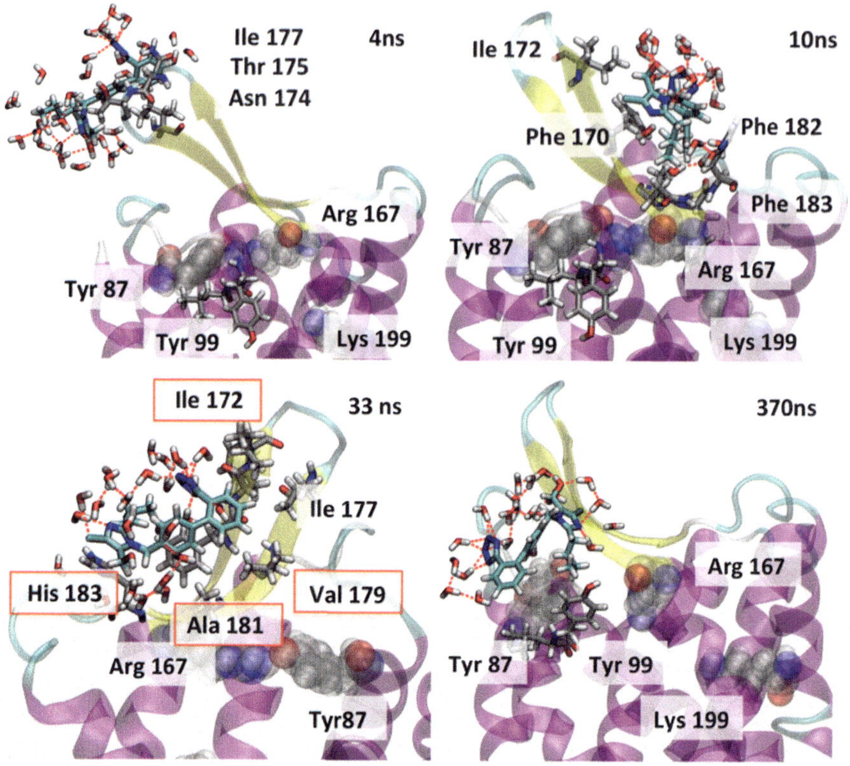

Fig. 11 Representative snapshots of a losartan molecule approaching the ECL2 of AT1 receptor. Water molecules are shown in licorice representation in red for oxygen and white for hydrogen atoms. Tyr87, Tyr99, Arg167 and Lys199 are shown in VDW spheres while other key residues as well as losartan are shown in licorice. The atoms are colored by atom type. The protein is shown in ribbons colored by secondary structure

for highly lipophilic molecules such as AT1 receptor antagonists to understand their mode of action. The question if they are first embedded in the lipidic core and then diffuse to the AT1 receptor is still not answered; (ii) the basic understanding of the simple system containing the lipid bilayers and the inserted drug through the use of biophysical techniques and MD is also crucial because more complicated systems can be studied containing various important components of lipid bilayers such as cholesterol; (iii) Studies seeking to examine the molecular basis of action of drugs can be extended to understand the formulation of drugs in deliverable vehicles. In the literature, there are already studies of the effects of cyclodextrins (CDs) in the drug encapsulation embedded in liposomes. The development of improved formulations is not an easy task as the drug must surpass an array of enemies ready to change its structural integrity. Drug-in-CD-in liposome (DCL) formulations are used to guide molecules to their target. The detailed knowledge of drug-membrane interactions in such systems will provide new potential in the formulations of low water-soluble

Table 2 Average number of hydrogen bonds

Hydrogen-bonded pairs	Number of hydrogen bonds	
	Current study	Zervou et al. [19]
Tetrazole—DPPC glycerol backbone	2.7 ± 1.6	3.0 ± 1.8
Hydroxyl—DPPC glycerol backbone	1.1 ± 1.0	1.0 ± 1.2
Hydroxyl—DPPC headgroup	1.7 ± 1.2 (PO_4) 1.3 ± 1.1 (NC_3)	4.0 ± 2.3
Tetrazole—DPPC headgroup	1.0 ± 1.0 (PO_4) 7.9 ± 3.2 (NC_3)	1.0 ± 1.2
Tetrazole—water	40.1 ± 6.0	9.2 ± 4.3
Hydroxyl—water	8.5 ± 2.2	8.0 ± 3.0

drugs such as AT1 receptor antagonists [85]. (iv) Comparative studies using MD calculations and experimental techniques must be performed in order to improve their predictability and develop more sophisticated systems simulating the biological environment; (v) In conclusion, the development of elegant experiments that promote the knowledge of the mode of drug-membrane interactions is desirable. Such experiments will aid the drug design and the development of safer drugs.

Acknowledgements This work was supported by CERIC funded programme (proposal number 20152002), for the promotion of the exchange and scientific cooperation between Greece and Germany IKYDA 2015 and by the Cy-Tera Project (ΝΕΑ ΥΠΟΔΟΜΗ/ΣΤΡΑΤΗ/0308/31), which is co-funded by the European Regional Development Fund and the Republic of Cyprus through the Research Promotion Foundation.

References

1. Laragh, J.H., Baer, L., Brunner, H.R., Buhler, F.R., Vaughan, J.E.: Renin, angiotensin and aldosterone system in pathogenesis and management of hypertensive vascular disease. Am. J. Med. **52**(5), 633–652 (1972). https://doi.org/10.1016/0002-9343(72)90054-X
2. Zhang, H., Unal, H., Gati, C., Han, G.W., Liu, W., Zatsepin, N.A., James, D., Wang, D., Nelson, G., Weierstall, U., Sawaya, M.R., Xu, Q., Messerschmidt, M., Williams, G.J., Boutet, S., Yefanov, O.M., White, T.A., Wang, C., Ishchenko, A., Tirupula, K.C., Desnoyer, R., Coe, J., Conrad, C.E., Fromme, P., Stevens, R.C., Katritch, V., Karnik, S.S., Cherezov, V.: Structure of the Angiotensin receptor revealed by serial femtosecond crystallography. Cell **161**(4), 833–844 (2015). https://doi.org/10.1016/j.cell.2015.04.011
3. Nirula, V., Zheng, W., Sothinathan, R., Sandberg, K.: Interaction of biphenylimidazole and imidazoleacrylic acid nonpeptide antagonists with valine 108 in TM III of the AT1 angiotensin receptor. Br. J. Pharmacol. **119**(8), 1505–1507 (1996)
4. Dascal, D., Nirula, V., Lawus, K., Yoo, S.E., Walsh, T.F., Sandberg, K.: Shared determinants of receptor binding for subtype selective, and dual endothelin-angiotensin antagonists on the

AT1 angiotensin II receptor. FEBS Lett. **423**(1), 15–18 (1998)
5. Ji, H., Zheng, W., Zhang, Y., Catt, K.J., Sandberg, K.: Genetic transfer of a nonpeptide antagonist binding site to a previously unresponsive angiotensin receptor. Proc. Natl. Acad. Sci. U S A **92**(20), 9240–9244 (1995)
6. Sakarellos, C., Lintner, K., Piriou, F., Fermandjian, S.: Conformation of the central sequence of angiotensin II and analogs. Biopolymers **22**(2), 663–687 (1983). https://doi.org/10.1002/bip.360220210
7. Balakumar, P., Jagadeesh, G.: Structural determinants for binding, activation, and functional selectivity of the angiotensin AT1 receptor. J. Mol. Endocrinol. **53**(2), R71–92 (2014). https://doi.org/10.1530/jme-14-0125
8. Leonis, G., Avramopoulos, A., Salmas, R.E., Durdagi, S., Yurtsever, M., Papadopoulos, M.G.: Elucidation of conformational states, dynamics, and mechanism of binding in human κ-opioid receptor complexes. J. Chem. Inf. Model. **54**(8), 2294–2308 (2014). https://doi.org/10.1021/ci5002873
9. Dror, R.O., Pan, A.C., Arlow, D.H., Borhani, D.W., Maragakis, P., Shan, Y., Xu, H., Shaw, D.E.: Pathway and mechanism of drug binding to G-protein-coupled receptors. Proc. Natl. Acad. Sci. U S A **108**(32), 13118–13123 (2011). https://doi.org/10.1073/pnas.1104614108
10. Shan, Y., Kim, E.T., Eastwood, M.P., Dror, R.O., Seeliger, M.A., Shaw, D.E.: How does a drug molecule find its target binding site? J. Am. Chem. Soc. **133**(24), 9181–9183 (2011). https://doi.org/10.1021/ja202726y
11. Rhodes, D.G., Sarmiento, J.G., Herbette, L.G.: Kinetics of binding of membrane-active drugs to receptor sites. Diffusion-limited rates for a membrane bilayer approach of 1,4-dihydropyridine calcium channel antagonists to their active site. Mol. Pharmacol. **27**(6), 612–623 (1985)
12. Kellici, T.F., Ntountaniotis, D., Kritsi, E., Zervou, M., Zoumpoulakis, P., Potamitis, C., Durdagi, S., Salmas, R.E., Ergun, G., Gokdemir, E., Halabalaki, M., Gerothanassis, I.P., Liapakis, G., Tzakos, A., Mavromoustakos, T.: Leveraging NMR and X-ray data of the free ligands to build better drugs targeting Angiotensin II Type 1 G-protein coupled receptor. Curr. Med. Chem. **23**(1), 36–59 (2016)
13. Kellici, T.F., Ntountaniotis, D., Leonis, G., Chatziathanasiadou, M., Chatzikonstantinou, A.V., Becker-Baldus, J., Glaubitz, C., Tzakos, A.G., Viras, K., Chatzigeorgiou, P., Tzimas, S., Kefala, E., Valsami, G., Archontaki, H., Papadopoulos, M.G., Mavromoustakos, T.: Investigation of the interactions of silibinin with 2-hydroxypropyl-beta-cyclodextrin through biophysical techniques and computational methods. Mol. Pharm. **12**(3), 954–965 (2015). https://doi.org/10.1021/mp5008053
14. Ntountaniotis, D., Kellici, T., Tzakos, A., Kolokotroni, P., Tselios, T., Becker-Baldus, J., Glaubitz, C., Lin, S., Makriyannis, A., Mavromoustakos, T.: The application of solid-state NMR spectroscopy to study candesartan cilexetil (TCV-116) membrane interactions. Comparative study with the AT1R antagonist drug olmesartan. Biochim. Biophys. Acta **1838**(10), 2439–2450 (2014). https://doi.org/10.1016/j.bbamem.2014.06.003
15. Kellici, T.F., Tzakos, A.G., Mavromoustakos, T.: Rational drug design and synthesis of molecules targeting the angiotensin II type 1 and type 2 receptors. Molecules **20**(3), 3868–3897 (2015). https://doi.org/10.3390/molecules20033868
16. Kellici, T., Ntountaniotis, D., Vrontaki, E., Liapakis, G., Moutevelis-Minakakis, P., Kokotos, G., Hadjikakou, S.G., Tzakos, A., Afantitis, A., Melagraki, G., Bryant, S., Langer, T., Di Marzo, V., Mavromoustakos, T.: Rational drug design paradigms: the Odyssey for designing better drugs. Comb. Chem. High Throughput Screen. **18**(3), 238–256 (2015)
17. Kellici, T.F., Ntountaniotis, D., Liapakis, G., Tzakos, A.G., Mavromoustakos, T.: The dynamic properties of angiotensin II type 1 receptor inverse agonists in solution and in the receptor site. Arab. J. Chem. (2016). https://doi.org/10.1016/j.arabjc.2016.11.014
18. Moschonas, I.C., Kellici, T.F., Mavromoustakos, T., Stathopoulos, P., Tsikaris, V., Magafa, V., Tzakos, A.G., Tselepis, A.D.: Molecular requirements involving the human platelet protease-activated receptor-4 mechanism of activation by peptide analogues of its tethered-ligand. Platelets, 1–10 (2017). https://doi.org/10.1080/09537104.2017.1282607

19. Zervou, M., Cournia, Z., Potamitis, C., Patargias, G., Durdagi, S., Grdadolnik, S.G.: Mavromoustakos T (2014) Insights into the molecular basis of action of the AT1 antagonist losartan using a combined NMR spectroscopy and computational approach. Biochim. Biophys. Acta **3**, 1031–1046 (1838). https://doi.org/10.1016/j.bbamem.2013.12.012
20. Zhang, H., Unal, H., Desnoyer, R., Han, G.W., Patel, N., Katritch, V., Karnik, S.S., Cherezov, V., Stevens, R.C.: Structural basis for ligand recognition and functional selectivity at angiotensin receptor. J. Biol. Chem. **290**(49), 29127–29139 (2015). https://doi.org/10.1074/jbc.M115.689000
21. Oliveira, T.R., Lamy, M.T., De Paula, U.M., Guimaraes, L.L., Toledo, M.S., Takahashi, H.K., Straus, A.H., Lindsey, C.J., Paiva, T.B.: Structural properties of lipid reconstructs and lipid composition of normotensive and hypertensive rat vascular smooth muscle cell membranes. Braz. J. Med. Biol. Res. **42**(9), 844–853 (2009)
22. Netticadan, T.J., Ashavaid, T.F., Nair, K.G.: Characterisation of the canine cardiac sarcolemma in experimental myocardial ischemia. Indian J. Clin. Biochem. **12**(1), 49–54 (1997). https://doi.org/10.1007/BF02867955
23. Vist, M.R., Davis, J.H.: Phase equilibria of cholesterol/dipalmitoylphosphatidylcholine mixtures: 2H nuclear magnetic resonance and differential scanning calorimetry. Biochemistry **29**(2), 451–464 (1990)
24. Marsh, D.: Liquid-ordered phases induced by cholesterol: a compendium of binary phase diagrams. Biochim. Biophys. Acta **1798**(3), 688–699 (2010). https://doi.org/10.1016/j.bbamem.2009.12.027
25. Lingwood, D., Simons, K.: Lipid rafts as a membrane-organizing principle. Science **327**(5961), 46–50 (2010). https://doi.org/10.1126/science.1174621
26. Zoumpoulakis, P., Daliani, I., Zervou, M., Kyrikou, I., Siapi, E., Lamprinidis, G., Mikros, E., Mavromoustakos, T.: Losartan's molecular basis of interaction with membranes and AT1 receptor. Chem. Phys. Lipids **125**(1), 13–25 (2003)
27. Fotakis, C., Christodouleas, D., Chatzigeorgiou, P., Zervou, M., Benetis, N.P., Viras, K., Mavromoustakos, T.: Development of a CP ^{31}P NMR broadline simulation methodology for studying the interactions of antihypertensive AT1 antagonist losartan with phospholipid bilayers. Biophys. J. **96**(6), 2227–2236 (2009). https://doi.org/10.1016/j.bpj.2008.11.057
28. Theodoropoulou, E., Marsh, D.: Interactions of angiotensin II non-peptide AT(1) antagonist losartan with phospholipid membranes studied by combined use of differential scanning calorimetry and electron spin resonance spectroscopy. Biochim. Biophys. Acta **1461**(1), 135–146 (1999)
29. Schneider, M.F., Marsh, D., Jahn, W., Kloesgen, B., Heimburg, T.: Network formation of lipid membranes: triggering structural transitions by chain melting. Proc. Natl. Acad. Sci. U S A **96**(25), 14312–14317 (1999)
30. Estep, T.N., Mountcastle, D.B., Biltonen, R.L., Thompson, T.E.: Studies on the anomalous thermotropic behavior of aqueous dispersions of dipalmitoylphosphatidylcholine-cholesterol mixtures. Biochemistry **17**(10), 1984–1989 (1978)
31. Bruggemann, E.P., Melchior, D.L.: Alterations in the organization of phosphatidylcholine/cholesterol bilayers by tetrahydrocannabinol. J. Biol. Chem. **258**(13), 8298–8303 (1983)
32. Mavromoustakos, T., Yang, D.P., Makriyannis, A.: Effects of the anesthetic steroid alphaxalone and its inactive delta 16-analog on the thermotropic properties of membrane bilayers. A model for membrane perturbation. Biochim. Biophys. Acta **1239**(2), 257–264 (1995)
33. O'Leary, T.J., Levin, I.W.: Raman spectroscopic study of an interdigitated lipid bilayer. Dipalmitoylphosphatidylcholine dispersed in glycerol. Biochim. Biophys. Acta **776**(2), 185–189 (1984)
34. Hsu, J.C., Yip, C.M.: Molecular dynamics simulations of indolicidin association with model lipid bilayers. Biophys. J. **92**(12), L100–102 (2007). https://doi.org/10.1529/biophysj.107.108050
35. Tian, X., Pavlopoulos, S., Yang, D.P., Makriyannis, A.: The interaction of cannabinoid receptor agonists, CP55940 and WIN55212-2 with membranes using solid state 2H NMR. Biochim. Biophys. Acta **1808**(9), 2095–2101 (2011). https://doi.org/10.1016/j.bbamem.2010.11.026

36. Ntountaniotis, D., Mali, G., Grdadolnik, S.G., Halabalaki, M., Skaltsounis, A.L., Potamitis, C., Siapi, E., Chatzigeorgiou, P., Rappolt, M., Mavromoustakos, T.: Thermal, dynamic and structural properties of drug AT1 antagonist olmesartan in lipid bilayers. Biochim. Biophys. Acta **1808**(12), 2995–3006 (2011). https://doi.org/10.1016/j.bbamem.2011.08.001
37. Mavromoustakos, T., Theodoropoulou, E., Yang, D.P.: The use of high-resolution solid-state NMR spectroscopy and differential scanning calorimetry to study interactions of anaesthetic steroids with membrane. Biochim. Biophys. Acta **1328**(1), 65–73 (1997)
38. Yang, D.P., Mavromoustakos, T., Beshah, K., Makriyannis, A.: Amphipathic interactions of cannabinoids with membranes. A comparison between Δ8-THC and its O-methyl analog using differential scanning calorimetry, X-ray diffraction and solid state 2H-NMR. BBA—Biomembr. **1103**(1), 25–36 (1992). https://doi.org/10.1016/0005-2736(92)90053-o
39. Mavromoustakos, T., Theodoropoulou, E.: A combined use of 13C-cross polarization/magic angle spinning, 13C-magic angle spinning and ^{31}P-nuclear magnetic resonance spectroscopy with differential scanning calorimetry to study cannabinoid-membrane interactions. Chem. Phys. Lipids **92**(1), 37–52 (1998). https://doi.org/10.1016/S0009-3084(98)00005-X
40. Mavromoustakos, T., Daliani, I.: Effects of cannabinoids in membrane bilayers containing cholesterol. Biochim. Biophys. Acta **1420**(1–2), 252–265 (1999)
41. Santos, J.S., Lee, D.K., Ramamoorthy, A.: Effects of antidepressants on the conformation of phospholipid headgroups studied by solid-state NMR. Magn. Reson. Chem. **42**(2), 105–114 (2004). https://doi.org/10.1002/mrc.1327
42. O'Leary, T.J., Ross, P.D., Levin, I.W.: Effects of anesthetic and nonanesthetic steroids on dipalmitoylphosphatidylcholine liposomes: a calorimetric and Raman spectroscopic investigation. Biochemistry **23**(20), 4636–4641 (1984)
43. Potamitis, C., Chatzigeorgiou, P., Siapi, E., Viras, K., Mavromoustakos, T., Hodzic, A., Pabst, G., Cacho-Nerin, F., Laggner, P.: Rappolt M (2011) Interactions of the AT1 antagonist valsartan with dipalmitoyl-phosphatidylcholine bilayers. Biochim. Biophys. Acta **6**, 1753–1763 (1808). https://doi.org/10.1016/j.bbamem.2011.02.002
44. Colthup, N.B., Daly, L.H., Wiberley, S.E.: Introduction to Infrared and Raman Spectroscopy, 3rd edn. Academic Press, Boston (1990)
45. Levin, I.W., Lewis, E.N.: Fourier transform Raman spectroscopy of biological materials. Anal. Chem. **62**(21), 1101A–1111A (1990)
46. Bista, R.K., Bruch, R.F., Covington, A.M.: Variable-temperature Raman spectro-microscopy for a comprehensive analysis of the conformational order in PEGylated lipids. J. Raman Spectrosc. **40**(4), 463–471 (2009). https://doi.org/10.1002/jrs.2156
47. Gaber, B.P., Peticolas, W.L.: On the quantitative interpretation of biomembrane structure by Raman spectroscopy. Biochim. Biophys. Acta **465**(2), 260–274 (1977)
48. Benetis, N.-P., Kyrikou, I., Zervou, M., Mavromoustakos, T.: Static CP ^{31}P NMR multilamellar bilayer broadlines in the absence and presence of the bioactive dipeptide β-Ala-Tyr or Glu. Chem. Phys. **314**(1–3), 57–72 (2005). https://doi.org/10.1016/j.chemphys.2005.01.028
49. Kyrikou, I., Hadjikakou, S.K., Kovala-Demertzi, D., Viras, K., Mavromoustakos, T.: Effects of non-steroid anti-inflammatory drugs in membrane bilayers. Chem. Phys. Lipids **132**(2), 157–169 (2004). https://doi.org/10.1016/j.chemphyslip.2004.06.005
50. Michel, M.C., Foster, C., Brunner, H.R., Liu, L.: A systematic comparison of the properties of clinically used angiotensin II type 1 receptor antagonists. Pharmacol. Rev. **65**(2), 809–848 (2013). https://doi.org/10.1124/pr.112.007278
51. Tosco, P., Rolando, B., Fruttero, R., Henchoz, Y., Martel, S., Carrupt, P.A., Gasco, A.: Physicochemical profiling of sartans: a detailed study of ionization constants and distribution coefficients. Helv. Chim. Acta **91**(3), 468–482 (2008). https://doi.org/10.1002/hlca.200890051
52. Kritsi, E., Potamitis, C., Durdagi, S., Zoumpoulakis, P., Golic Grdadolnik, S., Mavromoustakos, T.: Molecular insights into the AT1 antagonism based on biophysical and in silico studies of telmisartan. Med. Chem. Res. **22**(10), 4842–4857 (2013). https://doi.org/10.1007/s00044-012-0464-5
53. Liossi, A.S., Ntountaniotis, D., Kellici, T.F., Chatziathanasiadou, M.V., Megariotis, G., Mania, M., Becker-Baldus, J., Kriechbaum, M., Krajnc, A., Christodoulou, E., Glaubitz, C., Rappolt,

M., Amenitsch, H., Mali, G., Theodorou, D.N., Valsami, G., Pitsikalis, M., Iatrou, H., Tzakos, A.G.: Mavromoustakos T (2017) Exploring the interactions of irbesartan and irbesartan-2-hydroxypropyl-beta-cyclodextrin complex with model membranes. Biochim. Biophys. Acta **6**, 1089–1098 (1859). https://doi.org/10.1016/j.bbamem.2017.03.003
54. Bocskei, Z., Simon, K., Rao, R., Caron, A., Rodger, C.A., Bauer, M.: Irbesartan Crystal Form B. Acta Crystallogr Sect C: Cryst. Struct. Commun. **54**(6), 808–810 (1998). doi:https://doi.org/10.1107/s0108270197019884
55. Sadeghpour, A., Rappolt, M., Ntountaniotis, D., Chatzigeorgiou, P., Viras, K., Megariotis, G., Papadopoulos, M.G., Siapi, E., Mali, G.: Mavromoustakos T (2015) Comparative study of interactions of aliskiren and AT1 receptor antagonists with lipid bilayers. Biochim. Biophys. Acta **4**, 984–994 (1848). https://doi.org/10.1016/j.bbamem.2014.12.004
56. Ivetac, A., Andrew McCammon, J.: Mapping the druggable allosteric space of G-Protein coupled receptors: a fragment-based molecular dynamics approach. Chem. Biol. Drug Des. **76**(3), 201–217 (2010). https://doi.org/10.1111/j.1747-0285.2010.01012.x
57. Gkeka, P., Eleftheratos, S., Kolocouris, A., Cournia, Z.: Free energy calculations reveal the origin of binding preference for aminoadamantane blockers of influenza A/M2TM pore. J. Chem. Theory Comput. **9**(2), 1272–1281 (2013). https://doi.org/10.1021/ct300899n
58. Vanni, S., Neri, M., Tavernelli, I., Rothlisberger, U.: Predicting novel binding modes of agonists to β adrenergic receptors using all-atom molecular dynamics simulations. PLoS Comp. Biol. **7**(1) (2011). https://doi.org/10.1371/journal.pcbi.1001053
59. Gkeka, P., Sarkisov, L.: Interactions of phospholipid bilayers with several classes of amphiphilic α-helical peptides: insights from coarse-grained molecular dynamics simulations. J. Phys. Chem. B **114**(2), 826–839 (2010). https://doi.org/10.1021/jp908320b
60. Gkeka, P., Papafotika, A., Christoforidis, S., Cournia, Z.: Exploring a Non-ATP pocket for potential allosteric modulation of PI3 Kα. J. Phys. Chem. B **119**(3), 1002–1016 (2015). https://doi.org/10.1021/jp506423e
61. Cournia, Z., Allen, T.W., Andricioaei, I., Antonny, B., Baum, D., Brannigan, G., Buchete, N.V., Deckman, J.T., Delemotte, L., Del Val, C., Friedman, R., Gkeka, P., Hege, H.C., Henin, J., Kasimova, M.A., Kolocouris, A., Klein, M.L., Khalid, S., Lemieux, M.J., Lindow, N., Roy, M., Selent, J., Tarek, M., Tofoleanu, F., Vanni, S., Urban, S., Wales, D.J., Smith, J.C., Bondar, A.N.: Membrane protein structure, function, and dynamics: a perspective from experiments and theory. J. Membr. Biol. **248**(4), 611–640 (2015). https://doi.org/10.1007/s00232-015-9802-0
62. Periole, X., Huber, T., Marrink, S.-J., Sakmar, T.P.: G protein-coupled receptors self-assemble in dynamics simulations of model bilayers. J. Am. Chem. Soc. **129**(33), 10126–10132 (2007). https://doi.org/10.1021/ja0706246
63. Periole, X., Knepp, A.M., Sakmar, T.P., Marrink, S.J., Huber, T.: Structural determinants of the supramolecular organization of G protein-coupled receptors in bilayers. J. Am. Chem. Soc. **134**(26), 10959–10965 (2012). https://doi.org/10.1021/ja303286e
64. Filizola, M., Wang, S.X., Weinstein, H.: Dynamic models of G-protein coupled receptor dimers: indications of asymmetry in the rhodopsin dimer from molecular dynamics simulations in a POPC bilayer. J. Comput. Aided Mol. Des. **20**(7), 405–416 (2006). https://doi.org/10.1007/s10822-006-9053-3
65. Ísberg, V., Balle, T., Sander, T., Jørgensen, F.S., Gloriam, D.E.: G protein- and agonist-bound Serotonin 5-HT2A receptor model activated by steered molecular dynamics simulations. J. Chem. Inf. Model. **51**(2), 315–325 (2011). https://doi.org/10.1021/ci100402f
66. Kruse, A.C., Hu, J., Pan, A.C., Arlow, D.H., Rosenbaum, D.M., Rosemond, E., Green, H.F., Liu, T., Chae, P.S., Dror, R.O., Shaw, D.E., Weis, W.I., Wess, J., Kobilka, B.K.: Structure and dynamics of the M3 muscarinic acetylcholine receptor. Nature **482**(7386), 552–556 (2012)
67. Mavromoustakos, T., Durdagi, S., Koukoulitsa, C., Simcic, M., Papadopoulos, M.G., Hodoscek, M., Grdadolnik, S.G.: Strategies in the rational drug design. Curr. Med. Chem. **18**(17), 2517–2530 (2011)
68. Topiol, S., Sabio, M.: X-ray structure breakthroughs in the GPCR transmembrane region. Biochem. Pharmacol. **78**(1), 11–20 (2009). https://doi.org/10.1016/j.bcp.2009.02.012

69. Congreve, M., Langmead, C.J., Mason, J.S., Marshall, F.H.: Progress in structure based drug design for G protein-coupled receptors. J. Med. Chem. **54**(13), 4283–4311 (2011). https://doi.org/10.1021/jm200371q
70. Matsoukas, M.T., Potamitis, C., Plotas, P., Androutsou, M.E., Agelis, G., Matsoukas, J., Zoumpoulakis, P.: Insights into AT1 receptor activation through AngII binding studies. J. Chem. Inf. Model. **53**(11), 2798–2811 (2013). https://doi.org/10.1021/ci4003014
71. Oates, J., Watts, A.: Uncovering the intimate relationship between lipids, cholesterol and GPCR activation. Curr. Opin. Struct. Biol. **21**(6), 802–807 (2011). https://doi.org/10.1016/j.sbi.2011.09.007
72. Sengupta, D., Chattopadhyay, A.: Identification of cholesterol binding sites in the Serotonin1A receptor. J. Phys. Chem. B **116**(43), 12991–12996 (2012). https://doi.org/10.1021/jp309888u
73. Khelashvili, G., Grossfield, A., Feller, S.E., Pitman, M.C., Weinstein, H.: Structural and dynamic effects of cholesterol at preferred sites of interaction with rhodopsin identified from microsecond length molecular dynamics simulations. Proteins **76**(2), 403–417 (2009). https://doi.org/10.1002/prot.22355
74. Fillion, D., Cabana, J., Guillemette, G., Leduc, R., Lavigne, P., Escher, E.: Structure of the human angiotensin II Type 1 (AT(1)) receptor bound to angiotensin II from multiple chemoselective photoprobe contacts reveals a unique peptide binding mode. J. Biol. Chem. **288**(12), 8187–8197 (2013). https://doi.org/10.1074/jbc.M112.442053
75. Unal, H., Jagannathan, R., Bhatnagar, A., Tirupula, K., Desnoyer, R., Karnik, S.S.: Long range effect of mutations on specific conformational changes in the extracellular loop 2 of angiotensin II type 1 receptor. J. Biol. Chem. **288**(1), 540–551 (2013). https://doi.org/10.1074/jbc.M112.392514
76. Vauquelin, G., Packeu, A.: Ligands, their receptors and … plasma membranes. Mol. Cell Endocrinol. **311**(1–2), 1–10 (2009). https://doi.org/10.1016/j.mce.2009.07.022
77. Seddon, A.M., Casey, D., Law, R.V., Gee, A., Templer, R.H., Ces, O.: Drug interactions with lipid membranes. Chem. Soc. Rev. **38**(9), 2509–2519 (2009). https://doi.org/10.1039/b813853m
78. Matsoukas, M.T., Cordomi, A., Rios, S., Pardo, L., Tselios, T.: Ligand binding determinants for angiotensin II type 1 receptor from computer simulations. J. Chem. Inf. Model. **53**(11), 2874–2883 (2013). https://doi.org/10.1021/ci400400m
79. Hodzic, A., Zoumpoulakis, P., Pabst, G., Mavromoustakos, T., Rappolt, M.: Losartan's affinity to fluid bilayers modulates lipid-cholesterol interactions. Phys. Chem. Chem. Phys. **14**(14), 4780–4788 (2012). https://doi.org/10.1039/c2cp40134g
80. de Meyer, F.J.M., Benjamini, A., Rodgers, J.M., Misteli, Y., Smit, B.: Molecular simulation of the DMPC-cholesterol phase diagram. J. Phys. Chem. B **114**(32), 10451–10461 (2010). https://doi.org/10.1021/jp103903s
81. Bennett, W.F.D., Tieleman, D.P.: Computer simulations of lipid membrane domains. Biochim. Biophys. Acta **1828**(8), 1765–1776 (2013). https://doi.org/10.1016/j.bbamem.2013.03.004
82. Corvera, E., Mouritsen, O.G., Singer, M.A., Zuckermann, M.J.: The permeability and the effect of acyl-chain length for phospholipid bilayers containing cholesterol: theory and experiment. Biochim. Biophys. Acta **1107**(2), 261–270 (1992)
83. Haines, T.H.: Do sterols reduce proton and sodium leaks through lipid bilayers? Prog. Lipid Res. **40**(4), 299–324 (2001)
84. Grossfield, A.: Recent progress in the study of G protein-coupled receptors with molecular dynamics computer simulations. Biochim. Biophys. Acta **1808**(7), 1868–1878 (2011). https://doi.org/10.1016/j.bbamem.2011.03.010
85. Kellici, T.F., Liapakis, G., Tzakos, A.G., Mavromoustakos, T.: Pharmaceutical compositions for antihypertensive treatments: a patent review. Expert Opin. Ther. Pat. **25**(11), 1305–1317 (2015). https://doi.org/10.1517/13543776.2015.1086337

Biophysical Characterization of Polysialic Acid—Membrane Nanosystems

Karolina Sapoń, Teresa Janas and Tadeusz Janas

Abstract Polysialic acid (polySia) is a long, membrane-bound, polyanionic polymer (with the degree of polymerization, DP, up to 400) of negatively charged sialic acid monomers. Biological roles of polySia are based on its ability to modulate repulsive and attractive interactions, and its ability to modulate membrane surface charge density, pH at the membrane surface, and membrane potentials. PolySia is used in anti-bacterial and anti-cancer therapies, and in neural tissue repair. Hydrophobically-modified polySia chains can form nano-structures (micelles or liposomes) with high stability and low toxicity for drug delivery. The analysis, based on the Goldman-Hodgkin-Katz equation, of transmembrane potential changes resulting from trans-membrane translocation of polySia is described. Membrane electrical equivalent circuit can be applied to calculate polySia transmembrane fluxes. Both the surface potential (ψ_S) and the potassium diffusion potential ($\Delta\psi_{diff}$, calculated using the Nernst equation) can modulate polySia-mediated membrane interactions. Langmuir monolayer technique can be applied to determine the excess free energy of mixing, ΔG_{exc}, of lipid components in the presence of polySia. The effect of polySia on the van't Hoff enthalpy (ΔH_{VH}) of the phase transition of lipid bilayer can be determined using fluorescence spectroscopy. The dissociation constant, K_D, for polySia-liposome complex can be determined using polySia-to-membrane FRET spectroscopy.

Keywords Dissociation constant · FRET · Liposome · Membrane potential · Micelle · Nanodroplet · Nanoparticle · Nanovesicle · Phase transition · Polysialic acid · Van't hoff enthalpy

---

K. Sapoń · T. Janas · T. Janas (✉)
Department of Biotechnology and Molecular Biology, University of Opole,
Kominka 6, 45-032 Opole, Poland
e-mail: tadeusz.janas@uni.opole.pl

C. Demetzos and N. Pippa (eds.), *Thermodynamics and Biophysics of Biomedical Nanosystems*, Series in BioEngineering, https://doi.org/10.1007/978-981-13-0989-2_11

1 Introduction: Membrane Polysialic Acid (PolySia)

Polysialic acid (polySia) forms linear chains (Fig. 1) with degree of polymerization (DP) of 8–400, shorter polySia chains with DP of 2–7 are referred to as oligosialic acids (oligoSia) [36]. The degree of polymerization of polySia is defined as the number of sialic acid residues within a polySia chain.

PolySia is a highly hydrated polyanion because sialic acid monomers (Sia) are negatively charged at physiological pH due to the presence of a carboxylic group with a pKa of ~2.6 [17]. In humans, polySia chain consists of N-acetylneuraminic acid (Neu5Ac) and has a chemical structure of (α-2, 8-Neu5Ac)$_n$ which forms a helical configuration [5, 36]. ST8Sia family of sialyltransferases catalyzes the synthesis of α2, 8-linked sialic/polysialic acid chains, including processive synthesis of oligo-(in the case of ST8Sia III) and polySia (in the case of ST8Sia II & ST8Sia IV) chains [15]. PolySia is implicated in development and plasticity of the brain, pathophysiology of schizophrenic brains, cancer metastasis, neuroinvasive potential of pathogenic bacterial strains, and the immune response [35, 37].

1.1 Presence of PolySia on Membranes

PolySia chains are predominantly attached to the external surface of the plasma membrane, although nuclear localization has been also reported [41]. It was determined that polySia is present on (reviewed in [18]):

- CD36 glycoprotein in human fat globule outer membrane. PolySia in human milk may be important for neonatal development in terms of protection and nutrition because of the low Neu5Ac synthetic capacity and the high Neu5Ac requirement for neonatal brain development and cognition.
- Flagellasialin glycoproteins and on gangliosides in the rafts of plasma membrane of sea urchin sperm flagella. Flagellasialin may regulate intracellular calcium level, sperm motility, and fertilization.

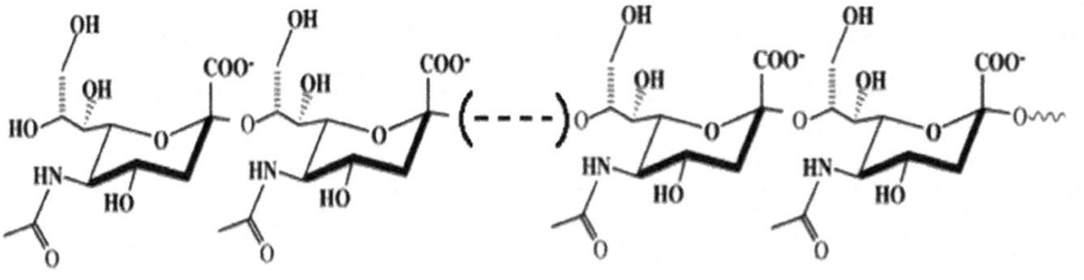

Fig. 1 Structure of polysialic acid (poly-α-2,8-N-acetylneuraminic acid). PolySia is usually conjugated to a biomembrane anchor (lipid or protein) through their reducing end (the right end of the chain)

- Megalin glycoprotein in plasma membranes of normal and carcinoma cells in rats. Megalin functions as a receptor for a multitude of ligands and may regulate calcium homeostasis.
- NCAM (neural cell adhesion molecule) in the plasma membrane of neural cells, neural stem cell–derived progenitors/precursors cells, cancer cells, and leukocytes. The major membrane protein carrier of polySia in mammalian cells is NCAM. There are several isoforms of NCAM, three of which can carry polySia: NCAM-180 and NCAM-140 (integral membrane isoforms), and NCAM-120, the isoform that is anchored to the plasma membrane via a glycosylphosphoinositol (GPI).
- Neuropilin-2 glycoprotein in the plasma membrane of human dendritic cells. PolySia chains on Neuropilin-2 can modulate dendritic cell–T lymphocyte interactions.
- Oligosialylated gangliosides in the outer leaflet of plasma membranes. Gangliosides are highly enriched in the vertebrate nervous system.
- Oligosialylated membrane proteins.
- Phospholipids in the outer membrane of some capsule-expressing bacteria. Bacteria express polySia with chemical and immunological properties identical to those of mammalian-produced polySia; therefore, polySia is used to mask the antigenicity of invading organisms from the immune system of the host.
- Polysialyltransferases PST and STX in Golgi membranes. PST and STX are integral membrane proteins of Golgi membranes, and their catalytic domains are located in the lumen of Golgi vesicles.
- Sodium and potassium channels in plasma membranes. PolySia may regulate electrophysiological properties of these channels.
- SynCAM (synaptic cell adhesion molecule) in synaptic plasma membranes of brain progenitors. SynCAM induces synaptic differentiation and organizes excitatory synapses through interactions across the nascent and mature synaptic cleft.

1.2 Biophysical Mechanisms of the Activity of Membrane Poly/OligoSia

The mechanisms of poly/oligoSia activity are based on a combination of interactions between polySia and another molecule (through repulsive and/or attractive interactions) and interactions between a membrane and a molecule (through modulation of pH at the membrane surface, surface charge density, and membrane potentials).

Repulsive electrostatic interactions mediated by membrane polySia chains can be observed in the case of (reviewed in [18]):

- heterophilic (i.e. between two molecules of different kind) *cis* interactions, e.g. between NCAM and NMDA (N-methyl-D-aspartate) receptor;
- heterophilic *trans* interactions;
- homophilic (i.e. between two molecules of the same kind) *cis* (i.e. in the plane of the host membrane) dimerization of NCAMs;

- homophilic *trans* (i.e. between two apposing membranes) interactions of NCAMs.

Attractive electrostatic interactions mediated by membrane polySia can be observed in the case of:

- anti-polySia antibodies;
- cytokine CCL21, thus regulating the migratory capacity of human dendritic cells;
- Endo-N neuraminidase, thus bacteriophages gain access to the bacterial outer membrane;
- extracellular histone H1, which may have a role in regeneration of the nervous system;
- heparan sulfate proteoglycans, thus modulating activity-dependent remodeling of synapses;
- other polysialic acid chains, in the presence of calcium;
- phospholipids in the lipid bilayer membrane;
- polybasic motifs of polysialyltransferases;
- positively-charged long-chain bases.

PolySia was also shown to bind a number of neurologically important molecules, such as brain-derived neurotropic factor (BDNF), basic fibroblast growth factor (FGF2), and neurotransmitters, such as dopamine, norepinephrine and epinephrine, thus regulating their functions through attractive interactions [5].

The binding to oligoSia chains (in oligosialylated gangliosides) occurs for:

- anti-ganglioside antibodies;
- bacterial neurotoxins, thus facilitating the uptake of bacteria through endocytosis;
- *Escherichia coli* K1 polysialyltransferase;
- sialic-acid-binding immunoglobuline-like lectins (Siglecs);
- viruses, thus facilitating virus entry into the cell.

The action of polySia on biomembrane-related phenomena can result from a combination of several factors:

- attractive interactions—between polySia chains and membrane lipids, positively-charged domains of membrane proteins, or soluble proteins/peptides/proteoglycans;
- repulsive interactions—between polySia chains, or between polySia chains and negatively-charged domains of membrane proteins;
- modulation of pH at the membrane surface—arising from the presence of carboxylic groups in polySia chains;
- modulation of surface charge density arising from the negative charge of the polySia monomers;
- modulation of membrane potentials (dipole potential, transmembrane potential and surface potential)—arising from polySia transmembrane translocation or from the binding of polySia chains to membranes, or from the incorporation of polySia chains into the polar part of the membrane.

PolySia chains at the outer membrane surfaces increase surface density of negative charges, decrease pH at the membrane surface, decrease the electric field inside the membrane, and decrease the absolute value of negative transmembrane potential. These modifications can further lead to changes in protein localization and function with impact on malignant transformation, in transmembrane channel (or membrane carrier) activities, and in mitochondrial release of pro-apoptotic factors. Therefore, the effect of polysialylated NCAM on membrane channel protein activity may not result from indirect membrane effects on the channel, but can result from modulation of local surface charge density, pH at the membrane surface, or/and membrane potentials by polySia. Although other polyanionic molecules (e.g. nucleic acids, other anionic polysaccharides) can have similar effects on membranes, polySia is unique since polySia seems to be the only polyanion permanently attached to membranes of eukaryotic cells.

1.3 Modes of Attachment of Poly/OligoSia Chains to the Membrane

The reducing end of polySia chain can be conjugated to the following biomembrane anchors (as shown in Fig. 2):

- protein attached to the membrane through a glycosylphosphatidylinositol (GPI) lipid anchor; studies suggest that polySia conjugation may occur through core

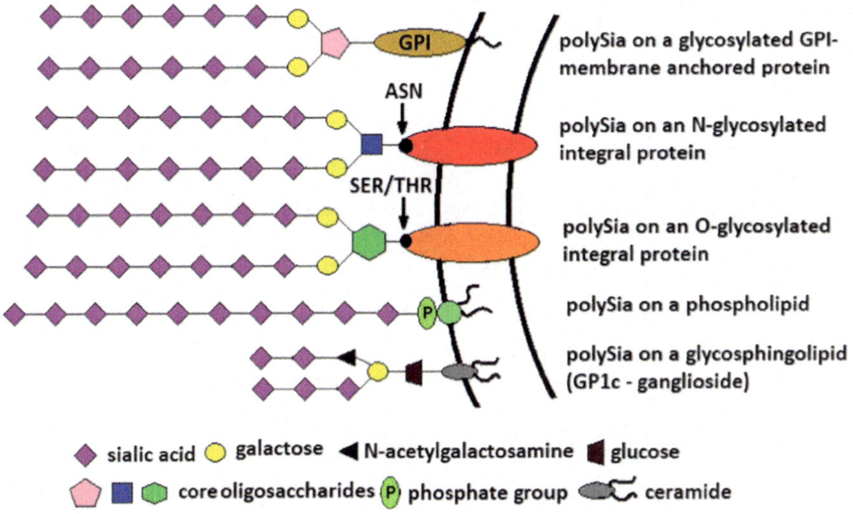

Fig. 2 Membrane anchors of polysialic acids. GPI—glycosylphosphatidylinositol, ASN—asparagine, SER—serine, THR—threonine

oligosaccharides O-linked to a serine or a threonine located in the polypeptide chain, e.g. NCAM-120, however there is not a direct evidence for the O-linked version of the GPI-anchored proteins;
- integral membrane protein—through core oligosaccharides N-linked to an asparagine located in the polypeptide chain, e.g. NCAM-180;
- integral membrane protein—through core oligosaccharides O-linked to a serine or a threonine located in the polypeptide chain, e.g. neuropilin 2, glycoprotein of sea urchin sperm, CD36 glycoprotein, NCAM;
- phospholipid—through the phosphate group in the case of polySia in bacterial K1 capsule;
- glycosphingolipid—through core oligosaccharide chains in a ceramide lipid (e.g. an oligosialylated ganglioside GP1c).

2 Nano-structures Containing PolySia Chains

PolySia chains are frequently used in preparations of nano-structures with potential medical applications. In Fig. 3 the major classes of polySia-containing nano-structures are represented.

2.1 *Nanoparticles Containing PolySia-Protamine Complexes*

The first polySia-containing type of nanoparticles is based on the electrostatic interactions between the positively-charged protamine and negatively-charged polySia

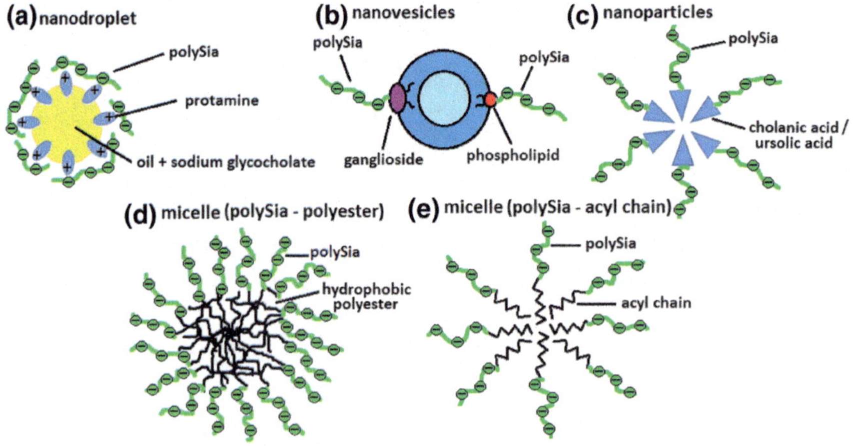

Fig. 3 Nano-structures containing polysialic acid (polySia) chains

(Fig. 3a). Single-layer protamine and double layer polySia/protamine nanocapsules were designed [38] to be used as carriers to facilitate the transport of macromolecules (insulin) across the intestinal epithelium. The oily core (esters of saturated coconut and palm kernel oil-derived caprylic and capric fatty acids and glycerin) of these nanocapsules contained also insulin together with the penetration enhancer, sodium glycocholate. The shell of the nanocapsules was made of protamine, which has the role of facilitating the interaction with the epithelium and PEG-stearate, which is supposed to facilitate the transport of the nanocapsules across the mucus layer. In addition, an extra layer of polySia was added to the nanocapsules with the aim of preventing their interaction with proteolytic enzymes followed by the degradation of the associated insulin. Different in vitro cell monolayer models simulating the different mucosal barriers of the intestinal epithelium were adopted. For cell uptake studies, fluorescent protamine nanocapsules were prepared with TAMRA-labelled protamine. TAMRA is the carboxylic acid of tetramethylrhodamine, and it is used as a fluorescent dye. The ultimate goal of this work was to evaluate the potential of protamine and polySia/protamine nanocapsules to transport peptide drugs such as insulin across the epithelial barrier using a model cell monolayer. The ability of the nanocapsules to cross the cell monolayer was estimated by quantifying the amount of fluorescent maker covalently linked to protamine, detected in the basolateral side compared to the initial amount added in the apical side of the monolayer. Caco-2 and Caco-2/Raji cells were seeded at a density of 5×10^5 cells/well onto Matrigel-coated polycarbonate inserts (12 mm insert diameter, 3 μm pore size). The double layer polySia-protamine nanocapsules exhibited a significantly higher transport rates across the cell monolayer compared to protamine nanocapsules without polySia. The apparent permeability coefficient (P_{app}, cm/s) of the cell monolayer for the drug-containing nanocapsules was calculated according to the following equation:

$$P_{app} = (dQ/dt)\ (1/AC_o) \quad (2.1)$$

where: dQ/dt is the drug transport rate (μg/s), C_o is the initial drug concentration on the apical side (μg/mL) and A is the surface area of the membrane filter (cm^2).

2.2 *Nano-structures Formed by Hydrophobically-Modified PolySia Chains*

The second polySia-containing type of nanoparticles is based on the chemical modification of polySia chain with hydrophobic molecules (glycosphingolipids and phospholipids, triterpenoids, long-chain bases, or lipophilic polymers).

2.2.1 Nano-vesicles Containing PolySia-Lipid Conjugate

Synthesis of polysialylated neoglycosphingolipids on the surface of nano-vesicles (Fig. 3b) was performed [3] by using the polysialyltransferase (polyST) from neuroinvasive *Escherichia coli* K1 and glycosphinglipids as exogenous sialyl acceptors. Inside-out membrane vesicles containing the polyST complex from *E. coli* K1 were prepared by disrupting bacterial cells by passage through an ice-cold French pressure cell at 8000 psi (1 psi = 6.9 kPa). To ensure that the glycosphinglipids were incorporated into the membrane vesicles, glycosphinglipids were first resuspended in a buffer containing Triton X-100, such that the final concentration of Triton X-100 in the incubation mixture was 0.017%. This concentration of Trition X-100 has no effect on the *E. coli* K1 polyST activity. Then, the glycosphinglipids were preincubated with vesicles for 20 min at 33 °C. Endo-N-acylneuraminidase was used to confirm the α2,8-specifc polysialylation of glycosphinglipids.

PolySia from *Escherichia coli* K1 (the K1-antigen), which is capped at its potential reducing end with phosphatidic acid acting as a lipid anchor group, was readily incorporated into liposomal membranes (Fig. 3b) of dimyristoylphosphatidylcholine (DMPC) in giant liposomes (diameter 5–50 μm) as model membranes for immunological studies [8]. The incorporation of this K1-antigen has been proven by immunofluorescence using a FITC-labeled monoclonal antibody mAb735 specifically recognizing poly-α-2,8-NeuAc. The giant vesicles were observed using an inverse microscope with phase contrast and epifluorescence equipment (phase contrast objectives 16 × and 40×). To distinguish whether antibody binding to liposomes is caused by incorporated Kl-antigen or by unspecific adsorption of antibody, control experiments were done using giant vesicles without Kl-antigen. No detectable fluorescence was observed when liposomes were incubated with antibody alone, assuring that unspecific adsorption of the antibody is not relevant in these experiments. It was suggested that the driving force for the insertion of the lipid residues (of the K1-antigen) into membranes is the gain in free energy resulting from the transition of a dynamic, micellar system (of the extracted K1-antigen) with partially solvent-exposed alkyl chains to a system in which the hydrophobic residues are situated in the hydrophobic interior of the membrane. The possibility for further manipulation of this model system has been shown by using a poly-α-2,8-NeuAc cleaving enzyme (endoneuraminidase). Two series of experiments were performed: addition of the endoneuraminidase to the Kl-antigen-bearing liposomes before and after binding of the mAb735. The function of the endoneuraminidase has been proven by showing no binding of the antibody after enzyme treatment of Kl-bearing liposomes as well as by rapid loss of fluorescence of a previously bound FITC-antibody. Thus, the insertion of the polySia-lipid conjugate (the Kl-antigen) in giant liposomes occurred spontaneously via its lipid residue.

PolySia chains were chemically conjugated to dipalmitoyl-phosphatidylethanolamine (DPPE) in liposomes through a series of chemical modifications [32], with final reaction between thiol-reactive-liposomes and thiol-containg polysialic acid-cysteamine. The liposomes were composed of dipalmitoyl-phosphatidylcholine (DPPC), cholesterol, and modified DPPE. The presence of liposomes enhanced the interfacial reaction between the reagents. A sonicated dilute solution of twin-lipid modified polySia gave a broad range of nano-particles; on filtration, however they became smaller with tighter distribution as evaluated by photon-correlation spectroscopy.

2.2.2 Nanoparticles Containing PolySia- Triterpenoid Conjugate

A novel pH-sensitive degradable micelle, containing polySia-ursolic acid conjugate (Fig. 3c), for targeted intracellular release of a drug paclitaxel (PTX) was constructed [44]. Results showed paclitaxel-loaded-micelles exhibited well-defined spherical shape and homogeneous distribution. The degree of substitution (DS, defined as the number of ursolic acids per 100 sugar residues of polySia polymer) was estimated to be 3.3. The critical micelle concentration (CMC) was determined using pyrene as a fluorescence probe for micelle formation, in which, the ratio I_{338}/I_{333} from pyrene excitation spectra, could be used as an index of micelle hydrophobicity. The intersection of the extrapolated straight line segments yielded the CMC value of 0.11 mg/ml for polySia-ursolic acid conjugate in water at room temperature. The drug loading content and encapsulation efficiency were calculated from the following equations:

$$\text{Drug loading content}\,(\%) = (\text{wt of the PTX in micelles/wt of the micelles}) \times 100\% \quad (2.2)$$

$$\text{Encapsulation efficiency}\,(\%) = (\text{wt of the PTX in micelles/wt of the feeding PTX}) \times 100\% \quad (2.3)$$

The drug-loading was 4.5% with an entrapment efficiency of 67.5%. The results indicated that the pH-responsive biodegradable PTX-loaded—polySia-ursolic acid micelles possess extracellular stability.

Nanoparticles containing polySia-triterpenoid conjugate were also constructed by coupling between N-deacetylated polySia and 5β-cholanic acid [28] (Fig. 3c). The physicochemical characteristics of these nanoparticles (zeta potential, morphology and size) were measured, and in vitro cytotoxicity and the cellular uptake of these nanoparticles loaded with doxorubicin were tested.

In the case of a nanoparticle (e.g. liposome, micelle) containing a surface charge (positive or negative), zeta potential is defined as the electric potential near the surface of this nanoparticle, within the interfacial double layer surrounding this nanoparticle, and at the location of the slipping plane that separates mobile water solution from molecules (water and ions) that remains attached to the surface. The value of the zeta potential can affect the migration rate of nanoparticles.

The sizes (ca. 600–800 nm) of the nanoparticles were determined at 25 °C using dynamic light scattering. The nanoparticles showed good deformability, with most of them easily passing through smaller-size filter membranes. The zeta-potential was from −3 to −5 mV depending on the molar ratio of polySia:cholanic acid during coupling reaction. The zeta potential value of polySia chains was −1.25 mV. The morphology of the particles was observed by transmission electron microscopy and the analysis demonstrated that polySia-cholanic acid nanoparticles were nearly spherical in shape.

2.2.3 Nanoparticles Containing PolySia-Hydrophobic Polyester Conjugate

Amphiphilic block copolymers have been prepared based on polySia as a hydrophilic block and polycaprolactone (PCL) as the hydrophobic block [9]. The block copolymers formed spherical micelles (Fig. 3d) with diameter in the range of 270–390 nm, depending on the length of polycaprolactone. The zeta potentials of these micelles (ca. −20 mV) were not dependent on the length of PCL. PolySia chains hold more water molecules than other synthetic, hydrophilic polymers. The poly(ethylene glycol) repeating unit forms a hydrogen bond with three water molecules, whereas polysaccharides form a hydrogen bonds with four to six water molecules per their repeating unit. Therefore polySia-PCL micelles may have potential as drug carriers, requiring prolonged circulation in the body. The in vitro cytotoxicity MTT assay showed that the polySia-PCL micelles are not toxic.

The cyclosporine A (CyA)-loaded polySia–polycaprolactone (PCL) micelles were constructed, characterized and tested for rheumatoid arthritis [42]. The critical micelle concentration of polySia-PCL conjugates was determined to be ca. 85 μg/ml, with a loading capacity and loading efficiency of 0.09 mg CyA/mg polySia–PCL and 29%, respectively. CyA loading resulted in a size increase from 74 to 108 nm. The fixed aqueous layer thickness (FALT) of the polySia–PCL micelles was determined to be 0.63 ± 0.02 nm, comparable to that obtained for traditionally utilized poly(ethylene glycol) coated liposomes. To measure the FALT, the polySia–PCL micelles were suspended in aqueous solution with increasing NaCl concentrations, and the zeta potentials $\psi(L)$ were measured. Assuming a spherical particle and a homogeneous charge distribution, Gouy-Chapman theory can be applied to give the zeta potential as a function of the Debye-Hückel parameter, κ:

$$\ln \psi(L) = \ln A - \kappa L \tag{2.4}$$

where: C is the molarity of the NaCl, A is a constant, $\kappa = C^{1/2}/0.3$. Plotting $\ln \psi(L)$ versus κ yields a straight line with a slope L equal to the thickness of the FALT in nm.

2.2.4 Nanoparticles Containing PolySia—Long-Chain Bases Conjugate

PolySia was modified with a long-chain hydrocarbon through reaction of the carboxylic acid side groups with N-decylamine and the resultant polySia-decylamine conjugates self-assembled into micelles (Fig. 3e) for encapsulation of hydrophobic drug molecules [1]. Poly(ethylene glycol) (PEG) conjugation has been used to improve the circulatory stability of free drugs, as well as nanosized drug delivery systems. However, PEG may not be the ideal solution for improving circulatory stability because there is evidence of continued uptake by the reticuloendothelial system, and concerns remain about the nondegradable nature of the synthetic polymer as well as potential immunogenicity [1]. PolySia is a natural, biodegradable carbohydrate polymer, which should confer stealth properties to associated molecules because of the lack of immunogenicity, in conjunction with the absence of receptors and high hydrophilicity. The critical micelle concentrations (CMCs) for polySia-decylamine prepared with various degrees of substitution were determined with steady-state fluorescence measurements using pyrene as a probe and equaled 50–100 μg/ml. The size of the polySia-decylamine nanoparticles was in the range of 25–140 nm depending on the degree of substitution, DS (of a polySia chain by long-chain amine); the micelles increased in size with higher degrees of substitution. Micelles were further characterized by zeta potential. For all DS values, the zeta potential varied between −30 and −40 mV. The zeta potential provides a gauge of the surface charge resultant from interaction with counterions in a specified medium. The values obtained are indicative of micelle stability because previous research has demonstrated that zeta potentials with absolute values >30 mV will permit electrostatic stabilization. Cytotoxicity of the nanoparticles was dependent on the degree of substitution of polySia chain with decylamine.

PolySia was chemically conjugated with octadecylamine (ODA), and the obtained polySia-ODA copolymer could self-assemble into micelle in aqueous solution with a 120 μg/ml critical micelle concentration, CMC [40]. The CMC of polySia-ODA copolymer was determined by fluorescence spectrophotometry using pyrene as a probe. PolySia—ODA micelles were loaded with a drug (a calmodulin antagonist) for therapy of vascular dementia. The polySia-ODA micelles could cross the blood—brain barrier mainly via active endocytosis by brain endothelial cells followed by transcytosis. The zeta potential of polySia chains was determined as −33.4 ± 2.56 mV. After modification with ODA, the zeta potential of the polySia-ODA micelles decreased to −21.2 ± 1.01 mV. The reason was that partial carboxyl groups were replaced by ODA chains. The relatively high zeta potential was a significant factor for increasing the stability of micelles in aqueous medium through repulsion interaction. PolySia-ODA micelle's morphology and size were evaluated by transmission electron microscopy and dynamic light scattering. TEM images showed a uniform distribution of both polySia-ODA micelle and these micelles loaded with the drug with sizes around 50 and 30 nm, respectively. The reduced size was mainly attributed to the enhanced hydrophobic interaction between ODA segments and the hydrophobic drug.

PolySia derivative conjugated with the long hydrophobic aliphatic chain octadecyl dimethyl betaine (BS18) has been synthesized [43]. PolySia-BS18 conjugate was synthesized by grafting the hydroxyl groups of polySia with the carboxylic group of the long hydrophobic aliphatic chain of BS18 through esterification. The resulting compound polySia-BS18 was attached to the liposomes and their pharmacokinetic behavior was studied in Wistar rats. A fluorescent probe DiR was used for liposome labeling for bio-distribution experiments, and the unentrapped DiR was removed by a Sephadex G-50 column, and the DiR in liposomes was assessed using a microplate reader fluorescence spectrophotometer. The mean particle size (110–135 nm in diameter) of liposomes was determined by dynamic light scattering. The zeta potential of liposomes (from −13 to −36 mV, depending on the liposome composition) was measured by electrophoretic light scattering. Although the tertiary ammonium salt of BS18 is positive charged and the ammonium group of BS18 can neutralize a part of negative charge of the carboxylic group on polySia, the BS18 substitution degree of polySia-BS18 was only 15%. Therefore there was still a large number of negative charged carboxylic groups which exist on the polySia-BS18. The drug epirubicin was loaded into liposomes using a pH transmembrane gradient ($\Delta pH = 3$). For another nano-formulation containing polySia—octadecyl dimethyl betaine, liposomes were replaced by Pluronic acid, which consists of a central hydrophobic block of polypropylene glycol flanked by two hydrophilic blocks of polyethylene glycol. PolySia—octadecyl dimethyl betaine—Pluronic F127 mixed micelles, which were loaded with an anticancer drug docetaxel, were prepared by a self-assembly method [45] in order to increase the solubility and stability of docetaxel and enhance its antitumor effect. The measured critical micelle concentration was 1.3 μM, and the micelles had a diameter of ca. 34 nm with a narrow polydispersity, the entrapment efficiency was ca. 99%, and the drug loading was ca. 1.4%.

3 Membrane Potentials in PolySia—Membrane Nanosystems

There are three major electrical potentials at the biomembranes (Fig. 4).

Transmembrane potential (or "membrane potential") ($\Delta\psi_m$, V_m) is the potential difference between the solutions at the inner and outer sides of the cell membrane; the values are typically from −10 to −100 mV (negative at the membrane cytoplasmic side). It arises from the transfer of charges (typically ions) between two

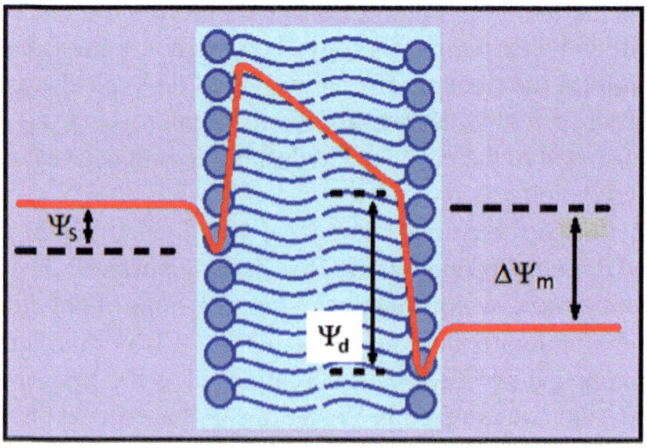

Fig. 4 The electrical potential profile across the lipid bilayer. $\Delta\psi_m$—transmembrane potential, ψ_s—surface potential of the membrane, ψ_d—dipole potential of the membrane

water compartments separated by the membrane; if the transmembrane transfer of ions is passive (without energy input), the transmembrane potential usually is called a "diffusion potential". The surface potential (ψ_s,) is localized at the membrane-solution interface and is generated by a net charge of membrane lipids, proteins or compounds adsorbed on the membrane surface. The dipole potential (ψ_d) is generated inside the polar regions of the membrane by the intramolecular dipole moments of lipids, proteins and water molecules, e.g. by the ester linkage of the hydrocarbon chains of phospholipids; the values are typically from +200 mV to +400 mV (positive inside the lipid bilayer). The superposition of the transmembrane potential, the dipole potentials, and the surface potentials gives rise to the potential inside the membrane (called the "intramembrane potential").

3.1 *Application of Goldman-Hodgkin-Katz Equation for Analysis of Transmembrane Potential Changes Resulting from Transmembrane Translocation of PolySia*

The analysis, based on the Goldman-Hodgkin-Katz equation, of transmembrane potential changes resulting from transmembrane translocation of polySia chains has been performed [19]. The flow of ions across a neutral membrane of thickness h, enclosing a cell or a subcellular vesicular structure, was considered. Membrane surfaces were placed at $x = 0$ and $x = h$. The outer compartment was for $x > h$, the inner compartment was for $x < 0$. The basic equations applied for this analysis were following:

(a) The Goldman-Hodgkin-Katz equation for flux of positive ions, J_+, across the membrane:

$$J_+ = z_+ P_+ v_m \left[c_{+,o} - c_{+,i} \exp(z_+ v_m)\right] / \left[1 - \exp(z_+ v_m)\right] \tag{3.1}$$

where: z_+ is the positive ions valence, P_+ is the membrane permeability coefficient for positive ions, v_m is the dimensionless membrane potential normalized in units of kT/e, $c_{+,o}$ is the concentration of positive ions in the outer compartment, $c_{+,i}$ is the concentration of positive ions in the inner compartment.

(b) The Goldman-Hodgkin-Katz equation for flux of negative ions, J_-, across the membrane:

$$J_- = z_- P_- v_m \left[c_{-,o} - c_{-,i} \exp(z_- v_m)\right] / \left[1 - \exp(z_- v_m)\right] \tag{3.2}$$

where: z_- is the negative ions valence, P_- is the membrane permeability coefficient for negative ions, v_m is the dimensionless membrane potential normalized in units of kT/e, $c_{-,o}$ is the concentration of negative ions in the outer compartment, $c_{-,i}$ is the concentration of negative ions in the inner compartment. The valence of polySia corresponds to its degree of polymerization.

(c) The total current, i, of ions passing through the membrane at the stationary state:

$$i = i_+ + i_- = 0 \tag{3.3}$$

where: i_+ is the current of positive ions, i_- is the current of negative ions.

(d) Electroneutrality requirements (for both sides of the membrane)

$$\sum |z_+|c_+ = \sum |z_-|c_- \tag{3.4}$$

where: $|z_+| = z_+$, $|z_-| = -z_-$.

Combining Eqs. 3.1–3.4, the final equation was obtained and solved using a Mathematica computational application. The relationships between the transmembrane potential and the degree of polymerization of polySia (DP up to 200), the temperature, the ratio of membrane permeability for cations and anions, and the inner/outer concentration ration of polySia have been plotted and discussed. The changes in the transmembrane potential, up to 118 mV, were obtained for the permeability ratio greater than 1. However, the temperature-dependent changes in $\Delta\psi_m$ were less than 7 mV in the temperature range 0–50 °C.

3.2 *Application of the Membrane Electrical Equivalent Circuit for Analysis of the PolySia Transmembrane Fluxes*

The analysis of polySia transmembrane translocation was performed using membrane electrical equivalent circuit [21, 22]. Hodgkin and Huxley used an equivalent circuit and Ohm's law to derive a relationship between total membrane current and the membrane capacitance, the membrane voltage, the time, ion and leakage conductance, the equilibrium potentials for each of the ions, which they applied to nerve signaling [14]. Modifications of the Hodgkin-Huxley equation have been applied to solve for membrane voltage in the presence of electrogenic pumps. The models, having the form of equivalent electrical circuits, can be subjected to standard electrical circuit analysis techniques to describe quantitatively the voltage and current relationships among the components. These fundamental techniques of circuit analysis include Ohm's law and Kirchhoff's laws. Pumps are special cases of porters and the input side of a pump can be driven by the free energy, the pump potential, expressed in volts of the chemical reaction.

The electrical equivalent circuit of the membrane is presented in Fig. 5. The left-hand branch of the circuit represents the passive diffusion of ions across the membrane; the membrane resistance G_d^{-1} is connected in series with the electromotive force E_d which is equivalent to the diffusion potential. The next branch represents the electrogenic pumps which actively transport ions out of the cell; G_p is the total

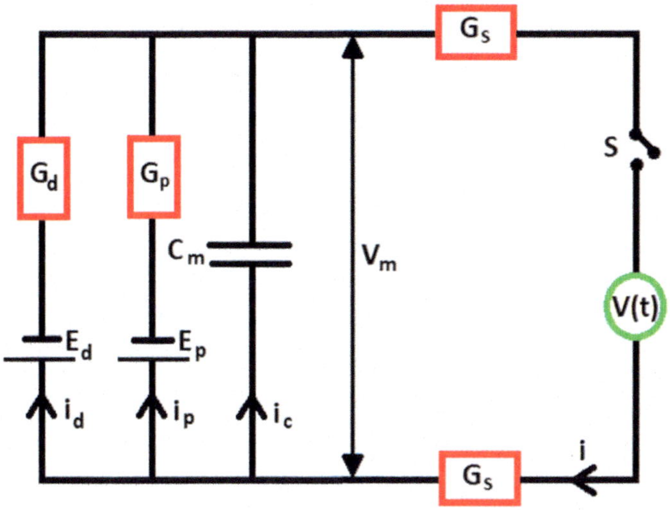

Fig. 5 The electrical equivalent circuit of the membrane. G_d—the electrical conductance of the membrane resulting from diffusion process, G_p—the electrical conductance of ion pumps, G_s—the electrical conductance of the bathing solution, E_d—the diffusion potential, E_p—the electromotive force of ion pumps, C_m—the capacitance of the membrane, V_m—the electrical transmembrane potential, V(t)—the external potential source, i—the total current through the membrane, i_d—the diffusive current, i_p—the current passing through the pumps, i_c—the capacitive current

conductance of the electrogenic pumps located in the membrane; E_p is the electromotive force of the electrogenic pumps. C_m, G_s and V(t) represent the membrane capacitance, the electrical conductance of the bathing solution and the electrodes, and the external potential source, respectively. The switch, S, closes the circuit at the time $t = 0$.

The basic equations applied for this analysis were following:

(a) The first Kirchhoff's law

$$i = i_d + i_p + i_c \tag{3.5}$$

where: i is the total current through the membrane, i_d is the diffusive current, i_p is the current passing through the pumps, i_c is the capacitive current (all currents are per unit area).

(b) The Ohm's law:

$$i_d = G_d(V_m - E_d) \tag{3.6}$$

$$i_p = G_p(V_m - E_p) \tag{3.7}$$

$$i_c = C_m(dV_m/dt) \tag{3.8}$$

where: t(s) is time, G_d ($\Omega^{-1}cm^{-2}$) is the electrical conductance of the membrane resulting from diffusion process, E_d (V) is the diffusion potential, G_p ($\Omega^{-1}cm^{-2}$) is the electrical conductance of ion pumps, E_p (V) is the electromotive force of ion pumps, C_m (Fcm^{-2}) is the capacitance of the membrane, V_m (V) is the electrical potential of the membrane.

(c) The second Kirchhoff's law:

$$V = V_m + 2V_s \tag{3.9}$$

where: V(V) is the external potential, V_s (V) is the drop of the electrical potential on the electrical resistance, R_s (Ω), of the bathing solution and electrodes.

Combining Eqs. 3.5–3.9, the final first order differential equation was obtained and solved for $V = V_0$: the electrical potential source, V(t), keeps the external potential, V, at the level V_0 for the initial condition, the flux $J(t = 0) = J_0$. The changes in the polyion flux were up to 88% after 1 ms. Both the increase of polyion chain length and the decrease of membrane conductance resulted in the diminution of this effect. Inversion of flux direction was observed as a result of external potential changes.

3.3 Determination of the Membrane Surface Potential and the PH at the Membrane Surface in the Presence of PolySia Using a Fluorescence Probe Fluorescein-PE

Adsorption of negatively charged polySia chains on the surface of phosphatidylcholine liposomes causes an increase of concentration of positively charged hydrogen ions, and this change of the pH at the liposomal surface was detected using a fluorescent probe fluorescein-PE [23]. The fluorescence of this probe increases with the increase of pH of the solution. According to Gouy–Chapman–Stern theory, the concentration of protons at the membrane surface, $[H^+]$, depends on the surface charge density, σ, and the surface potential, ψ_o, at the distance $h = 0$ from the surface:

$$\psi_o = \psi_s(h = 0) \tag{3.10}$$

$$[H^+] = [H^+]_o \exp(-F\psi_o/RT) \tag{3.11}$$

$$\psi_o = s/\varepsilon\varepsilon_o\kappa \tag{3.12}$$

$$\kappa = (2e^2z^2Nc/\varepsilon\varepsilon_o kT)^{1/2} \tag{3.13}$$

where: $[H^+]_o$ is the bulk concentration of $[H^+]$, F is the Faraday constant, R is the gas constant, T is the absolute temperature, ε is a dielectric constant of the medium, ε_o is the permittivity of free space, c is the total concentration of monovalent electrolyte in the bulk aqueous solution, z is the valance of the ion, e is the elementary charge, N is the Avogadro number, and k is the Boltzmann constant.

The surface charge density of lipid bilayer, after adsorption of polySia, was calculated assuming that the mean DP of polySia chains equals 50 and each sialyl residue bears one negative charge. The calculated surface density is 0.0257 e/nm^2, which corresponds to ca. 16 polySia chains per one liposome. Defining the degree of adsorption as the molar ratio of polySia chains on liposomes and in bulk solution, this coefficient was calculated as 0.04%.

3.4 *Determination of the Membrane Diffusion Potential in the Presence of PolySia Using a Fluorescence Probe Oxonol-V and the Nernst Equation*

The effect of the transmembrane potential on polySia binding to the liposomal surface was investigated using the fluorescent probe Oxonol-V (OX-V) [33], which is sensitive to the value of the transmembrane potential. The maximum of the emission fluorescence spectra of OX-V is shifted from 625 to 650 nm in the presence of phospholipid vesicles. When a valinomycin/potassium ion-induced transmembrane gradient (valinomycin-induced potassium diffusion potential, $\Delta\psi_{diff}$) is created in lipid vesicles (positive inside the vesicles), an increase in fluorescence at 650 nm is observed in the case of a low OX-V/phospholipid molar ratio. The relationship between the generated transmembrane potential (calculated using the Nernst equation) and the increase in the OX-V fluorescence in DOPC liposomes was measured. Both the transmembrane potential and the surface potential modulated polySia-mediated membrane interactions (Fig. 6).

Procedure 1—Determination of the membrane diffusion potential in the presence of polySia using a fluorescence probe Oxonol-V and the Nernst Equation

1. *Measurement of the relationship between the generated transmembrane potential and the increase in the Oxonol-V fluorescence in liposomes.*

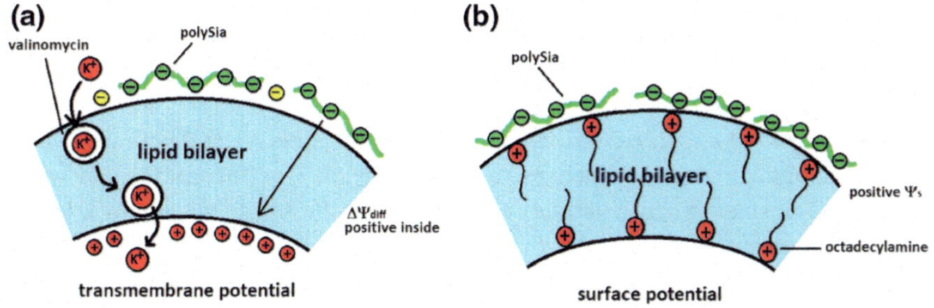

Fig. 6 The potassium diffusion potential, $\Delta\Psi_{diff}$, (A) and the surface potential, Ψ_s, (B) as driving forces for binding of polySia chains to the surface of lipid membranes

(a) ***Preparation of unilamellar liposomes***: DOPC (dioleoylphosphatidylcholine) was dissolved in chloroform/methanol (2/1). Then, the solvents were evaporated under a stream of nitrogen gas and the remainder was desiccated under a vacuum for at least 2 h. The lipid was resuspended in 50 mM TRIS (pH 7.5) and 0.5 mM KCl. Multilamellar liposomes were formed by gentle vortexing. The suspension underwent several freeze-thaw cycles, achieved by repeated immersion in liquid nitrogen followed by warming in 60 °C water. Large unilamellar vesicles were prepared by extrusion at 60 °C using an Avanti MiniExtruder with a filter pore diameter of 100 nm.

(b) ***Generation of the transmembrane potential, $\Delta\psi_m$, at the liposomal membrane using valinomycin***: large unilamellar vesicles (13 μM DOPC) containing Oxonol-V (130 nM) were prepared in 50 mM TRIS (pH 7.5) and 0.5 mM KCl. The external KCl concentration was increased by adding concentrated KCl. Upon the addition of valinomycin (1 nM), a diffusion potential for potassium ions was generated (positive inside), and then detected by measuring Oxonol-V fluorescence at 650 nm. To prevent osmotic effects, the vesicles were prepared in a buffer containing an appropriate concentration of choline chloride. Experiments were also performed with equal internal and external concentrations of KCl (in the absence and in the presence of a transmembrane potential).

(c) ***Calculation of the value of transmembrane potential, $\Delta\psi_m$, using the Nernst equation***:

$$\Delta\psi_m = (RT/zF)\ln\left([K^+]_{out}/[K^+]_{in}\right) \tag{3.14}$$

where: R is the gas constant, T is the absolute temperature, F is Faraday constant, z is the valance of the potassium ion (z = + 1), $[K^+]_{out}$ is the concentration of potassium ions outside liposomes, $[K^+]_{in}$ is the concentration of potassium ions inside liposomes.

The transmembrane potential was generated due to facilitated diffusion of potassium ions using valinomycin (a naturally occurring dodecadepsipeptide, a neutral ionophore) as the transmembrane carrier.

(d) ***Measurements of Oxonol-V fluorescence as a function of $\Delta\psi_m$***: Oxonol-V fluorescence was measured at excitation $\lambda_{ex} = 610$ nm, and emission $\lambda_{em} = 650$ nm. The relative change in Oxonol-V fluorescence was calculated as $(F-F_o)/F_o$, where F_o is the Oxonol-V fluorescence at zero transmembrane potential (in the absence of a transmembrane potential, equal internal and external concentrations of KCl).

(e) ***Drawing of the calibration curve***: the obtained experimental data can be fit to a straight line using linear regression analysis. The increase in the value of the transmembrane potential (positive inside the liposomes) gives a linear increase in OX-V fluorescence at 650 nm, with the slope of the line equal to $(1.45 \pm 0.11) \times 10^{-3}\ mV^{-1}$ (Fig. 7). The overall ionic strength was kept constant to avoid its influence on OX-V fluorescence.

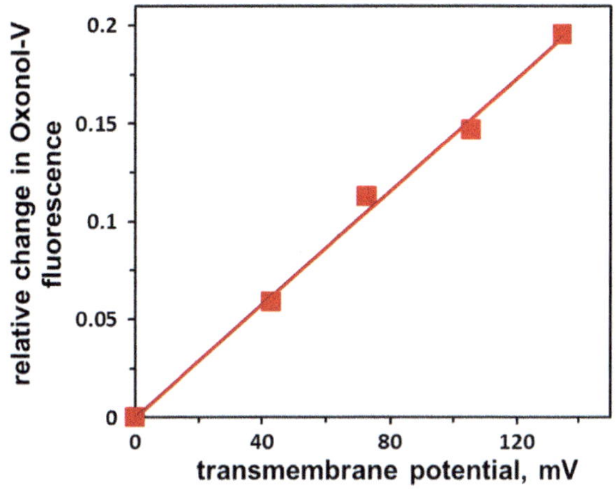

Fig. 7 Calibration curve: the relative change in Oxonol-V fluorescence as a function of transmembrane potential

2. *Measurement of the relative change in Oxonol-V fluorescence as a function of polySia concentration in the absence and in the presence of a transmembrane potential (positive inside liposomes).*

The effect of polySia at different concentrations on the relative change in OX-V fluorescence in DOPC liposomes was tested both in the absence and presence of a transmembrane potential. PolySia increases the OX-V fluorescence in liposomal membranes in the presence of a transmembrane potential, which suggests an increase in the membrane potential (positive inside) upon the binding of negatively charged polySia chains to the external surface of the liposomal membrane. In the absence of a transmembrane potential, polySia decreases the OX-V fluorescence, which can reflect a competition between negatively charged polySia chains and negatively charged OX-V for the same binding sites on the liposomal surface.

4 Thermodynamics of PolySia-Membrane Nanosystems

4.1 Determination of Excess Free Energy of Mixing, ΔG_{exc}, of Lipid Components in the Presence of PolySia Using the Langmuir Monolayer Technique

Langmuir monolayer technique was applied to study the role of a simple long-chain base, octadecylamine (ODA), in both *cis* and *trans* interactions mediated by polySia in model membranes composed of ODA and dioleoylphospatidycholine (DOPC) [24].

The monolayers were deposited and then they were equilibrated at zero pressure for 5 min to allow evaporation of chloroform. The experiments were performed both in the presence of polySia (0.05 mg/ml) and in the absence of polySia in the subphase. The excess area per molecule, ΔA_{ex}, was calculated according to the equation:

$$\Delta A_{ex} = A_{12} - (A_1 X_1 + A_2 X_2) \tag{4.1}$$

where: A_{12} is the mean area per molecule in a mixed monolayer at a constant surface pressure, A_1 and A_2 are the molecular areas of a single component at the same surface pressure, and X_1 and X_2 are the mole fractions of components 1 and 2 in the mixed film.

The interactions of two mixed monolayer components can be studied from the point of view of miscibility of the components, based on the calculations of ΔA_{ex}, as a function of composition. According to the additivity rule, a mixed monolayer can show non-ideal behavior as a consequence of the interactions between the two components. When the two components are immiscible or form an ideal mixture, ΔA_{ex} equals zero.

The excess free energy of mixing, ΔG_{exc}, was calculated according to the equation:

$$\Delta G_{exc} = \int_{\Pi o}^{\Pi} (A_{12} - X_1 A_1 - X_2 A_2) d\Pi \tag{4.2}$$

where: A_{12} is the area per molecule in the mixed film, A_1 and A_2 are molar areas in pure films, X_1 and X_2 are the mole fractions of components 1 and 2 in the mixed film, and Π is the surface pressure. The integrals correspond to the areas under the Π-A isotherms, and Π_o is defined as the surface pressure where the monolayer components are ideally miscible. It is generally assumed that Π_o is close to zero. In practice, Π_o is commonly set to the lowest measurable surface pressure, and in this work Π_o was set to 0.4 mN/m.

ΔG_{exc} gives information on whether interactions between membrane components are energetically favored (ΔG_{exc} negative) or not (ΔG_{exc} positive) as compared to an ideal mixture. The positive values of ΔG_{exc} indicate a tendency of molecules to interact preferentially with molecules of the same kind, suggesting a formation of aggregates. In addition, a minimum in the plot of ΔG_{exc} versus molar ratio suggests a formation of a complex between molecules of different species. Negative values of ΔG_{exc} indicate more attractive interaction between membrane components as compared to those in a pure monolayer. Moreover, lower values of ΔG_{exc} indicate that the mixed film is more stable.

DOPC isotherms underwent the usual behavior observed for unsaturated phospholipids during compression. First, a transition from a gaseous state to a liquid expanded state was observed with a small increase in surface pressure. The subsequent large increase in surface pressure indicated the formation of a liquid condensed state of the monolayer. When collapse occurs, a fraction of the molecules are forced out of the monolayer. The collapse pressure is determined from the change of the slope of the isotherms occurring during a further compression of the film. The limiting molecular area represents the beginning of observable intermolecular forces between adjacent molecules in the monolayer. When added free to an aqueous solution, polySia increased the collapse pressure of ODA/DOPC monolayers, reduced the effect of ODA on the limiting molecular area, and inversed the values of excess area per molecule and of excess free energy of mixing from positive to negative.

4.2 Determination of the Van't Hoff Enthalpy, ΔH_{VH}, of the Phase Transition of Lipid Bilayer in the Presence of PolySia Using Fluorescence Spectroscopy

The effect of polySia on the van't Hoff enthalpy, ΔH_{VH}, of the phase transition of lipid bilayer was investigated using the fluorescent probe RH-421, which is sensitive to the lipid bilayer fluidity [4]. The ratiometric method was applied, i.e., the ratio of the fluorescence intensities detected at two excitation wavelengths on the blue and red flanks of the excitation spectrum was measured. In the case of RH-421, the excitation wavelengths chosen were 440 and 540 nm. The ratio was measured at an emission wavelength of 670 nm. The relationship between the value of the parameter R and the increasing temperature in the range of 34–54 °C was measured.

Procedure 2. Determination of the van't Hoff enthalpy, ΔH_{VH}, of the phase transition of lipid bilayer in the presence of polySia.

1. *Measurements of the fluorescence of the RH-421probe in lipid bilayer as a function of temperature.*

(a) ***preparation of unilamellar liposomes***: DPPC (dipalmitoylphosphatidylcholine) and the fluorescent probe RH-421 (0.2 mol% of lipids) were dissolved in chloroform/methanol (2/1) and mixed. Then, the solvents were evaporated under a stream of nitrogen gas and the remainder was desiccated under a vacuum for at least 2 h. The buffer was added and multilamellar liposomes were formed by gentle vortexing. The suspension underwent several freeze-thaw cycles, achieved by repeated immersion in liquid nitrogen followed by warming in 60 °C water. Large unilamellar vesicles were prepared by extrusion at 60 °C using an Avanti MiniExtruder with a filter pore diameter of 100 nm.

(b) ***measurements of RH-421 fluorescence as a function of temperature***: steady-state fluorescence spectroscopy experiments on RH-421 in 100 nm DPPC liposomes (2:1000 mol ratio, probe:lipid) were performed as a function of temperature. Excitation and emission wavelengths were 440 and 670 nm, and also 540 and 670 nm, respectively. Liposomes were added to polysialic acid solution in buffer. For control experiments the same volume of buffer was used without polySia.

(c) ***calculation of the parameter R for the fluorescence probe RH-421***: the parameter R was calculated as:

$$R = \frac{fluorescence\ emission\ at\ 670\,\text{nm},\quad when\ excitation\ is\ at\ 440\,\text{nm}}{fluorescence\ emission\ at\ 670\,\text{nm},\quad when\ excitation\ is\ at\ 540\,\text{nm}} \tag{4.3}$$

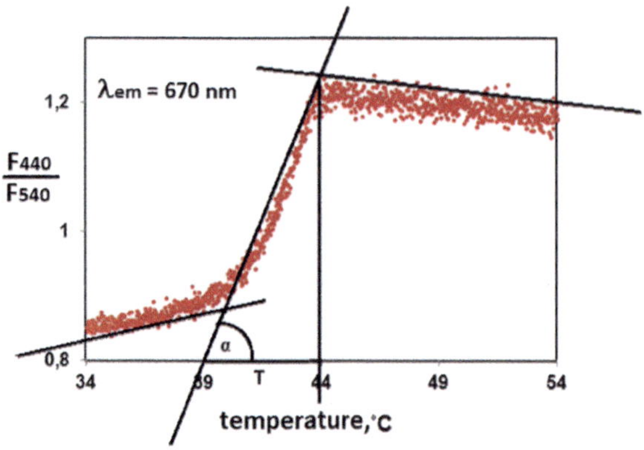

Fig. 8 The temperature scan, showing the phase transition of DPPC bilayer, using the fluorescence probe RH-421

2. *Calculations of the van't Hoff enthalpy.*

The van't Hoff enthalpy (ΔH_{VH}) can be determined from the sharpness of the phase transition [2], as measured by the value of parameter R as a function of temperature (Fig. 8):

$$\left(\Delta H_{VH} = 4R_g T_m^2 \left(\frac{dR}{dT}\right)_{T_m}\right) \tag{4.4}$$

where:

ΔH_{VH}	the van't Hoff enthalpy (¢HVH) of the phase transition, [kJ/mol],
R_g	the ideal gas constant = 8,31 [J/mol·K],
T_m	the phase transition temperature [K],
$(dR/dT)_{Tm} = tg\alpha$	the slope of the scan at the phase transition temperature [1/K]
T	temperature [K].

Other membrane fluorescent probes, sensitive to lipid bilayer fluidity, can be used for the determination of the van't Hoff enthalpy for membrane phase transition, e.g. in the case of DPH probe, the anisotropy, r, can be measured as a function of temperature [2].

3. *Measurement of the value of the van't Hoff enthalpy for phase transition of the DPPC bilayer as a function of polySia concentration.*

The effect of polySia at different concentrations on the value of the van't Hoff enthalpy for phase transition of the DPPC bilayer was tested. The presence of polySia caused the decrease of the value of the van't Hoff enthalpy in the case of liposomal membranes.

5 Binding of PolySia to Liposomes

5.1 Determination of the Amount of PolySia Bound to Liposomes Using the Toluidine Blue (TB) Absorption Spectroscopy

Toluidine blue (TB) is a basic dye that is used to determine the surface charge density of plasma membrane. It has a maximum absorption at 630 nm. The binding of TB to the negative charges of polySia (or any other polyanionic molecule) results in a metachromatic shift in its maximum absorption to 560 nm. As the concentration of polyanionic polySia increases, the absorption at 630 nm (A630) decreases and the absorption at 560 nm (A560) increases [33]. Therefore, the difference (A630–A560) and the ratio A630/A560 are inversely related to the concentration of polySia in the liposomal fractions. The smaller the absolute value of the difference (or the ratio), the higher the concentration of polySia bound to the liposomes.

First, measurements were performed to determine the relationship between the toluidine blue absorbance spectra and the increase in concentration of polySia in the solution. Next, measurement of the relationship between the percentage of positively-charged octadecylamine (ODA) in liposomal membrane and the ratio of the absorbance difference (R) were performed. The presence of ODA in the lipid bilayer generates the positive surface charge density.

PolySia in the buffer at a final concentration of 0.05 mg/ml was mixed with DOPC or ODA/DOPC (molar ratio of 0.05) liposomes prepared in the buffer with a final concentration before gel filtration of 10 mg/ml. After incubation for 5 min at room temperature, the suspension was applied to a 1 ml Sephacryl S-1000 column that voids liposomes but retains polyanions, and eluted with the buffer. The samples were collected as ca. 30 μl fractions. Co-elution of polyanions and liposomes indicates binding. The presence of polySia in the liposomal fractions was detected by immediate titration of the liposomal fraction (liposome concentration: 1 mg/ml) into a solution of the cationic dye toluidine blue (TB). Upon binding of TB to the polySia chain, the absorption maximum shifts from 630 to 560 nm.

The ratio of the absorbance difference (R) was determined as:

$$R = \frac{(A630 - A560)\ for\ the\ sample\ with\ ODA/DOPC\ liposomes}{(A630 - A560)\ for\ the\ sample\ with\ DOPC\ liposomes} \quad (5.1)$$

If the binding of polySia to ODA/DOPC liposomal membranes is greater than the binding to pure DOPC membranes, the ratio is less than 1. The ratio R decreased linearly with increasing volume of the liposomal fraction added to the TB solution indicating a higher affinity of polySia chains to liposomes with a positive surface charge than to liposomes with a zero net surface charge (DOPC-only liposomes). It can be estimated from the TB experiments that with a polySia concentration of 0.05 mg/ml, ca. 30% of polySia chains are bound to the ODA/DOPC liposomes.

5.2 Determination of the Dissociation Constant, K_D, for PolySia-Liposome Complex Using FRET Fluorescence Spectroscopy

To determine the dissociation constant for the polySia-liposomal lipid complex, fluorescence and FRET (Forster Resonance Energy Transfer) spectroscopy measurements between free polySia chains labeled with the intercalated green dye, YOYO-1, and membranes labeled with the intercalated red dye, Nile Red were performed [25]. Nile Red shows almost no fluorescence in water solution. However, upon incorporation into the hydrophobic interior of a lipid bilayer, the dye undergoes a ~40-fold fluorescence enhancement. Because FRET is highly distance-dependent, it becomes efficient only when the polySia binds the membrane, bringing the two fluorophores together.

Procedure 3. Determination of the dissociation constant, K_D, for polySia-liposome complex.

1. *Measurements of the polySia-to-membrane FRET.*

(a) ***Preparation of unilamellar liposomes***: DOPC (or DPPC) and the fluorescent probe Nile Red (0.2 mol% of lipids) were dissolved in chloroform/methanol (2/1) and mixed. Then, the solvents were evaporated under a stream of nitrogen gas and the remainder was desiccated under a vacuum for at least 2 h. The buffer was added and multilamellar liposomes were formed by gentle vortexing. The suspension underwent several freeze-thaw cycles, achieved by repeated immersion in liquid nitrogen followed by warming in 60 °C water. Large unilamellar vesicles were prepared by extrusion at 60 °C using an Avanti MiniExtruder with a filter pore diameter of 100 nm.

(b) ***Measurements of the decrease of YOYO-1 fluorescence as a function of the increase in Nile-Red-labeled liposome concentration***: polySia-to-membrane FRET between polySia-bound YOYO-1 and membrane-bound Nile Red was measured at room temperature by exciting YOYO-1 and monitoring the decrease of its emission intensity in the presence of FRET. In detail, polySia (5 mg/ml in buffer) was mixed with the cationic cyanide fluorescence dye YOYO-1 iodide (final concentration 1.0 μM). The liposomes were added and polySia-to-membrane FRET was monitored as decrease in YOYO-1 emission intensity, $\lambda_{ex} = 491$ nm and $\lambda_{em} = 508$ nm.

2. *Calculations of the dissociation constant, K_D, for the polySia-liposome complex.*

The dissociation constant, K_D, for the polySia-liposome complex can be determined from the Langmuir equation [20]:

$$\Delta F = \Delta F_{max}[L/(K_D + L)] \tag{5.2}$$

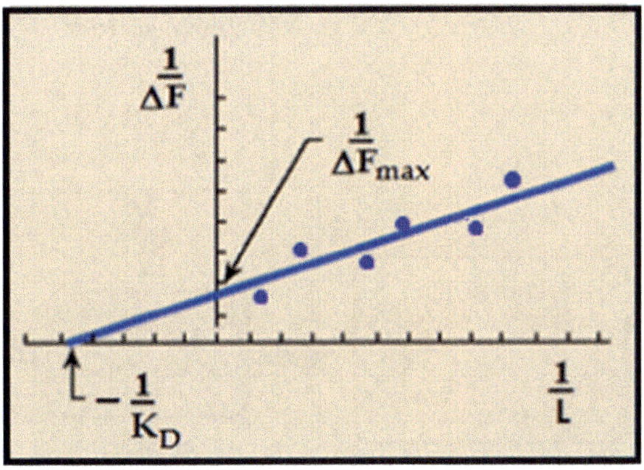

Fig. 9 Determination of K_D from the Langmuir equation using the double reciprocal plot. K_D—the dissociation constant for the polySia-liposome complex, L—concentration of liposomal lipids, ΔF—the fluorescence change, ΔF_{max}—the maximal fluorescence change

where: ΔF_{max} is the calculated maximal fluorescence change, L is the liposomal lipid concentration, and K_D is the equilibrium dissociation constant for polySia-liposome complex.

The K_D values can be determined from the Langmuir equation using mathematical software or using a standard double reciprocal plot (Fig. 9).

3. *Measurement of the value of the dissociation constant, K_D, for the polySia-liposome complex in the case of liposomes made from DOPC or DPPC.*

The value of the dissociation constant, K_D, for the polySia-liposome complex was determined. The K_D value in the case of liposomal membranes made from DOPC molecules was higher in comparison to the value of K_D for liposomes made form DPPC molecules.

6 Concluding Remarks: Medical Applications of PolySia

The medical applications of polySia include: N-propionyl-polySia immunization; polySia-containing nano-carriers of drugs; polysialylation of peptides, proteins, and low molecular weight drugs; and polySia-mediated neural tissue repair after neural injuries. These applications are summarized in Fig. 10, and they are also described below.

6.1 N-Propionyl-PolySia Immunization

Some of bacterial polySia chains retain a phosphoglycolipid on their reducing termini, and although small, this lipid component causes the individual chains to aggregate resulting in the higher molecular weight nanoparticles. Despite this aggregation,

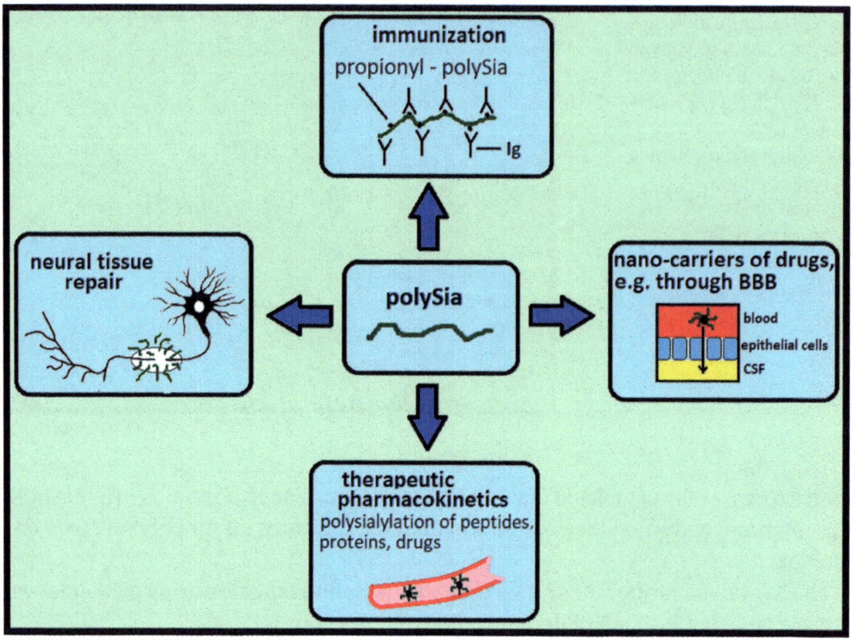

Fig. 10 Medical applications of polysialic acid. Ig—immunoglobulin (antibody), BBB-blood-brain-barrier, CSF—cerebrospinal fluid

polySia is still poorly immunogenic [27]. The poor immunogenicity of the group B meningococcal polysaccharide and the structurally identical *E. coli* K1 capsular polysaccharide (i.e. polySia) preclude their use as vaccines against meningitis caused by group B meningococcal and *E. coli* K1. Therefore additional chemical manipulation of the basic structure of polySia was attempted, on the premise that a synthetically derived artificial antigen might be capable of modulating the immune response in mice [26]. N-acetyl groups of the sialic acid residues were replaced by N-propionyl groups, followed by conjugation of the resultant N-propionylated polySia to tetanus toxoid, which yielded an antigen that when injected in mice induced in them high levels of cross-reactive IgG antibodies. This procedure was also applied in patients with small cell lung cancer, SCLC [30, 31] because polySia bound to the neural cell adhesion molecule is extensively expressed on the surface of SCLC cells. Robust antibody response was noted in patients with SCLC after vaccination with 30 μg of keyhole limpet hemocyanin-conjugated N-propionylated polySia, but peripheral neuropathy and ataxia were detected in several vaccinated patients. It was found that vaccination with the 10 μg dose was significantly more immunogenic than the 3 μg dose. Self-limited grade 3 ataxia of unclear etiology was seen only in 1 of 18 patients. Sera from all patients in the 10 μg dose group also had bactericidal activity against group B meningococci with rabbit complement. Thus the lowest optimal dose was established with confirmation of the safety of the induction of antibodies against polySia with the N-propionylated polySia vaccine. Further evidence was provided [27] that extended epitopes have the ability to break immune tolerance associated with the polysialic acid capsule of these pathogens.

6.2 PolySia-Containing Nano-carriers

Several polySia-containing nano-structures were tested for potential medical applications. Insulin-loaded polySia/protamine nanocapsules administered intra-jejunally to healthy rats resulted in a moderate reduction of the glucose levels [38]. In vitro cytotoxicity assay showed that micelles, containing polySia-ursolic acid conjugate and loaded with an antitumor drug paclitaxel, retained anti-tumor activity with a cell viability of ca. 50% indicating that these micelles could efficiently release paclitaxel into the tumor cells for cancer chemotherapy [44]. In vivo cancer therapy with doxorubicin-loaded polySia-cholanic acid nanoparticles in mice showed antitumor effects that resembled those of the free anticancer drug doxorubicin [28]. Moreover, these nanoparticles had low toxicity toward other organs, reflecting their tumor-targeting property. Therefore, polySia-cholanic acid nanoparticles were considered a potential nanocarrier for anticancer agents.

The ability of the polySia-decylamine micelles to encapsulate hydrophobic therapeutics was demonstrated with Cyclosporine A [1], an immunosuppressant to prevent organ rejection following transplantation, which is also regularly used as a disease-modifying antirheumatic drug in the treatment of rheumatoid arthritis. Sizes and zeta potentials of the loaded and unloaded micelles indicated that the drug carriers had acceptable properties for passive targeting of diseased regions. The polySia-octadecylamine (polySia-ODA) micelles loaded with a calmodulin antagonist were tested for therapy of vascular dementia [39, 40]. These micelles could cross the blood—brain barrier (BBB) mainly via active endocytosis by brain endothelial cells followed by transcytosis. In a water maze test for spatial learning, these micelles significantly reduced the escape latencies of right unilateral common carotid arteries occlusion mice with dosage significantly reduced versus free drug. In addition, the decrease of hippocampal phospho-CaMKII and phospho-synapsin I was partially restored in mice following calmodulin antagonist loaded polySia-ODA micelle treatment. Therefore it was concluded, that these micelles might promote the development of therapeutic approaches for improving the efficacy of brain-targeted drug delivery and have great potential for vascular dementia treatment. In addition, this nanodrug carrier system improved the cognitive deficit by inhibiting the nitrosative stress and inflammasome activation. In particular, polySia-ODA micelles encapsulated reduced dosage of the drug, which is important for therapy of neurovascular diseases.

PolySia-modified liposomes, containing polySia-octadecyl dimethyl betaine conjugate, for efficient delivery of an anticancer drug epirubicin, has been tested [43]. The pharmacokinetic study showed that polySia-modified liposomes prolonged the residence time of the epirubicin in the blood compared with that observed from common liposomes. Biodistribution results obtained from tumor-bearing mice demonstrated that polySia-modified liposomes increased the accumulation of modified liposomes in tumors compared with that of liposomes without polySia. To assess whether polySia-modified liposomes possessed the advantage of a long circulation lifetime in vivo, the formulations were administered in rats intravenously. In the antitumor efficacy study, polySia-modified liposomes showed antitumor and life-prolonging effects, indicating that epirubicin-loaded polySia-modified liposomes might have potential as an effective approach for anticancer therapy. PolySia—octadecyl dimethyl betaine—Pluronic F127 mixed micelles loaded with docetaxel [45] accumulated at tumor sites as demonstrated by in vivo imaging study. The in vivo antitumor activity of these mixed micelles against a tumor xenograft model showed a significant higher inhibition and a lower toxicity compared with micelles without polySia, thus suggesting that the mixed polymeric micelles containing polySia may be a promising strategy for antitumor delivery of docetaxel.

The cyclosporine A (CyA)-loaded polySia–polycaprolactone (PCL) micelles had a negligible effect on the viability of a synovial fibroblast cell line [42]. CyA is used in the treatment of rheumatoid arthritis. Fluorescent microscopy was utilized to demonstrate cellular uptake of the drug-loaded micelles by the synovial fibroblasts through a non-receptor mediated form of endocytosis, release of the CyA from the micelles, and partitioning of CyA into the membrane.

6.3 Polysialylation of Peptides, Proteins, and Low Molecular Weight Drugs

Covalent chemical attachment of compounds such as polySia to therapeutic proteins can overcome three major issues in protein therapeutics: in vitro and in vivo instability, immunogenicity and short half-lives [34]. The oxidized polySia was then coupled to catalase by reductive amination [11]. Enzyme kinetics studies revealed an increase in the apparent K_m (the Michaelis constant) of the polysialylated enzyme indicating a reduction of enzyme affinity for the substrate. This enzyme was however much more stable in the presence of specific proteinases. Another enzyme, asparaginase, was polysialylated using similar method, and polysialylated constructs contained from 4 to 8 polySia chains per molecule of enzyme [12]. Such constructs retained most of the initial asparaginase activity and also maintained the K_m values of the native enzyme. In contrast to the native enzyme, polysialylated asparaginase constructs exhibited resistance to proteolysis. In vivo experiments with intravenously injected mice revealed a significant increase in the half-life of the polysialylated asparaginase. The pharmacological activity of polysialylated insulin constructs was compared with that of intact insulin in normal mice injected subcutaneously [16]. Polysialylated insulin exerted a more prolonged reduction of blood glucose levels. It was concluded that polySia can be used as a means to increase the circulatory half-life of proteins and thus serve as an alternative to the non-biodegradable monomethoxy-poly(ethylene glycol).

In addition to the therapeutic enzymes and insulin, modulation of antibody pharmacokinetics by chemical polysialylation was also tested [6]. Polysialylation generally increased Fab fragment blood half-life resulting in higher tumor uptake in a human tumor xenograft mouse model without significant cross-reaction with normal tissues. This supported the premise that polySia is an inert polymer with no natural receptor. Antibody fragments such as single-chain Fvs (scFvs) are a choice antibody format for human therapeutic antibody discovery [7]. Because of their smaller size and lower molecular complexity, they can be easily selected from combinatorial libraries and more readily expressed in a bacterial expression system, leading to rapid and inexpensive production. For therapy in vivo, the smaller size allows for faster tissue penetration; however, the targeted absolute uptake of such species may be limited by the subsequent faster clearance rate through glomerular filtration. It was shown that site-specific polysialylation of an antitumor single-chain Fv fragment led to antibody fragments with full immunoreactivity and increased blood half-life causing improved, up to 30-fold, tumor uptake while displaying favorable tumor/normal tissue specificity [7].

6.4 PolySia-Mediated Neural Tissue Repair After Neural Injuries

Induced polySia expression has been used as a strategy for promoting tissue repair involving both replacement of cells and rebuilding of neural connections [10]. Glial scars that form at CNS (Central Nervous System) injury sites block axon regeneration. However, transfection of scar astrocytes by a viral vector encoding a polysialyltransferase led to sustained expression of high levels of polySia. This treatment promoted growth of corticospinal axon processes through the lesion site and increased the number of progenitors available to the injury site, suggesting that this procedure could serve as part of a therapy for recovery from injury-induced paralysis. In vivo data have been presented on the complementation of synthetic nerve transplants with exogenous soluble polySia, isolated from *Escherichia coli* K1 (K1-polySia) capsules [13]. After establishing stable contacts nerves lose polySia expression and the molecule reappears after nerve injury accounting for selective re-innervation of motor targets. Regeneration across sciatic nerve gaps in rat was studied after reconnecting the transected nerve stumps with differentially filled silicon tubes (in the absence or presence of Schwann cells and exogenous K1-polySia). The new data on the biocompatibility of exogenous K1-polySia in the in vivo regeneration system demonstrated polySia potency to serve as innovative building block to develop biodegradable nerve conduits. A hypothesis was tested that permanent synthesis of polySia in Schwann cells impaired functional recovery of lesioned peripheral nerves, using transgenic mice expressing polysialyltransferase ST8SiaIV under control of a glial-specific promoter [29]. Both impaired re-myelination of regenerated axons at the lesion site and the proportion of successfully re-innervated motor endplates were significantly increased. The combined negative and positive effects of a continuous polysialyltransferase overexpression observed during peripheral nerve regeneration suggested that an optimized control of polysialyltransferase expression in Schwann cells might improve recovery after peripheral nerves injury.

Acknowledgements This work was supported by National Science Centre, Poland, Grant no. 2015/17/N/NZ1/00029.

References

1. Bader, R.A., Silvers, A.L., Zhang, N.: Polysialic acid-based micelles for encapsulation of hydrophobic drugs. Biomacromol **12**, 314–320 (2011)
2. Bothun, G.D., Knutson, B.L., Strobel, H.J., Nokes, S.E.: Liposome fluidization and melting point depression by pressurized CO_2 determined by fluorescence anisotropy. Langmuir **21**, 530–536 (2005)
3. Cho, J.W., Troy, F.A.: Polysialic acid engineering: synthesis of polysialylated neoglycosphingolipids by using the polysialyltransferase from neuroinvasive *Escherichia coli* K1. Proc. Natl. Acad. Sci. U.S.A. **91**, 11427–11431 (1994)
4. Clarke, R.L.: Effect of lipid structure on the dipole potential of phosphatidylcholine bilayers. Biochim. Biophys. Acta **1327**, 269–278 (1997)
5. Colley, K.J., Kitajima, K., Sato, C.: Polysialic acid: biosynthesis, novel functions and applications. Crit. Rev. Biochem. Mol. Biol. **49**, 498–532 (2014)
6. Constantinou, A., Epenetos, A.A., Hreczuk-Hirst, D., Jain, S., Deonarain, M.P.: Modulation of antibody pharmacokinetics by chemical polysialylation. Bioconjug. Chem. **19**, 643–650 (2008)
7. Constantinou, A., Epenetos, A.A., Hreczuk-Hirst, D., Jain, S., Wright, M., Chester, K.A., Deonarain, M.P.: Site-specific polysialylation of an antitumor single-chain Fv fragment. Bioconjug. Chem. **20**, 924–931 (2009)
8. Decher, G., Ringsdorf, H., Venzmer, J., Bitter-Suermann, D., Weisgerber, C.: Giant liposomes as model membranes for immunological studies: spontaneous insertion of purified Kl-antigen (poly-α-2,8-NeuAc) of *Escherichia coli*. Biochim. Biophys. Acta **1023**, 357–364 (1990)
9. Deepagan, V.G., Thambi, T., Ko, H., Kang, Y.M., Park, J.H.: Amphiphilic polysialic acid derivatives: synthesis, characterization, and in-vitro cytotoxicity. J. Nanosci. Nanotechnol. **13**, 7312–7318 (2013)
10. El Maarouf, A., Petridis, A.K., Rutishauser, U.: Use of polysialic acid in repair of the central nervous system. Proc. Natl. Acad. Sci. U.S.A. **103**, 16989–16994 (2006)
11. Fernandes, A., Gregoriadis, G.: Synthesis, characterization and properties of sialylated catalase. Biochim. Biophys. Acta **1293**, 90–96 (1996)
12. Fernandes, A.I., Gregoriadis, G.: Polysialylated asparaginase: preparation, activity and pharmacokinetics. Biochim. Biophys. Acta **1341**, 26–34 (1997)
13. Haastert-Talini, K., Schaper-Rinkel, J., Schmitte, R., Bastian, R., Muhlenhoff, M., Schwarzer, D., Draeger, G., Su, Y., Scheper, T., Gerardy-Schahn, R., Grothe, C.: *In vivo* evaluation of polysialic acid as part of tissue-engineered nerve transplants. Tissue Eng A **16**, 3085–3098 (2010)
14. Hodgkin, A.L., Huxley, A.F.: A quantitative description of membrane current and its application to conduction and excitation in nerve. J. Physiol. **117**, 500–544 (1952)
15. Huang, R.B., Cheng, D., Lu, B., Liao, S.M., Troy II, F.A., Zhou, G.P.: The intrinsic relationship between structures and functions of the sialyltransferase ST8Sia family members. Curr. Topics. Medic, Chem. **17**, 2359–2369 (2017)

16. Jain, S., Hreczuk-Hirst, D.H., McCormack, B., Mital, M., Epenetos, A., Laing, P., Gregoriadis, G.: Polysialylated insulin: Synthesis, characterization and biological activity in vivo. Biochim. Biophys. Acta **1622**, 42–49 (2003)
17. Janas, T., Janas, T.: Polysialic acid: structure and properties. In: Dumitriu, S. (ed.) Polysaccharides: Structural Diversity and Functional Versality, 2nd edn, pp. 707–727. Marcel Dekker, New York (2005)
18. Janas, T., Janas, T.: Membrane oligo- and polysialic acids. Biochim. Biophys. Acta—Biomembranes **1808**, 2923–2932 (2011)
19. Janas, T., Janas, T., Krajiński, H.: Membrane transport of polysialic acid chains: modulation of transmembrane potential. Eur. Bioph. J. **29**, 507–514 (2000)
20. Janas, T., Janas, T., Yarus, M.: A membrane transporter for tryptophan composed of RNA. RNA **10**, 1541–1549 (2004)
21. Janas, T., Krajiński, H., Janas, T.: Electromigration of polyion homopolymers across biomembranes: a biophysical model. Biophys. Chem. **87**, 167–178 (2000)
22. Janas, T., Krajiński, H., Timoszyk, A., Janas, T.: Translocation of polysialic acid across model membranes: kinetic analysis and dynamic studies. Acta Biochim. Polon. **48**, 163–173 (2001)
23. Janas, T., Nowotarski, K., Janas, T.: Polysialic acid can mediate membrane interactions by interacting with phospholipids. Chem. Phys. Lipids **163**, 286–291 (2010)
24. Janas, T., Nowotarski, K., Janas, T.: The effect of long-chain bases on polysialic acid-mediated membrane interactions. Biochim. Biophys Acta—Biomembranes **1808**, 2322–2326 (2011)
25. Janas, T., Yarus, M.: Visualization of membrane RNAs. RNA **9**, 1353–1361 (2003)
26. Jennings, H.J., Roy, R., Gamian, A.: Induction of meningococcal group B polysaccharide-speciWc IgG antibodies in mice by using an N-propionylated B polysaccharide-tetanus toxoid conjugate vaccine. J. Immunol **137**, 1708–1713 (1986)
27. Johal, A.R., Jarrell, H.C., Letts, J.A., Khieu, N.H., Landry, R.C., Jachymek, W., Yang, Q., Jennings, H.J., Brisson, J.R., Evans, S.V.: The antigen-binding site of an N-propionylated polysialic acid-specific antibody protective against group B meningococci is consistent with extended epitopes. Glycobiol **23**, 946–954 (2013)
28. Jung, B., Shim, M.K., Park, M.J., Jang, E.H., Yoon, H.Y., Kim, K., Kim, J.H.: Hydrophobically modified polysaccharide-based on polysialic acid nanoparticles as carriers for anticancer drugs. Int. J. Pharm. **520**, 111–118 (2017)
29. Jungnickel, J., Eckhardt, M., Haastert-Talini, K., Claus, P., Bronzlik, P., Lipokatic-Takacs, E., Maier, H., Gieselmann, V., Grothe, C.: Polysialyltransferase overexpression in Schwann cells mediates different effects during peripheral nerve regeneration. Glycobiol **22**, 107–115 (2012)
30. Krug, L.M., Ragupathi, G., Ng, K.K., Hood, C., Jennings, H.J., Guo, Z., Kris, M.G., Miller, V., Pizzo, B., Tyson, L., Baez, V., Livingston, P.O.: Vaccination of small cell lung cancer patients with polysialic acid or N-propionylated polysialic acid conjugated to keyhole limpet hemocyanin. Clin. Cancer Res. **10**, 916–923 (2004)
31. Krug, L.M., Ragupathi, G., Ng, K.K., et al.: Immunization with N-propionyl polysialic acid–KLH conjugate in patients with small cell lung cancer is safe and induces IgM antibodies reactive with SCLC cells and bactericidal against group B meningococci. Cancer Immunol. Immunother. **61**, 9–18 (2012)
32. Matthews, I., Petrak, K.: The preparation and characterization of a twin-tailed, lipid-linked polysialic acid. Biopolymers **33**, 453–457 (1993)
33. Nowotarski, K., Sapoń, K., Kowalska, M., Janas, T., Janas, T.: Membrane–potential dependent binding of polysialic acid to lipid membranes. Cell. Mol. Biol. Lett. **18**, 579–594 (2013)
34. Pisal, D.S., Kosolski, M.P., Balu-Iyer, S.V.: Delivery of therapeutic proteins. J. Pharm. Sci. **99**, 2557–2575 (2010)
35. Rutishauser, U.: Polysialic acid in the plasticity of the developing and adult vertebrate nervous system. Nat. Rev. Neurosci. **9**, 26–35 (2008)
36. Sato, C., Kitajima, K.: Disialic, oligosialic and polysialic acids: distribution, functions and related disease. J. Biochem. **154**, 115–136 (2013)
37. Schnaar, R.L., Gerardy-Schahn, R., Hildebrandt, H.: Sialic acids in the brain: gangliosides and polysialic acid in nervous system development, stability, disease, and regeneration. Physiol. Rev. **94**, 461–518 (2014)

38. Thwala, L.N., Beloqui, A., Csaba, N.S., Gonzalez-Touceda, D., Tovar, S., Dieguez, C., Alonso, M.J., Preat, V.: The interaction of protamine nanocapsules with the intestinal epithelium: a mechanistic approach. J. Control. Release **243**, 109–120 (2016)
39. Wang, R., Yin, Y.X., Mahmood, Q., et al.: Calmodulin inhibitor ameliorates cognitive dysfunction via inhibiting nitrosative stress and NLRP3 signaling in mice with bilateral carotid artery stenosis. CNS Neurosci Therapeut **23**, 818–826 (2017)
40. Wang, X.J., Gao, Y.P., Lu, N.N., et al.: Endogenous polysialic acid based micelles for calmodulin antagonist delivery against vascular dementia. ACS Appl. Mater. Interfaces. **8**, 35045–35058 (2016)
41. Westphal, N., Kleene, R., Lutz, D., Theis, T., Schachner, M.: Polysialic acid enters the cell nucleus attached to a fragment of the neural cell adhesion molecule NCAM to regulate the circadian rhythm in mouse brain. Mol. Cell. Neurosci. **74**, 114–127 (2016)
42. Wilson, D.R., Zhang, N., Silvers, A.L., Forstner, M.B., Bader, R.A.: Synthesis and evaluation of cyclosporine A-loaded polysialic acid-polycaprolactone micelles for rheumatoid arthritis. Eur. J. Pharm. Sci. **51**, 146–156 (2014)
43. Zhang, T., Zhou, S., Hu, L., et al.: Polysialic acid-modifying liposomes for efficient delivery of epirubicin, in-vitro characterization and in-vivo evaluation. Int. J. Pharm. **515**, 449–459 (2016)
44. Zhang, W., Dong, D., Li, P., et al.: Novel pH-sensitive polysialic acid based polymeric micelles for triggered intracellular release of hydrophobic drug. Carbohydr. Polym. **139**, 75–81 (2016)
45. Zhang, W., Zhou, S., Liu, Y., et al.: Polysialic acid and pluronic F127 mixed polymeric micelles of docetaxel as new approach for enhanced antitumor efficacy. Drug Dev. Ind. Pharm. **43**, 1827–1835 (2017)

Self-assembling of Thermo-Responsive Block Copolymers: Structural, Thermal and Dielectric Investigations

A. Kyritsis, A. Laschewsky and C. M. Papadakis

Abstract Plethora of amphiphilic polymers and copolymers have been synthesized that form self-assembled structures in aqueous media, resembling the assemblies of biopolymers invented by nature. Such polymeric systems serve as stimuli-responsive materials, i.e. they respond to small external changes in the environmental conditions, which is a common process for biopolymers in living organisms. Temperature is the most widely used stimulus in environmentally responsive polymer systems. Thermoresponsive polymers have attracted much research interest because of their potential applications, which include rheological control additives, thermal affinity separation, controlled drug release, gene therapy and regenerative medicine. On the other hand, they represent model systems for many biological systems, for example for the investigation of the interaction between peptide-like groups and solvents and, thus, for the study of protein stability in aqueous solutions. In this chapter, we provide a comprehensive view on recent investigations on the micellar aggregation and the thermoresponsive behavior of amphiphilic model polymers. Firstly, we will present general characteristics of the thermoresponsive behavior of macromolecules and discuss in more detail their applications with biomedical interest. Next, we will focus on the experimental investigation of thermoresponsive polymers and present, briefly, research outcomes concerning the properties of the well-studied poly(N-isopropylacrylamide) (PNIPAM) polymer. Then, we will present results with respect to the thermoresponsive behavior of a rather new class of polymers based on the non-ionic poly(methoxy diethylene glycol acrylate) (PMDEGA) polymer. Copolymers with various architectures, namely diblock, triblock and star block copolymers are studied, as well as a PMDEGA homopolymer as reference. To that aim, complemen-

A. Kyritsis (✉)
Physics Department, National Technical University of Athens,
Iroon Polytechneiou 9, Zografou Campus, 15780 Athens, Greece
e-mail: akyrits@central.ntua.gr

A. Laschewsky
Fraunhofer Institut Für Angewandte Polymerforschung,
Geiselbergstr. 69, 14476 Potsdam-Golm, Germany

C. M. Papadakis
Physik-Department, Fachgebiet Physik Weicher Materie, Technische Universität München,
James-Franck-Str. 1, 85748 Garching, Germany

C. Demetzos and N. Pippa (eds.), *Thermodynamics and Biophysics of Biomedical Nanosystems*, Series in BioEngineering, https://doi.org/10.1007/978-981-13-0989-2_12

tary methods were applied, such as small-angle X-ray (SAXS), differential scanning calorimetry (DSC) and broadband dielectric spectroscopy (BDS). Seeking for understanding of fundamental aspects of the macromolecular thermoresponsive behavior, we present, in a comparative way, results obtained on PNIPAM- and PMDEGA-based systems. Characteristic differences between the two series of polymeric solutions are worked out, concerning the self-organization, the width and hysteresis of the transition, and the chain conformations during the demixing phase transition.

1 Introduction

Numerous supramolecular structures of varying complexity are formed in nature upon self-assembly of biomacromolecules via non-covalent interactions in aqueous media. Many of such molecules are amphiphilic, i.e. they consist of hydrophilic and hydrophobic moieties. Also synthetic amphiphilic macromolecules form self-assembled structures in aqueous media. The spontaneous formation of self-assembled structures is the outcome of a delicate balance between attractive and repulsive forces, each one with its characteristic strength. For the purpose of self-assembly, the attractive interactions do not have to be covalent, i.e. "permanent". Macromolecular self-associations are driven by noncovalent weak and complementary interactions: Coulombic, hydrogen bonding, van der Waals, exchange, repulsive, and hydrophobic interactions.

Macromolecules that can form self-assembled structures can, often, also respond to stimuli, which is a common process for biopolymers in living organisms. Stimuli-responsive polymers are defined as polymers that undergo relatively large and abrupt, physical or chemical changes in response to small external changes in the environmental conditions. A simplified design of the stimuli-responsive association concept in aqueous polymer solution is shown in Fig. 1 [1, 2]. Various names have been given to 'stimuli-responsive' polymers like stimuli-sensitive, intelligent, smart, or environmentally sensitive polymers [3]. There are many different stimuli to modulate the response of polymer systems. These stimuli could be classified as either physical or chemical stimuli. Chemical stimuli, such as pH, ionic factors and chemical agents, will change the interactions between polymer chains or between polymer chains and solvents at the molecular level. The physical stimuli, such as temperature, electric or magnetic fields, and mechanical stress, will affect the level of various energy sources and alter molecular interactions at critical onset points.

Temperature is the most widely used stimulus in environmentally responsive polymer systems. Temperature-responsive (thermoresponsive) polymers undergo reversible conformational or phase changes in response to negligible variations of temperature. One of the unique properties of thermoresponsive polymers is the presence of a critical solution temperature. The critical solution temperature is the temperature at which the phase of polymer and solvent mixture is discontinuously changed, with its value depending on the composition of the mixture [3–5]. When a polymer is molecularly dissolved in a suitable solvent, it may become insoluble upon increase or

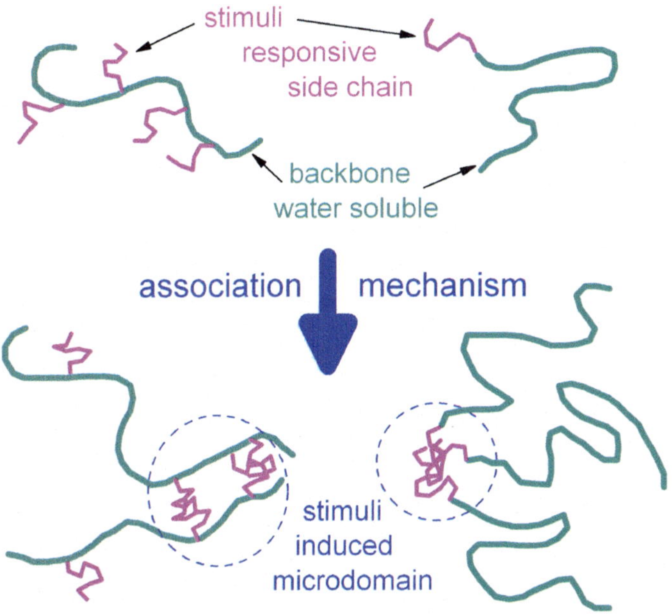

Fig. 1 A simplified design of the stimuli-responsive association concept in aqueous polymer solution

decrease in temperature and, thus, precipitate from the solution. In other words, the binary polymer/solvent mixture undergoes a temperature-induced phase separation from a one-phasic towards a bi-phasic system due to the existence of a miscibility gap in the phase diagram [6–8]. If elevation of temperature results in phase separation, the system exhibits lower critical solution temperature (LCST) behavior. To be more precise, the polymer does not simply precipitate from the solution, but two phases are formed in equilibrium, whereby one phase has a high polymer concentration and the other one has a low polymer concentration. As shown in Fig. 2a, the LCST is defined as the temperature at the minimum of the binodal (or the coexistence curve) of the phase diagram. The corresponding concentration is the lower critical solution concentration (LCSC). The reverse case, where phase separation occurs upon decreasing temperature, is called upper critical solution temperature (UCST) behavior (Fig. 2b).

Most of the studies discussed in this chapter use polymers that present an LCST, first described by Heskins and Guillet [9]. Poly(N-isopropylacrylamide) (PNIPAM) has been by far the most studied thermoresponsive polymer in materials science. The chemical structure of PNIPAM is shown in Fig. 3a. This synthetic polymer exhibits a LCST around 32 °C in aqueous medium and is therefore very useful for preparing smart materials for biological applications [5, 10, 11]. However, it should be noted that the main reason for its biomedical popularity is not really the fact that the LCST of PNIPAM is close to body temperature (other polymers exhibit LCST values even closer to 37 °C) but rather the fact that its sharp phase

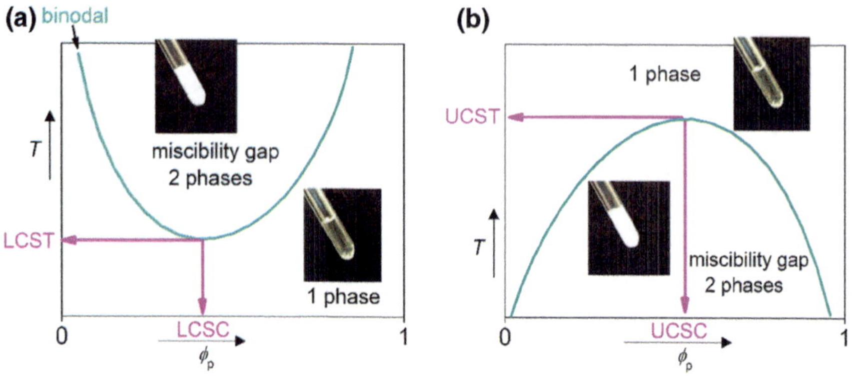

Fig. 2 Phase diagram for a binary polymer-solvent mixture exhibiting LCST (**a**) and UCST (**b**) behavior

Fig. 3 Chemical structure of PNIPAM (**a**) and PMDEGA (**b**) homopolymers

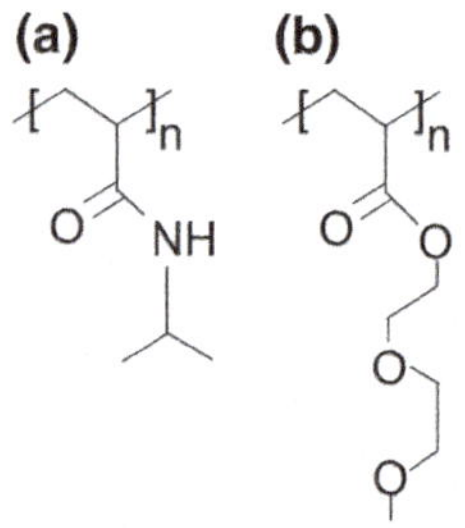

transition is relatively insensitive to slight variations of environmental conditions, like pH, concentration, or chemical environment and to changes of molar mass, as well, and due to its versatility in terms of copolymer architecture variation [12]. Adjustment of the LCST of PNIPAM has been achieved by copolymerizing with hydrophilic or hydrophobic monomers rendering the overall hydrophilicity of the polymer higher or lower respectively [10, 13]. Other polymers with thermoresponsive properties include poly(N,N-diethylacrylamide) (PDEAM) with an LCST over the range of 25–32 °C, poly(N-vinylcaprolactam) (PVCL) with an LCST between 25 and 35 °C, poly(vinyl methyl ether) (PVME) with an LCST at 33.8 °C, poly[2-(dimethylamino) ethyl methacrylate] (PDMAEMA) with an LCST of around 50 °C, methylcellulose (MC) with LCST ~50 °C, poly(2-ethyl-2-oxazoline) (PEOZ) ~62 °C and poly(ethylene glycol) (PEG), also called poly(ethylene oxide) (PEO) whose LCST is around 85 °C [4, 14, 15].

A major limitation for the use of PNIPAM in biomedical applications is that the unreacted NIPAM monomer has shown evidence of neurotoxicity [4, 16]. Thus, thermoresponsive copolymers that do not contain acrylamides are gaining increasing attention, and they have proven a very promising area for research. Among them, thermoresponsive polymers containing short oligo(ethylene glycol) side chains have gained rapid success in fundamental polymer science but also in applied materials research [12, 17–21]. The chemical structure of the member of this family with the shortest side chain, poly(methoxy diethylene glycol acrylate) (PMDEGA), is shown in Fig. 3b. These polymers combine the advantages of PEG (i.e., biocompatibility) and of thermoresponsive polymers (i.e., LCST behavior in water) in a single macromolecular structure. Moreover, they have inherent advantages as compared to traditional PNIPAM, such as: (i) an excellent bio-repellency below the LCST (i.e., anti-fouling behavior), (ii) reversible phase transitions (i.e., no marked hysteresis), and (iii) bio-inert properties (i.e., no specific interactions with biological materials). Furthermore, since oligo(ethylene glycol) pendant chains can be connected to various types of polymeric backbones, the polymerization of oligo(ethylene glycol) (meth)acrylates is surely the most effective strategy for preparing thermoresponsive polymers [15, 18].

As the stimulus-responsive behavior occurs in aqueous solutions, thermoresponsive polymers that exhibit LCST behavior are becoming increasingly attractive for biotechnology and medicine, for their use, among others, as basic components of biosensors and actuators, thermally modulated drug and gene deliveries, affinity precipitation, thermoresponsive bioconjugated proteins and enzymes, thermoresponsive cell culture substrates for fabricating cell sheets, tissue engineering and artificial muscles [4, 8, 11, 14, 22–30].

On the other hand, the interest in this coil-to-globule transition is not only due to its importance as a fundamental concept in polymer physics and solution dynamics, but also due to its relevance for many biological systems. The close similarity of the cooperative hydration of PNIPAM and polypeptide backbone indicates the relevance of this study for the thermal stability of hydrogen bonded water networks enveloping such biological polymers [5, 10, 31]. Plenty of experimental results indicate that knowledge of the balance of interactions between water and the hydrophilic and hydrophobic groups at the coil-to-globule transition in the PNIPAM aqueous solution is essential for understanding complicated molecular mechanisms in biological systems, such as the stabilization of protein conformation and self-association of lipids. Moreover, the response to stimuli is a common process for biopolymers in living organisms. In the past, the concept of cooperative interactions between the functional segments of biopolymer has led to the invention of novel synthetic thermoresponsive polymer systems.

2 Characteristics of the Thermoresponsive Behavior

It is worth stressing here that thermal sensitivity is a general phenomenon for polymers in solution: the solubility of all polymers in any solvent depends on temperature. For that reason, Allan Hoffman defined intelligent stimuli-responsive polymers as polymers that respond to a small physical or chemical stimulus with large property changes [1, 32]. The coil-to-globule transition is a typical polymer response to a change in its solution temperature. Non-ionic polymers can also undergo a coil-globule transition in aqueous solutions. Thermoresponsive homopolymers may have a repeat unit that contains hydrophilic and hydrophobic moieties, and one of the best known examples of a synthetic thermoresponsive polymer is poly(N-isopropylacrylamide) (PNIPAM). It is worth mentioning that their transition significantly differs from the transition of polymers in organic media. Hydrogen bonds and hydrophobic and hydrophilic interactions contribute much more to the solubility of a polymer in water than do short range van der Waals interactions, which prevail in solutions of polymers in organic solvents. The majority of non-ionic water-soluble polymers undergo phase separation upon heating. The phase separation of these polymers can be described by a phase diagram with an LCST (Fig. 2a) [9], which reflects a local structural transition involving water molecules surrounding specific segments of the polymer in solution [5]. As in nature, the bulk response of the thermosensitive polymer is usually due to multiple cooperative interactions such as loss of hydrogen bond (HB) binding, which, although individually weak, ultimately evokes a large structural change in the material when summed up over the whole polymer [2].

2.1 Thermodynamics of Homo-Polymer Solutions

The behavior of a polymer in a given medium reflects the balance of like and unlike interactions among its own segments and the surrounding molecules. In the case of aqueous solutions, the solvent-solvent interaction in water is particularly strong and indicated by its partially ordered structure. Ordering of solutes such as PNIPAM in aqueous solution results from specific orientations required to hydrogen bonding with the already somewhat arranged water molecules.

The solution process at temperature T is governed by the Gibbs free energy of mixing

$$\Delta G_m(T) = \Delta H_m(T) - T \cdot \Delta S_m(T) \tag{1}$$

As long as the free energy of the mixing is lower than the sum of the respective phases, the phase separation will not occur. Thus, negative values of $\Delta G_m(T)$ lead to homogeneous mixing of solute and solvent. Specific interactions between solute and solvent determine the enthalpic, ΔH_m, and entropic, ΔS_m, contribution to the Gibbs free energy. At the phase transition, the fragile balance between these two terms is destroyed and polymer chain conformations and water hydrogen bonds network change dramatically, always in order to minimize the changes in Gibbs free energy.

In more detail, the water molecules trapped by hydrogen bonds between the hydrophilic polymer groups and bulk water form a thin shell of ordered structure around the hydrophilic part of the polymer. Formation of this hydration shell gives a negative enthalpy for the solution process $\Delta H_m(T_0)$ at a temperature T_0, at which the hydrogen bonds stay substantially unbroken. At the same time, water molecules must reorient around nonpolar regions of solutes, being unable to form hydrogen bonds with them. Therefore, another hydrated shell of ordered ice-like structure is formed upon the hydrogen bonds between water molecules themselves around the hydrophobic part of the polymer [33]. These have been claimed to be clathrate-like structures and the involved water molecules are called "cage" water molecules [34, 35]. This ordered ice-like structure of "cage" water molecules results in a large negative entropy change of mixing $\Delta S_m(T_0)$. This phenomenon, known as the hydrophobic effect, may be considered as a complex result of the tendency of hydrophobic molecules to minimize the interface with water molecules and of the consecutive entropy change of water molecules around the hydrophobic group. Thus, according to Eq. (1), the solvation of amphiphilic macromolecules at $\boldsymbol{T_0}$ results from the domination of enthalpy of mixing.

In the case of LCST polymers, increasing temperature tends to break the hydrogen bonds and, as a result, the hydrated shells around the hydrophobic part of the polymers are disordered. Water molecules are expelled into the bulk, which increases $\Delta H_m(T_0)$ by δH_w and $\Delta S_m(T_0)$ by δS_w. As a consequence, the hydrophobic groups of the polymer may approach each other, forming gradually new intra- and intermolecular bonds (hydrophobing bonding). These intermolecular interactions suppress conformation changes of the polymer and, assuming that do not disturb seriously the ordered structure of "cage" water molecules, lead to a decrease in $\Delta S_m(T_0)$ by δS_{hp} ($\delta S_{hp} < 0$) and in $\Delta H_m(T_0)$ by δH_{hp} ($\delta H_{hp} < 0$). Hence, $\Delta H_m(T_0)$ increases by $\delta H = \delta H_w + \delta H_{hp}$ as a result of hydrogen bond breaking (strong positive contribution) and hydrophobic bonding (weaker negative contribution). Both effects also lead to a slight increase in $\Delta S_m(T_0)$ given by $\delta S = \delta S_w + \delta S_{hp}$ (assuming that polymer-polymer interactions are very weak at temperatures close to T_o). Then, the above considerations allow the Gibbs free energy of mixing in Eq. (1) for the solution process at temperature T to be expressed by

$$\Delta G_m(T) = \Delta H_m(T_0) + \delta H(T) - T[\Delta S_m(T_0) + \delta S(T)] \quad (2)$$

where,

$$\delta H(T) = \delta H_w(T) + \delta H_{hp}(T) \quad (3)$$

$$\delta S(T) = \delta S_w(T) + \delta S_{hp}(T) \quad (4)$$

The $\delta H(T)$ term is positive and increases with T to make $\Delta H_m(T) = \Delta H_m(T_0) + \delta H(T)$ positive at a certain temperature. The entropy change $\delta S(T)$ in Eq. (4) is positive, but decreases with temperature due to the increasing effect of hydrophobic bonding of the polymer chains. The term $-T\,[\Delta S_m(T_0) + \delta S(T)]$ in Eq. (2) is therefore also taken as positive. As a consequence, the sum of the second and third terms on the right-hand-side of Eq. (2) increases monotonically with increasing T, mainly due to water-polymer HB breaking, and the negative $\Delta H_m(T_0)$ is increasingly compensated by the sum of these terms, finally resulting in a positive $\Delta G_m(T)$. This predicts that, as the temperature is raised, the system becomes increasingly unstable, and phase separation takes place above a critical temperature. During phase separation, hydrophobic bonding can take place between the same and between different polymer chains. The former process manifests itself as single chain collapse (intra-chain contraction/coil-to-globule transition [36]) and the latter as intermolecular aggregation (inter-chain association/phase transition). At higher temperatures, the entropy term ($-$ T $\Delta S_m(T)$) dominates the otherwise exothermic enthalpy of the hydrogen bonds formed between the polymer polar groups and water molecules ($\Delta H_m(T)$) that is the initial driving force for dissolution. A high gain in the entropy change comes, apart from the release of hydrogen bonded water molecules into the bulk, also from the disruption of the clathrate-like structure of water around the non-polar groups of the polymer. Thus, when a homopolymer in solution encounters a situation in which the thermodynamic quality of the solvent is poor, individual chains of the homopolymer undergo a coil-to-globule collapse. The globules associate immediately, and macroscopic phase separation seems unavoidable. If the concentration of the polymer is high enough, this replacement of polymer-water contacts with polymer-polymer and water-water contacts is manifested by precipitation.

Worth noticing here, that it has been reported that a number of polymers in water or in organic solvents form equilibrium globules, i.e. single chain globules that remain isolated in solution without immediate association and precipitation. When multimolecular aggregates are formed they are called "*mesoglobules*" [5]. Mesoglobules of thermosensitive polymers that are formed beyond CP are spherical in shape and monodispersed in size and typically have a radius on the order of 50–200 nm. Various thermoresponsive polymers and their derivatives form colloidally stable suspensions instead of the expected macrophase separation upon heating of their dilute aqueous solutions above CP, including homopolymers such as PNIPAM, PVLC, and poly(methylvinyl ether), PMVEth [5].

In order to study the phase separation of a polymer solution and to create the phase diagram, the Gibbs free energy must be defined as a function of the volume fraction ϕ_{pol} of the polymer at various temperatures. In Fig. 4 a schematic representation of possible phase diagram of a polymer that exhibits LCST behavior is shown. In general, the LCST-type phase diagram of homogeneous polymer solutions is characterized by stable, metastable, and unstable regions. These regions are separated by two phase lines, namely the binodal and spinodal lines, which meet at the critical point [9, 10, 37–39]. The binodal line marks the crossover from the stable, homogeneous low-temperature (LT) phase to the metastable range. The upper limit of stability of the LT phase is given by the spinodal line of the phase diagram, above which a spinodal decomposition of the homogeneous solution definitely occurs into the high-temperature (HT) phase. During experiments involving moderate heating rates, the phase separation presumably often starts in the metastable range via a nucleation and growth process. From the structural viewpoint, the macromolecules undergo a coil-to-globule transition. Polymer-rich agglomerates grow on the time scale of seconds to hours within the phase-separating solution. Different, sometimes subsequent growth processes, which can in part be described by colloidal aggregation, are given in the literature [40]. Regarding the related terminology, as seen in Fig. 4, the lowest point of the binodal line, which is also the point at which the binodal and the spinodal line meet, is termed LCST. Hence, for each given system, one

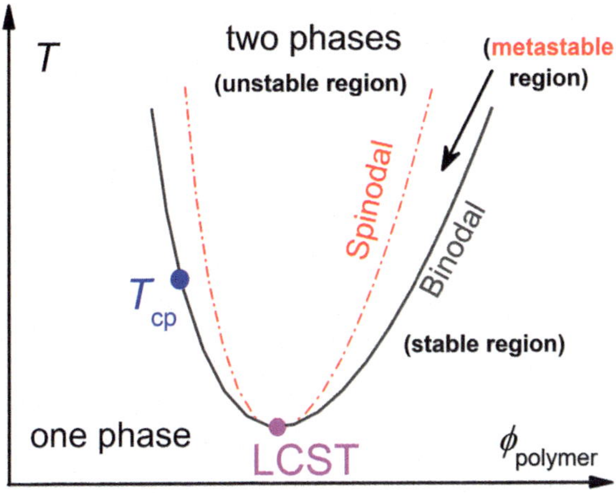

Fig. 4 Phase diagram of a polymer that exhibits LCST behavior. The stable, metastable and unstable regions are also indicated

LCST value is expected. Any other point on the binodal term is called cloud point (T_{cp} or CP), a term originating from the transition from a transparent solution below CP to a turbid (cloudy) one above CP. Depending on the method of detection, CP can also be expressed as demixing temperature T_{dem}. Often, the term LCST is incorrectly used instead of CP, especially in the case of PNIPAM where the phase diagram is relatively flat as a function of polymer concentration (ϕ_{pol}) [41]. In a LCST system, the right-hand branch of the curve is characterized by a positive slope, indicating that the polymer (or gel) will precipitate (collapses) as the temperature increases.

2.2 *Classification of Thermoresponsive Polymers*

2.2.1 Classification According to Their Miscibility with Water

Two types of thermoresponsive polymers are distinguished [5]: those which can act only as an hydrogen bond (HB) donor or as an acceptor, such as poly(ethylene oxide) (PEO), and those which can act both as a HB donor and acceptor, such as poly(N-isopropyl acrylamide) (PNIPAM), and can therefore form hydrogen bonds among themselves. In the first case, the water molecules are believed to adsorb onto the chain independently from each other, whereas in the latter case, they are expected to adsorb in sequences and are released collectively at the cloud point [42, 43]. This results in different solvation processes and, consequently, in different thermoresponsive behavior of the macromolecules.

Berghmans and Van Mele proposed the following phenomenological classification of thermoresponsive polymers according to their miscibility with water:

Type I polymers (e.g. poly(N-vinylcaprolactam), PVCL) are species that follow the classic Flory–Huggins behavior [44, 45]: their LCST (i.e. the absolute minimum in the phase diagram) shifts upon increasing the polymer molar mass towards lower polymer concentrations.

Type II polymers (e.g. PNIPAM [37, 46]) are polymers for which the minimum of the demixing curves is hardly affected by chain length. For Type II polymers, the architecture has a negligible effect (e.g. the LCST of star PNIPAM is similar to that of linear polymer), except for polymers with hydrophobic or hydrophilic end-groups and polymers with a high number of arms or spherical brushes [5].

Type III polymers (e.g. poly(methylvinylether), PMVEth) exhibit a bimodal phase diagram, presenting two critical points for low and high polymer concentrations corresponding to the Type I and Type II behaviors, respectively.

2.2.2 Classification According to Their Physical Form

Mimicking biological systems, several stimuli-responsive systems have been developed, with the majority of studies dealing with polymeric **solutions**, **gels**, **surfaces and interfaces**, and to some extent, **polymeric solids** (Fig. 5). Therefore thermoresponsive polymers (and in general smart polymers) could be classified in response to their physical form into three main classes [1, 25, 47], i.e.

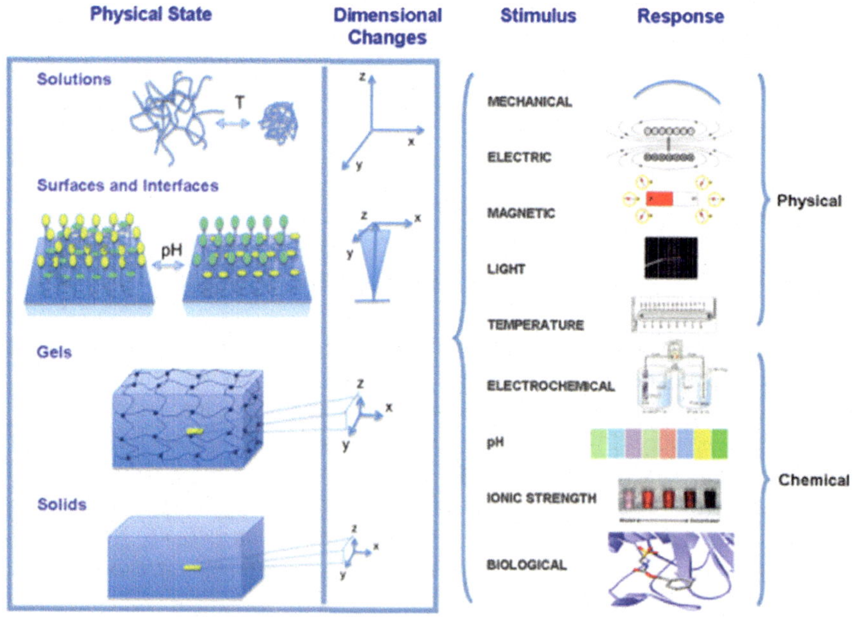

Fig. 5 Schematic representation of dimensional changes in polymeric solutions, at surfaces and interfaces, in polymeric gels, and polymer solids resulting from physical or chemical stimuli (Reproduced from [47], with permission from Elsevier)

(i) linear free chains in solutions, where the polymer undergoes a reversible collapse after an external stimulus is applied,
(ii) chains adsorbed or grafted on a surface, where the polymer reversibly adsorbs or collapses on a surface, respectively, converting the interface from hydrophilic to hydrophobic, once a specific external parameter is modified and
(iii) covalently cross-linked gels and reversible or physical gels, which can be either micro- or macroscopic networks and on which swelling behavior is environmentally triggered.

The states of matter shown in Fig. 5 impose different degrees of restrictions on the mobility of polymeric segments or chains, thus making dimensional responsiveness easily attainable for the systems with a higher solvent content and minimal energy inputs. Significantly greater challenges exist when designing chemically or physically crosslinked gels and solid polymeric networks that require maintaining their mechanical integrity. Restricted mobility within the network results from significant spatial limitations, thus imposing limits on obtaining stimuli-responsiveness. While going from a solution phase to surfaces and interfaces, or gels to solid state phases, the segmental mobilities of polymer chains decrease due to significant spatial restrictions manifested by smaller displacement vectors in all spatial directions. Consequently, energetic requirements for responses to temperature, mechanical, electro-magnetic irradiation, or electrochemical, pH, ionic strength, or bioactive species will be different for each physical state [48]. Examples of responses are depicted in Fig. 5 and are classified into physical and chemical categories, where multiple stimuli may result in one or more responses, or one stimulus may result in more than one response [47].

Since the first **hydrogel** was synthesized from copolymers of 2-hydroxyethyl methacrylate with ethylene dimethacrylate, hydrogels have been successfully utilized in commercial applications, such as contact lenses and implant materials, and have proven biocompatibility in their clinical uses [3]. The tremendous interest in hydrogels was increased by introducing stimuli responsive properties. Of these properties, temperature responsive hydrogels have attracted strong interest. By means of cross-linking, the drastic change of the molecular conformation at the phase transition temperature is translated into macroscopically collapsing materials. Indeed, when hydrogels are prepared by crosslinking thermoresponsive polymers, the temperature sensitivity in water results in changes in the degree of hydration of the polymers. Below the transition temperature, the polymer swells up to equilibrium hydration degree, being in an expanded state. When increasing the temperature above the transition, the hydrogel de-swells to a collapsed state. This process is generally reversible and can be applied in a pulsatile manner, making the polymer to behave as an on-off system, when the stimulus is applied or removed. Gels of small dimension with a radius in the range from tens of nanometers to about 1 μm often are colloidally stable even in the collapsed state and, compared to macroscopic gels, respond to the temperature change more quickly [49]. The transition temperature for these types of gels is termed volume phase transition temperature (VPTT, instead of LCST), because of a considerable change in size, while no macroscopic phase separation occurs (except the expelling of solvent). This makes such so-called microgels outstandingly well suitable for studying the LCST phenomenon, which might even give insight into the denaturation of proteins, but furthermore gives rise to applications relying on triggered release [50].

Surface modification with thermoresponsive polymers leads to the preparation of **responsive interfaces** which may show a behavior in response to one of these small changes in the environmental parameters which is very different from the one in the bulk. A temperature responsive surface or interface is another route to biomedical applications such as temperature-modulated membranes, chromatography, and cell culture dish [3, 29]. Two main categories can be considered for **interfaces** designed with immobilized temperature responsive polymers (broadly all stimuli responsive polymers); to modulate pores of a porous matrix as temperature responsive gates [51, 52], and to control the wettability of nonporous matrix surfaces [53, 54]. For example, the modification of a membrane surface with temperature responsive polymer chains can modulate the diffusion profiles [3].

2.2.3 Classification According to Combination of Co-monomers

For block and graft copolymer structures, there are two effects that contribute to the thermoresponsive behavior: it may be given more by thermoresponsivity of the thermoresponsive part of the molecule alone (this is the case of, e.g., most poly(N-isopropylacrylamide)-based systems) **or** by the overall hydrophobicity/hydrophilicity balance of the whole molecule (this is the case for, e.g., triblock poly(2-alkyl-2-oxazoline)-based systems, poly(ethylene oxide)-block-poly(propylene oxide)-block-poly(ethylene oxide)-based systems or, generally, systems with shorter blocks) [23].

Thermoresponsive copolymers combining two or more co-monomers, where at least one of these would give a thermoresponsive homopolymer, enable convenient tuning of the LCST as well as incorporation of additional functionality. For example, they can incorporate charge or biodegradability within a thermoresponsive chain. They can also enable grafting of the copolymers to surfaces and particles. Thermoresponsive copolymers provide an excellent opportunity to finely tune the LCST and thermoassociative properties. It is through copolymerization that the properties of thermoresponsiveness can be transferred to other systems. A number of copolymer architectures can be prepared (Fig. 6) [4].

Self-assembled structures, such as macromolecular micelles, vesicles or aggregates, can be designed to take up a range of activities. Biodegradability can be readily incorporated. Their ability to form reversible gels on their own or with other dispersions provides potential applications as injectable soft biomaterials [4].

A series of amphiphilic block copolymers were reported to exhibit thermoresponsive behavior by having temperature responsive **micellization behavior** and by forming hydrogels above a critical gelation temperature [55]. For example, block copolymers based on PEO–PPO sequences are a family of commercially available triblock copolymers (e.g. Pluronics® or Poloxamer, Tetronics®) which exhibit a sol-gel transition below or close to the physiological temperature and a gel–sol transition

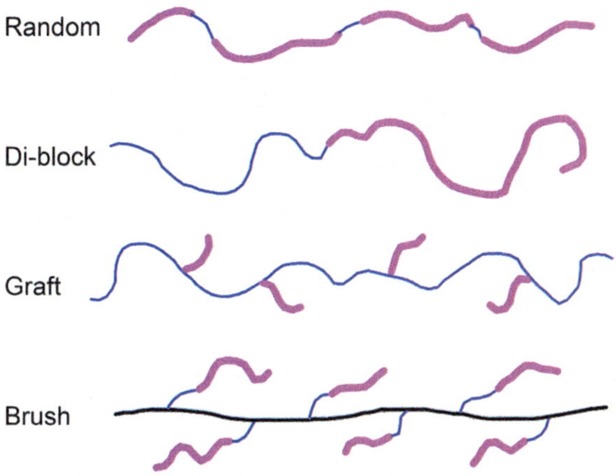

Fig. 6 Thermoresponsive copolymer architectures. The copolymers are depicted at temperatures below the LCST. The purple (and thicker) segments are thermoresponsive

around 50 °C. The gelation responds by a 3-dimensional packing of micelles due to the hydrophilic–hydrophobic balance, increasing the micelle volume and provoking micelle packaging when the sol-gel transition takes place [3, 55]. This balance can be modulated by incorporating different side chains with hydrophilic or hydrophobic segments [29]. PEO, PPO, and PEO–PPO–PEO block copolymers have their respective LCSTs (e.g. cloud point). However, the sol–gel transition of PEO–PPO–PEO should be separated from their LCSTs, because the temperature responsive gelation occurs by the three dimensional packing of micelles, which are formed by the hydrophilic-hydrophobic balance [56]. Thermo-responsiveness is one of the fundamental and most accessible properties of supramolecular polymeric materials, as the weak noncovalent interactions are inherently susceptible to a thermal stimulus. Therefore, thermo-responsive supramolecular materials have attracted broad interest.

3 Applications with Biomedical Interest

With respect to biomedical applications, thermoresponsive polymers have been successfully used, mainly, for drug/gene delivery applications and for cell sheet engineering.

Polymers with LCST have been tested in controlled **drug delivery** matrices and in on-off release profiles in response to a stepwise temperature change. It has been shown that thermoresponsive polymers are especially effective in terms of delivering a pharmaceutical payload in drug delivery systems. There are two general drug delivery strategies: passive and active targeting [57, 58]. Both delivery mechanisms rely upon the particles resisting adsorption of proteins within the body. The ability of the delivery vehicles to circulate within the body (e.g. plasma) depends on a number of factors, including particle size. Particles which are much less than 100 nm have an advantage in this regard. Macromolecular micelles formed by thermoresponsive

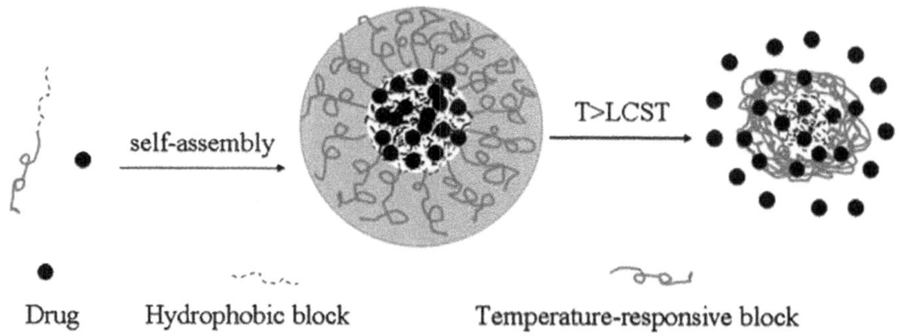

Fig. 7 Depiction of thermoresponsive amphiphilic copolymer used for drug delivery and controlled release (Reproduced from [4], with permission from Springer Nature) (pending)

copolymers have sizes that are well within this range. These ***micelles*** can solubilize hydrophobic compounds and enable triggered release at target sites within the body. Amphiphilic block copolymers featuring a thermoresponsive component can self-assemble into macromolecular micelles in aqueous solutions. Hydrophobic polymer chains, such as PMMA or PS [4, 15], can form the core, and the thermoresponsive chains compose a responsive hydrophilic shell. During the process of the formation of the micelles, hydrophobic drug molecules can be absorbed within the hydrophobic cores. The macromolecular micelles were stable to aggregation at room temperature but collapsed at the CP, **triggering drug release**. The process is depicted in Fig. 7. For thermoresponsive delivery systems, a trigger that is being pursued is hyperthermia. To release drugs site specifically upon local heating requires the drugs to be loaded by using thermoresponsive polymers with an LCST between 37 and 42 °C.

In the case of gene delivery, two types of gene carriers have been developed [3]; one is a soluble temperature responsive polymeric carrier, which can bind anionic DNA or hydrophobic drugs at body temperature and dissociate them at hyperthermic temperatures. The other is a temperature responsive micelle or nano-hydrogel. Thermoresponsive block copolymers can also be used for protein delivery.

Surfaces with thermoresponsive properties have been demonstrated to be effective in controlling cell attachment and detachment [59]. The application of temperature responsive polymers to modified surfaces exploits the fact that most proteins show significantly greater adsorption on hydrophobic surfaces than in hydrophilic ones. Above the LCST the thermo-sensitive polymer will adsorb peptides and proteins from a solution and these biomolecules can be desorbed by decreasing the temperature. It has been shown that the surface of tissue culture polystyrene grafted with PNIPAM allows cells to adhere and proliferate above the LCST of the polymer whereas a cell detachment was detected at temperatures below LCST [54]. Moreover, this type of application has been tested on biodegradable polymers such as poly(L-lactic acid) and chitosan exhibiting a similar behavior than that performed on polystyrene culture plates [29]. Developments in the field of stimuli-responsive polymers have been rapid, and concomitantly a number of novel applications of responsive surfaces in the biomedical field have been demonstrated. Thermoresponsive polymers have been successfully used for engineering cell sheets for reconstruction of functional tissues. In recent years, cell sheet engineering without scaffolds has attracted significant attention in tissue engineering, mainly owing to its advantage in maintaining cell-cell interactions and in mimicking the microarchitecture of native tissue [60]. One common approach for generating cell sheets is utilizing thermo-responsive polymers, such as poly(N-isopropylacrylamide, PNIPAM).

In recent years progress has been made also in the synthesis and development of polymer protein **bioconjugates** [61, 62] useful for affinity separations, biosensors, diagnostics, enzyme processes, and targeted delivery of drugs or chemical agents, labels and other signals. Two different kinds of bioconjugates including stimuli-responsive polymers have been prepared by:

1. Random polymer conjugation to lysine amino groups of a protein.
2. Site-specific conjugation of the polymer to genetically engineered specific amino acid sites. The placement of stimuli-sensitive polymers near the active place of a recognition protein can provide a highly environmental-sensitive system [29].

Several investigators have used thermoresponsive polymer properties in **bio-separations**. A typical application is thermoresponsive chromatography [63, 64]. Such systems use a thermoresponsive-polymer-modified stationary phase and an all-aqueous mobile phase. Bioactive compounds, peptides and proteins have been separated through the hydrophobic interactions between analytes and a thermoresponsive-polymer modified stationary phase. Thermoresponsive polymers are also used for cell separation. Different cells have different adhesion or detachment properties on thermoresponsive-polymer-modified surfaces. These properties can also be modulated based on the properties of the grafted thermoresponsive polymer, e.g., graft chain length, graft density, and graft amount [30].

4 Experimental Characterization of Thermoresponsive Behavior

The LCST, being a thermodynamic quantity, has a precisely determined value that is not dependent on the measurement technique. Several experimental techniques can be employed to study the critical transition behavior of these polymers. Distinctions must be made between the techniques based on the underlying phenomena that are being probed. Studies in the literature typically prepare samples of fixed composition and then measure them in a setup in which the temperature-dependent change of a physical quantity is observed. Experimentally, the LCST can be estimated from the cloud point using concentration-dependent, temperature-resolved turbidity measurements or visual observations. Moreover, thermodynamic susceptibilities like the specific heat capacity, the compressibility, the dielectric and magnetic susceptibilities and the volume expansion coefficient yield valuable information about changes in structure morphology and molecular dynamics during the critical transition.

4.1 Experimental Techniques

4.1.1 Differential Scanning Calorimetry (DSC)

Differential scanning calorimetry (DSC) is a convenient technique to study the critical transition of thermoresponsive polymers. DSC is a thermal analysis technique, conventionally employed for studying thermal transitions of materials and comprises probably the most widely used materials characterization technique [65, 66]. Calorimetry is based on the following equation

$$\delta Q = c \cdot \Delta T = C \cdot m \cdot \Delta T \tag{5}$$

where δQ is the heat exchange, ΔT the temperature change caused by exchanged heat, c the heat capacity, $C = c/m$ the specific heat capacity and m the sample mass. Assuming that neither the sample mass nor the specific heat capacity do not depend on time, t, Eq. (5) becomes

$$\frac{\delta Q}{dt} = \Phi = c \cdot \frac{dT}{dt} = C \cdot m \cdot \beta \tag{6}$$

where Φ is the heat flow and β the scan rate (heating or cooling). DSC is the main technique for measuring heat capacity at elevated temperatures in a reasonably short time. Depending on the specific task, different modes of DSC may be employed, such as scanning at constant rate, isothermal, temperature modulated, and fast-scanning calorimetry [67].

The most common operation mode of DSC is heating or cooling at constant rate, often in combination of both. Data are typically presented as heat flow versus temperature (Fig. 8), e.g. in milliWatts per gram of the sample, mW/g, as taken from the DSC instrument, or, after normalizing the heat flow with the heating/cooling rate (K/min), as heat capacity (actually specific heat capacity, C_p, e.g. in J/gK). The rate of the DSC scan is an important experimental parameter for fundamental studies, for example the cooling rate when studying the kinetics of phase transitions in polymers [67]. It is well-known that the enthalpy change (ΔH) resulting from water dissociation during the coil-to-globule transition is large enough to be detected via conventional DSC [37, 68–71].

Figure 8 schematically shows DSC thermograms, cooling and heating scans, for an aqueous solution of polystyrene–*b*–poly(N-isopropylacrylamide) diblock copolymer (PS-*b*-PNIPAM) with a polymer concentration of 25 wt% [72], in the temperature range of its cloud point. During the heating ramp, the demixing phase transition is observed as an endothermic peak; thus, we witness a first-order phase transition. In a typical cycle, the transition temperature observed during the heating ramp may be higher than the one in the cooling ramp (Fig. 8). This is the manifestation of the hysteresis effect. The calorimetrically obtained transition temperature of PNIPAM-based polymers is often defined as the onset of the DSC transition endotherm, T_{onset} (the intersection of the baseline and the leading edge of the endotherm). The temper-

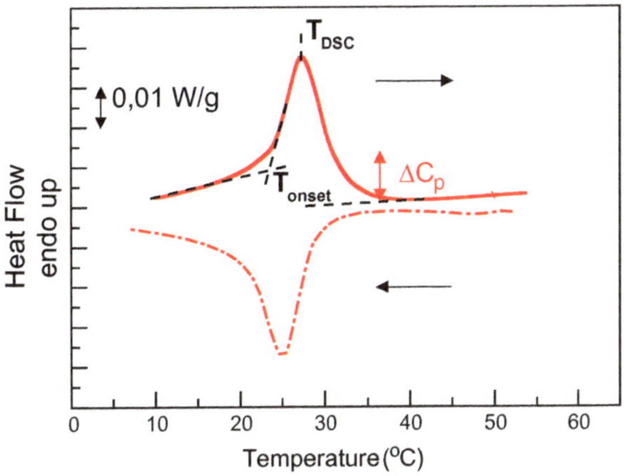

Fig. 8 DSC heating (—) and cooling (- · · -) thermograms obtained on an aqueous PS-*b*-PNIPAM solution having a polymer concentration of 25 wt%. The heating/cooling rate was 3 °C/min

ature of the maximum heat capacity, T_{DSC}, is also used to characterize the transition. The differential heat flow in polymer solutions provides information not only with respect to critical temperature and enthalpy of the transition, but also about the transitional heat capacity increment and the kinetics of the chain collapse. Furthermore, it was shown that calorimetric measurements offer the best way to study the degree of cooperativity of the temperature-induced transition in the thermoresponsive systems [36, 73]. Moreover, due to the fact that the samples are mounted in hermetically closed pans, it is more suitable to characterize the phase transition by re-iteration of several heating–cooling sequences. The dependence of the DSC thermograms on polymer concentration and heating rate has been widely studied in various polymer solutions.

Temperature modulated DSC (TMDSC) technique has been also used for the investigation of the phase transition of thermoresponsive systems regarding the mechanism of the demixing process [45, 68, 74]. In the temperature modulated mode (temperature modulated DSC—TMDSC or MDSC [75, 76]), a sinusoidal component is added to the constant heating rate which further increases the calorimetric capabilities [67]

$$T(t) = T_0 + \beta_0 t + A_T \sin(\omega t) \tag{7}$$

where T_0 is the start temperature, β_0 underlying heating (cooling) rate, A_T the amplitude of temperature perturbation, and $\omega = 2\pi/T_p$, the angular frequency with T_p the modulation period. Specifically, by properly setting the amplitude and the period of the oscillating component, TMDSC can provide a higher resolution and sensitivity compared to classic DSC, since it can achieve an instant heating rate as high as 10 °C/min while the average heating rate is only 1 °C/min. The response signal is analyzed by Fourier transformation, which provides the possibility to uncover transitions ruled by kinetic effects. Furthermore, the analysis of the oscillating signal can give information about the molecular dynamics (heat capacity spectroscopy covering more than six decades). In this way it is possible to determine the complex heat capacity susceptibility where the temperature is the external dynamic variable [77].

4.1.2 Small Angle X-Ray Scattering (SAXS)

Small-angle X-ray scattering allows structural investigations on the mesoscopic length scale, i.e. ~1–100 nm [78]. The scattering intensity is measured in dependence of the momentum transfer, q, which is defined as $q = 4\pi \sin(\theta/2)/\lambda$. θ is the scattering angle and λ the X-ray wavelength. At a CuKα laboratory source, the wavelength is $\lambda = 0.154$ nm. Experiments can also be carried out at synchrotrons, where the photon flux is orders of magnitude higher, which allows measurements of more dilute polymer solutions than at a laboratory source and time-resolved measurements. Sample cells for polymer solutions are usually capillaries from quartz glass, which have a diameter of 1–2 mm. The scattering intensity from the solvent-filled cell is measured as well and is subtracted from the data of the polymer solution. Data analysis may be carried out by fitting structural models of the $I(q)$ data. Depending on the complexity of the system, different models may be used.

In semi-dilute solutions of homopolymers, for instance, the scattering intensity is usually described by the Ornstein-Zernike structure factor:

$$I(q) = \frac{I_0}{1 + q^2\xi^2} \tag{8}$$

which comprises the correlation length of concentration fluctuations, ξ, as well as the intensity at $q = 0$, I_0 [79]. For PNIPAM solutions below the cloud point, both quantities show critical behavior as the cloud point is approached [79, 80]. Above the cloud point, the polymers collapse; however, since the collapse goes along with aggregation, it is difficult to be observed. The data are dominated by scattering from large, compact aggregates. These are usually described by the Porod law [81]:

$$I(q) = \frac{2\pi(\Delta\rho)^2\, {}^{S}\!/_{V}}{q^4} \tag{9}$$

$(\Delta\rho)^2$ denotes the difference of scattering length densities of the polymers and the solvent and S/V the specific interface of the aggregates.

For block copolymers having a thermoresponsive block and one or two hydrophobic blocks, micelle formation is expected in aqueous solutions below the cloud point. Using SAXS, their size, shape and inner structure can be determined. If the hydrophobic blocks are short, it is likely that the di- or triblock copolymers form spherical micelles, where the hydrophobic blocks form the core, and the thermoresponsive blocks form the shell. Their SAXS data can often be modelled by spheres having a core-shell structure, and from fitting, among others, the core radius and the thickness of the shell can be determined. At the collapse transition, the latter decreases drastically, whereas the former stays constant [82, 83]. To describe the concentration fluctuations in the micellar shell, the Ornstein-Zernike structure factor is often used, and the correlation length, ξ, decreases at the collapse transition. For high polymer concentration, the positions of the micelles are correlated, and this interaction may be modeled by the Percus-Yevick structure factor, which is based on the hard-sphere interaction [84]. It comprises the hard-sphere radius, i.e. half the average distance between the positions of the micelles, and the volume fraction of correlated micelles. Upon collapse of the micellar shell, the hard-sphere radius decreases discontinuously, especially in the case of triblock copolymers with hydrophobic end blocks and thermoresponsive end blocks [82].

4.1.3 Broadband Dielectric Spectroscopy (BDS)

The dielectric function (also dielectric permittivity and dielectric "constant") $\varepsilon^*(\omega)$ describes the material response to the application of an alternating electric field $E(\omega)$. For small electric field strengths, a linear relationship holds between E and the polarization P

$$P(\omega) = (\varepsilon^*(\omega) - 1)\,\varepsilon_0\, E(\omega) \tag{10}$$

where ε_0 is the permittivity in vacuum. $\varepsilon^*(\omega)$ is related by the theory of dielectric relaxation to the correlation function $\Phi(t)$ of the polarization fluctuations [85–87],

$$\frac{\varepsilon^*(\omega) - \varepsilon_\infty}{\varepsilon_s - \varepsilon_\infty} = \int_0^\infty \left[-\frac{d\Phi(t)}{dt}\right] \exp(-i\omega t)dt \tag{11}$$

and

$$\Phi(t) = \frac{< \Delta P(t)\,\Delta P(0) >}{< \Delta P(0)^2 >} \tag{12}$$

where ΔP denotes a fluctuation of the polarization around its equilibrium value and the brackets denote the averaging over an ensemble or time t. The dielectric function is measured by dielectric techniques, which are a powerful tool for studying molecular dynamics in various materials. The main advantage of dielectric techniques over other techniques of measuring molecular dynamics is the extremely broad frequency range covered, which extends from about 10^{-5} to about 10^{11} Hz [85–87]. Obviously, this broad frequency range cannot be covered by a single technique and measurements are performed in the frequency domain, the time domain or the temperature domain. In most cases, measurements are performed in the frequency domain, i.e. with variation of the frequency of the applied electric field (broadband dielectric spectroscopy, BDS).

BDS has been increasingly used to study polymer solutions and smart gels [88, 89]. Due to broad frequency range covered by BDS, processes of different time and length scales can be studied by considering the molecular mobility and/or the conductivity as a probe for the polymer structure. Dielectric measurements on polymer solutions provide valuable information that includes, in addition to the electrical properties of each domain in the heterogeneous solution, the mobility of the whole polymer chain or of its side chains, the diffusion of ions over different lengths, up to the size of microgel particles, and the degree of association of protons to the polymer chains. For aqueous solutions of thermoresponsive polymers, their demixing transition and the subsequent assembly process were monitored by means of dielectric spectroscopy measurements, usually in the high frequency domain (1 MHz–50 GHz) [88–97]. In this frequency range, the investigated underlying molecular mechanisms include the polarization of either condensed counterions or freely mobile ions, the polarization of the ionic atmosphere, long-range solvent-ordering effects and, at higher frequencies, the orientational polarization of the water molecules themselves, possibly modified by the interaction with the solute. Recent measurements in the THz frequency range provided very interesting insight into the dynamics of individual water molecules, related to the hydrogen bond interactions they are involved in, either with the polymer chain or with other water molecules [31]. Moreover, dielectric measurements at lower frequencies, in the range of 10^{-2}–10^{6} Hz, were also performed on solutions of thermoresponsive polymers, mainly based on PNIPAM [96–102]. Such measurements allow for the study of the evolution of micellar aggregates, of changes in microgel structures and of their charge accumulation capacity during the demixing phase transition, on the basis, mainly, of monitoring interfacial polarization processes that are activated due to long-range transport of ions and their accumulation at interfaces between the aggregates and the aqueous medium.

4.2 PNIPAM and PNIPAM-Based Systems

One of the most studied thermoresponsive polymers is poly(N-isopropylacrylamide) (PNIPAM), which features an LCST of around 32 °C in aqueous media. The thermoresponsive properties of PNIPAM are presented in the excellent reviews by Schild [103], Pelton [49], Gil and Hudson [3], Asevey et al. [5], Halperin et al. [10]. Apart the numerous applications, the thermodynamic properties of aqueous PNIPAM solutions have been thoroughly investigated in order to understand complicated molecular mechanisms in biological systems. The phase transition of PNIPAM solutions, which is accompanied by the formation of inter- and/or intra-chain HBs above the CP, spans over a very narrow temperature range, and, in a wide range, the CP has only a weak dependence on the degree of polymerization and the concentration of the solutions. The sharpness of the transition is usually assigned to a highly cooperative dehydration of PNIPAM [10, 42, 43, 104].

The molecular architecture of PNIPAM consists of a hydrophobic backbone and a hydrophilic amide group capped by a hydrophobic iso-propyl moiety, as shown in Fig. 9. In Fig. 9a, b, the monomeric unit of PNIPAM (chemical formula) and a schematic presentation of the trimmer unit are shown and in Fig. 9c the monomeric unit with the water structures around the PNIPAM chains below the LCST [31, 68].

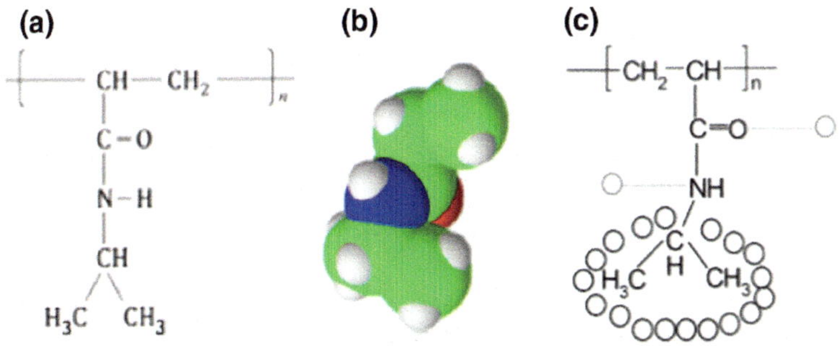

Fig. 9 Chemical structure of PNIPAM (**a**), schematic of the NIPAM trimer unit (**b**) and water structures around the PNIPAM chain below LCST (**c**) (Reprinted with permission from [31]. Copyright 2015 American Chemical Society)

The water around the isopropyl group of the PNIPAM residue (Fig. 9c) is known to have a well-ordered structure, called "water cages" or "water clathrates" [34, 35, 68].

PNIPAM has been suggested to have two types of bound water [31, 68]. Below CP, where hydrophilic regions are exposed to the solvent, hydrophilic amide (–NH) and carbonyl (–CO) groups form HBs with proximal water molecules driven by an enthalpic term (hydrophilic hydration). At the same time, the "cage-like" rigid ice-like HB network of water is found around the isopropyl groups and is called hydrophobic hydration. Thus, below CP, the macromolecules are enveloped by hydration shells. The hydration number, defined as the number of H_2O molecules bound to one NIPAM monomer, including both, hydrophilic and hydrophobic hydrated water, is assumed to depend on the concentration, architecture, and topology of the PNIPAM chains in the aqueous solution [90–92, 105]. Remarkably, the hydration number of NIPAM monomers dissolved in water amounts to 5 [91], whereas that of linear PNIPAM chains in semidilute aqueous solutions lies at about 11–13 [31, 90, 92]. A cooperative effect concerning the orientation of the side chains of PNIPAM during hydration is responsible for the more voluminous hydration shells of PNIPAM in comparison to the one of its monomer. Support to these claims comes from recent experimental findings that indicate the crucial role of chain backbone mobility in PNIPAM self-assembly properties and in the reorganization of the water HB network around the macromolecules [106]. It is unclear in how far the hydration number also depends on the concentration of aqueous PNIPAM solutions because of the entanglement of the macromolecules.

Numerous published reports suggest that the origin of the demixing phase transition of PNIPAM at CP is as follows (Fig. 10): once the temperature exceeds CP, the entropic term dominates the enthalpic one, leading to an unfavorable Gibbs free energy state, and the hydrophobic groups start to aggregate together (so-called hydrophobic interaction) to compensate for the entropic penalty. According to this hydrophobic interaction, the extended coil-like conformation of PNIPAM with the hydrophilic region exposed to the solvent changes into a swollen

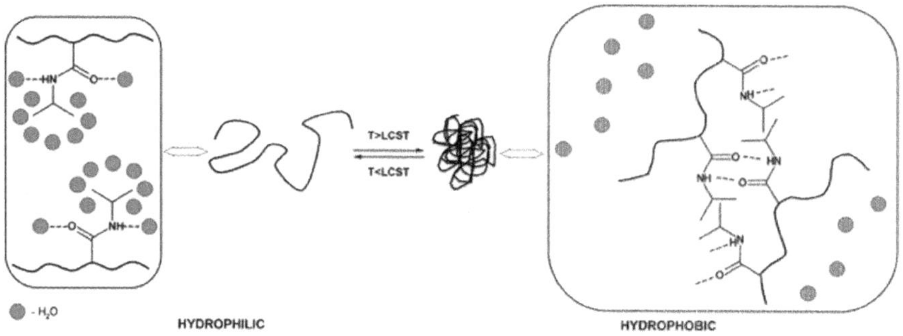

Fig. 10 Illustration of temperature induced PNIPAM phase transition (Reproduced from [2], with permission from Elsevier)

globule-like structure. During this process, the hydrophilic–NH and–CO groups are partially re-oriented into the inner part of the PNIPAM molecule forming intra-molecular–NH· · ·OC–hydrogen bonds, with the hydrophobic isopropyl residue also being modified because of a cooperative decrease in the number of hydrophobic hydration water [31]. Recent experimental findings suggest that a gradual dehydration starts at CP because of coil-to-globule transition, and a reduction of the hydration number around 35% (from 10 below CP to 6.5 above CP) is estimated to occur. That means that the collapsed state of PNIPAM still contains a large amount of water [31, 105, 107]. Predominantly, the dehydration above CP was identified as the loss of hydrophobic hydrated water, according to the minimization of the solvent-exposed hydrophobic surface area driven by hydrophobic interactions [31, 107]. Furthermore, the collapse at higher temperatures is coupled with the large translational entropy gain of the water molecules that overcomes the loss of polymer conformational entropy upon collapse. This is the well-established entropy driven mechanism of LCST polymer collapse [108, 109]. In addition, recent molecular dynamics (MD) simulation showed that the drastic change in the conformational state of hydrophobic isopropyl group serves as a dynamic trigger for PNIPAM coil-to-globule transition, including the hydrophilic–hydrophobic transition and HB transformation [110]. Grinberg et al. [69] suggested that the phase transition of the PNIPAM hydrogel involves three different heats: the heat from apolar dehydration, the one from polar dehydration, and the residual heat caused by the strong interactions (van der Waals or H-bonding interactions) of the residues of the PNIPAM chain (intra- and/or inter-molecular interactions among the side chains of PNIPAM [111]).

4.2.1 Tuning PNIPAM Thermoresponsive Behavior

Several factors can influence the critical temperature of an aqueous PNIPAM solution. For example, an increased hydrophobicity of the end groups lowers the cloud point [112]. Furthermore, with increasing molecular weight, the critical temperature decreases, until the effect of the end groups is negligible and a plateau of CP is reached at a critical molecular weight [113]. In this context, the coil-to-globule transition and the influence of factors, such as molecular weights, temperatures, co-solvents, and/or ions, to mention a few, have been extensively investigated both experimentally as well as using computer simulations [109]. The effect of PNIPAM chain length and PNIPAM concentration was also investigated. A clear conformational transition was found either for chains with at least 30 monomers or for high oligomer concentrations. Furthermore, computational studies have addressed several systems, starting from a single PNIPAM chain in solution to gels and networks, as well as micro- and nano-gels. In the case of PNIPAM gels, the effect of degree of cross-linking on the local structure of water around polar and apolar groups is rather small. However, they present remarkable global conformational effects due to their geometrical constraints [109].

The subtleties on the behavior of these smart polymers increase even further when dealing with complex molecular architectures. In this context, it has been shown that the LCST of a PNIPAM chain can be tuned by changing its tacticity and/or incorporating a few different hydrophilic or hydrophobic monomeric centers along its backbone [10, 109]. An increase in the concentration of such hydrophobic moieties can result in PNIPAM co-polymers with water insoluble properties. Thus, these co-polymers can combine the characteristics of each block, creating a new polymer with unique capabilities. However, one of the biggest obstacles in controlling the macroscopic properties of these co-polymers is the lack of understanding of their atomic-level structure and network of proximal solvent molecules. Even with the advances in experimental techniques it is almost impossible to probe the dynamics of atomic-level structure of these co-polymers and proximal water. In addition to the exact atomic origin of the LCST, the effect of the nature (hydrophilic vs. hydrophobic) of the co-monomer and chain length of the co-polymer on the LCST of PNIPAM is unknown. The structure of water at the junction (where two polymers are attached) of these co-polymers is inaccessible by experiments [114].

Furthermore, the effect of different salts on the CP of PNIPAM has been studied extensively. The strength of the effect has been found to follow the Hofmeister series [115]. More similar to drug molecular structures, other studies have found strong effects of small aromatic cosolutes on the phase transition of PNIPAM, even at low concentrations. So far, the mechanism of interaction has not been explained, and even structurally very similar cosolutes may differ in their influence on CP by several degrees centigrade [113].

From a general point of view, addition of water-miscible organic solvents to aqueous LCST-type polymer solutions has a similar effect as salt, namely the reduction of the CP. This effect is called co-nonsolvency [33, 116]. For example, PNIPAM is soluble at room temperature in water and also in ethanol, but it is not soluble in a mixture water/ethanol; the minimum of solubility occurs at 40 vol.% ethanol content. One explanation put forward is that the co-nonsolvency effect is caused by the dehydration of polymer chains due to competition of hydrophobic interaction between the polymer and the alcohol. This is a cooperative effect characterized by a sharp phase transition when the ethanol content is increased—once the hydration shell of the polymer is broken, the entire polymer is quickly dehydrated and collapses [11, 23, 117].

Though PNIPAM has been in the focus of the research for quite a long time, a lot of interesting results appear still each year. The last decade has seen the publication of hundreds of papers involving poly(N-isopropylacrylamide), as homopolymers and as constituents of copolymers, gels, microgels and surface layers, shedding light on various aspects of the self-assembling properties of PNIPAM macromolecules.

4.3 PMDEGA-Based Systems

Recently, an alternative to PNIPAM type of thermoresponsive copolymers based on poly (diethylene glycol) (acrylate)s blocks was introduced [18]. They belong to the class of comb polymers composed of an acrylate backbone and oligo(ethylene glycol) side-chains [17] which undergo the demixing phase transition through thermally induced dehydration, but without formation of inter- and/or intra-chain hydrogen bonds in the polymer-rich phase. Within this family of polymers, poly(methoxy diethylene glycol acrylate) (PMDEGA) has the shortest side chain to achieve water solubility (Fig. 11) [18, 118] and thus, is structurally the most alike to PNIPAM. Furthermore, the CP of aqueous solutions of PMDEGA homopolymers having intermediate to high molar masses is around 35–45 °C, i.e., similar to the CP of PNIPAM. Yet, as PMDEGA shows LCST behavior of type I (Flory–Huggins type) while PNIPAM follows type II behavior, characteristic differences exist also between solutions of block copolymers featuring PNIPAM or PMDEGA blocks, concerning the self-organization, the width and hysteresis of the transition as well as the switching kinetics. These novel thermosensitive macromolecules are very promising for biomedical applications since they are principally composed of biocompatible oligo(ethylene glycol) segments. Indeed, poly(ethylene glycol) (PEG) is an uncharged, water-soluble, nontoxic, nonimmunogenic polymer and therefore the most applied synthetic polymer in the biomedical field.

PMDEGA features a diethylene glycol side chain which is responsible for the thermoresponsivity in aqueous media (Fig. 11). In dilute solution, the CP of the homopolymer PMDEGA is in the physiologically particularly interesting range of 35–42 °C for number average degrees of polymerization N > 100, as measured using turbidimetry. The observed transition is narrow (width below 2 K) and shows nearly no hysteresis upon cooling.

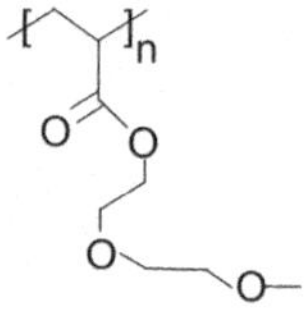

Fig. 11 Chemical structure of PMDEGA homopolymer

4.3.1 Effects of Molecular Architecture

By block copolymerizing PMDEGA with short polystyrene (PS) end blocks, new diblock, triblock or star-like architectures can be achieved [15]. The transitions from clear to opaque solutions are also sharp, but a certain hysteresis of the CP was observed in tri- and starblock copolymers. Moreover, the CPs decrease in the order diblock > triblock > starblock copolymer. Both phenomena have been attributed to topological effects, such as the tethering of the PMDEGA block to the PS core of the micelles.

A number of experiments were designed and carried out focusing on the **effects** of the **molecular architecture** on the CP, the micelle formation and the aggregation behavior above the CP. PMDEGA homopolymers, P(S-*b*-MDEGA) diblock and P(S-*b*-MDEGA-*b*-S) triblock copolymers as well as three-arm starblock copolymer P(MDEGA-*b*-S)$_3$ were investigated in a comparative way. The various block copolymers were synthesized by two successive reversible addition-fragmentation chain transfer (RAFT) polymerizations, starting from a mono-functional, symmetrically di-functional, or tri-functional trithiocarbonate chain transfer agent, respectively [18]. The structure of the polymers investigated is shown in Fig. 12 [102]. The PS blocks are in all cases short (N = 8–11), thus enabling direct dissolution in water, while the PMDEGA blocks are long (N = 172–665). A number of methods to characterize the **structural**, **mechanical**, **thermal**, and **dielectric** properties were used, and a wide concentration range was covered [102].

The critical micellization concentrations (CMC) were detected using fluorescence correlation spectroscopy (FCS). At this, identical fluorescence-labeled polymers were used as tracers. The micellar size was monitored as a function of temperature using dynamic light scattering (DLS), and the CPs were determined from the onset of mesoglobule formation. The CMC values as determined from the appearance of micelles and the estimated radii r_h of the micelles are listed in Table 1. The CMCs were found at $1.4-7 \times 10^{-8}$ mol/L (which corresponds to $0.43–5.9 \times 10^{-3}$ g/L) whereas the estimated r_h values were located in the range 10.2–12.8 nm for the homopolymer and the three block copolymers. Our studies have shown that neither the hydrodynamic radii nor the unimer and micellar hydrodynamic radii depend strongly on the polymer architecture [102].

DSC measurements on highly concentrated solutions reveal characteristics of both the glass transition of the polymer and the (thermoresponsive) collapse transition. Figure 13a shows DSC thermograms obtained on a PMDEGA homopolymer solution, with a polymer concentration of 25 wt% (or a water fraction $h_w = 0.75$), and on a P(S-d$_8$)$_9$-*b*-PMDEGA$_{353}$-*b*-P(S-d$_8$)$_9$ triblock copolymer solution with $h_w = 0.80$, focusing on the temperature range of the glass transition of the polymers. The curves clearly show the endothermic step related to the glass transition of the polymeric chains. In Table 2, the glass transition temperatures of the copolymer, T_g, and the corresponding heat capacity increment, ΔCp, are compiled. T_g is nearly independent of the polymer architecture, whereas significant changes occur in the heat capacity increment, where upon tethering PMDEGA block on PS cores, ΔCp is reduced approximately to the half of the values corresponding to free chains.

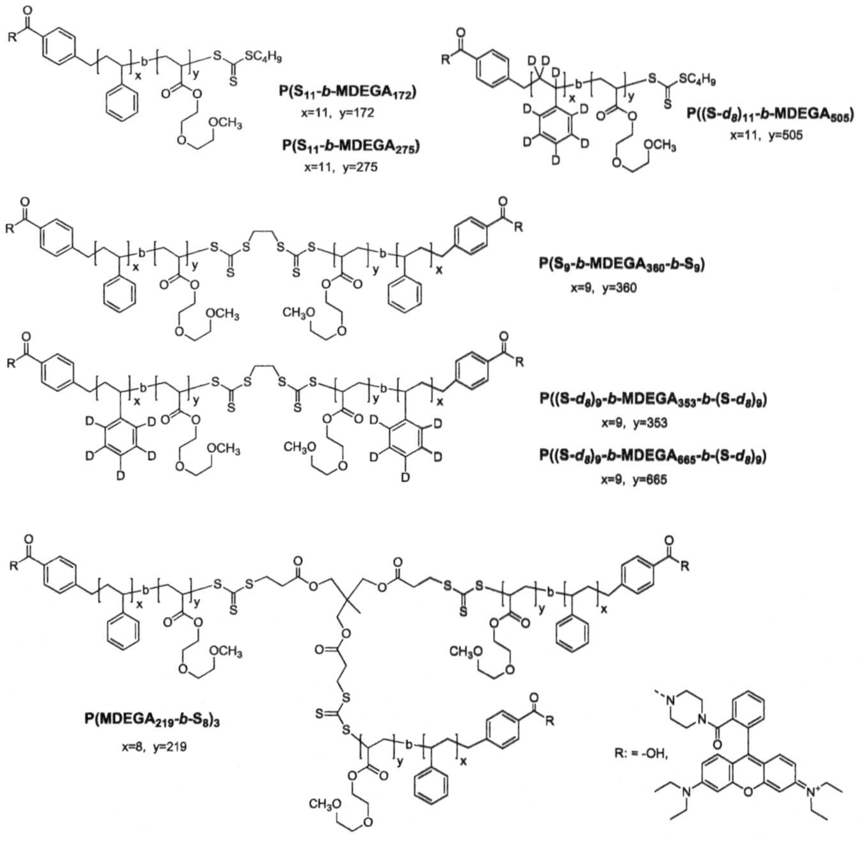

Fig. 12 Chemical structure of the polymers investigated (Reproduced from [102], with permission from Springer Nature)

Table 1 Results from FCS and DLS measurements on PMDEGA-based copolymers with varying architecture

Sample	FCS[a]			DLS	
	CMC (10^{-8} mol/l)	$r_{\mathrm{h,uni}}$ (nm)	$r_{\mathrm{h,mic}}$ (nm)	$r_{\mathrm{h,mic}}^{\mathrm{a,b}}$ (nm)	CP[b] (°C)
PS_{11}-*b*-$PMDEGA_{172}$	1.4 ± 0.3	0.6 ± 0.5	12.5 ± 0.4		
PS_{11}-*b*-$PMDEGA_{275}$				10.9 ± 0.1	41 ± 1
PS_9-*b*-$PMDEGA_{360}$-*b*-PS_9	7 ± 2	1.1 ± 0.2	10.2 ± 0.2	10.3 ± 0.1	39.0 ± 0.5
$(PMDEGA_{219}$-*b*-$PS_8)_3$	5 ± 2	1.0 ± 0.8	12.8 ± 0.5	11.7 ± 0.1	38.0 ± 0.5

[a]at 30–35 °C; [b]at concentration 2 g/L

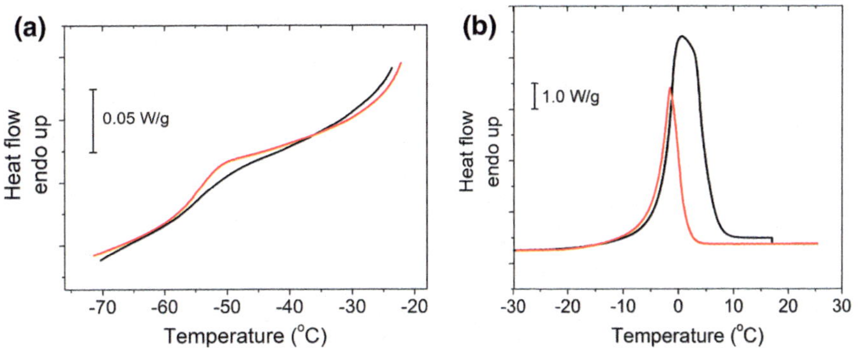

Fig. 13 DSC thermograms of solutions of homopolymer PMDEGA ($h_w = 0.75$, red line) and PS–*b*–PMDEGA–*b*–PS copolymer ($h_w = 0.80$, black line) in glass transition (**a**) and ice melting region (**b**). A heating rate of 10 °C/min was applied

Table 2 Parameters obtained from DSC with a 10 °C/min rate applied

Polymer	Glass transition		Ice melting		
	T_g (°C)	ΔC_p (J/g °C)	T_m (°C)	ΔH (J/g)	X_{UCW} (%)
$PMDEGA_{153}$	−54.9	0.21	−1.5	188	25
$PS(d\text{-}_8)_9$-b-$PMDEGA_{353}$-b-$P(S\text{-}d_8)_9$	−55.0	0.12	1.0	213	21

Figure 13b shows the ice melting peak appearing in the heating thermograms at slightly higher temperatures. The temperatures of the melting peaks, T_m, and the calculated enthalpies of melting, ΔH_m, are given in Table 2. Taking into account the corresponding enthalpy of melting for bulk hexagonal ice, $\Delta H_{water} = 333.55$ J/gK, the fraction of uncrystallized water, X_{UCW}, in the two solutions reads:

$$X_{UCW} = \frac{\Delta H_{water} - \Delta H^m/h^w}{\Delta H_{water}} \quad (13)$$

X_{UCW} is found to decrease slightly with increase of structure complexity, i.e. from 0.25 for the homopolymer down to 0.21 for the triblock copolymer. Thus, with respect to the effects of water molecules on the polymeric segmental dynamics and to their crystallization ability, the experimental findings suggest a remarkable influence of the polymer architecture [102].

Figure 14 shows the part of the DSC thermograms related to the thermoresponsive behavior, obtained during heating and cooling with a rather high (10 °C/min) heating/cooling rate. We observe that the curves are very asymmetric, presenting a shoulder after the peak during heating. Also, during heating, the signal drops under the (extrapolated) baseline after the endothermic peak has reached a plateau after completion of the transition, which indicates that the water- and polymer-rich phases

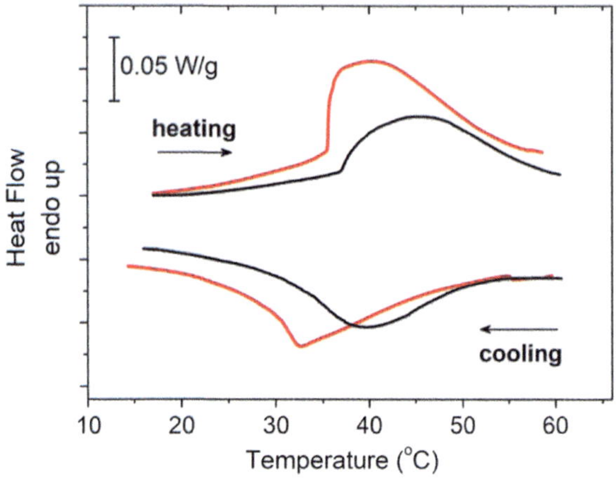

Fig. 14 Heating and cooling thermograms obtained on solutions of homopolymer PMDEGA (h_w = 0.75, red line) and PS–*b*–PMDEGA–*b*–PS copolymer (h_w = 0.80, black line) at a heating/cooling rate of 10 °C/min

Table 3 Parameters related to the thermoresponsive transition obtained from DSC studies

Sample	Heating			Cooling		
	T_{DSC} (°C)	T_{on} (°C)	ΔH(j/g)	T_{DSC} (°C)	T_{on} (°C)	ΔH (J/g)
$PMDEGA_{153}$	40	36	5.59	33	50	5.44
$PS(d\text{-}_8)_9$-b-$PMDEGA_{353}$-b-$P(S\text{-}d_8)_9$	45	37	4.66	39	52	4.39

of the solution are in equilibrium. Hysteresis is observed between heating and cooling thermograms. In addition, for the copolymer, the transition is shifted to higher temperatures. The parameters resulting from an analysis of these thermograms are given in Table 3. Both calorimetrically obtained transition temperatures of the thermoresponsive polymers, i.e. T_{on}, defined by the intersection of the baseline and the leading edge of the endotherm and T_{DSC}, determined by the maximum of the endothermic peak, (Sect. 4.1.1) are given in Table 3.

By comparing the CP measured using turbidimetry (Table 1) with the values of T_{on} and T_{DSC}, (Table 3) it is observed that for the PS–*b*–PMDEGA–*b*–PS triblock copolymer, T_{on} is closer to CP, whereas for the PMDEGA homopolymer, the CP value (38.3 °C) is closer to the peak maximum, T_{DSC}, i.e. homopolymer solutions undergo the transition earlier in DSC measurements.

Dielectric measurements on the same solutions provide information with respect to the formation and re-dissolution of large aggregates at the CP. Figure 15 shows the real part of dielectric permittivity, ε', and ac conductivity, σ'_{ac}, of the PMDEGA (Fig. 15a, c) and PS-*b*-PMDEGA-*b*-PS (Fig. 15b, d) solutions, measured isothermally during heating in the temperature region around CP. Only the high frequency

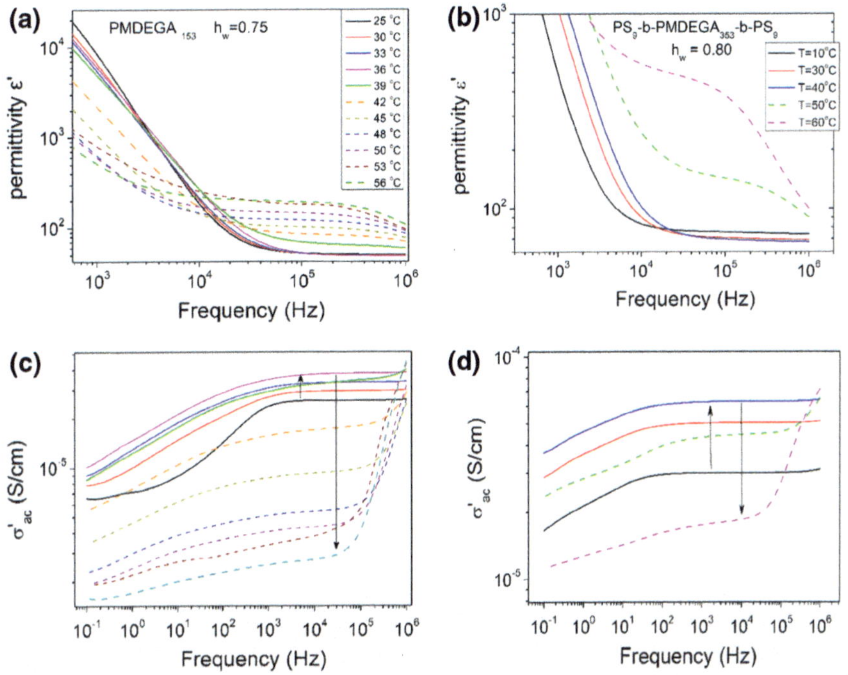

Fig. 15 Real part of dielectric permittivity, ε′, and ac conductivity, σ'_{ac}, versus frequency measured isothermally at temperatures indicated in the plots on the PMDEGA (**a**, **c**) and the PS-*b*-PMDEGA-b-PS (**b**, **d**) solutions. Dashed lines mark the responses recorded at temperatures above the T_{DSC} of each solution. The arrows in the conductivity spectra show the evolution of the spectra with increasing temperature

part of the spectra is shown here. Concerning the dielectric permittivity, we observe that below T_{DSC} of each solution (40 °C for the PMDEGA and 45 °C for the PS-*b*-PMDEGA-*b*-PS solution in the quasi-static case), all spectra exhibit a plateau at high frequencies along with a steep increase of the ε′ values at frequencies lower than ~10^4 Hz. The constant values at high frequencies correspond to dipolar polarization processes relaxing at frequencies higher than 1 MHz due to the reorientation of water molecules with possible contributions from fast local chain polarization processes. The steep increase of ε′ at lower frequencies is attributed to ionic polarization effects usually observed in aqueous solutions [89, 119]. The high frequency dielectric permittivity, ε′, decreases with increasing temperature, as expected for orientational polarization processes. Regarding the conductivity data, the curves in Fig. 15c, d exhibit at frequencies higher than about 1 kHz the conductivity plateau related to the bulk dc conductivity of the solution. The drop of the σ'_{ac} values at low frequencies is attributed to electrode polarization effect. We observe that below T_{DSC} the dc conductivity increases with increasing temperature for both solutions.

Interestingly, for both solutions, the dielectric response changes significantly, when crossing T_{DSC}. More specifically, the measured real part of dielectric permittivity increases in the whole frequency range (Fig. 15a, b), and the spectra reveal the appearance of a new polarization process at temperatures above CP, in the frequency range of 0.1–1 MHz, i.e. having a time scale of 10^{-5} s. Concerning the real part of conductivity, during the temperature rise around T_{DSC}, the values of σ'_{ac} in the kHz region decrease, indicating that a strong polarization process is activated at lower frequencies, which results in the decrease of the measured real part of the conductivity in the kHz region. It is important to note here that the dielectric behavior of both solutions is thermally reversible.

Summarizing, the BDS results suggest that, immediately above the CP, aggregates are formed providing the medium for motion and accumulation of charge carriers that are detached from the polymers during the phase transition. The analysis of the data suggests that, initially, the size of these aggregates is around 20 nm and increases with temperature, reaching values close to 50 nm in saturation (approximately 20 °C above CP). Comparing with the finding from SAXS, that 2 r_g of the aggregates right above the CP is about 23 nm, we conclude that BDS and SAXS provide the same order of magnitude regarding the size of the aggregates [72].

4.4 Comparison Between PNIPAM- and PMDEGA-Based Systems

In order to gain more insight in the temperature-dependent self-organization of the thermoresponsive polymers PNIPAM and PMDEGA, two series of block copolymers were prepared, which are based on their use as hydrophilic and thermo-responsive building blocks. The comparative study of these copolymers reveal that both polymers show characteristic differences with respect to their thermo-responsive behavior despite their very close LCST values and thus ought to be considered as complementary rather than equivalent options for implementing "intelligent" hydrogel systems [15].

In these studies, the macromolecular structure of the model systems were broadly varied (Fig. 16), by varying the molecular architecture (BA diblock vs. BAB symmetrical triblock vs. $(BA)_{3\times}$ 3-arm star block copolymers), as well as the relative and absolute sizes of the thermoresponsive A and the permanently hydrophobic B blocks [15 and references therein]. For comparison, these systematic variations were complemented by PNIPAM and PMDEGA homopolymer references. For the various amphiphilic block copolymers, the rather long thermo-responsive blocks were end capped by—mostly short—hydrophobic blocks, with a particular emphasis on PS as model hydrophobic polymer.

Turbidimetry of dilute aqueous solutions was used to map the general trends of thermo-responsive behavior of the various polymer structures. Already at this level, a major difference between the PNIPAM-based and the PMDEGA-based polymers became evident. For the PNIPAM-based polymers, a nearly constant cloud point of 30–32 °C is observed, independent of the molar mass of the polymers, the nature

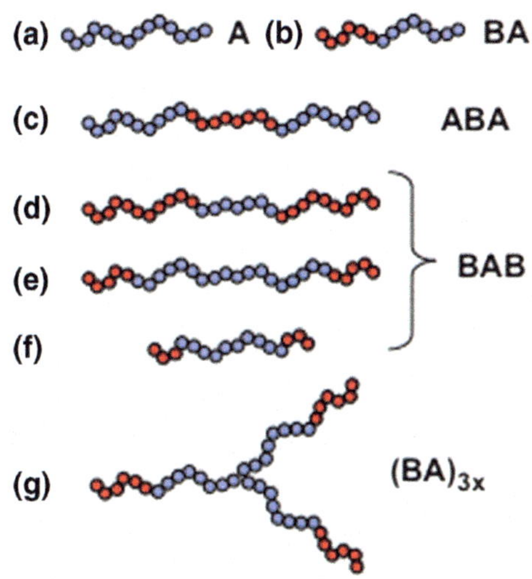

Fig. 16 Schematic architectures of the thermo-responsive polymers studied: **a** homopolymer references, **b** amphiphilic BA diblock copolymer, **c** amphiphilic ABA symmetrical triblock copolymer, **d–f** amphiphilic BAB symmetrical triblock copolymers of varying relative and absolute block sizes, **g** amphiphilic $(BA)_{3x}$ 3-arm star block copolymer (with light blue color is the thermoresponsive block A and with red color is the hydrophobic block B) (Reproduced from [15], with permission from Springer Nature) (pending)

of the end groups (defined by the specific RAFT agent used), length and nature of the hydrophobic blocks, or the type of copolymer architecture (homopolymer, BA diblock as well as BAB and ABA triblock copolymers), as illustrated in Fig. 17 [120]. Also, variation of the polymer concentration between 0.2 and 40 wt% only marginally affected the position of the cloud points [120]. This is in good agreement with the literature, in particular also with reports on the missing effect of hydrophobic end groups or attached blocks, respectively, if the PNIPAM block surpasses a minimum molar mass (>100 g/mol), and if the hydrophobic end groups are equivalent to a hexadecyl chain or larger. This effect has been attributed to local micro-phase separation in the aqueous environment, and to the resulting efficient shielding of the associated hydrophobic components from contact with water [121, 122].

In striking contrast, PMDEGA-based polymers show a very pronounced effect of the molar mass even up to high molar masses (Fig. 17). Moreover, the effect of the molar mass is counter-intuitive, as even for hydrophilic end groups, the cloud point apparently increases with increasing molar mass. Furthermore, the cloud points are very sensitive to the chemical nature of low or high molar mass end groups as well as to the polymer architecture. Diblock, triblock and star block copolymers even of identical ratio of thermoresponsive to hydrophobic block sizes exhibit significantly varying phase transition temperatures (Fig. 17). At the same time, the cloud points of dilute as well as concentrated aqueous solutions of PMDEGA-based polymers display a marked dependence on the polymer concentration [18, 123]. With increasing

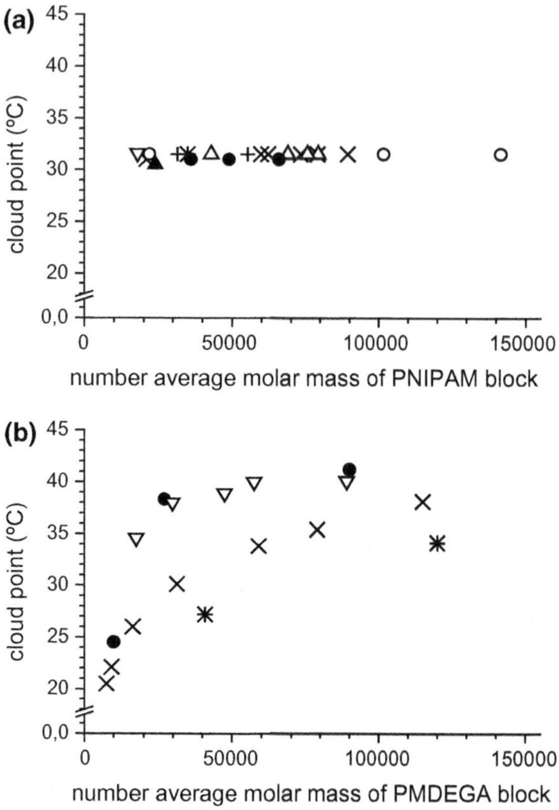

Fig. 17 Comparison of the LCST-type behavior of amphiphilic block copolymers of different structures and architectures in dilute aqueous solution, in dependence on the nature and absolute length of the thermo-responsive block (Data from Refs. [18, 120, 123, 124]). **a** PNIPAM-based polymers: (•) homopolymer; (∇) BA diblock copolymers with B = PS; (×) BAB triblock copolymers with B = PS; (+) BAB triblock copolymers with B = PtbS; (✱) BAB triblock copolymers with B = DBBA; (Δ) BAB triblock series with B = PEHA; (O) BAB triblock series with B = PODA (variable size of the B blocks, 1.0 g·L^{-1} polymer in water, heating rate 1 °C min^{-1}). **b** PMDEGA-based polymers: (•) homopolymer (using RAFT agent 1,2-bis[4-(*tert*-butoxycarbonyl)benzyl sulfanylthiocarbonyl sulfanyl] ethane); (∇) BA diblock copolymers with B = PS_{11}; (×) BAB triblock copolymers with B = PS_8; (✱) $(BA)_3$ 3-arm star block copolymers with B = PS_8 (3.0 g·L^{-1} polymer in water, heating rate 1 °C·min^{-1}) (Reproduced from [15], with permission from Springer Nature) (pending)

concentration, the cloud point passes through a minimum, which shifts to lower concentrations with increasing molar mass, in agreement with predictions from simple thermodynamics.

4.4.1 Kinetics of the Collapse or Re-Swelling Processes

Another striking difference is the different kinetics of the collapse or re-swelling processes of the two different thermoresponsive blocks. It is worth noticing at this point, that different from the homopolymers PNIPAM and PMDEGA, the aqueous solutions of the amphiphilic blocks copolymers become highly viscous with increasing concentration and eventually form a hydrogel [120, 123].

Turbidimetry studies of the PMDEGA-based polymers show virtually no difference between the heating and cooling curves when slow rates were applied (as 0.1 °C/min), i.e., the collapse and re-swelling of the polymer coils do not show a notable thermal hysteresis [18]. In contrast, PNIPAM-based polymers exhibit characteristic differences between the heating and cooling curves in turbidimetry [82, 120]. The notable hysteresis was putatively attributed to a reorganization process of the PNIPAM coils above the cloud point, where part of the secondary amide moieties form hydrogen bonds between neighboring repeat units instead of bonds with water, as below the phase transition. These extra HBs must be broken, and the reorganization must be reversed when cooling the dispersions below the phase transition temperature. The PMDEGA chains, in contrast, cannot establish hydrogen bonds between themselves, and thus may react faster.

The collapse and re-swelling process during the phase transition has been investigated also by employing DSC measurements, and distinct differences were recorded between the two thermoresponsive systems. Figure 18 presents DSC thermograms obtained on PS_{62}-*b*-$PNIPAM_{115}$ and PS_{11}-*b*-$PMDEGA_{331}$ solutions with polymer concentrations of 25 wt%, focused on the temperature range of the phase transition. The applied heating/cooling rate was 3 °C/min. It is obvious that the response of the two systems is significantly different. For the PS-*b*-PNIPAM solution, the peaks are more symmetric and narrower in both, heating and cooling thermograms. During heating, the signal drops below the baseline after the completion of the endothermic peak, reaching a plateau value. This effect, indicating that heat capacity decreases slightly at temperatures above CP, is considered as the onset of partial vitrification due to the dissociation of water molecules and polymer units [72]. On the contrary, for the PS-*b*-PMDEGA solution, the signal meets the (extended) baseline after the transition.

It is worth to notice here, that in the case of the PS-*b*-PNIPAM solution, the DSC measurements point to T_{DSC} values (27–28 °C) which are lower than the CP estimated by turbidimetry (31.6 °C), whereas for the PS-*b*-PMDEGA solution, the T_{DSC} values (41–44 °C) are higher than the CP estimated by turbidimetry (40.0 °C). The breadth $\Delta T_{breadth}$ of the transition, which can be defined as $\Delta T_{breadth} = T_{\mathrm{end}} - T_{\mathrm{on}}$, was found to be about 11 °C for the PS-*b*-PNIPAM and 22 °C for PS-*b*-PMDEGA solution, at the heating rate of 3 °C/min. Summarizing, DSC measurements show that, during the demixing process, the hydration number per oligo(ethylene oxide) side chain in PMDEGA blocks may decrease gradually, and that the whole processes is completed only about 20 °C above the CP. On the contrary, the sharp, narrow endothermic peaks recorded on the PS-*b*-PNIPAM solution point to an abrupt, stepwise dehydration

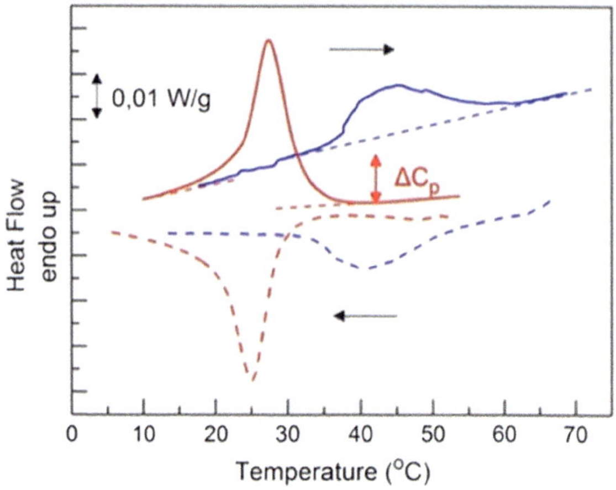

Fig. 18 DSC heating (—) and cooling (- - -) thermograms obtained on 25 wt% PS_{62}-*b*-$PNIPAM_{115}$ (red) and PS_{11}-*b*-$PMDEGA_{331}$ (blue) solutions. The heating/cooling rate was 3 °C/min (Reprinted with permission from [72]. Copyright 2018 American Chemical Society)

behavior of PNIPAM at CP, which has been suggested to be of cooperative character [42, 43, 104].

4.4.2 Aggregation and Segmental Dynamics in Solution

The transition behavior of diblock and triblock copolymers is mainly determined by the chain architecture: The absence or presence of bridging between micelles seems to be the most important feature. All copolymers studied form spherical core-shell micelles in aqueous solution: star-like micelles for PS-*b*-PNIPAM and PS-*b*-PMDEGA diblock copolymers and flower-like micelles with a certain degree of bridging for PS-*b*-PNIPAM-*b*-PS and PS-*b*-PMDEGA-*b*-PS triblock copolymers [15].

For instance, the collapse and aggregation behavior at the cloud point is shown in Fig. 19 for the diblock copolymer PS_{50}-*b*-$PNIPAM_{160}$ and the $P(S\text{-}d_8)_{10}$-*b*-$PNIPAM_{390}$-*b*-$P(S\text{-}d_8)_{10}$ and PS_8-*b*-$PMDEGA_{452}$-*b*-PS_8) triblock copolymers, as studied by DLS. The diblock copolymers PS_{50}-*b*-$PNIPAM_{160}$ form micelles in dilute aqueous solutions which shrink slightly as temperature is increased towards the cloud point (Fig. 19a). At the cloud point, their hydrodynamic radius decreases drastically, i.e., the micellar shell has collapsed. At higher temperatures, large aggregates ("clusters") are observed which, interestingly, are stable in size at ~80 nm.

Two triblock copolymers $Pd8S_{10}$-$PNIPAM_{390}$-$Pd8S_{10}$ and PS_8-$PMDEGA_{452}$-PS_8 having very short PS block as well as thermoresponsive blocks which are approximately twice as long as the ones of the PS_{50}-$PNIPAM_{160}$ diblock copolymer were chosen for comparison. In contrast to the diblock copolymer, both triblock copolymers immediately form large clusters at the cloud point, and single collapsed micelles cannot be observed (Fig. 19b, c). The cluster size depends on the nature of the thermoresponsive block: For $Pd8S_{10}$-$PNIPAM_{390}$-$Pd8S_{10}$, the size is ~70 nm, i.e., similar

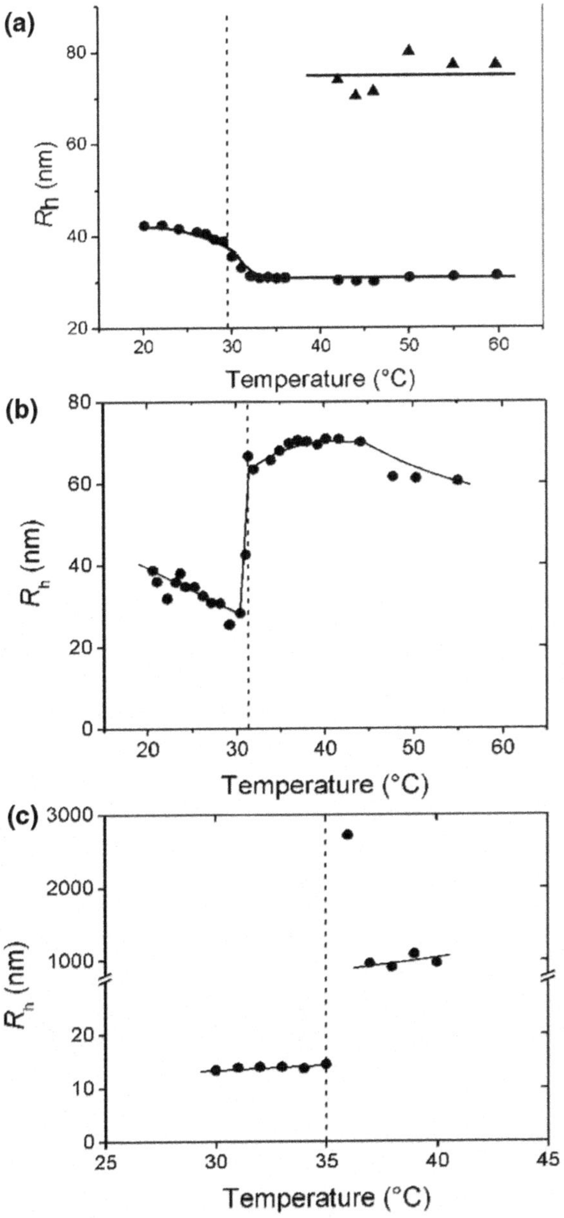

Fig. 19 Hydrodynamic radii of aqueous solutions PS_{50}-$PNIPAM_{160}$ at 0.20 g L^{-1} [124] (**a**), $Pd8S_{10}$-$PNIPAM_{390}$-$Pd8S_{10}$ at 0.2 g L^{-1} [82] (**b**), and PS_8-$PMDEGA_{452}$-PS_8 at 3.0 g L^{-1} [123] as determined using DLS (**c**) (Reproduced from [15], with permission from Springer Nature) (pending)

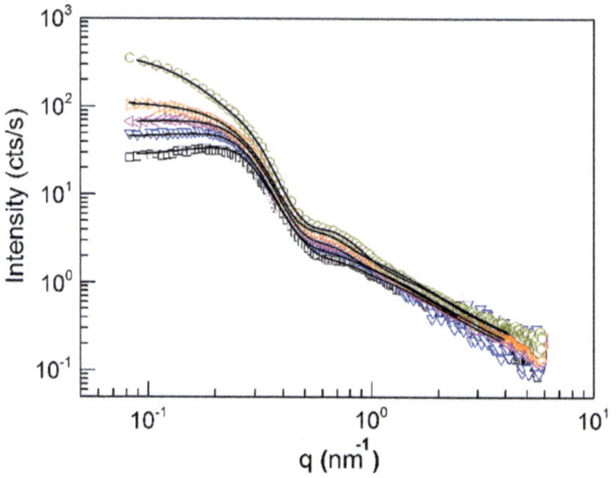

Fig. 20 Representative SAXS curves of a 25 wt% PS-*b*-PMDEGA solution in dependence on temperature. Black squares: 32 °C, blue triangles: 36 °C, pink triangles: 38 °C, orange triangles: 40 °C, dark yellow circles: 42 °C. The lines are fits, see text. Only every second point is presented for clarity (Reprinted with permission from [72]. Copyright 2018 American Chemical Society)

to the diblock copolymer (Fig. 19a), whereas the ones in PS_8-$PMDEGA_{452}$-PS_8 are much bigger with ~1000 nm in radius. It is assumed that PNIPAM becomes glassy when water is released (dry $T_g = 133$ °C), which hampers the growth of the clusters, in agreement with the literature [122]. In contrast, PMDEGA (dry $T_g < -40$ °C) stays liquid-like, enabling the growth of very large clusters. In all cases, the micellar structure is preserved in the clusters.

Regarding the aggregation behavior of concentrated PS-*b*-PNIPAM and PS-*b*-PMDEGA solutions (25–30 wt% polymer concentration), synchrotron SAXS measurements [124] and small-angle neutron scattering (SANS) measurements [83] revealed also different thermoresponsive behavior. Representative SAXS curves of PS-*b*-PMDEGA solutions at 32–42 °C, i.e. below and above the CP from turbidimetry are given in Fig. 20. At higher temperatures, the polymer precipitated. At 32 °C, a maximum is observed at q ~ 0.2 nm^{-1} which is attributed to the correlation between the positions of the micelles. With increasing temperature, the correlation peak becomes weaker and, at the same time, the forward scattering below 0.1 nm^{-1} increases, implying the formation of larger aggregates. At 42 °C, the maximum is not discernible any longer because of the increased forward scattering. Equations (8) and (9) was fitted, and the fits obtained are good in the entire q-range (Fig. 20). The resulting fitting parameters are given in Fig. 21.

The data shown in Fig. 21 indicate a smooth demixing transition around the CP for the PS-b-PMDEGA solution, where individual micelles start to form aggregates only several degrees centigrade above the CP. On the contrary, PS-*b*-PNIPAM forms core-shell micelles with a peculiar behavior around the CP [83]: At temperatures up to 26 °C, the PNIPAM shell is swollen, however, less than expected for a good or a theta solvent. At 27–28 °C, i.e., just below the CP, the micellar shell starts to

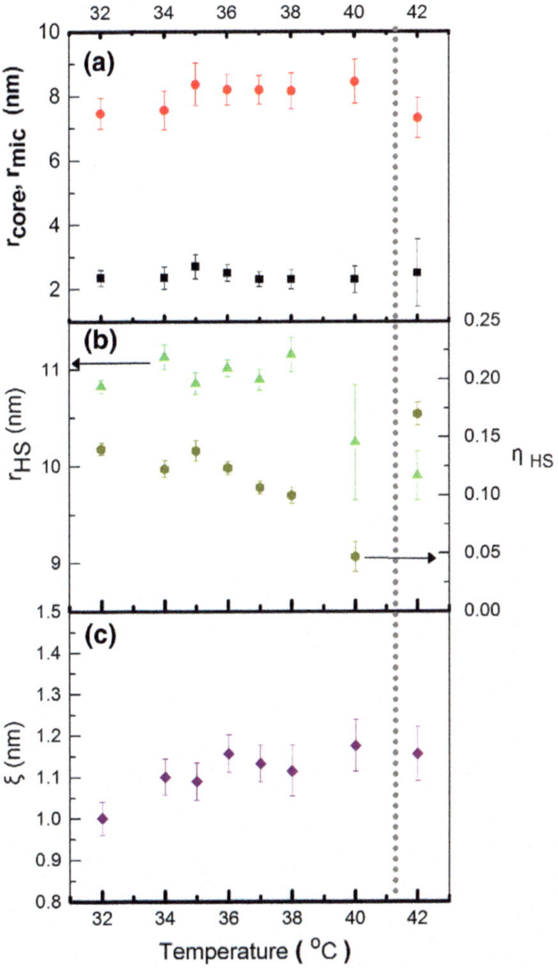

Fig. 21 Parameters resulting from model fitting to the SAXS curves of PS-*b*-PMDEGA solution. **a** Core radius, r_{core} (black squares) and micellar radius, r_{mic} (red circles). **b** Hard-sphere radius, r_{HS} (green triangles, left axis), and volume fraction η_{HS} (dark yellow circles, right axis). **c** Correlation length of the fluctuations, ξ. The grey dashed line indicates the cloud point from DSC (Reprinted with permission from [72]. Copyright 2018 American Chemical Society)

collapse, and the micelles form already small aggregates. Between 28 and 31 °C, the number of aggregates increases. Above 31 °C, the micellar shells are fully collapsed with the micellar core-shell structure being preserved. The distance between the micelles is minimal, and their liquid-like correlation becomes stronger again. The aggregates grow as temperature is increased. Obviously, the experiments indicate that for PS-b-PNIPAM solutions, the phase separation consists of discrete stages, the first stage taking place below the CP where aggregation of the individual micelles occurs. Figure 22 summarizes the behavior observed for three of the PNIPAM- and PMDEGA-based systems.

Neutron spin-echo spectroscopy revealed a number of dynamic processes in concentrated solutions of PS_{50}-$PNIPAM_{160}$ and Pd_8S_{10}-$PNIPAM_{390}$-Pd_8S_{10} below the cloud point: the segmental dynamics of PS, (in case of protonated PS), the so-called

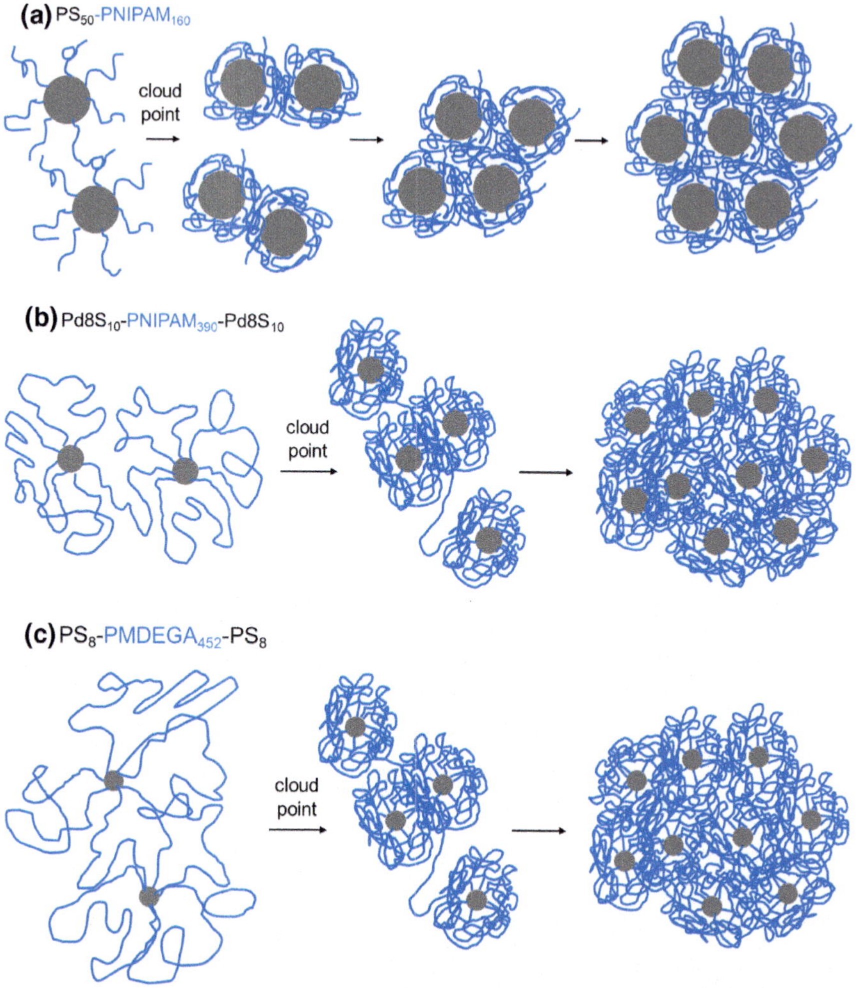

Fig. 22 Model of the phase behavior of the three systems studied in solution. For simplicity, the micellar cores of PS are sketched as homogeneous circles. The blue lines denote the thermo-responsive PNIPAM or PMDEGA blocks (Reproduced from [15], with permission from Springer Nature)

breathing mode of PNIPAM, i.e., the collective dynamics of the swollen PNIPAM shell, and the diffusion of entire micelles [15]. Interestingly, for both, diblock and triblock copolymers, the segmental dynamics in the collapsed state are independent of polymer concentration and faster than in the swollen state. This finding was attributed to a "freezing in" of the segmental dynamics of the PNIPAM shell except a few strands that stick out into the solution. The segmental dynamics thus reflects the micellar architecture: Because of the relatively high grafting density of shell blocks on the core surface, the breathing mode of the shell blocks dominates and is influenced by back folding and bridging.

Recently dielectric measurements shed light in the vitrification process occurred in PNIPAM-based diblock copolymer in the collapsed state [72]. BDS results implied that dielectric permittivity ε' (at 1 MHz) is a sensitive probe of the so-called m-process (related to the segmental mobility of PNIPAM) in PS-*b*-PNIPAM solutions. In Fig. 23, we present the temperature dependence of the real part of dielectric permittivity, ε', and ac conductivity, σ'_{ac}, measured isochronally at $f = 1$ MHz on the PS-*b*-PNIPAM (Fig. 23a) and PS-*b*-PMDEGA (Fig. 23b) diblock copolymer solutions, during heating/cooling with a rate of 3 °C/min. We observe that, in the PS-*b*-PNIPAM solution (Fig. 23a), a steep increase of ε' is recorded at a critical temperature T_{cH}, which is somehow related to the critical phase transition temperatures recorded by turbidimetry (CP) and DSC (T_{DSC}). However, crossing T_{cH} (and remaining below T_{DSC} and CP), ε' increases rapidly with temperature and start to decrease again with increasing temperature, exhibiting thus a peak in the ε'(T) curve and reaching finally saturation values lower than the initial one, at temperatures as high as 30 °C above CP. This ε'(T) behavior is reversible, as seen from the values recorded during cooling shown in Fig. 23a. It has been suggested that the sudden increase of ε' at temperatures just below the CP (Fig. 23a) is related with enhanced PNIPAM chain polarization process (m-process) [72]. In agreement with the proposed water detaching mechanism, this enhancement of the local chain polarization process indicates that enhanced local chain motions occur at the first stages of the demixing transition [111]. This finding may suggest that just prior (or more accurately, at the first stages of) the transition, water re-organization takes place, resulting in a temporal increase of the local chain mobility, which in terms of BDS spectroscopy, results in the temporal increase of the strength of the m-process. However, in parallel with this water re-organization process, the solution enters into the metastable phase of the transition, and some of the amide (−NH) and the carbonyl (−CO) groups prefer to form intra-molecular HBs. The establishment of these intra- and/or inter-chain HBs together with the existing local chain mobility alters dramatically the conformations of the hydrophobic isopropyl groups, and consequently, the "cage like" rigid HB network of the water molecules around them [31, 107]. The recorded gradual decrease of the dielectric strength of the m-process with increasing temperature above the CP reflects a gradual decrease of local chain mobility, implying, thus, that the number of intra- and/or inter-chain HBs gradually increases, leading probably to a decrease in the hydrophobic surface that is exposed to water. This scenario is fully supported by small-angle neutron scattering (SANS) measurements on concentrated solutions (70 wt% water) of the similar diblock copolymer PS_{50}-*b*-$PNIPAM_{160}$ [83], as mentioned above. By comparing the thermoresponsive behavior of the two solutions in Fig. 23, we observe that the temperature dependence of ε' at and above CP is totally different for the two copolymers. For the PS-*b*-PMDEGA solution, contrary to the behavior of PS-*b*-PNIPAM solution, ε' increases rapidly with temperature, reaching a plateau at temperatures 30 °C above CP (Fig. 23b). This polarization process is also evident in the σ'_{ac}(T) curve in Fig. 23b, where a remarkable drop of conductivity occurs at T_{cH}.

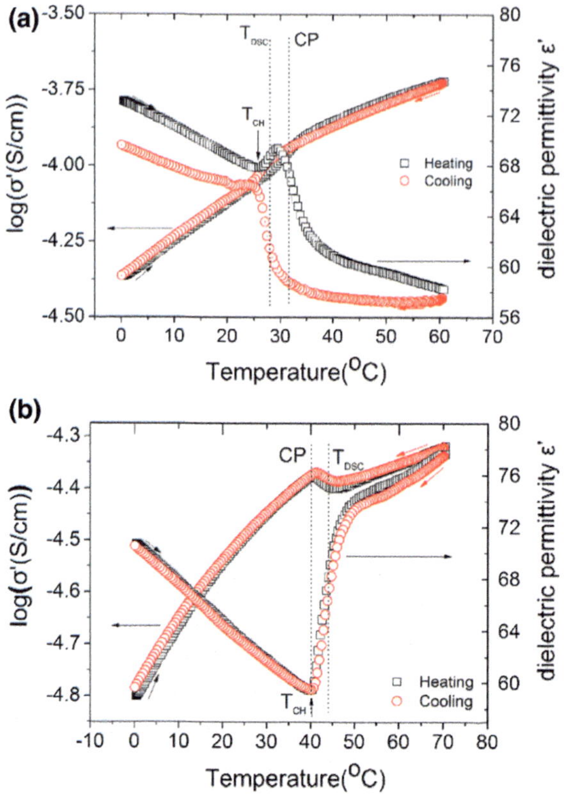

Fig. 23 Real part of dielectric permittivity, ε', and ac conductivity, σ'_{ac}, measured at 1 MHz, as a function of temperature for PS_{62}-*b*-$PNIPAM_{115}$ (**a**) and PS_{11}-*b*-$PMDEGA_{331}$ (**b**) copolymer solutions with 25 wt% polymer concentration. The dotted lines indicate the cloud point temperature as measured by turbidimetry (CP) and by DSC (T_{DSC}). The critical temperature, T_{cH}, where ε' suddenly changes during heating, is indicated for both solutions (Reprinted with permission from [72]. Copyright 2018 American Chemical Society)

Thus, the dielectric susceptibility is very sensitive to morphological changes occurring in the solutions during the demixing transition and to the assembly process within the two-phase high temperature region. Especially its real part, the dielectric permittivity ε′, proves to be a well-suited probe for monitoring both, the reversible changes in molecular dipolar bond polarizabilities and the long-range ionic interfacial polarization during the phase transition.

5 Conclusions and Perspectives

The studies of the bulk behavior have thus revealed a number of differences between PNIPAM- and PMDEGA-based thermoresponsive systems, concerning the transition temperature, the width and hysteresis of the transition, the collapse and aggregation behavior of micellar solutions and the segmental dynamics. Numerous studies have revealed several factors that affect the themoresponsive behavior of macromolecules, however, despite the huge accumulation of data, the underlying molecular mechanisms have not been explained so far. For applications, the choice of a specific polymer for implementing supramolecular and responsive systems is not simply a matter of the exact value of the phase transition temperature. Clearly, the detailed chemical nature of the thermoresponsive polymers is of extreme importance for their behavior in aqueous solution. On the other hand, this behavior cannot be described and predicted by a simplified model considering just size, volume fraction and distribution of the various thermoresponsive and hydrophobic polymer fragments within the macromolecules. Unfortunately, the self-organization in aqueous solutions of each of them cannot be deduced solely on the basis of our current understanding of basic self-assembly principles.

Understanding of self-organization principles for ***homopolymers*** is the first step to the further understanding of more complex systems. To accomplish this target, it is critical to improve our knowledge along two main directions: First, to clarify the polymer-water interactions and the dynamics of water molecules around the macromolecules during the transition (dehydration process). Modern experimental methods are available now, e.g. THz dielectric spectroscopy, for studying selectively the dynamics of hydration and bulk water molecules in aqueous solutions of amphiphilic macromolecules. Second, to get insight into the role of macromolecular chain conformations and segmental mobility. The cooperative effects related with self-organization properties seem to be closely related with chain connectivity issues. Then, the study of ***more complex molecular structures*** (copolymers, gels, surface adsorbed chains) and thin thermoresponsive films is necessary in order to transfer the new knowledge to systems with practical applications. Real progress in our understanding of the fundamental aspects of behavior of complex multisensitive macromolecular systems and in the rational design of new functional materials is possible only by applying a multi-disciplinary approach bringing together chemistry and physics, experiment and theory.

Finally, it is worth mentioning that the close similarity of the cooperative hydration of PNIPAM and of polypeptide backbones indicates the relevance of the investigation of the self-assembling properties of the former for understanding the thermal stability of hydrogen bonded water networks enveloping such biological polymers. Plenty of experimental results indicate that knowledge of the balance of interactions between water and the hydrophilic and hydrophobic groups at the coil-to-globule transition in the PNIPAM aqueous solution is essential for understanding complicated molecular mechanisms in biological systems, such as the stabilization of protein conformation and self-association of lipids. The response to stimuli is a common process for biopolymers in living organisms. In the past, the concept of cooperative interactions between the functional segments of biopolymer has led to the invention of novel synthetic thermoresponsive polymer systems. Therefore, the interest in self-organization properties of macromolecules is not only due to its importance as a fundamental concept in polymer physics and solution dynamics, but also due to its relevance to many biological systems.

Acknowledgements The authors would like to thank Dionysia Aravopoulou (NTUA), Achille M. Bivigou-Koumba, Anna Miasnikova (Universität Potsdam), Joseph Adelsberger, Konstantinos Kyriakos, Andreas Meier-Koll, Peter Müller-Buschbaum and Kordelia Troll (TU München) who were or are involved in studies of thermoresponsive polymers in our groups. Financial support was provided by Deutsche Forschungsgemeinschaft (DFG) via the priority program SPP 1259 "Intelligente Hydrogele" (grants LA611/7, MU1487/8 and PA771/4) and by the German Academic Exchange Service (DAAD) within the program "Hochschulpartnerschaften mit Griechenland" (ResComp).

References

1. Hoffman, A.S.: "Intelligent" polymers in medicine and biotechnology. Artif. Organs **19**, 458–467 (1995)
2. Dimitrov, I., Trzebicka, B., Müller, A.J.E., Dworak, A., Tsvetanov, C.B.: Thermosensitive water-soluble copolymers with doubly responsive reversibly interacting entities. Prog. Polym. Sci. **32**, 1275–1343 (2007)
3. Gil, E.S., Hudson, S.M.: Stimuli-responsive polymers and their bioconjugates. Prog. Polym. Sci. **29**, 1173–1222 (2004)
4. Liu, R., Fraylich, M., Saunders, B.R.: Thermoresponsive copolymers: from fundamental studies to applications. Colloid Polym. Sci. **287**, 627–643 (2009)
5. Aseyev, V., Tenhu, H., Winnik, F.: Non-ionic thermoresponsive polymers in water. Adv. Polym. Sci. **242**, 29–89 (2011)
6. Koningsveld, R., Staverman, A.J.: Liquid–liquid phase separation in multicomponent polymer solutions. I. Statement of problem and description of methods of calculation. J. Polym. Sci. A2 **6**, 305–323 (1968)
7. Koningsveld, R., Staverman, A.J.: Liquid–liquid phase separation in multicomponent polymer solutions. 2. Critical state. J. Polym. Sci. A2 **6**, 325–347 (1968)
8. Weber, C., Hoogenboom, R., Schubert, U.S.: Temperature responsive bio-compatible polymers based on poly(ethylene oxide) and poly(2-oxazoline)s. Prog. Polym. Sci. **37**, 686–714 (2012)
9. Heskins, M., Guillet, J.E.: Solution properties of poly (N-isopropylacrylamide). J. Macromol. Sci., Chem. **2**, 1441–1455 (1968)
10. Halperin, A., Kröger, M., Winnik, F.M.: Poly(N-isopropylacrylamide) phase diagrams: fifty years of research. Angew. Chem. Int. Ed. **54**, 15342–15367 (2015)
11. Lanzalaco, S., Armelin, E.: Poly(N-isopropylacrylamide) and copolymers: a review on recent progresses in biomedical applications. Gels **3**, 36 (2017) (31 pages)
12. Lutz, J.-F., Akdemir, O., Hoth, A.: Point by point comparison of two thermosensitive polymers exhibiting a similar LCST: is the age of poly(NIPAM) over? J. Am. Chem. Soc. **128**, 13046–13047 (2006)
13. Hacker, M.C., Klouda, L., Ma, B.B., Kretlow, J.D., Mikos, A.G.: Synthesis and characterization of injectable, thermally and chemically gelable, amphiphilic poly(N-isopropylacrylamide)-based macromers. Biomacromol **9**, 1558–1570 (2008)
14. Ward, M.A., Georgiou, T.K.: Thermoresponsive polymers for biomedical applications. Polymers **3**, 1215–1242 (2011)

15. Laschewsky, A., Müller-Buschbaum, P., Papadakis, C.M.: Thermo-responsive amphiphilic Di- and triblock copolymers based on poly(N-isopropylacrylamide) and poly(methoxy diethylene glycol acrylate): aggregation and hydrogel formation in bulk solution and in thin films. In: Sadowski, G., Richtering, W. (eds.) Intelligent Hydrogels, Progress in Colloid and Polymer Science, vol. 140, pp. 15–34. Springer International Publishing, Switzerland (2013)
16. Tanii, H., Hashimoto, K.: In vitro neurotoxicity study with dorsal root ganglia for acrylamide and its derivatives. Toxicol. Lett. **58**, 209–213 (1991)
17. Lutz, J.-F.: Polymerization of oligo(ethylene glycol) (meth)acrylates: toward new generations of smart biocompatible materials. J. Polym. Sci., Part A: Polym. Chem. **46**, 3459–3470 (2008)
18. Miasnikova, A., Laschewsky, A.: Influencing the phase transition temperature of poly(methoxy diethylene glycol acrylate) by molar mass, end groups, and polymer architecture. J. Polym. Sci., Part A: Polym. Chem. **50**, 3313–3323 (2012)
19. Zhong, Q., Wang, W., Adelsberger, J., Golosova, A., Bivigou-Koumba, A.M., Laschewsky, A., Funari, S.S., Perlich, J., Roth, S.V., Papadakis, C.M., Müller-Buschbaum, P.: Collapse transition in thin films of poly(methoxydiethylenglycol acrylate). Colloid Polym. Sci. **289**, 569–581 (2011)
20. Zhong, Q., Metwalli, E., Rawolle, M., Kaune, G., Bivigou-Koumba, A.M., Laschewsky, A., Papadakis, C.M., Cubitt, R., Wang, J., Müller-Buschbaum, P.: Influence of hydrophobic polystyrene blocks on the rehydration of polystyrene-block-poly(methoxy diethylene glycol acrylate)-block-polystyrene films investigated by in situ neutron reflectivity. Macromolecules **49**, 317–326 (2016)
21. Zhong, Q., Metwalli, E., Rawolle, M., Kaune, G., Bivigou-Koumba, A.M., Laschewsky, A., Papadakis, C.M., Cubitt, R., Wang, J., Müller-Buschbaum, P.: Vacuum induced dehydration of swollen poly(methoxy diethylene glycol acrylate) and polystyrene-block-poly(methoxy diethylene glycol acrylate)-block-polystyrene films probed by in-situ neutron reflectivity. Polymer **124**, 263–273 (2017)
22. Chen, T., Ferris, R., Zhang, J., Ducker, R., Zauscher, S.: Stimulus-responsive polymer brushes on surfaces: transduction mechanisms and applications. Prog. Polym. Sci. **35**, 94–112 (2010)
23. Hrubý, M., Filippov, S.K., Štěpánek, P.: Smart polymers in drug delivery systems on crossroads: which way deserves following? Eur. Polym. J. **65**, 82–97 (2015)
24. Gupta, P., Vermani, K., Garg, S.: Hydrogels: from controlled release to pH-responsive drug delivery. Drug Discov. Today **7**, 569–579 (2002)
25. Jeong, B., Gutowska, A.: Lessons from nature: stimuliresponsive polymers and their biomedical applications. Trends Biotechnol. **20**, 305–311 (2002)
26. Qiu, Y., Park, K.: Environment-sensitive hydrogels for drug delivery. Adv. Drug Deliv. Rev. **53**, 321–339 (2001)
27. Yokoyama, M.: Gene delivery using temperature-responsive polymeric carriers. Drug Discov. Today **7**, 426–432 (2002)
28. Galaev, L.Y., Mattiasson, B.: 'Smart' polymers and what they could do in biotechnology and medicine. Trends Biotechnol. **17**, 335–340 (2000)
29. Aguilar, M.R., Elvira, C., Gallardo, A., Vázquez, B., Román, J.S.: Smart polymers and their applications as biomaterials. In: Ashammakhi, N., Reis, R., Chiellini, E. (eds.) Topics in Tissue Engineering, vol. 3, Chap. 6 (2007)
30. Nagase, K., Okano, T.: Thermoresponsive-polymer-based materials for temperature-modulated bioanalysis and bioseparations. J. Mater. Chem. B **4**, 6381–6397 (2016)
31. Shiraga, K., Naito, H., Suzuki, T., Kondo, N., Ogawa, Y.: Hydration and hydrogen bond network of water during the coil-to-globule transition in poly(N isopropyl-acrylamide) aqueous solution at cloud point temperature. J. Phys. Chem. B **119**, 5576–5587 (2015)
32. Chen, G., Hoffman, A.S.: Graft copolymers that exhibit temperature-induced phase transitions over a wide range of pH. Nature **373**, 49–52 (1995)
33. van der Vegt, N.F.A., Nayar, D.: The hydrophobic effect and the role of cosolvents. J. Phys. Chem. B **121**, 9986–9998 (2017)
34. Otake, K., Inomata, H., Konno, M., Saito, S.: Thermal analysis of the volume phase transition with N-isopropylacrylamide gels. Macromolecules **23**, 283–289 (1990)

35. Bae, Y.H., Okano, T., Kim, S.W.: Temperature dependence of swelling of crosslinked poly(N,N′-alkyl substituted acrylamides) in water. J. Polym. Sci., Polym. Phys. Ed. **28**, 923–936 (1990)
36. Ding, Y., Ye, X., Zhang, G.: Can coil-to-globule transition of a single chain be treated as a phase transition? J. Phys. Chem. B **112**, 8496–8498 (2008)
37. Afroze, F., Nies, E., Berghmans, H.: Phase transitions in the system poly(N-isopropylacrylamide)/water and swelling behaviour of the corresponding networks. J. Mol. Struct. **554**, 55–68 (2000)
38. Aleksandrova, R., Philipp, M., Müller, U., Riobóo, R.J., Ostermeyer, M., Sanctuary, R., Müller-Buschbaum, P., Krüger, J.K.: Phase instability and molecular kinetics provoked by repeated crossing of the demixing transition of PNIPAM solutions. Langmuir **30**, 11792–11801 (2014)
39. Taylor, M.J., Tomlins, P., Sahota, T.S.: Thermoresponsive gels. Gels **3**, 4 (2017) (31 pages)
40. Zaccone, A., Crassous, J.J., Ballauff, M.: Colloidal gelation with variable attraction energy. J. Chem. Phys. **138**, 104908–104908 (2013)
41. Kyriakos, K.: Cononsolvency of PNIPAM in water/alcohol mixtures—A neutron scattering study, Ph.D. Thesis, Technical University of Munich (2015)
42. Okada, Y., Tanaka, F.: Cooperative hydration, chain collapse, and flat LCST behavior in aqueous poly(N-isopropylacrylamide) Solutions. Macromolecules **38**, 4465–4471 (2005)
43. Tanaka, F., Koga, T., Winnik, F.M.: Competitive hydrogen bonds and cononsolvency of poly(N-isopropylacrylamide)s in mixed solvents of water/methanol. Prog. Colloid Polym. Sci. **136**, 1–7 (2009)
44. Meeussen, F., Nies, E., Berghmans, H., Verbrugghe, S., Goethals, E., Du Prez, F.E.: Phase behaviour of poly(N-vinyl caprolactam) in water. Polymer **41**, 8597–8602 (2000)
45. Van Durme, K., Verbrugghe, S., Du Prez, F.E., Van Mele, B.: Influence of poly(ethylene oxide) grafts on kinetics of LCST behavior in aqueous Poly(N-vinylcaprolactam) solutions and networks studied by modulated temperature DSC. Macromolecules **37**, 1054–1061 (2004)
46. Van Durme, K., Van Assche, G., Van Mele, B.: Kinetics of demixing and remixing in poly(N-isopropylacrylamide)/water studied by modulated temperature DSC. Macromolecules **37**, 9596–9605 (2004)
47. Liu, F., Urban, M.W.: Recent advances and challenges in designing stimuli-responsive polymers. Prog. Polym. Sci. **35**, 3–23 (2010)
48. Urban, M.W.: Stratification, stimuli-responsiveness, self-healing, and signaling in polymer networks. Prog. Polym. Sci. **34**, 679–687 (2009)
49. Pelton, R.: Temperature-sensitive aqueous microgels. R. Adv. Colloid Interface Sci. **85**, 1–33 (2000)
50. Hofmann, C.H., Plamper, F.A., Scherzinger, C., Hietala, S., Richtering, W.: Cononsolvency revisited: solvent entrapment by N-isopropylacrylamide and N, N-diethylacrylamide microgels in different water/methanol mixtures. Macromolecules **46**, 523–532 (2013)
51. Geismann, C., Ulbricht, M.: Photoreactive functionalization of poly(ethylene terephthalate) tracketched pore surfaces with "smart" polymer systems. Macromol. Chem. Phys. **206**, 268–281 (2005)
52. Li, S.K., D'Emanuele, A.: On-off transport through a thermoresponsive hydrogel composite membrane. J. Control. Release **75**, 55–67 (2001)
53. Lupitsky, R., Roiter, Y., Minko, S., Tsitsilianis, C.: From smart polymer molecules to responsive nanostructured surfaces. Langmuir **21**, 8591–8593 (2005)
54. Yamato, M., Konno, C., Utsumi, M., Kikuchi, A., Okano, T.: Thermally responsive polymer-grafted surfaces facilitate patterned cell seeding and co-culture. Biomaterials **23**, 561–567 (2002)
55. Bromberg, L.E., Ron, E.S.: Temperature-responsive gels and thermogelling polymer matrices for protein and peptide delivery. Adv. Drug Deliv. Rev. **31**, 197–221 (1998)
56. Zhang, K., Khan, A.: Phase behavior of poly(ethyleneoxide)–poly(propylene oxide)–poly(ethylene oxide) triblock copolymers in water. Macromolecules **28**, 3807–3812 (1995)

57. Sweta, M., Jay Prakash, J., Domb, A.J., Neeraj, K.: Exploiting EPR in polymer drug conjugate delivery for tumor targeting. Curr. Pharm. Des. **12**, 4785–4796 (2006)
58. Marcucci, F., Lefoulon, F.: Active targeting with particulate drug carriers in tumor therapy: fundamentals and recent progress. Drug Discov. Today **9**, 219–228 (2004)
59. Neoh, K.G., Kang, E.T.: Responsive surfaces for biomedical applications. MRS Bull. **35**, 673–681 (2010)
60. Alghunaim, A., Brink, E.T., Newby, B.-Z.: Surface immobilization of thermo-responsive poly(N-isopropylacrylamide) by simple entrapment in a 3-aminopropyl-triethoxysilane network. Polymer **101**, 139–150 (2016)
61. Hoffman, A.S.: Bioconjugates of intelligent polymers and recognition proteins for use in diagnostics and affinity separations. Clin. Chem. **46**, 1478–1486 (2000)
62. Hoffman, A.S., Stayton, P.S., Press, O., Murthy, N., Lackey, C.A., Cheung, C., Black, F., Campbell, J., Fausto, N., Kyriakides, T.R., Bornstein, P.: Design of "smart" Polymers that can direct intracellular drug delivery. Polym. Advan. Technol. **13**, 992–999 (2002)
63. Kikuchi, A., Okano, T.: Intelligent thermoresponsive polymeric stationary phases for aqueous chromatography of biological compounds. Prog. Polym. Sci. **27**, 1165–1193 (2002)
64. Kobayashi, J., Kikuchi, A., Sakai, K., Okano, T.: Aqueous chromatography utilizing hydrophobicity-modified anionic temperature-responsive hydrogel for stationary phases. J. Chromatogr. A **958**, 109–119 (2002)
65. Wunderlich, B.: Thermal analysis. Academic Press, New York (1990)
66. Hatakeyama, T., Quinn, F.X.: Thermal analysis, fundamentals and applications to polymer science. Wiley, Chichester (1994)
67. Schick, C.: Differential scanning calorimetry (DSC) of semicrystalline polymers. Anal. Bioanal. Chem. **395**, 1589–1611 (2009)
68. Cho, E.C., Lee, J., Cho, K.: Role of bound water and hydrophobic interaction in phase transition of poly(N-isopropylacrylamide) aqueous solution. Macromolecules **36**, 9929–9934 (2003)
69. Grinberg, V.Y., Dubovik, A.S., Kuznetsov, D.V., Grinberg, N.V., Grosberg, A.Y., Tanaka, T.: Studies of the thermal volume transition of poly(N-isopropylacrylamide) hydrogels by high-sensitivity differential scanning microcalorimetry. 2. Thermodynamics functions. Macromolecules **33**, 8685–8692 (2000)
70. Ding, Y., Ye, X., Zhang, G.: Microcalorimetric investigation on aggregation and dissolution of poly(N-isopropylacrylamide) chains in water. Macromolecules **38**, 904–908 (2005)
71. Maeda, T., Yamamoto, K., Aoyagi, T.: Importance of bound water in hydration-dehydration behavior of hydroxylated poly(N-isopropylacrylamide). J. Colloid Interface Sci. **302**, 467–474 (2006)
72. Aravopoulou, D., Kyriakos, K., Miasnikova, A., Laschewsky, A., Papadakis, C.M., Kyritsis, A.: Comparative investigation of the thermoresponsive behavior of two diblock copolymers comprising PNIPAM and PMDEGA blocks. J. Phys. Chem. B **122**, 2655–2668 (2018)
73. Tiktopulo, E.I., Bychkova, V.E., Ricka, J., Ptitsyn, O.B.: Cooperativity of the coil-globule transition in a homopolymer: microcalorimetric study of poly(N-isopropylacrylamide). Macromolecules **27**, 2879–2882 (1994)
74. Swier, S., Van Durme, B., Van Mele, B.: Modulated-temperature differential scanning calorimetry study of temperature-induced mixing and demixing in poly(vinylmethylether)/water. J. Polym. Sci.: Part B: Polym. Phys. **41**, 1824–1836 (2003)
75. Schick, C.: Temperature modulated differential scanning calorimetry (TMDSC)—basics and applications to polymers. In: Gallagher P.K. (Ed.) Handbook of Thermal Analysis and Calorimetry, vol. 3. Elsevier Science, Amsterdam, Lausanne, New York, Oxford, Shannon, Singapore, Tokyo (2002)
76. Reading, M., Hourston, D.J.: Modulated temperature differential scanning calorimetry: theoretical and practical applications in polymer characterisation. Springer, Berlin (2006)
77. Müller, U., Philipp, M., Thomassey, M., Sanctuary, R., Krüger, J.K.: Temperature modulated optical refractometry: a quasi-isothermal method to determine the dynamic volume expansion coefficient. Thermochim. Acta **555**, 17–22 (2013)

78. Roe, R.-J.: Methods of X-Ray and neutron scattering in polymer science. Topics in Polymer Science, Oxford University Press (2000)
79. Shibayama, M., Tanaka, T., Han, C.C.: Small angle neutron scattering study on poly(N-isopropyl acrylamide) gels near their volume-phase transition temperature. J. Chem. Phys. **97**, 6829–6841 (1992)
80. Meier-Koll, A., Pipich, V., Busch, P., Papadakis, C.M., Müller-Buschbaum, P.: Phase separation in semidilute aqueous poly(N-isopropylacrylamide) solutions. Langmuir **28**, 8791–8798 (2012)
81. Porod, G.: Die Röntgenkleinwinkelstreuung von dichtgepackten kolloiden Systemen. Kolloid Z. Z. Polym. **125**, 51–57 (1952)
82. Adelsberger, J., Kulkarni, A., Jain, A., Wang, W., Bivigou-Koumba, A., Busch, P., Pipich, V., Holderer, O., Hellweg, T., Laschewsky, A., Müller-Buschbaum, P., Papadakis, C.M.: Thermoresponsive PS-*b*-PNIPAM-*b*-PS Micelle: aggregation behavior, segmental dynamics, and thermal response. Macromolecules **43**, 2490–2501 (2010)
83. Adelsberger, J., Meier-Koll, A., Bivigou-Koumba, A., Busch, P., Holderer, O., Hellweg, T., Laschewsky, A., Müller-Buschbaum, P., Papadakis, C.M.: The collapse transition and the segmental dynamics in concentrated micellar solutions of P(S-*b*-NIPAM) diblock copolymers. Colloid Polym. Sci. **289**, 711–720 (2011)
84. Percus, J.K., Yevick, G.: Analysis of classical statistical mechanics by means of collective coordinates. Phys. Rev. **110**, 1–13 (1958)
85. McGrum, N.G., Read, B.E., Williams, G.: Anelastic and dielectric effects in polymeric solids. Wiley, New York (1967)
86. Runt, J., Fitzgerald, J.J. (eds.): Dielectric Spectroscopy of Polymeric Materials. American Chemical Society, Washington, DC (1997)
87. Kremer, F., Schoenhals, A. (eds.): Broadband Dielectric Spectroscopy. Springer, Berlin (2003)
88. Yang, M., Zhao, K.: Anomalous volume phase transition temperature of thermosensitive semi-interpenetrating polymer network microgel suspension by dielectric spectroscopy. J. Phys. Chem. B **119**, 13198–13207 (2015)
89. Mohanty, P.S., Nöjd, S., Bergman, M.J., Nägele, G., Arrese-Igor, S., Alegria, A., Roa, R., Schurtenberger, P., Dhont, J.K.G.: Dielectric spectroscopy of ionic microgel suspensions. Soft Matter **12**, 9705–9727 (2016)
90. Ono, Y., Shikata, T.: Hydration and dynamic behavior of poly(N-isopropylacrylamide)s in aqueous solution: a sharp phase transition at the lower critical solution temperature. J. Am. Chem. Soc. **128**, 10030–10031 (2006)
91. Ono, Y., Shikata, T.: Contrary hydration behavior of N-isopropylacrylamide to its polymer, P(NIPAm), with a lower critical solution temperature. J. Phys. Chem. B **111**, 1511–1513 (2007)
92. Satokawa, Y., Shikata, T., Tanaka, F., Qiu, X., Winnik, F.M.: Hydration and dynamic behavior of a cyclic poly(N-isopropylacrylamide) in aqueous solution: effects of the polymer chain topology. Macromolecules **42**, 1400–1403 (2009)
93. Masci, G., Cametti, C.: Dielectric properties of thermo-reversible hydrogels: the case of a dextran copolymer grafted with poly(N-isopropylacrylamide). J. Phys. Chem. B **113**, 11421–11428 (2009)
94. Masci, G., De Santis, S., Cametti, C.: Dielectric properties of micellar aggregates due to the self-assembly of thermoresponsive diblock copolymers. J. Phys. Chem. B **115**, 2196–2204 (2011)
95. Füllbrandt, M., Ermilova, E., Asadujjaman, A., Hölzel, R., Bier, F.F., von Klitzing, R., Schönhals, A.: Dynamics of linear poly(N isopropylacrylamide) in water around the phase transition investigated by dielectric relaxation spectroscopy. J. Phys. Chem. B **118**, 3750–3759 (2014)
96. Su, W., Zhao, K., Wei, J., Ngai, T.: Dielectric relaxations of poly (N-isopropyl-acrylamide) microgels near the volume phase transition temperature: impact of crosslinking density distribution on the volume phase transition. Soft Matter **10**, 8711–8723 (2014)
97. Su, W., Yang, M., Zhao, K., Ngai, T.: Influence of charged groups on the structure of microgel and volume phase transition by dielectric analysis. Macromolecules **49**, 7997–8008 (2016)

98. Gómez-Galván, F., Lara-Ceniceros, T., Mercado-Uribe, H.: Device for simultaneous measurements of the optical and dielectric properties of hydrogels. Meas. Sci. Technol. **23**, 025602 (2012) (6 pp)
99. Zhou, J., Wei, J., Ngai, T., Wang, L., Zhu, D., Shen, J.: Correlation between dielectric/electric properties and cross-linking/charge density distributions of thermally sensitive spherical PNIPAM microgels. Macromolecules **45**, 6158–6167 (2012)
100. Füllbrandt, M., von Klitzing, R., Schönhals, A.: Probing the phase transition of aqueous solutions of linear low molecular weight poly(N-isopropylacrylamide) by dielectric spectroscopy. Soft Matter **8**, 12116–12123 (2012)
101. Füllbrandt, M., von Klitzing, R., Schönhals, A.: The dielectric signature of poly(N-isopropylacrylamide) microgels at the volume phase transition: dependence on the crosslinking density. Soft Matter **9**, 4464–4471 (2013)
102. Kyriakos, K., Aravopoulou, D., Augsbach, L., Sapper, J., Ottinger, S., Psylla, C., Rafat, A., Benitez-Montoya, C.A., Miasnikova, A., Di, Z., Laschewsky, A., Müller-Buschbaum, P., Kyritsis, A., Papadakis, C.M.: Novel thermoresponsive block copolymers having different architectures—structural, mechanical, thermal and dielectric investigations. Colloid Polym. Sci. **292**, 1757–1774 (2014)
103. Schild, H.G.: Poly(N-isopropylacrylamide): experiment, theory and application. Prog. Polym. Sci. **17**, 163–249 (1992)
104. Choi, H.S., Yui, N.: Design of rapidly assembling supramolecular systems responsive to synchronized stimuli. Prog. Polym. Sci. **31**, 121–144 (2006)
105. Philipp, M., Kyriakos, K., Silvi, L., Lohstroh, W., Petry, W., Krüger, J.K., Papadakis, C.M., Müller-Buschbaum, P.: From molecular dehydration to excess volumes of phase-separating pnipam solutions. J. Phys. Chem. B **118**, 4253–4260 (2014)
106. Futscher, M.H., Philipp, M., Müller-Buschbaum, P., Schulte, A.: The role of backbone hydration of poly(N-isopropyl acrylamide) across the volume phase transition compared to its monomer. Sci. Rep. **7**, 17272–7 (2017)
107. Bischofberger, I., Calzolari, D.C.E., De Los Rios, P., Jelezarov, I., Trappe, V.: Hydrophobic hydration of poly-N-isopropyl acrylamide: a matter of the mean energetic state of water. Sci. Rep. **4**, 4377 (2014) (7 pages)
108. de Gennes, P.G.: Scaling Concepts in Polymer Physics. Cornell University Press, London (1979)
109. de Oliveira, T.E., Mukherji, D., Kremer, K., Netz, P.A.: Effects of stereochemistry and copolymerization on the LCST of PNIPAm. J. Chem. Phys. **146**, 034904 (2017) (10 pages)
110. Liu, S.M., Taylor, C., Chong, B., Liu, L., Bilic, A., Terefe, N.S., Stockmann, R., Thang, S.H., De Silva, K.: Conformational transitions and dynamics of thermal responsive poly(N-isopropylacrylamide) polymers as revealed by molecular simulation. Eur. Polym. J. **55**, 153–159 (2014)
111. Katsumoto, Y., Tanaka, T., Ihara, K., Koyama, M., Ozaki, Y.: Contribution of intramolecular $C=O\cdots H–N$ hydrogen bonding to the solvent-induced reentrant phase separation of poly(N-isopropylacrylamide). J. Phys. Chem. B **111**, 12730–12737 (2007)
112. Xia, Y., Burke, N.A.D., Stover, H.D.H.: End group effect on the thermal response of narrow-disperse poly(N-isopropylacrylamide) prepared by atom transfer radical polymerization. Macromolecules **39**, 2275–2283 (2006)
113. Termühlen, F., Kuckling, D., Schönhoff, M.: Isothermal titration calorimetry to probe the coil-to-globule transition of thermoresponsive polymers. J. Phys. Chem. B **121**, 8611–8618 (2017)
114. Singh, R., Deshmukh, S.A., Kamath, G., Sankaranarayanan, S.K.R.S., Balasubramanian, G.: Controlling the aqueous solubility of PNIPAM with hydrophobic molecular units. Comput. Mater. Sci. **126**, 191–203 (2017)
115. Zhang, Y., Furyk, S., Bergbreiter, D.E., Cremer, P.S.: Specific ion effects on the water solubility of macromolecules: PNIPAM and the hofmeister series. J. Am. Chem. Soc. **127**, 14505–14510 (2005)

116. Scherzinger, C., Schwarz, A., Bardowb, A., Leonhard, K., Richtering, W.: Cononsolvency of poly-N-isopropyl acrylamide (PNIPAM): Microgels versus linear chains and macrogels. Curr. Opin. Colloid Interface Sci. **19**, 84–94 (2014)
117. Kyriakos, K., Philipp, M., Adelsberger, J., Jaksch, S., Berezkin, A.V., Lugo, D.M., Richtering, W., Grillo, I., Miasnikova, A., Laschewsky, A., Müller-Buschbaum, P., Papadakis, C.M.: Cononsolvency of water/methanol mixtures for PNIPAM and PS-b-PNIPAM: pathway of aggregate formation investigated using time-resolved SANS. Macromolecules **47**, 6867–6879 (2014)
118. Hua, F., Jiang, X., Li, D., Zhao, B.: Well-Defined thermosensitive, water-soluble polyacrylates and polystyrenics with short pendant oligo(ethylene glycol) groups synthesized by nitroxide-mediated radical polymerization. J. Polym. Sci., Part A: Polym. Chem. **44**, 2454–2467 (2006)
119. Bordi, F., Cametti, C., Colby, R.H.: Dielectric spectroscopy and conductivity of polyelectrolyte solutions. J. Phys.: Condens. Matter **16**, R1423–R1463 (2004)
120. Bivigou-Koumba, A.M., Görnitz, E., Laschewsky, A., Müller-Buschbaum, P., Papadakis, C.M.: Thermoresponsive amphiphilic symmetrical triblock copolymers with a hydrophilic middle block made of poly(N-isopropylacrylamide): synthesis, self-organization, and hydrogel formation. Colloid. Polym. Sci. **288**, 499–517 (2010)
121. Kujawa, P., Segui, F., Shaban, S., Diab, C., Okada, Y., Tanaka, F., Winnik, F.M.: Impact of end-group association and main-chain hydration on the thermosensitive properties of hydrophobically modified telechelic poly(N-isopropylacrylamides) in water. Macromolecules **39**, 341–348 (2006)
122. Koga, T., Tanaka, F., Motokawa, R., Koizumi, S., Winnik, F.M.: Theoretical modeling of associated structures in aqueous solutions of hydrophobically modified telechelic PNIPAM based on a neutron scattering study. Macromolecules **41**, 9413–9422 (2008)
123. Miasnikova, A., Laschewsky, A., De Paoli, G., Papadakis, C.M., Müller-Buschbaum, P., Funari, S.S.: Thermoresponsive hydrogels from symmetrical triblock copolymers poly(styrene-block-(methoxy diethylene glycol acrylate)-block-styrene). Langmuir **28**, 4479–4490 (2012)
124. Troll, K., Kulkarni, A., Wang, W., Darko, C., Bivigou-Koumba, A.M., Laschewsky, A., Müller-Buschbaum, P., Papadakis, C.M.: The collapse transition of poly(styrene-b-(N-isopropyl acrylamide)) diblock copolymers in aqueous solution and in thin films. Colloid Polym. Sci. **286**, 1079–1092 (2008)

Sol-Gel Synthesis and Characterization of Hybrid Materials for Biomedical Applications

Michelina Catauro and Stefano Vecchio Ciprioti

Abstract "Biomaterials" are able to interact with human tissue and body fluids to treat, improve, or replace anatomical elements of the human body, without adverse reactions in the host organism. However, it is not currently possible to avoid the early failure of implants, because a good combination of properties (mechanical, chemical, and tribological) and biocompatibility of the materials has not been yet achieved. Bioactive glasses constitute a promising class of bioactive materials for bone repair and substitution. They have the capability of bonding with the living bone by forming a hydroxyapatite layer on their surface with a composition equivalent to that of the mineral phase of bones. The biological properties of bioglasses are influenced by their composition and by the synthesis method as well. An ideal method to prepare a bioglass is the sol-gel technique, a versatile synthesis process. The process involves the transition of a system from mostly colloidal liquid ('sol') into a solid 'gel'. The sol-gel method was employed to synthesize many materials widely used in biomedical implants such as hip and knee prostheses, heart valves. Among the products of this synthesis process are ceramic and glassy materials as carriers for drug release in many biomedical applications has been extensively reported. Nanoparticles are interesting products prepared by the sol-gel method for biomedical applications, with particular reference to the cancer therapy.

M. Catauro
Dipartimento di Ingegneria, Università degli Studi della Campania Luigi Vanvitelli,
Via Roma 21, 81031 Aversa, Italy
e-mail: michelina.catauro@unicampania.it

S. Vecchio Ciprioti (✉)
Dipartimento di Scienze di Base ed Applicate per l'Ingegneria,
Sapienza Università di Roma, Via del Castro Laurenziano 7, 00161 Rome, Italy
e-mail: stefano.vecchio@uniroma1.it

C. Demetzos and N. Pippa (eds.), *Thermodynamics and Biophysics of Biomedical Nanosystems*, Series in BioEngineering, https://doi.org/10.1007/978-981-13-0989-2_13

1 Introduction

The human body is considerate a machine beautifully engineered, but it has a limited ability to repair its major tissues and organs when a seriously damage occurs such as trauma or different diseases. In the last decades, it was possible to overcome these problems with the development of new materials able to replace tissues or body parts.

The early failure of the implants can be caused often by several reasons and many of them are quite preventable. In many patients, the implants failure can be determined from the rejection of it, due to a reaction by the immune system, but also, it could be causes by mechanical, chemical, tribological, surgical, and biocompatibility problems. For example, in dentistry and orthopedics, after implantation, are recorded inflammation phenomena and frequently infection processes that lead to an unsuccessful osteointegration. Many study reported that the high oxidative stress due to inflammatory process causes an extensive formation of granulation tissue and fibrous capsule around the prostheses, thus inducing an osteolysis process [35, 100]. For this purpose, new materials, called "biomaterials", were developed [91]. The biomaterials are able to interact with human tissue and body fluids to treat, improve, or replace anatomical elements of the human body, without adverse reactions in the host organism [91].

The ideal biomaterials should present a porous structure able to exhibit high tolerability and integration capability to allow regeneration of bone defects. Therefore, a good combination of chemical and mechanical properties, and biocompatibility are required. Various classes have been investigated to produce durable implants, and among them bioactive glasses constitute a promising class of bioactive materials for bone repair and substitution [4, 68]. They have the ability of bonding with the living bone by forming a hydroxyapatite layer on their surface with a composition equivalent to that of the mineral phase of bones [60].

The first bioglass (45S5) was discovered in 1971 by Henchet al. [44] and since then many other glass compositions were explored [90]. Moreover, the results reported in the literature [55, 68] show that, the biological properties of bioglasses are influenced by their composition and the synthesis method as well.

An ideal method to prepare a bioglass is the sol-gel technique. It is a colloidal chemistry technology, which offers the possibility to produce various materials with novel, predefined properties in a simple process. It is a versatile synthesis process used to produce ceramics and glasses materials at low temperatures. The process involves the transition of a system from mostly colloidal liquid ('sol') into a solid 'gel' [7]. The sol-gel method has many advantages, such as the purity of products and, among the most important one, the possibility to incorporate thermolabile molecules. In fact, it is possible to obtain organic-inorganic hybrids materials [124], in which the organic and inorganic phase are bonded together at the nanometer to sub-micrometer scales [15, 120]. In the recent years, glasses, ceramics or drug delivery systems obtained by incorporating polymers or drugs in the inorganic matrix [124], or nanocomposites derived by hybrid materials, all prepared by this method, were proposed to be interesting biomaterials for several pharmaceutical and biomedical applications [27,

78]. They were used, for example, as hip and knee prostheses or heart valves [14, 23, 24].

Many studies have shown that glasses prepared via the sol–gel technique are more bioactive than those with a similar composition but obtained by melting methods [62, 110]. Their bioactivities can be ascribed to the presence of –OH groups on their surface, which are able to induce hydroxyapatite nucleation.

The sol–gel method can be coupled with coating techniques, e.g., dip, spin or spray coating, as in the case of the sol–gel dip-coating technique for biomedical applications, where this procedure was applied to modify the properties of the implants surfaces, thus causing a reduction of corrosion [14] and ion release, and improving bioactivity and biocompatibility [16, 75]. In fact, another event that cause the implant failure is the tendency to release ions (toxic) from metallic materials, when they come in contact with body fluids, determining the lack of integration in bone [23].

Among the products of this synthesis process, ceramic and glassy materials are used as carriers for drug delivery [47, 57, 113]. They have been proposed as implantable therapeutic systems to release drugs (as anti-inflammatory agents, antibiotics and antitumor) to decrease inflammatory processes or infections that can result from their implantation [18, 51, 103]. In particular, the antitumor drugs incorporated in the sol-gel materials are object of many studies. The main problem found in the cancer therapy is the non-selective action of antitumor drugs. In fact, chemotherapy drugs may not kill only cancer cells but they could also give side effects on healthy tissues and organs. Cancer is a complex disease and is the second cause of death in the world. Although in the last decade many works and advances are focused on cancer research, it remains a major public health problem yet. Nanoparticles loaded with antitumor drugs prepared by sol-gel method represent new and more specific cancer treatments. Nanoparticles size, adjustable by controlling the synthesis parameters, plays a crucial role in improving the deep penetration and local accumulation of the drug in the tumor tissues [32, 80].

2 Sol-Gel Method

The sol-gel method is an interesting technique to prepare materials with controlled structures and properties. In fact, it is possible to modulate the operative conditions of the synthesis to obtain different shape materials (Fig. 1).

The starting point to synthesize the sol-gel materials is to understand the chemistry behind this method. In the synthesis process molecular precursors, such as metal alkoxides M(OR)x (where M = Si, Sn, Ti, Zr, Al, Mo, V, W, Ce) are used, which are involved in two important reactions: hydrolysis and polycondensation.

The typical synthesis procedure used to produce bulk gels, film, powers, and nanoparticles, starts when the precursor is added in a solvent, such as alcohol. In the initial stage of synthesis process, the formation of a sol, a stable suspension of colloidal particles in a liquid, is obtained by the first hydrolysis reaction (Eq. 1) [7].

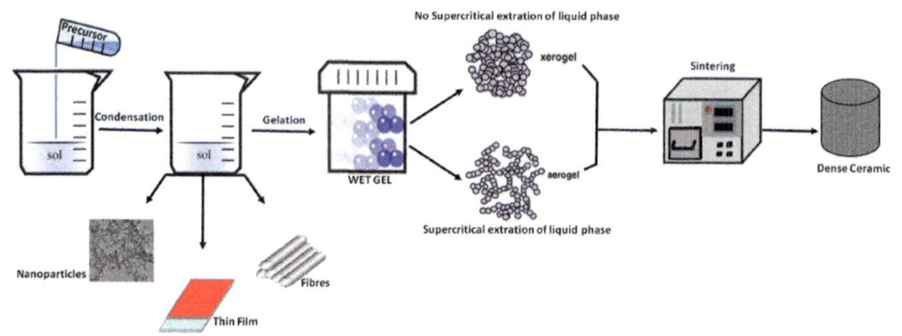

Fig. 1 Scheme of the sol-gel processing

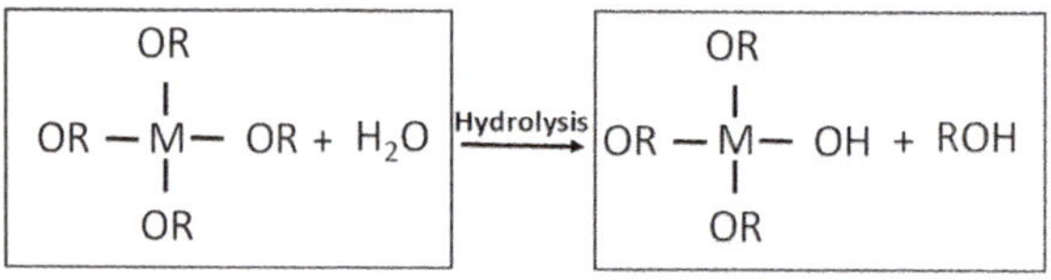

(1)

In this reaction the alkoxide group is replaced with hydroxyl group. Furthermore, condensation starts (Eqs. 2–3) before hydrolysis is completed. The different reactions involve the transition of a system from mostly colloidal liquid ('sol') into a solid 'gel', and it provokes the formation of solid network caused by covalent bonds among the sol particles.

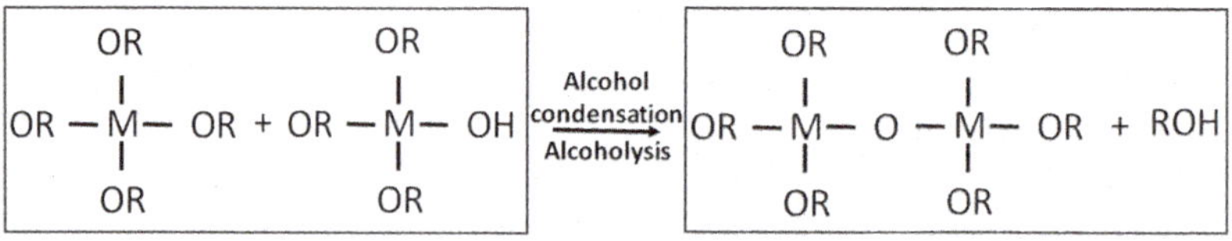

(2)

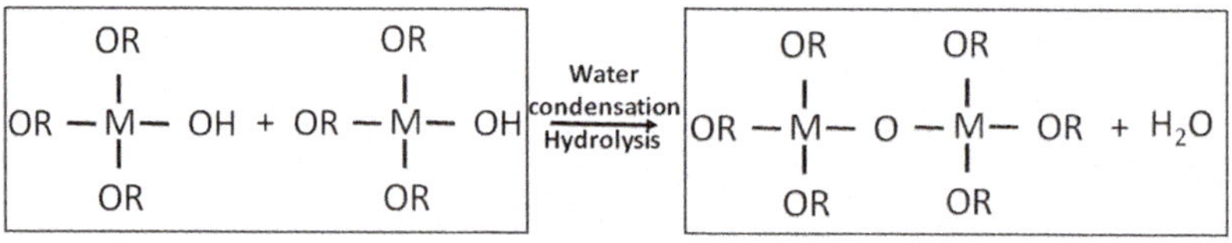

(3)

Many metal alkoxides show a high reactivity toward water and, in some cases, it could lead to the spontaneous formation of precipitates, instead of gels [98]. Generally, to overcome this limit, acetic acid, KOH, amines, KF, HF, titanium alkoxides, and vanadium alkoxides and oxides are used as catalysts. In particular, mineral acids are more effective than the equivalent base concentrations as catalysts.

Furthermore, many parameters involved in the synthesis via sol-gel can affect hydrolysis and condensation rates, thus influencing the structural properties of the synthesized material, such as the kind of precursors, catalysts, molar ratios of reactants, solvent composition and the aging and drying conditions [76].

The solvent, normally, is used as a homogenizing agent, because water and some precursors are immiscible and because it facilitates hydrolysis. Precursors are the building blocks for the final materials, and the increase of their chains length and its branching allows the hydrolysis rate to be decreased. The molar ratio of reactants affects both hydrolysis and condensation reactions. In fact, increasing its value causes the promotion of hydrolysis reaction, its high reaction rate, and lead to completion of hydrolysis reaction before condensation has started. The molar ratio of reactants influences both the time of gelation and the pore size. The higher is the degree of aggregation and condensation, the higher is the viscosity of the sol. The gel point is reached when a continuous network is formed. Before the gel point has been reached, the colloidal dispersion behaves like a more or less viscous liquid. At the gel point, the viscosity increases sharply, and a form-stable, elastic gel body is obtained. The particles grow by aggregation or condensation until they collide to give clusters of particles. The clusters become bigger and bigger by repeated collisions. When the gel is formed, many clusters will be present in the sol phase, but they are not attached to the spanning cluster. Later, they progressively become connected to the network. The gel appears when the last link is formed between two large clusters to create the spanning cluster. It is worth noting that the bond resulting in the formation of the spanning cluster is not different from the previously formed bonds; that is, gelation is not a special thermodynamic event. Actually, it is structurally related to the glass-forming processes. It is very important to realize that the chemical reactions leading to network formation are not finished with gelation, and structural rearrangements take place in the wet gels. This phenomenon increases the stiffness of the gels, and is called aging.

The drying stage is a critical part of the whole sol-gel process, which involves the removal of the liquid phase from the wet gel. As evaporation occurs even drying stresses can cause the cracking of bulk materials. During the drying process, the gel shrinks by the volume that was previously occupied by the liquid that flows from the internal of the gel body to its surface. Upon shrinkage, OH groups at the internal surface approach each other and can react with each other and new M-O-M bridges are formed. As drying proceeds, the network becomes increasingly stiffer and the surface tension in the liquid rises correspondingly, because the pore radii become smaller. Furthermore, it is possible to obtain a xerogel or an aerogel, respectively, using different drying conditions at ambient pressure or in supercritical conditions (Fig. 1).

3 Hybrid Materials

The hybridization of different materials produces new materials with enhanced properties [83, 86]. Organic and inorganic hybrids are materials composed of intimately distributed organic and inorganic phases. Particle size of each phase of the hybrid materials ranges from 1 to 100 nm, with characteristic properties that result in different advantages and limitations [41]. The key aspect to synthesize hybrid materials is keep or enhances the best properties of each of the components, while eliminating or reducing their particular limitations. Although the synthesis of hybrid materials, seems to be recent, a primitive precursor of hybrid composites was found: *Adobe* [41] is a mixture of clay and straw that has been used to make walls in arid regions throughout history, obtaining a material with synergic properties. During the past decades, polymer science has strongly been entering the field of organic–inorganic hybrids, aiming at combining the properties of polymers with those of inorganic components. These hybrid materials based on combinations of polymers with various inorganic hosts, such as inorganic clay compounds [102], metal oxo clusters [9], oligosilsesquioxanes and derivatives [48, 49], mesoporous silica (zeolite) [69], calcium carbonate [123] and metal [40], or inorganic nano-particles [64], have been prepared by different chemical routes and have been exploited in different applications [61].

The simplest method for producing hybrid materials is solution blending. In this method, a polymer is first dissolved in a good solvent and then inorganic nanoparticles have been added, thus producing a dispersion via mechanical stirring or ultrasonication. Although this method is simple and of low cost, it has some limits due to the strong tendency of the inorganic nanoparticles to agglomerate, so as to reduce their surface energy. Nanoparticle agglomeration phenomena lead to materials characterized by a high inhomogeneity and worst physical, chemical, and mechanical properties. For this reason, inorganic nanoparticles are usually introduced in the polymer solution after being functionalized with either organic molecules or macromolecules (Fig. 2) [61].

Another synthesis procedure is the Melt blending that does not require the use of organic solvents, thus rendering it more environmentally friendly compared to the solution blending method. It is performed using the thermal treatment and shearing at temperatures above the polymer-softening point. Many works are reported in literature describing the synthesis of polymer-based hybrids, by using the melt-blending technique, in which the polymers are combined with different types of inorganic components [61].

Moreover, polymers with poor solubility in organic solvents or with high melting temperatures cannot be co-processed with inorganic components, as described in previous techniques [61]. For this reason, another interesting technique is the solid-state processes, which does not require any solvent. Castrillo et al. [10] have been used the high-energy ball milling (HEBM) method to synthesize polymer–inorganic hybrid materials. The authors shown that during the HEBM process, the agglomerates of the filler particles break up into smaller one, obtaining a homogeneous dispersion

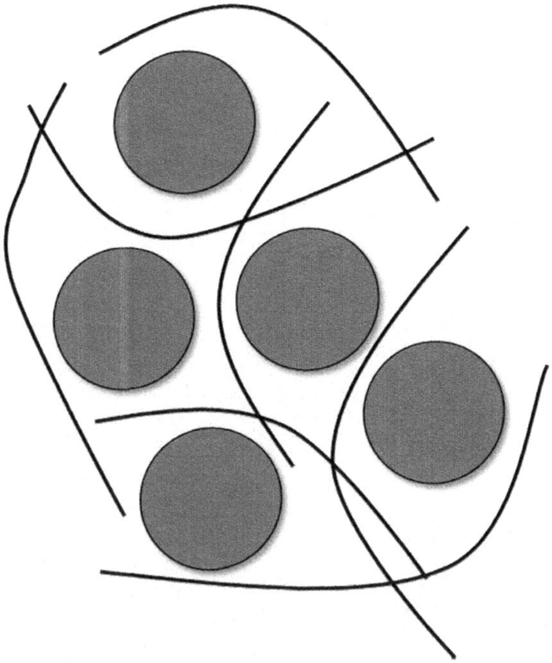

Fig. 2 Inorganic nanoparticles in the polymer solution

inside the polymer matrix. The final material retains the properties of the individual components, and new properties may arise due to synergistic effects between the filler particles and the polymer matrix.

Finally, the sol–gel method, which dates back more than two decades, is still used for the preparation of organic–inorganic hybrids. The main advantages of the sol–gel techniques for preparation of hybrid materials are the low processing temperature and pressure, versatility and flexibility. This technique has been widely employed in the formation and application of hybrid materials [2]. The temperatures used during the synthesis are low and, consequently, the thermal decomposition of organic material itself and any entrapped species is minimized and high purity and stoichiometry can be achieved [121]. The homogeneous sol solutions and highly porous materials are easily achieved. Moreover, it is possible to control the particle size and pore size, porosity and pore wall chemistry of the final material by a chemical modification of the precursors (Fig. 3) [34].

Numerous organic–inorganic hybrids including silica, alumina, titania, germania, and vanadia [50, 126] have been prepared by this method, since the sol-gel process offers a versatile approach to design new organic–inorganic hybrid materials [70, 115]. The obtained hybrid materials show not only both organic and inorganic properties, but also attractive novel properties resulted from hybridization at a nanometer level [83]. They are considered as innovative advanced materials and promisingly

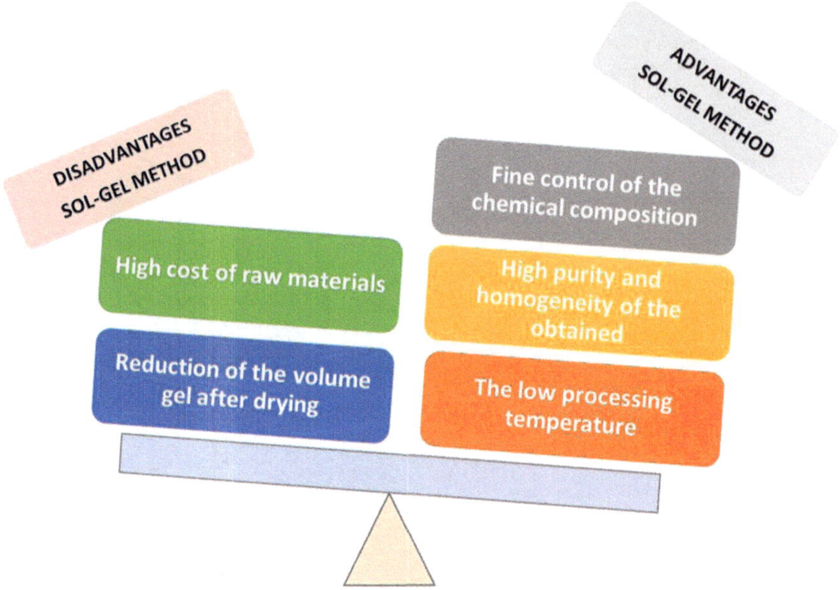

Fig. 3 Disadvantages and advantages of the sol–gel technique

applied for many fields such as optics, coatings, electronics, mechanics, sensors, membranes, and biology [94].

Material properties of hybrids strongly depend on the nature and strength of the interactions between the organic and inorganic components [84]. Judenstein and Sanchez [53] have proposed the classification of the hybrid materials according to the nature of their interactions. The hybrid materials have been divided in two classes (Fig. 4).

Class I indicates hybrids in which no covalent or ionic-covalent bond is formed between the organic and inorganic phases, but each component interacts by only weak bonds, such as hydrogen or van der Waals bond, π-π interactions or electrostatic forces. Class I hybrids can be produced by several approaches. In particular, Catauro et al. has synthesized Class I hybrid materials by the sol gel method [15, 17, 21]. In this class of materials the organic and inorganic phases are linked by hydrogen bonds between the hydroxyl group of the inorganic matrix and amide, carboxyl or carbonyl groups that exist in the polymer chains (Fig. 5). The organic components used were poly-ε-caprolactone (PCL) or polyethylene glycol (PEG), while silicon, titanium and zirconium dioxides (SiO_2, TiO_2 and ZrO_2 respectively) as inorganic matrixes. Gill et al. and Wojcik et al. have produced hybrids with interpenetrating inorganic and organic substance networks [38, 116]. The interpenetrating 3D networks are formed by simultaneous gelation of organic and inorganic components. Through this approach, elastic hydrogel, flexible rubber and hard glass with mesoporous networks have been obtained, accompanied by high biocompatibility [37].

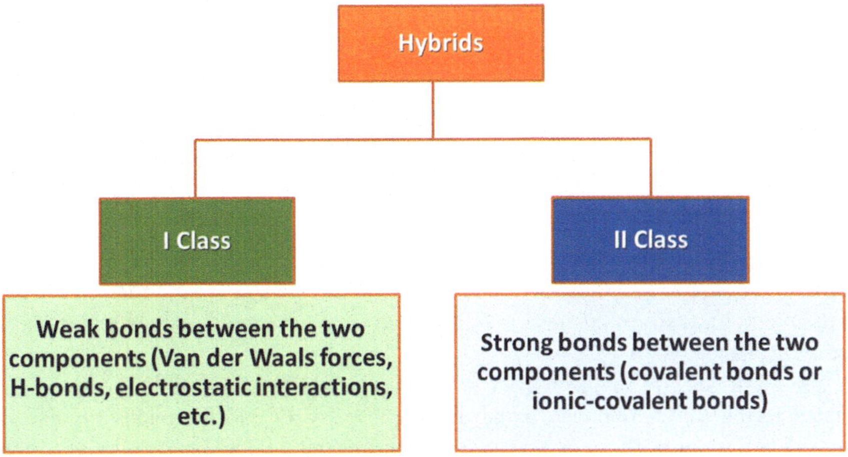

Fig. 4 Flow chart of the hybrid synthesis by the sol-gel method

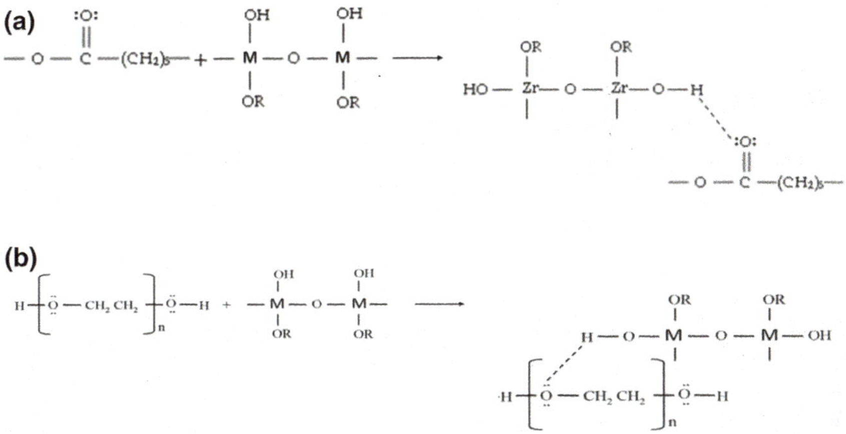

Fig. 5 Hydrogen bonds between the inorganic and organic components **a** PCL and **b** PEG

In Class II hybrid materials, covalent or ionic-covalent bonding connects the organic and inorganic components. This class of materials could be produced using a telechelic polymer such as hydroxyl-terminated poly(dimethylsiloxane) (PDMS) [101, 106].

Hybrid materials could be used in many areas such as optics, electronics, ionics, mechanics, energy, environment, biology, medicine for example as membranes and separation devices, fuel and solar cells, catalysts, sensors and functional smart coatings [2]. For this reason, the properties of inorganic and organic components have been investigated on the basis of their potential and actual applications.

3.1 Chemical Characterization of the Organic-Inorganic Hybrid Materials

Hybrid properties are strongly influenced by the composition of their components, the particles size, interfacial interactions, etc. [122]. For this reason a depth characterization of the chemical structure, microstructure and morphology, as well as of their thermal behaviour of the material is necessary.

The chemical structure of the organic-inorganic hybrids and the interaction between the organic and the inorganic components are generally identified by means of the Fourier transform infrared (FTIR) spectroscopy and solid-state Nuclear Magnetic Resonance (NMR) [3, 17].

The FTIR spectra of the hybrid materials, generally, show the typical signals related to the M–O–M bonds in the inorganic network and the bands typical of the bonds formed in the polymer chains. The intensity of the bands related to the each phases is a function of their relative amount in the hybrids. Moreover, when the condensation reaction is not complete, also the signal of the residue M–OH bonds in the inorganic matrix is visible. FTIR spectra can also provide evidence of the existence of hydrogen or covalent bonds between the organic and inorganic components. Many studies reports the formation of hydrogen bonds between the polymer and the residual M–OH of the inorganic network in the hybrids, proved by FTIR spectra [66, 67].

Solid-state NMR spectra can confirm the results obtained by FTIR analysis and can give further information on the structure of the metal oxide network and on its condensation rate. This technique requires a special equipment made by a magic-angle spinning (MAS), cross polarization (CP), etc. [65]. In the corresponding spectra, the peaks related to the network forming atomic nucleus are generally denoted by the symbol Qn to show the substitution degree of the structural units. For examples, for a ^{29}Si CP MAS NMR spectrum of a silica-based hybrid, *n* represents the number of bridging oxygen bonds formed by the SiO_4 tetrahedral units. Therefore, a sample showing a chemical environment consisting of only Q4 would possess a full condensed metal oxide network phase (in a silica-based hybrid, this case corresponds to stoichiometric SiO_2) [1], whereas the presence of Q3 and Q2 structures reflects that an incomplete condensation occurred.

X-ray techniques and neutron scattering can provide information about microstructure of the hybrid materials, whereas the materials morphology can be analyzed using some microscopy techniques.

In particular, X-Ray Diffraction (XRD), Wide-angle X-ray diffraction (WAXD) or Wide-angle X-ray scattering (WAXS), are widely used in the characterization of hybrid materials. In particular, XRD is an efficient analytical technique to investigate the presence of crystalline phases in the materials and eventually to identify them [22]. On the other hand, WAXD was commonly performed to analyse the degree of crystallinity, which is an important parameter because profoundly influences the physical, mechanical and biological properties of materials [46].

The morphology of the hybrids can be successfully observed using the following powerful microscopy techniques: the transmission electron microscope (TEM), the scanning electron microscope (SEM) and the Atomic Force Microscope (AFM).

TEM is a microscopy technique able to provide information about the inner structure of the analysed specimen. It is based on the transmission of a beam of electrons through an ultrathin specimen. When samples interact weakly with the electron beam, stains such as phosphotungstic acid and RuO_4 can be used to improve the quality of the recorded images. Sometimes, the electron beam can cause the decomposition of the organic components of the sample. To overcome this drawback cryogenic microscopy (cryo-TEM) can be used, where the sample is analysed at liquid nitrogen or liquid helium temperatures in a frozen state [112].

SEM is an electron microscope that allows to record 3D samples images and, thus, is successfully used to analyse the surface structure of the sample [117]. SEM uses a source of electrons once emitted strike the samples surface and are inelastically scattered by atoms in the sample. The detection of such emitted electrons allows creating the sample image. Moreover, in addition to the emitted electrons, also X-rays are produced by the interaction of electrons with the sample. The SEM equipped for energy-dispersive X-ray (EDX) spectroscopy, can provides sample surface microanalysis by detecting the emitted X-rays [58].

AFM is a microscope useful to acquire information about the morphology of the sample surface. It consists of a sharp tip (10–20 nm diameter) attached to a stiff cantilever. The tip is brought closer to the surface and the sample is scanned under the tip. The tip moves in response to a tip-surface interaction, and this movement is measured by focusing a laser beam onto the back of the cantilever, followed by the detection of the position of the reflected beam with a photodiode. AFM is an effective and widely used tool to characterize nanocomposites [77, 92].

X-ray photoelectron spectroscopy (XPS) is an analytical technique used for evaluating surface compositions of a specimen. The sample is placed under high vacuum and is bombarded with X-rays, which penetrate (approximately some nanometers) into the top layer of the sample and exciting the electrons. Some of these electrons from the upper layer are emitted from the sample and can be detected. The electron binding energy is a function of the chemical environment of the atom. Therefore, by using XPS it is possible to identify the elements present on the sample surface of organic-inorganic hybrids [8, 82, 87].

3.2 *Thermal Characterization of the Organic-Inorganic Hybrid*

The structure (porosity, crystallinity degree, etc.) and, thus, the properties of the sol-gel materials are considerably influenced by drying and heating conditions used for the post-synthesis treatments of the obtained gels. During those processes the polycondensation reactions proceed, leading to the formation of further M–O–M bonds and the removal of –OH groups residues in the gel network (dehydroxylation) [7]. In addition, the elimination of the water, solvents and organic modifiers (e.g. catalysts) takes place. Therefore, the drying and heating conditions (e.g. heat treatment temperature, heating rate, etc.) can be modulated to obtain materials with a specific microstructure and, thus, definite properties.

Many studies were reported in literature where it was explained that the heating conditions (that generally results in a different crystallization degree of the final materials) affect optoelectronic [93], magnetic [118], luminescence [6] properties, photocatalytic efficiencies [99], nanoparticle size [97] and many others features of the obtained sol-gel materials.

In particular, the different degree of crystallization influences extensively the biological performance of those materials [5, 20, 25, 105]. Cell adhesion and material bioactivity (the ability of inducing hydroxyapatite nucleation on their surface, if soaked in a simulated body fluid [60], which is related with their osseointegration ability are influenced by the surface topography and the ion release capability [63, 73, 89], which, in turn, depend on the materials structure. Dissolution rate of bioactive glasses plays a key role in their usage as implants because can cause the alterations in the mechanical properties of the materials, as well as the variation in the material surface charge, leading to a modification in the protein adsorption and hydroxyapatite nucleation ability.

Drying and heating conditions take on particular importance when organic-inorganic hybrid materials are synthesized. Those processes, indeed, can cause the degradation of the organic component, modifying irreversibly their properties.

Therefore, the in-depth knowledge of the thermal behavior of the sol-gel materials is essential and needed for the determination of the optimum condition for the calcination of the sol-gel materials [11, 19, 107–109]. This characterization can be effectively obtained by using thermal analysis techniques, which include differential scanning calorimetry (DSC), differential thermal analysis (DTA), thermogravimetry (TG), thermomechanical analysis (TMA), dynamic mechanical analysis/dynamic mechanical thermal analysis (DMA/DMTA), dielectric thermal analysis, etc. In some cases, some techniques are used simultaneously (TG/DTA and TG/DSC) to collect information on the change of more than one of its physical property when the sample is subjected to a specific temperature program under a controlled inert or reacting atmosphere. Simultaneous TG/DTA and TG/DSC apparatuses are recognized to be preferably used to determine the thermal properties of polymer nanocomposites. TG and DTG (first-order derivative of TG data) usually provide the onset degradation temperature, commonly used to assess a thermal stability scale in a series of materials undergoing the same process (described by the same reaction mechanism) [30, 31, 71, 72]. DTA and DSC techniques can be efficiently used to determine (from a qualitative or quantitative point of view, respectively) the thermal effects due to physical or chemical processes (phase transitions or dehydration and thermal decomposition, respectively).

3.3 Biomedical Applications of Hybrid Materials

Biomedical applications of hybrid materials are becoming more and more important as average life expectancy continues to increase and we are outliving our body parts. The sol–gel chemistry offers new and interesting possibilities for encapsulating heat-sensitive and fragile biomolecules such as enzyme, protein, antibody, because it takes advantage from an inherent low temperature and biocompatible process [74], but in aqueous solutions biomolecules lose their functionality. These problems can be minimized by direct immobilization of active biological substances in porous metal oxide carriers by sol–gel processes. This is due to its simplicity of preparation, low-temperature encapsulation, easy immobilization, chemical inertness, tunable porosity, optical transparency, mechanical stability and negligible swelling behaviour [52, 96].

Silica has been widely used as an inert and stable matrix for biological systems immobilization, particularly enzymes, antibodies and proteins. The approach to sol-gel silica loaded with biomolecules has produced important implementations in biotechnology. Betancor et al. [95] describe the advantages of silica matrix for enzyme immobilization, due to its high specific surface areas and controllable pore diameters, which can be tailored to the dimension of a specific enzyme.

However, the sol-gel technique presents a poor loading efficiency and enzyme leakage. The problem can be overcame by designing protocols for the preparation of matrices with a pore size that is adequate to allow the flow of substrates and products, but small enough to prevent the elution of the entrapped biomolecules [39].

In addition to the enzymes, the proteins have also been encapsulated in sol–gel derived silicate materials. Entrapped proteins typically reside in pores of size comparable to that of the protein. Thus, it is important to observe whether entrapped proteins maintain their native conformation during and after the entrapment, and whether they are able to undergo changes in conformation once entrapped. This aspect is very important, because in many cases the binding of an analyte to a protein requires that the protein is able to undergo conformational changes [33]. When the proteins are entrapped in the sol-gel matrix, they should remain functional, because their function depends on their conformational structure. The function of an entrapped protein depends on a number of factors such as the protein location and protein structure. Das et al. [29] indicated that such proteins tend to retain a native conformation immediately upon the entrapment, although some proteins, such as myoglobin, may undergo substantial conformational changes during the process.

Another example of bioencapsulation of biomolecules by the sol-gel technique is the antibodies encapsulated. Generally antibodies are high molecular weight proteins, if antigen is also high molecular weight compound, so that the interactions between antibody and antigen are difficult through the small pores of the matrix. Yang et al. [119] reported that the addition of PEG did not affect the encapsulation efficiencies of the antibody, but determined a strong effect on the binding activity of the encapsulated antibody. In fact, a high percentage of antibodies encapsulated with PEG is bound to the column compared to the antibody without PEG.

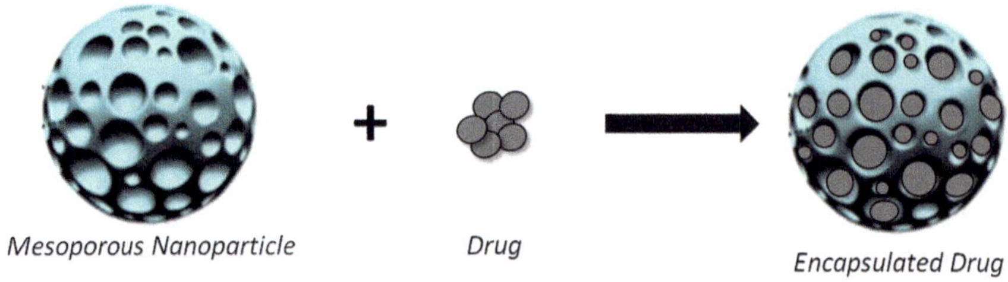

Fig. 6 Scheme of the drug encapsulation in an inorganic matrix

The versatility of the sol gel method allows loading also anti-thrombotic, antiproliferative, anti-inflammatory, anti migratory, and anesthetic agents, anti-coagulants, cell growth promoters, cell growth inhibitors and other pharmaceutical drugs in an inorganic matrix, such as mesoporous silica-based materials. The drug encapsulations in the hybrid organic-inorganic matrices are influenced by few parameters, such as porosity degree and specific surface of the pores, the organic component amount in the synthesis and the aging/drying conditions, (Fig. 6) [79].

The goal of an ideal drug-delivery system is to deliver the therapeutic amount of a drug to the proper site in the body to achieve promptly, and to maintain, the desired drug concentration [111]. The interactions between drugs and the sol-gel system can affect the release rate. The strength of these interactions is given by the functional groups of the main component of the system [88]. The release kinetics in the sol-gel system could follow the Higuchi model [42, 81]. This model is used both for water- and low-soluble drugs incorporated in solid/semi-solid matrix. If the matrix includes biodegradable polymers in biological medium, governed by the local enzymes, the drug release may occur via a degradation mechanism [59]. When the drug is released due to osmotic swelling, the kinetics release fits to Fick's law. In this case, the water penetrates into the drug carrier system, inducing relaxation of polymer chain and thus, the drug is released outside. An advantage in drug release is due to a porous morphological aspect, obtained by controlled drying or by using a modified catalyst molar ratio during the synthesis [56].

During the past decades, many newly designed drugs are based on biomolecules such as peptides, proteins, and DNA. These molecules are chemically unstable, thus encapsulating and protecting them can be useful in order to deliver drug molecules to the target tissues [114]. Furthermore, many cancer drugs have poor solubility and instability Therefore, the development of delivery systems that can carry a high payload of drug, protect the drug from degradation, facilitate cellular uptake and target specific cell populations is necessary for the clinical applicability of many drugs [43].

The majority of studies have focused on the delivery of anticancer drugs such as doxorubicin (DOX). Lee et al. [67] delivered DOX to tumor sites and observed the apoptosis ex vivo in the tumor cells. Hillegass et al. [45] applied similar sol–gel based silica nanoparticles as carriers for DOX to treat malignant mesothelioma.

DOX-loaded and nanoparticles without drug were injected both subcutaneously and directly into the tumor or intraperitoneally in an in vivo mouse model [45].

Finally, sol-gel materials can be used to realize medical diagnostic tools that require biocompatibility, nontoxicity, bioactivity and/or drug delivery properties. The medical devices must be without inflammatory tissue responses or immune reactions, if they are used like implant materials such as surgical and/or orthopedic devices, artificial heart-valves [66]. Some results shows that the use of polymers as substrates improves the properties of devices, because the polymeric part can hold the drug onto the surface of implantable medical devices, and controlled drug release occurs via degradation of the polymer or diffusion into the liquid or living tissue [85].

4 Case Studies: Details on Hybrid Materials Prepared and Characterized

4.1 Sol-Gel Synthesis of Organic–Inorganic Hybrids

Organic/inorganic hybrid materials are attracting considerable attention in the biomedical area. Catauro et al. [12, 15, 17, 21, 25, 107, 109] have prepared several hybrid materials by the sol-gel method with different inorganic matrices and different polymers. In particular, they have used a variety of oxides: silicon, titanium or zirconium dioxides (SiO_2, TiO_2 or ZrO_2), in which PCL or PEG have been incorporated at different concentrations.

Interesting results have been obtained with PEG/ZrO_2 hybrids [17, 24]. PEG has been incorporated with different percentage (6, 12, 24, 50, 60 and 70 wt%) in the zirconia matrix (Table 1).

A zirconium(IV) propoxide solution ($Zr(OC_3H_7)_4$ 70 wt% in n-propanol, Sigma Aldrich) and PEG (purchased by Sigma-Aldrich with average molar mass of about 400 g mol^{-1}) were, respectively, used as precursors of the inorganic and organic phases. The sol-gel synthesis starts when the precursor of zirconia was added to

Table 1 Variation in the gelation time controlled by changing the concentration of PEG

Inorganic phases	Organic phases	Gelation time (20–25 °C) (days)
ZrO_2	–	16
ZrO_2	6%wt PEG	16
ZrO_2	12%wt PEG	14
ZrO_2	24%wt PEG	12
ZrO_2	50%wt PEG	10
ZrO_2	60%wt PEG	9
ZrO_2	70%wt PEG	7

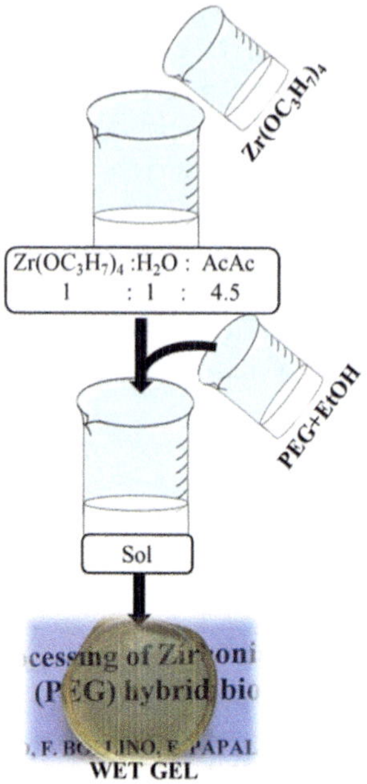

Fig. 7 Flow chart of the hybrid synthesis by the sol-gel method

a mixture of ethanol, water and acetylacetone (AcAc, Sigma-Aldrich) with molar ratios $Zr(OC_3H_7)_4$:H_2O:AcAc = 1:1:4.5. Acetylacetone was used to control the hydrolytic activity of zirconium alkoxide. Successively, PEG, previously dissolved in ethanol, was added to the solution under vigorous stirring to obtain a uniform and homogeneous sol. After gelation the wet gels were air-dried at 45 °C for 48 h to remove the residual solvent. Figure 7 shows the flow chart of the hybrid synthesis by the sol-gel method.

4.2 *Thermal Behavior Study and FTIR Identification of Degradation Mechanisms*

The thermal behavior of the pure ZrO_2, HAp and composites was investigated by using a simultaneous Mettler Toledo TG/DSC 2950 instrument, and processing the experimental raw data by a STARe software. The instrument has been equipped with two identical crucibles, one for the reference filled with alumina in a powder form and one for the sample, containing about 20–25 mg of solid in order to uniformly cover the bottom surface area of the crucible. The TG experiments were carried out under a purging nitrogen atmosphere at a flow rate of 60 mL min^{-1} from ambient temperature to 800 °C at a heating rate of 10 °C min^{-1}. The sample temperature was calibrated by performing identical experiments using very pure indium and zinc reference materials (purity higher than 99.998%), and assuming a final average uncertainty u(T) = ±1 K over the whole temperature range explored.

The mechanisms of thermally stimulated processes occurring during the TG/DSC experiments were identified by a SETARAM 92-16.18 TG apparatus coupled by a Thermofisher Scientific Nicolet iS10 Spectrophotometer. The apparatus has been equipped with 250 μL alumina crucibles, filled with about 100–150 mg of sample to obtain the minimum amount of gaseous species to be analyzed by FTIR using the Calisto software. A preliminary blank experiment was performed using empty crucibles under the same experimental conditions of the samples tested (argon purging gas at 40 mL min^{-1}, in the temperature range between 25 and 1200 °C at a constant heating rate of 10 °C min^{-1}). The gas mixture evolved during the TG experiments were conveyed to the FTIR unit through a heated transfer line kept at 200 °C. The instrument is able to collect a spectrum each 11 s, being eight scans performed at 0.5 cm^{-1} intervals with a resolution of 4 cm^{-1}.

The results of thermal behavior of pure ZrO_2 and the six ZP hybrid materials collected from TG and DTA experiments were summarized in Table 2.

The first step for all the materials is ascribed to dehydration. The highest amount of water is found in Z, followed by the PEG-poor hybrids, ZP6 and ZP12, while lower content was found for the PEG-rich hybrids (ZP24, ZP50, ZP60 and ZP70). After dehydration Z undergoes dehydroxylation, in which water is released through condensation of hydroxyl surface groups, similarly to what has been recently observed for pure SiO_2 [109]. Under inert atmosphere this process is very slow and takes place in a very wide temperature range from 200 to about 600 °C [125]. By contrast, all ZP hybrid materials undergo PEG decomposition between 200 and 350 °C. This reaction takes place mainly in one step of mass loss, and is simultaneously accompanied by some endo-exothermic effects. The thermal degradation of pure PEG under inert atmosphere was found in previous studies to take place at about 423 K [104]. As PEG interacts with the ZrO_2 matrix, higher temperatures are required for its degradation when it is embedded in ZP materials.

The thermal degradation process takes place through pyrolysis reactions, which follow a radical chain mechanism initiated by random scissions of C–C or C–O bonds. A third step of mass loss is found after decomposition of PEG, with percentages ranging from about 7 (for sample Z) to 15 (for sample ZP70). The onset temperature is shifted toward lowed values with increasing the PEG content. The exothermic effect associated to this process proves that this mass loss step is not imputable to PEG degradation, but to a decomposition/oxidation of AcAc [36], which leads to the formation of both gaseous and non-volatile products. Therefore, a limited mass loss is found in this range because a low amount of AcAc was used in the sol-gel synthesis and because a solid residue is formed. However, the higher the content of PEG the higher the mass loss recorded. When the AcAc forms coordination complexes with metals, indeed, its enol form is favored over the diketone one, so the –OH terminal groups of PEG chains can bond the alpha carbons of the AcAc molecules [54]. Therefore, the thermal degradation of this PEG-AcAc compound leads to products with molecular weight higher than those formed in the PEG-free sample (Z). The

Table 2 Mass loss percentages and temperature corresponding each step of mass loss obtained from the TG curves

Material	T range/°C	Δm/%	Thermal effect	Onset T/°C
Z	15–200 200–400 400–600	19.7 9.9 6.6	Dehydration/Endo Dehydroxylation/Endo Decomposition-oxidation of AcAc/Exo	82 255 461
ZP6	16–191 200–350 400–600	13.8 16.9 8.2	Dehydration/Endo Decomposition of PEG/Endo + Exo Decomposition-oxidation of AcAc/Exo	79 299 488
ZP12	15–188 200–350 400–600	13.5 19.0 8.1	Dehydration/Endo Decomposition of PEG/Endo + Exo Decomposition-oxidation of AcAc/Exo	77 296
ZP24	15–189 200–350	11.4 28.3	Dehydration/Endo Decomposition of PEG/Endo + Exo	74 291
ZP50	16–188 200–350	11.5 40.8	Dehydration/Endo Decomposition of PEG/Endo + Exo	74 287
ZP60	15–195 200–350	11.7 48.5	Dehydration/Endo Decomposition of PEG/Endo + Exo	72 274
ZP70	16–188 200–350	11.6 56.1	Dehydration/Endo Decomposition of PEG/Endo + Exo	81 278

exothermic effects observed in this temperature range are ascribable also to the formations of tetragonal zirconia nanocrystals upon heating.

Furthermore, in order to establish how PEG may resist to its degradation when the polymer interacts with the inorganic ZrO_2 matrix, the onset degradation temperature (Tonset) of the ZP materials has been also reported in Table 2, with the view to find a correlation with the increasing content of PEG, similarly to what has been recently observed in other PEG/clay nanotubes composites (at least starting from a given amount of PEG) [26]. As in other PEG-based organic-inorganic hybrids [109], a decreasing trend of the T_{onset} versus PEG content is observed. This result could be probably explained by the fact that when a high amount of polymer is added into the sol-gel matrix, part of it cannot form H-bonds with the matrix because a sufficient number of –OH groups is not present to bond all PEG chains. Therefore, the PEG non-linked to the inorganic matrix is not affected by the corresponding stabilizing

Table 3 Masses of the residues obtained from the TG curves at 800 °C

Material	Residual mass/%
Z	70.5
ZP6	67.8
ZP12	64.5
ZP24	53.9
ZP50	34.7
ZP60	28.0
ZP70	18.3

effect and degradation occurs at lower temperature [109]. Table 3 shows the sample mass of the residue after the TG experiments for the different content of PEG.

Except for pure zirconia in which PEG is absent, a decreasing trend of the sample masses of the residue (reasonably represented by ZrO_2) for all hybrids is observed after the complete degradation of the polymer with increasing the PEG content.

4.3 Morphological and Structural Characterization of Organic–Inorganic Hybrids

Figure 8 shows the FTIR spectra of PEG/ZrO_2, pure zirconia and PEG-based hybrid materials. The spectrum of ZP6 (PEG/ZrO_2 system with 6 wt% of PEG) is very similar to the spectrum of pure zirconia, because the concentration of polymer is extremely low. In fact, when the content of PEG increases, the recorded spectra are more similar to the spectrum of pure PEG. In particular, the signal of the ethereal C–O–C stretching at 1104 cm^{-1} is visible already in ZP12 and ZP24, while the C–H bending of the polymer at 1454 cm^{-1} [28] appears together with the bands due to alcohols (C–O stretching at 1250 cm^{-1} in the samples containing higher amount of PEG. In the spectra of ZP60 and ZP70 (containing 60 and 70 wt% of PEG, respectively) the intensities of the bands at 1585 and 1529 cm^{-1} decrease and appear slightly up shifted compared with that of the spectrum of ZP50. This observation was ascribed to the formation of interactions between PEG and AcAc, which increase with the polymer content. Furthermore, the formation of H bonds between the hydroxyl group of the inorganic matrix and the PEG chains is suggested by the change in the shape of the band assigned to Zr–OH stretching at 654 cm^{-1} and of the broader band of the –OH at 3440 cm^{-1} [107].

For this reason the obtained materials can be classified as class I hybrids [53]. The NMR analysis (Fig. 9) confirms the FTIR data. The small shoulder at about 72 ppm is due to the weak interactions between the carbons chain and the inorganic component, while the peak at about 71 ppm is attributed to the ether carbons belonging to the polymer chains, and the smaller peak at 61 ppm is due to the carbons connected to the –OH terminal groups [17].

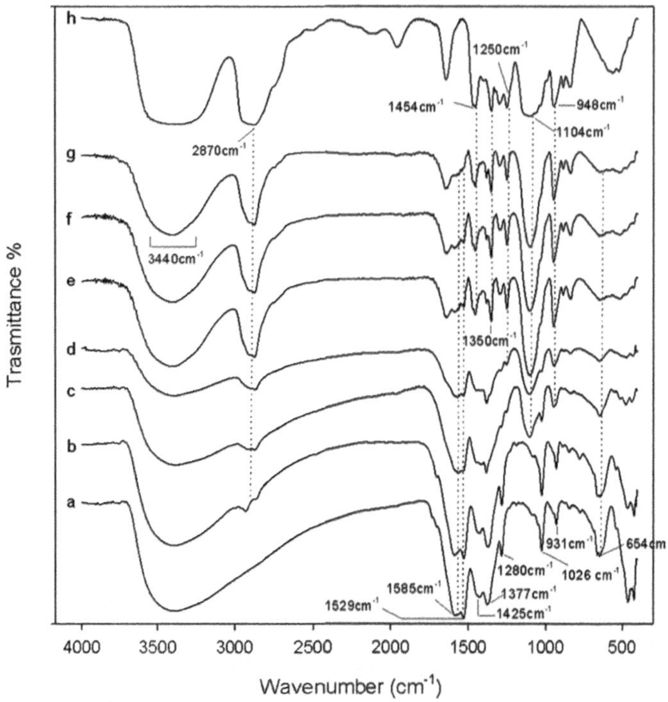

Fig. 8 FT-IR spectra of Z (**a**), ZP6 (**b**), ZP12 (**c**), ZP24 (**d**), ZP50 (**e**), ZP60 (**f**), ZP70 (**g**) and pure PEG (**h**)

Fig. 9 ^{1}H-^{13}C CPMAS NMR spectrum of the hybrid materials

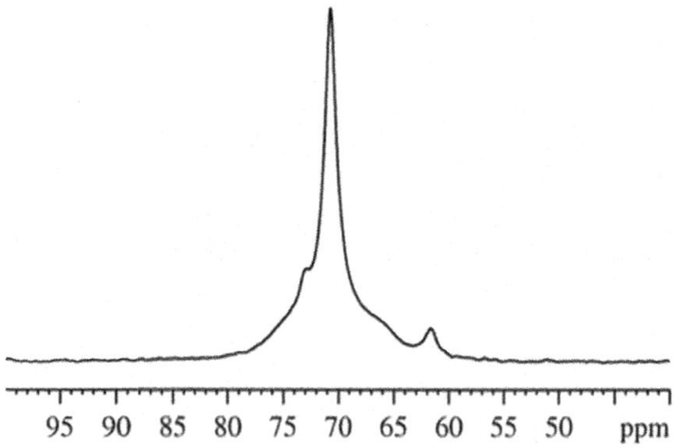

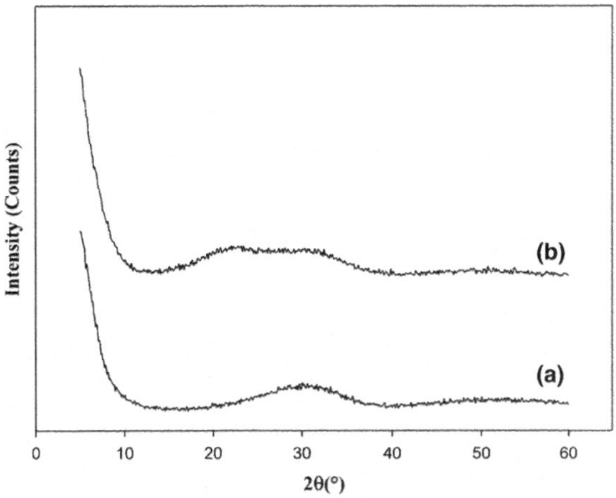

Fig. 10 XRD spectra of **a** ZrO_2 and **b** ZrO_2/70 wt% PEG

The XRD spectra of pure zirconia and ZP70 are reported in Fig. 10. The pure zirconia (curve a) exhibits a broad hump, which is characteristic of amorphous materials, while the same broad hump of pure zirconia is visible in that of ZP70 (curve b). These results suggest that peaks related to the crystals of PEG are not detected in the spectra of all hybrid materials regardless the amount of polymer [13, 17].

The SEM micrographs of the cross section of the hybrid materials show that the PEG-poor hybrids (ZP6, ZP12) and pure zirconia had similar morphologies and appeared as homogenous materials (Fig. 6). By contrast, the PEG-rich hybrids, namely ZP60 and ZP70, appear different from those with low concentration of PEG: the fiber-like structures disappear and they appear very rough, suggesting that modulation of the polymer content can be a tool to modify the microstructure of such hybrids. However, both kind of samples appears homogeneous and no phase separation is visible (Fig. 11).

The biological properties of the hybrid materials were also evaluated. To study their bioactivity, the hybrids were soaked in a simulated body fluid (SBF) with ion concentrations, nearly equal to those of the human blood plasma. During soaking the temperature was kept constant at 37 °C for 7, 14 and 21 days. Formation of the apatite layer was monitored by SEM/EDX microscopy observations on the materials after soaking in SBF [24, 60].

After 21 days of incubation, all the hybrid materials were observed with SEM and exhibited a similar deposition of apatite on their surface (Fig. 12). The EDX analysis of crystals showed an atomic ratio Ca/P equal to 1:6, which agrees with the chemical formula of hydroxyapatite [$Ca_{10}(PO_4)_6(OH)_2$]. The formation of the hydroxyapatite on the materials can be explained by the presence of surface Zr–OH groups on the hybrids, which interact with the Ca^{2+} ions present in the SBF. The Ca^{2+} ions combine with the negative charge of the phosphate ions to form amorphous phosphate, which is spontaneously converted into hydroxyapatite. These results suggest that the PEG/ZrO_2 hybrids are bioactive, regardless the polymer content.

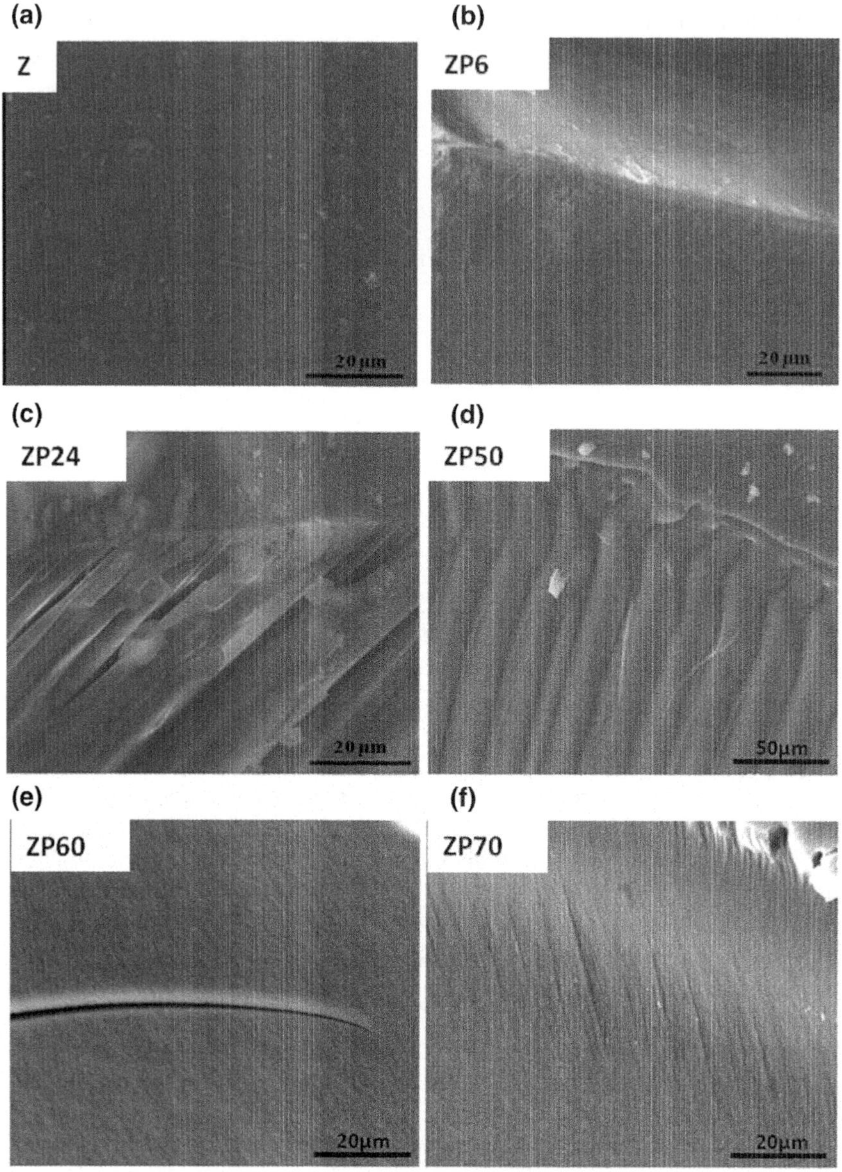

Fig. 11 Comparison of SEM images of Z (**a**), ZP50 (**b**), ZP60 (**c**), ZP70 (**d**)

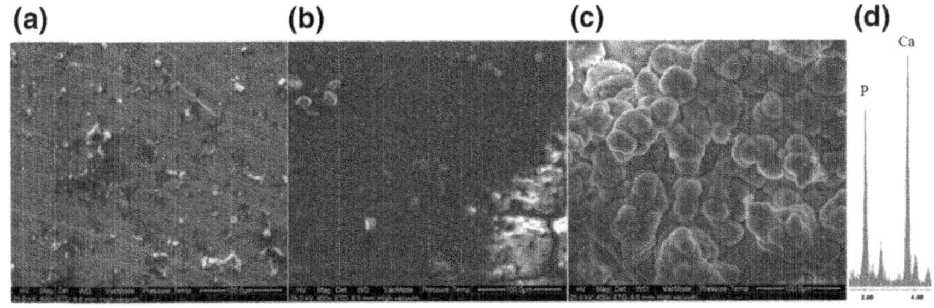

Fig. 12 SEM micrographs of samples after soaking in SBF for **a** 7 days, **b** 14 days, **c** 21 days. Panel D: EDX analysis

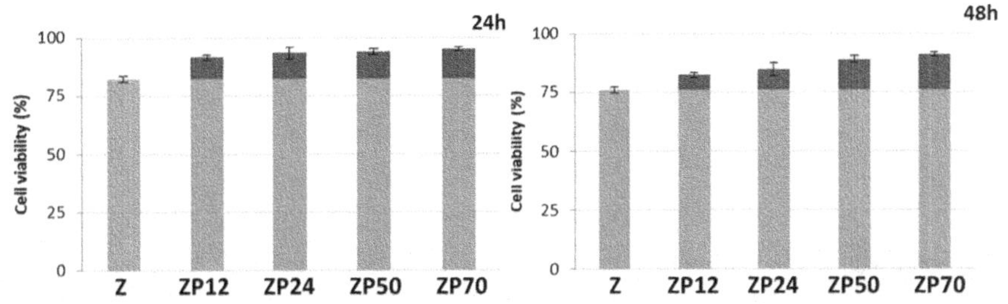

Fig. 13 Effects of the increasing amount of PEG contained in the hybrid materials (12, 24, 50, 70 wt%) on cell viability towards NIH 3T3 cell line at 24, 48 h exposure times (100% were considered as an untreated control cells)

Biocompatibility was evaluated on NIH-3T3 murine fibroblast cell line by conventional MTT assay. Cell viability was expressed as percentage of mitochondrial redox activity of the cells treated with the materials compared to an untreated control. Fibroblasts were grown for 24 and 48 h in absence and presence of the pure ZrO_2 and PEG/ZrO_2 hybrid materials, each containing different percentages of polymer [24].

Figure 13 shows that all the hybrids tested were able to induce an increase in the cell viability, compared to what observed in pure zirconia when it was tested in the non-hybrid form. The data allowed observing that the effects were dose- and time-dependent. The presence of different amounts of polyethylene glycol has promoted the preparation of biomaterials with improved biocompatibility.

5 Conclusions

The sol–gel technique has proved to be a valid system to synthesize hybrids materials. The materials synthesized and presented in this chapter show good qualities such as to be used in the biomedical field. The structure and microstructure of PEG/ZrO_2 hybrid materials, with different percentage of polymer, have been investigated by means of spectroscopic techniques (FT-IR and solid state NMR), XRD and SEM microscopy. It has been shown that amorphous class I organic–inorganic hybrid materials have been obtained. Indeed, FT-IR and solid state NMR have ascertained that the polymer (PEG) was incorporated into the inorganic network by weak bonds (likely H-bonds) between the organic polymer and the hydroxyl groups of the inorganic matrix. Moreover, the formation of a layer of hydroxyapatite on the surface after soaking samples in SBF, which is crucial for implant osteointegration in vivo, shown by SEM and related EDX, indicates that the PEG/ZrO_2 can be considered a bioactive materials.

The presence of different amounts of polyethylene glycol in hybrid materials showed a positive behavior on cell growth and proliferation of NIH 3T3 cells. Their effects were superior to ZrO_2, whose ceramics are already widely used for wear applications in joint replacements, it is due to the dependence between cell proliferation and the different percentages of the polymer in the several hybrid materials. These results suggest that the polymer incorporated in the inorganic matrix favors the preparation of biomaterials with improved biocompatibility.

References

1. Akhter, T., Saeed, S., Siddiqi, H.M., Park, O.O., Ali, G.: Synthesis and characterization of novel coatable polyimide-silica nanocomposites. J. Polym. Res. **21**(1), 332 (2013). https://doi.org/10.1007/s10965-013-0332-0
2. Amiri, S., Rahimi, A.: Hybrid nanocomposite coating by sol–gel method: a review. Iran. Polym. J. **25**(6), 559–577 (2016)
3. Atal, M.K., Saini, A., Jat, S.K., Rathore, K.S., Dhayal, V.: Synthesis and characterization of oxime-modified phenylimido vanadium(V) isopropoxide and their hydrolytic study. J. Sol-Gel. Sci. Technol. **83**, 281–290 (2017). https://doi.org/10.1007/s10971-017-4423-y
4. Bizari, D., Rabiee, M., Moztarzadeh, F., Tahriri, M., Alavi, S.H., Masaeli, R.: Synthesis, characterization and biological evaluation of sol-gel derived nanomaterial in the ternary system 64% SiO_2—31% CaO—5% P_2O_5 as a bioactive glass: in vitro study. Ceram.-Silik. **57**, 201–209 (2013)
5. Bollino, F., Armenia, E., Tranquillo, E.: Zirconia/hydroxyapatite composites synthesized via sol-gel: influence of hydroxyapatite content and heating on their biological properties. Materials **10**(7) (2017). https://doi.org/10.3390/ma10070757
6. Boukerika, A., Guerbous, L.: Annealing effects on structural and luminescence properties of red Eu^{3+}-doped Y_2O_3 nanophosphors prepared by sol–gel method. J. Lumin. **145**, 148–153 (2014). https://doi.org/10.1016/j.jlumin.2013.07.037
7. Brinker, C., Scherer, G.: Sol-Gel Science: The Physics and Chemistry of Sol-Gel Processing. Academic press, San Diego (1989)
8. Brusciotti, F., Snihirova, D.V., Xue, H., Montemor, M.F., Lamaka, S.V., Ferreira, M.G.S.: Hybrid epoxy–silane coatings for improved corrosion protection of Mg alloy. Corros. Sci. **67**, 82–90 (2013). https://doi.org/10.1016/j.corsci.2012.10.013
9. Carraro, M., Gross, S.: Hybrid materials based on the embedding of organically modified transition metal oxoclusters or polyoxometalates into polymers for functional applications: a review. Materials **7**(5), 3956–3989 (2014). https://doi.org/10.3390/ma7053956
10. Castrillo, P., Olmos, D., Amador, D., González-Benito, J.: Real dispersion of isolated fumed silica nanoparticles in highly filled PMMA prepared by high energy ball milling. J. Colloid Interface Sci. **308**(2), 318–324 (2007)

11. Catauro, M., Bollino, F., Dell'Era, A., Ciprioti, S.V.: Pure $Al_2O_3 \cdot 2SiO_2$ synthesized via a sol-gel technique as a raw material to replace metakaolin: chemical and structural characterization and thermal behavior. Ceram. Int. **42**(14), 16303–16309 (2016). https://doi.org/10.1016/j.ceramint.2016.07.179
12. Catauro, M., Bollino, F., Giovanardi, R., Veronesi, P.: Modification of Ti6Al4V implant surfaces by biocompatible TiO_2/PCL hybrid layers prepared via sol-gel dip coating: structural characterization, mechanical and corrosion behavior. Mater. Sci. Eng. C **74**, 501–507 (2017)
13. Catauro, M., Bollino, F., Papale, F.: Biocompatibility improvement of titanium implants by coating with hybrid materials synthesized by sol–gel technique. J. Biomed. Mater. Res. Part A **102**(12), 4473–4479 (2014)
14. Catauro, M., Bollino, F., Papale, F., Giovanardi, R., Veronesi, P.: Corrosion behavior and mechanical properties of bioactive sol-gel coatings on titanium implants. Mater. Sci. Eng. C **43**, 375–382 (2014). https://doi.org/10.1016/j.msec.2014.07.044
15. Catauro, M., Bollino, F., Papale, F., Marciano, S., Pacifico, S.: TiO_2/PCL hybrid materials synthesized via sol–gel technique for biomedical applications. Mater. Sci. Eng. C **47**, 135–141 (2015)
16. Catauro, M., Bollino, F., Papale, F., Mozetic, P., Rainer, A., Trombetta, M.: Biological response of human mesenchymal stromal cells to titanium grade 4 implants coated with PCL/ZrO_2 hybrid materials synthesized by sol-gel route: in vitro evaluation. Mater. Sci. Eng. C **45**, 395–401 (2014). https://doi.org/10.1016/j.msec.2014.09.007
17. Catauro, M., Bollino, F., Papale, F., Mozzati, M.C., Ferrara, C., Mustarelli, P.: ZrO_2/PEG hybrid nanocomposites synthesized via sol–gel: characterization and evaluation of the magnetic properties. J. Non-Cryst. Solids **413**, 1–7 (2015)
18. Catauro, M., Bollino, F., Papale, F., Pacifico, S.: Modulation of indomethacin release from ZrO_2/PCL hybrid multilayers synthesized via sol–gel dip coating. J. Drug Deliv. Sci. Technol. **26**, 10–16 (2015). https://doi.org/10.1016/j.jddst.2014.12.004
19. Catauro, M., Bollino, F., Papale, F., Vecchio Ciprioti, S.: Investigation on bioactivity, biocompatibility, thermal behavior and antibacterial properties of calcium silicate glass coatings containing Ag. J. Non-Cryst. Solids **422**, 16–22 (2015). https://doi.org/10.1016/j.jnoncrysol.2015.04.037
20. Catauro, M., Bollino, F., Renella, R.A., Papale, F.: Sol-gel synthesis of SiO_2–CaO–P_2O_5 glasses: influence of the heat treatment on their bioactivity and biocompatibility. Ceram. Int. **41**(10), 12578–12588 (2015). https://doi.org/10.1016/j.ceramint.2015.06.075
21. Catauro, M., Bollino, F., Veronesi, P., Lamanna, G.: Influence of PCL on mechanical properties and bioactivity of ZrO_2-based hybrid coatings synthesized by sol–gel dip coating technique. Mater. Sci. Eng. C **39**, 344–351 (2014)
22. Catauro, M., Dell'Era, A., Vecchio Ciprioti, S.: Synthesis, structural, spectroscopic and thermoanalytical study of sol-gel derived SiO_2–CaO–P_2O_5 gel and ceramic materials. Thermochim. Acta **625**, 20–27 (2016). https://doi.org/10.1016/j.tca.2015.12.004
23. Catauro, M., Papale, F., Bollino, F.: Coatings of titanium substrates with $xCaO(1{-}x)SiO_2$ sol-gel materials: characterization, bioactivity and biocompatibility evaluation. Mater. Sci. Eng. C Mater. Biol. Appl. **58**, 846–851 (2016). https://doi.org/10.1016/j.msec.2015.09.033
24. Catauro, M., Papale, F., Bollino, F., Gallicchio, M., Pacifico, S.: Biological evaluation of zirconia/PEG hybrid materials synthesized via sol-gel technique. Mater. Sci. Eng. C **40**, 253–259 (2014). https://doi.org/10.1016/j.msec.2014.04.001
25. Catauro, M., Renella, R., Papale, F., Ciprioti, S.V.: Investigation of bioactivity, biocompatibility and thermal behavior of sol–gel silica glass containing a high PEG percentage. Mater. Sci. Eng. C **61**, 51–55 (2016)
26. Cavallaro, G., Lazzara, G., Milioto, S.: Sustainable nanocomposites based on halloysite nanotubes and pectin/polyethylene glycol blend. Polym. Degrad. Stab. **98**(12), 2529–2536 (2013)
27. Chellappa, M., Thejaswini, B., Vijayalakshmi, U.: Biocompatibility assessment of SiO_2-TiO_2 composite powder on MG63 osteoblast cell lines for orthopaedic applications. IET Nanobiotechnol. **11**, 77–82 (2017)
28. Coates, J.: Interpretation of infrared spectra, a practical approach. Encycl. Anal. Chem. (2000)

29. Das, T.K., Khan, I., Rousseau, D.L., Friedman, J.M.: Preservation of the native structure in myoglobin at low pH by sol−gel encapsulation. J. Am. Chem. Soc. **120**(39), 10268–10269 (1998)
30. De Angelis Curtis, S., Kubiak, M., Kurdziel, K., Materazzi, S., Vecchio, S.: Crystal structure and thermoanalytical study of a cadmium(II) complex with 1-allylimidazole. J. Anal. Appl. Pyrol. **87**(1), 175–179 (2010). https://doi.org/10.1016/j.jaap.2009.11.007
31. De Angelis Curtis, S., Kurdziel, K., Materazzi, S., Vecchio, S.: Crystal structure and thermoanalytical study of cobalt(II) and nickel(II) complexes with 2,2′-bis-(4,5-dimethylimidazole). Thermochim. Acta **510**(1–2), 75–81 (2010). https://doi.org/10.1016/j.tca.2010.06.025
32. de Oliveira, L.F., Bouchmella, K., Goncalves, KdA, Bettini, J., Kobarg, J., Cardoso, M.B.: Functionalized silica nanoparticles as an alternative platform for targeted drug-delivery of water insoluble drugs. Langmuir **32**, 3217–3225 (2016). https://doi.org/10.1021/acs.langmuir.6b00214
33. dos Barbosa, A.S., de Silva M.A.O., Carvalho, N.B., Mattedi, S., Iglesias, M.A., Fricks, A.T., Lima, A.S., Franceschi, E., Soares, C.M.F.: Immobilization of lipase by encapsulation in silica aerogel. Quim Nova **37**, 969–976 (2014). https://doi.org/10.5935/0100-4042.20140155
34. Druart, M.E., Recloux, I., Thai, T.T., Ershov, S., Snyders, R., Olivier, M.G.: Impact of the addition of cerium salts (Ce(III) and Ce(IV)) on formation and ageing of a silica sol-gel layer. Surf. Coat. Technol. **304**, 40–50 (2016). https://doi.org/10.1016/j.surfcoat.2016.07.006
35. Gallo, J., Goodman, S.B., Konttinen, Y.T., Wimmer, M.A., Holinka, M.: Osteolysis around total knee arthroplasty: a review of pathogenetic mechanisms. Acta Biomater. **9**(9), 8046–8058 (2013). https://doi.org/10.1016/j.actbio.2013.05.005
36. Georgieva, I., Danchova, N., Gutzov, S., Trendafilova, N.: DFT modeling, UV-Vis and IR spectroscopic study of acetylacetone-modified zirconia sol-gel materials. J. Mol. Model. **18**(6), 2409–2422 (2012)
37. Gill, I.: Bio-doped nanocomposite polymers: sol−gel bioencapsulates. Chem. Mater. **13**(10), 3404–3421 (2001)
38. Gill, I., Ballesteros, A.: Bioencapsulation within synthetic polymers (Part 1): sol–gel encapsulated biologicals. Trends Biotechnol. **18**(7), 282–296 (2000)
39. Gill, J., Orsat, V., Kermasha, S.: Optimization of encapsulation of a microbial laccase enzymatic extract using selected matrices. Process Biochem. (Oxford, U K) **65**, 55–61 (2018). https://doi.org/10.1016/j.procbio.2017.11.011
40. Golabiewska, A., Lisowski, W., Jarek, M., Nowaczyk, G., Michalska, M., Jurga, S., Zaleska-Medynska, A.: The effect of metals content on the photocatalytic activity of TiO_2 modified by Pt/Au bimetallic nanoparticles prepared by sol-gel method. Mol Catal **442**, 154–163 (2017). https://doi.org/10.1016/j.mcat.2017.09.004
41. Gomez-Romero, P.: Hybrid organic–inorganic materials—in search of synergic activity. Adv. Mater. **13**(3), 163–174 (2001)
42. Grassi, M., Grassi, G.: Mathematical modelling and controlled drug delivery: matrix systems. Curr. Drug Deliv. **2**(1), 97–116 (2005)
43. Hakeem, A., Zahid, F., Zhan, G., Yi, P., Yang, H., Gan, L., Yang, X.: Polyaspartic acid-anchored mesoporous silica nanoparticles for pH-responsive doxorubicin release. Int. J. Nanomed. **13**, 1029 (2018)
44. Hench, L.L., Splinter, R.J., Allen, W.C., Greenlee, T.K.: Bonding mechanism at interface of ceramic prosthetic materials. J. Biomed. Mater. Res. **2**, 117–141 (1972)
45. Hillegass, J.M., Blumen, S.R., Cheng, K., MacPherson, M.B., Alexeeva, V., Lathrop, S.A., Beuschel, S.L., Steinbacher, J.L., Butnor, K.J., Ramos-Niño, M.E.: Increased efficacy of doxorubicin delivered in multifunctional microparticles for mesothelioma therapy. Int. J. Cancer **129**(1), 233–244 (2011)
46. Hong, S.-G., Huang, S.-C.: Crystallization properties of polyhydroxybutyrate with modified silicas. J. Polym. Res. **22**(4), 61 (2015). https://doi.org/10.1007/s10965-015-0706-6
47. Imam, S.S., Bukhari, S.N.A., Ahmad, J., Ali, A.: Formulation and optimization of levofloxacin loaded chitosan nanoparticle for ocular delivery: In-vitro characterization, ocular tolerance and antibacterial activity. Int. J. Biol. Macromol. **108**, 650–659 (2018)

48. Imoto, H., Katoh, R., Honda, T., Yusa, S.-i., Naka, K.: Self-association behavior of amphiphilic molecules based on incompletely condensed cage silsesquioxanes and poly(ethylene glycol)s. Polym J (Tokyo, Jpn): Ahead of Print (2018). https://doi.org/10.1038/s41428-017-0021-7
49. Imoto, H., Wada, S., Naka, K.: Rh-catalyzed direct arylation of a polyhedral oligomeric silsesquioxane. Dalton Trans. **46**(19), 6168–6171 (2017). https://doi.org/10.1039/c7dt01106g
50. Iwamura, T., Akiyama, K., Hakozaki, T., Shino, M., Adachi, K.: Synthesis of cellulose/silica gel polymer hybrids via in-situ hydrolysis method. Polym. Bull. (Heidelberg, Ger.) **74**(12), 4997–5009 (2017). https://doi.org/10.1007/s00289-017-2000-8
51. Jiménez-Flores, Y., Suárez-Quezada, M., Rojas-Trigos, J.B., Lartundo-Rojas, L., Suárez, V., Mantilla, A.: Characterization of Tb-doped hydroxyapatite for biomedical applications: optical properties and energy band gap determination. J. Mater. Sci., 1–11 (2017). https://doi.org/10.1007/s10853-017-1201-8
52. Jin, R., Liu, Z., Bai, Y., Zhou, Y., Chen, X.: Effective control of enzyme activity based on a subtle nanoreactor: a promising strategy for biomedical applications in the future. ACS Appl. Nano Mater. **1**, 302–309 (2018). https://doi.org/10.1021/acsanm.7b00152
53. Judeinstein, P., Sanchez, C.: Hybrid organic–inorganic materials: a land of multidisciplinarity. J. Mater. Chem. **6**(4), 511–525 (1996)
54. Karakhanov, E., Kardasheva, Y.S., Maksimov, A., Predeina, V., Runova, E., Utukin, A.: Macrocomplexes on the basis of functionalized polyethylene glycols and copolymers of ethylene oxide and propylene oxide: synthesis and catalysis. J. Mol. Catal. A: Chem. **107**(1–3), 235–240 (1996)
55. Kaur, G., Pandey, O.P., Singh, K., Homa, D., Scott, B., Pickrell, G.: A review of bioactive glasses: their structure, properties, fabrication and apatite formation. J. Biomed. Mater. Res. Part A **102**(1), 254–274 (2014)
56. Kempahanumakkagari, S., Kumar, V., Samaddar, P., Kumar, P., Ramakrishnappa, T., Kim, K.-H.: Biomolecule-embedded metal-organic frameworks as an innovative sensing platform. Biotechnol Adv. Ahead of Print (2018). https://doi.org/10.1016/j.biotechadv.2018.01.014
57. Khamsehashari, N., Hassanzadeh-Tabrizi, S., Bigham, A.: Effects of strontium adding on the drug delivery behavior of silica nanoparticles synthesized by P123-assisted sol-gel method. Mater. Chem. Phys. **205**, 283–291 (2018)
58. Khan, A., Asiri, A.M., Rub, M.A., Azum, N., Khan, A.A.P., Khan, S.B., Rahman, M.M., Khan, I.: Synthesis, characterization of silver nanoparticle embedded polyaniline tungstophosphate-nanocomposite cation exchanger and its application for heavy metal selective membrane. Compos. B Eng. **45**(1), 1486–1492 (2013). https://doi.org/10.1016/j.compositesb.2012.09.023
59. Kiwilsza, A., Milanowski, B., Druzbicki, K., Jenczyk, J., Jarek, M., Mielcarek, J., Lulek, J., Pajzderska, A., Wasicki, J.: Molecular dynamics and the dissolution rate of nifedipine encapsulated in mesoporous silica. Microporous Mesoporous Mater. **250**, 186–194 (2017). https://doi.org/10.1016/j.micromeso.2017.05.019
60. Kokubo, T.: Bioactive glass ceramics: properties and applications. Biomaterials **12**(2), 155–163 (1991)
61. Krasia-Christoforou, T.: Organic–inorganic polymer hybrids: synthetic strategies and applications. In: Hybrid and Hierarchical Composite Materials. Springer, pp. 11–63 (2015)
62. Kumar, A., Murugavel, S., Aditya, A., Boccaccini, A.R.: Mesoporous 45S5 bioactive glass: synthesis, in vitro dissolution and biomineralization behavior. J. Mater. Chem. B **5**, 8786–8798 (2017). https://doi.org/10.1039/c7tb01738c
63. Kumar, R., Münstedt, H.: Polyamide/silver antimicrobials: effect of crystallinity on the silver ion release. Polym. Int. **54**(8), 1180–1186 (2005). https://doi.org/10.1002/pi.1828
64. Kumar, S.K., Ganesan, V., Riggleman, R.A.: Perspective: Outstanding theoretical questions in polymer-nanoparticle hybrids. J. Chem. Phys. **147**(2), 020901/020901–020901/020918 (2017b). https://doi.org/10.1063/1.4990501
65. Kurczewska, J., Ryczkowski, J., Pasieczna-Patkowska, S., Schroeder, G.: Photoacoustic infrared spectroscopic studies of silica gels with organically functionalized surface. Spectrosc. Lett. **49**, 529–534 (2016). https://doi.org/10.1080/00387010.2016.1213298

66. Lee, E.J., Huh, B.K., Kim, S.N., Lee, J.Y., Park, C.G., Mikos, A.G., Choy, Y.B.: Application of materials as medical devices with localized drug delivery capabilities for enhanced wound repair. Prog. Mater Sci. **89**, 392–410 (2017). https://doi.org/10.1016/j.pmatsci.2017.06.003
67. Lee, J.E., Lee, N., Kim, H., Kim, J., Choi, S.H., Kim, J.H., Kim, T., Song, I.C., Park, S.P., Moon, W.K.: Uniform mesoporous dye-doped silica nanoparticles decorated with multiple magnetite nanocrystals for simultaneous enhanced magnetic resonance imaging, fluorescence imaging, and drug delivery. J. Am. Chem. Soc. **132**(2), 552–557 (2009)
68. Li, H.C., Wang, D.G., Hu, J.H., Chen, C.Z.: Influence of fluoride additions on biological and mechanical properties of Na_2O–CaO–SiO_2–P_2O_5 glass-ceramics. Mater. Sci. Eng. C **35**, 171–178 (2014). https://doi.org/10.1016/j.msec.2013.10.028
69. Li, W., Hui, L., Xue, B., Dong, C., Chen, Y., Hou, L., Jiang, B., Wang, J., Yang, Y.: Facile high-temperature synthesis of weakly entangled polyethylene using a highly activated Ziegler-Natta catalyst. J. Catal. **360**, 145–151 (2018). https://doi.org/10.1016/j.jcat.2018.01.024
70. Loy, D.A., Shea, K.J.: Bridged polysilsesquioxanes. Highly porous hybrid organic-inorganic materials. Chem. Rev. **95**(5), 1431–1442 (1995)
71. Materazzi, S., Vecchio, S., Wo, L.W., Curtis, S.D.A.: Thermoanalytical studies of imidazole-substituted coordination compounds: Mn(II)-complexes of bis(1-methylimidazol-2-yl)ketone. J. Therm. Anal. Calorim. **103**(1), 59–64 (2011). https://doi.org/10.1007/s10973-010-1137-6
72. Materazzi, S., Vecchio, S., Wo, L.W., De Angelis, Curtis S.: TG-MS and TG-FTIR studies of imidazole-substituted coordination compounds: Co(II) and Ni(II)-complexes of bis(1-methylimidazol-2-yl)ketone. Thermochim. Acta **543**, 183–187 (2012). https://doi.org/10.1016/j.tca.2012.05.013
73. Mavropoulos, E., Costa, A.M., Costa, L.T., Achete, C.A., Mello, A., Granjeiro, J.M., Rossi, A.M.: Adsorption and bioactivity studies of albumin onto hydroxyapatite surface. Colloids Surf. B **83**(1), 1–9 (2011). https://doi.org/10.1016/j.colsurfb.2010.10.025
74. Meng, S., Mansouri, J., Ye, Y., Chen, V.: Effect of templating agents on the properties and membrane distillation performance of TiO_2-coated PVDF membranes. J. Membr. Sci. **450**, 48–59 (2014). https://doi.org/10.1016/j.memsci.2013.08.036
75. Midha, S., Kim, T.B., van den Bergh, W., Lee, P.D., Jones, J.R., Mitchell, C.A.: Preconditioned 70S30C bioactive glass foams promote osteogenesis in vivo. Acta Biomater. **9**, 9169–9182 (2013). https://doi.org/10.1016/j.actbio.2013.07.014
76. Milea, C., Bogatu, C., Duta, A.: The influence of parameters in silica sol-gel process. Bull. Transilvania Univ Brasov **4**, 53 (2011)
77. Mohseni, M., Bastani, S., Jannesari, A.: Influence of silane structure on curing behavior and surface properties of sol–gel based UV-curable organic–inorganic hybrid coatings. Prog. Org. Coat. **77**(7), 1191–1199 (2014). https://doi.org/10.1016/j.porgcoat.2014.04.008
78. Montazerian, M., Yekta, B.E., Marghussian, V.K., Bellani, C.F., Siqueira, R.L., Zanotto, E.D.: Bioactivity and cell proliferation in radiopaque gel-derived CaO–P_2O_5–SiO_2–ZrO_2 glass and glass–ceramic powders. Mater. Sci. Eng. C **55**, 436–447 (2015). https://doi.org/10.1016/j.msec.2015.05.065
79. Nadgir, M.M., Coffey, A., Murari, B.M.: Modified sol-gel processed silica matrix for gel electrophoresis applications. J. Sol-Gel. Sci. Technol. **83**(1), 155–164 (2017). https://doi.org/10.1007/s10971-017-4401-4
80. Naghibi, S., Madaah Hosseini, H.R., Faghihi Sani, M.A., Shokrgozar, M.A., Mehrjoo, M.: Mortality response of folate receptor-activated, PEG-functionalized TiO_2 nanoparticles for doxorubicin loading with and without ultraviolet irradiation. Ceram. Int. **40**, 5481–5488 (2014). https://doi.org/10.1016/j.ceramint.2013.10.136
81. Naghipoor, J., Rabczuk, T.: A mechanistic model for drug release from PLGA-based drug eluting stent: a computational study. Comput. Biol. Med. **90**, 15–22 (2017). https://doi.org/10.1016/j.compbiomed.2017.09.001
82. Oberdisse, J., Hellweg, T.: Structure, interfacial film properties, and thermal fluctuations of microemulsions as seen by scattering experiments. Adv. Colloid Interface Sci. **247**, 354–362 (2017). https://doi.org/10.1016/j.cis.2017.07.011

83. Ogoshi, T., Chujo, Y.: Organic–inorganic polymer hybrids prepared by the sol-gel method. Compos. Interfaces **11**(8–9), 539–566 (2005)
84. Owens, G.J., Singh, R.K., Foroutan, F., Alqaysi, M., Han, C.-M., Mahapatra, C., Kim, H.-W., Knowles, J.C.: Sol–gel based materials for biomedical applications. Prog. Mater Sci. **77**, 1–79 (2016)
85. Pantelidis, D., Bravman, J.C., Rothbard, J., Klein, R.L.: Bioactive material delivery systems comprising sol-gel compositions. US Patent Application 20070071789 (2012)
86. Pérez-Fonseca, A.A., Robledo-Ortíz, J.R., Ramirez-Arreola, D.E., Ortega-Gudiño, P., Rodrigue, D., González-Núñez, R.: Effect of hybridization on the physical and mechanical properties of high density polyethylene–(pine/agave) composites. Mater. Des. **64**, 35–43 (2014). https://doi.org/10.1016/j.matdes.2014.07.025
87. Qian, X., Song, L., Hu, Y., Yuen, R.K.K.: Thermal degradation and flammability of novel organic/inorganic epoxy hybrids containing organophosphorus-modified oligosiloxane. Thermochim. Acta **552**, 87–97 (2013). https://doi.org/10.1016/j.tca.2012.11.010
88. Qu, H., Bhattacharyya, S., Ducheyne, P.: Silicon oxide based materials for controlled release in orthopedic procedures. Adv. Drug Deliv. Rev. **94**, 96–115 (2015). https://doi.org/10.1016/j.addr.2015.05.015
89. Radev, L.: Influence of thermal treatment on the structure and in vitro bioactivity of sol-gel prepared CaO–SiO_2–P_2O_5 glass-ceramics. Process. Appl. Ceram. **8**(3), 155–166 (2014)
90. Radev, L., Hristov, V., Michailova, I., Fernandes, H.M.V., Salvado, M.I.M.: In vitro bioactivity of biphasic calcium phosphate silicate glass-ceramic in CaO–SiO_2–P_2O_5 system. Process. Appl. Ceram. **4**(1), 15–24 (2010)
91. Raghavendra, G.M., Varaprasad, K., Jayaramudu, T.: Chapter 2—Biomaterials: design, development and biomedical applications A2—Thomas, Sabu. In: Grohens, Y., Ninan, N. (eds) Nanotechnology Applications for Tissue Engineering. William Andrew Publishing, Oxford, pp. 21–44 (2015). doi:https://doi.org/10.1016/B978-0-323-32889-0.00002-9
92. Rahimi, H., Mozaffarinia, R., Hojjati Najafabadi, A.: Corrosion and wear resistance characterization of environmentally friendly Sol–gel hybrid nanocomposite coating on AA5083. J. Mater. Sci. Technol. **29**(7), 603–608 (2013). https://doi.org/10.1016/j.jmst.2013.03.013
93. Samet, L., Ben Nasseur, J., Chtourou, R., March, K., Stephan, O.: Heat treatment effect on the physical properties of cobalt doped TiO_2 sol–gel materials. Mater. Charact. **85**, 1–12 (2013). https://doi.org/10.1016/j.matchar.2013.08.007
94. Sanchez, C., Julián, B., Belleville, P., Popall, M.: Applications of hybrid organic–inorganic nanocomposites. J. Mater. Chem. **15**(35–36), 3559–3592 (2005)
95. Sanchez, C., Lebeau, B., Chaput, F., Boilot, J.P.: Optical properties of functional hybrid organic–inorganic nanocomposites. Adv. Mater. **15**(23), 1969–1994 (2003)
96. Savage, T.J., Dunphy, D.R., Harbaugh, S., Kelley-Loughnane, N., Harper, J.C., Brinker, C.J.: Influence of silica matrix composition and functional component additives on the bioactivity and viability of encapsulated living cells. ACS Biomater. Sci. Eng. **1**(12), 1231–1238 (2015). https://doi.org/10.1021/acsbiomaterials.5b00261
97. Scalera, F., Gervaso, F., Sanosh, K.P., Sannino, A., Licciulli, A.: Influence of the calcination temperature on morphological and mechanical properties of highly porous hydroxyapatite scaffolds. Ceram. Int. **39**(5), 4839–4846 (2013). https://doi.org/10.1016/j.ceramint.2012.11.076
98. Schubert, U.: Chemistry and fundamentals of the sol–gel process. In: The Sol-Gel Handbook-Synthesis, Characterization, and Applications: Synthesis, Characterization and Applications, 3-Volume Set, pp. 1–28 (2015)
99. Shao, G.N., Imran, S.M., Jeon, S.J., Engole, M., Abbas, N., Salman Haider, M., Kang, S.J., Kim, H.T.: Sol–gel synthesis of photoactive zirconia–titania from metal salts and investigation of their photocatalytic properties in the photodegradation of methylene blue. Powder Technol. **258**, 99–109 (2014). https://doi.org/10.1016/j.powtec.2014.03.024
100. Steinbeck, M.J., Jablonowski, L.J., Parvizi, J., Freeman, T.A.: The role of oxidative stress in aseptic loosening of total hip arthroplasties. J. Arthroplasty **29**(4), 843–849 (2014)

101. Sur, G., Mark, J.: Elastomeric networks cross-linked by silica or titania fillers. Eur. Polym. J. **21**(12), 1051–1052 (1985)
102. Tang, Z., Wei, Q., Lin, T., Guo, B., Jia, D.: The use of a hybrid consisting of tubular clay and graphene as a reinforcement for elastomers. Rsc. Adv. **3**(38), 17057–17064 (2013)
103. Timin, A.S., Muslimov, A.R., Lepik, K.V., Okilova, M.V., Tcvetkov, N.Y., Shakirova, A.I., Afanasyev, B.V., Gorin, D.A., Sukhorukov, G.B.: Intracellular Breakable and Ultrasound-Responsive Hybrid Microsized Containers for Selective Drug Release into Cancerous Cells. Part Part Syst. Charact. **34**:n/a (2017). https://doi.org/10.1002/ppsc.201600417
104. Tiwari, A., Raj, B.: Reactions and Mechanisms in Thermal Analysis of Advanced Materials. Wiley, New York (2015)
105. Tredwin, C.J., Young, A.M., Abou Neel, E.A., Georgiou, G., Knowles, J.C.: Hydroxyapatite, fluor-hydroxyapatite and fluorapatite produced via the sol–gel method: dissolution behaviour and biological properties after crystallisation. J. Mater. Sci. Mater. Med. **25**(1), 47–53 (2014). https://doi.org/10.1007/s10856-013-5050-y
106. Uilk, J.M., Mera, A.E., Fox, R.B., Wynne, K.J.: Hydrosilation-cured poly (dimethylsiloxane) networks: intrinsic contact angles via dynamic contact angle analysis. Macromolecules **36**(10), 3689–3694 (2003)
107. Vecchio Ciprioti, S., Bollino, F., Tranquillo, E., Catauro, M.: Synthesis, thermal behavior and physicochemical characterization of ZrO_2/PEG inorganic/organic hybrid materials via sol–gel technique. J. Therm. Anal. Calorim. **130**(1), 535–540 (2017). https://doi.org/10.1007/s10973-017-6318-0
108. Vecchio Ciprioti, S., Catauro, M.: Synthesis, structural and thermal behavior study of four Ca-containing silicate gel-glasses: activation energies of their dehydration and dehydroxylation processes. J. Therm. Anal. Calorim. **123**(3), 2091–2101 (2016). https://doi.org/10.1007/s10973-015-4729-3
109. Vecchio Ciprioti, S., Catauro, M., Bollino, F., Tuffi, R.: Thermal behavior and dehydration kinetic study of SiO_2/PEG hybrid gel glasses. Polym. Eng. Sci. **57**(6), 606–612 (2017). https://doi.org/10.1002/pen.24561
110. Vichery, C., Nedelec, J.-M.: Bioactive glass nanoparticles: from synthesis to materials design for biomedical applications. Materials **9**(4), 288 (2016)
111. Wang, C., He, C., Tong, Z., Liu, X., Ren, B., Zeng, F.: Combination of adsorption by porous $CaCO_3$ microparticles and encapsulation by polyelectrolyte multilayer films for sustained drug delivery. Int. J. Pharm. **308**(1–2), 160–167 (2006)
112. Wang, X., Wang, L., Su, Q., Zheng, J.: Use of unmodified SiO_2 as nanofiller to improve mechanical properties of polymer-based nanocomposites. Compos. Sci. Technol. **89**, 52–60 (2013). https://doi.org/10.1016/j.compscitech.2013.09.018
113. Wang, X., Zhang, Y., Lin, C., Zhong, W.: Sol-gel derived terbium-containing mesoporous bioactive glasses nanospheres: in vitro hydroxyapatite formation and drug delivery. Colloids Surf. B **160**, 406–415 (2017)
114. Wang, Y., Cheetham, A.G., Angacian, G., Su, H., Xie, L., Cui, H.: Peptide–drug conjugates as effective prodrug strategies for targeted delivery. Adv. Drug Deliv. Rev. **110–111**, 112–126 (2017). https://doi.org/10.1016/j.addr.2016.06.015
115. Wen, J., Wilkes, G.L.: Organic/inorganic hybrid network materials by the sol−gel approach. Chem. Mater. **8**(8), 1667–1681 (1996)
116. Wojcik, A., Klein, L.: Organic-inorganic gels based on silica and multifunctional acrylates. J. Sol-Gel. Sci. Technol. **2**(1–3), 115–120 (1994)
117. Wu, W., Wu, Z., Yu, T., Jiang, C., Kim, W.-S.: Recent progress on magnetic iron oxide nanoparticles: synthesis, surface functional strategies and biomedical applications. Sci. Technol. Adv. Mater. **16**(2), 023501 (2015)
118. Yaghtin, M., Taghvaei, A.H., Hashemi, B., Janghorban, K.: Effect of heat treatment on magnetic properties of iron-based soft magnetic composites with Al_2O_3 insulation coating produced by sol–gel method. J. Alloy. Compd. **581**, 293–297 (2013). https://doi.org/10.1016/j.jallcom.2013.07.008

119. Yang, H.-H., Zhu, Q.-Z., Qu, H.-Y., Chen, X.-L., Ding, M.-T., Xu, J.-G.: Flow injection fluorescence immunoassay for gentamicin using sol-gel-derived mesoporous biomaterial. Anal. Biochem. **308**(1), 71–76 (2002)
120. Youssef, A.M., El-Nahrawy, A.M., Abou Hammad, A.B.: Sol-gel synthesis and characterizations of hybrid chitosan-PEG/calcium silicate nanocomposite modified with ZnO-NPs and (E102) for optical and antibacterial applications. Int. J. Biol. Macromol. **97**, 561–567 (2017). https://doi.org/10.1016/j.ijbiomac.2017.01.059
121. Zadeh, M.A., Van Der Zwaag, S., Garcia, S.: Routes to extrinsic and intrinsic self-healing corrosion protective sol-gel coatings: a review. Self-Healing Mater. **1**, 1–18 (2013)
122. Zanoaga, M., Tanasa, F.: Antimicrobial reagents as functional finishing for textiles intended for biomedical applications. I. Synthetic organic compounds. Chem. J. Mold. **9**(1), 14–32 (2014). https://doi.org/10.19261/cjm.2014.09(1).02
123. Zeng, Y., Cao, J., Wang, Z., Guo, J., Lu, J.: Formation of amorphous calcium carbonate and its transformation mechanism to crystalline $CaCO_3$ in laminar microfluidics. Cryst. Growth Des. Ahead of Print (2018). https://doi.org/10.1021/acs.cgd.7b01634
124. Zhao, L., Wu, Y., Chen, S., Xing, T.: Preparation and characterization of cross-linked carboxymethyl chitin porous membrane scaffold for biomedical applications. Carbohyd. Polym. **126**, 150–155 (2015). https://doi.org/10.1016/j.carbpol.2015.02.050
125. Zhuravlev, L.: The surface chemistry of amorphous silica. Zhuravlev model. Colloids Surf. A: Physicochem. Eng. Aspects **173**(1–3), 1–38 (2000)
126. Zou, H., Wu, S., Shen, J.: Polymer/silica nanocomposites: preparation, characterization, properties, and applications. Chem. Rev. **108**(9), 3893–3957 (2008)

Printed by Printforce, the Netherlands